The Interfacial Interactions
in Polymeric Composites

NATO ASI Series

Advanced Science Institutes Series

A Series presenting the results of activities sponsored by the NATO Science Committee, which aims at the dissemination of advanced scientific and technological knowledge, with a view to strengthening links between scientific communities.

The Series is published by an international board of publishers in conjunction with the NATO Scientific Affairs Division

A Life Sciences	Plenum Publishing Corporation
B Physics	London and New York
C Mathematical	Kluwer Academic Publishers
and Physical Sciences	Dordrecht, Boston and London
D Behavioural and Social Sciences	
E Applied Sciences	
F Computer and Systems Sciences	Springer-Verlag
G Ecological Sciences	Berlin, Heidelberg, New York, London,
H Cell Biology	Paris and Tokyo
I Global Environmental Change	

NATO-PCO-DATA BASE

The electronic index to the NATO ASI Series provides full bibliographical references (with keywords and/or abstracts) to more than 30000 contributions from international scientists published in all sections of the NATO ASI Series.
Access to the NATO-PCO-DATA BASE is possible in two ways:

– via online FILE 128 (NATO-PCO-DATA BASE) hosted by ESRIN,
Via Galileo Galilei, I-00044 Frascati, Italy.

– via CD-ROM "NATO-PCO-DATA BASE" with user-friendly retrieval software in English, French and German (© WTV GmbH and DATAWARE Technologies Inc. 1989).

The CD-ROM can be ordered through any member of the Board of Publishers or through NATO-PCO, Overijse, Belgium.

Series E: Applied Sciences - Vol. 230

The Interfacial Interactions in Polymeric Composites

edited by

Güneri Akovalı

Department of Chemistry,
Polymer Science and Technical Program,
Middle East Technological University,
Ankara, Turkey

Springer Science+Business Media, B.V.

Proceedings of the NATO Advanced Study Institute on
The Interfacial Interactions in Polymeric Composites
Antalya/Kemer, Turkey
15–26 June 1992

Library of Congress Cataloging-in-Publication Data

```
The Interfacial interactions in polymeric composites / edited by
  Güneri Akovali.
      p.   cm. -- (NATO ASI series.  Series E, Applied sciences ; no.
  230)
    Includes index.
    ISBN 978-94-010-4717-3    ISBN 978-94-011-1642-8 (eBook)
    DOI 10.1007/978-94-011-1642-8
    1. Polymeric composites.  2. Surface chemistry.  3. Polymeric
  composites--Surfaces.   I. Akovali, Güneri.  II. Series.
  TA418.9.C6I545   1992
  620.1'92--dc20                                          92-41725
```

ISBN 978-94-010-4717-3

Printed on acid-free paper

TABLE OF CONTENTS

PREFACE

Polymer composites represent materials of great and of continuously growing importance. Their potential for application appears to be limitless. They have been the subject of numerous studies both at academic and industrial levels. Much progress has been made in the incisive formulation of composites; sophisticated methods of property evaluation have been developed in the past decade and many, largely empirical solutions have been proposed to resolve the problem of their long-term performance under typical conditions of use (i.e. the use of silane or titane coupling agents to enhance adhesion within composite materials).

Assuredly one of the most essential factors in the performance of these systems is the condition of the interface and interphase among the constituents of a given system. It has become clear that it is the interface/interphase, and the interactions which take place in this part of a system, which determine to a significant degree the initial properties of the material. In order to achieve leadership in the formulation and application of polymer composites, it is evident that in depth understanding of interfacial and interphase phenomena becomes a prerequisite. Included in that understanding is, interalia, a grasp of thermodynamic, dispersion-force and non-dispersion-force interactions; adhesion phenomena at interfaces; the morphological and mechanical characteristics of interfaces and interphases; the time dependent variations in these characteristics; state-of-the science approaches to modifying, controllably, key interactions through the medium

of surface modification by chemical and especially by
electrical discharge methods; diagnostic methods capable of
yielding quantitative information on surface and interface
chemistry. Fortunately, in response to the evident
importance of the factors listed here, intensive research
activity has taken and is taking place in the Universities
of the Nato countries. Resident in these locations, and at
certain industrial sites, are experts, who are able to
disseminate information of high value to a wide number of
scientists and engineers, whose task is to evaluate further
the technology and the applications of material composites.
This Nato-ASI meeting is proposed as an outstanding vehicle
for congregating leading workers in the field, with the view
of meeting the targets of incisive information transfer to a
critical and critically involved audience. Therefore it
functions both as a means of direct transfer of information
to concerned parties and as a means of publishing a
compendium of information.

The Institute considered the interfacial interactions
in polymeric composites and for this, first a
differentiation between adhesion, interfaces and interphases
is made; which included the discussion of the
adsorption-mechanical (hooking)-electrostatic and diffusion
theories of adhesion, as well as the rheological theories of
adhesive joints. The concept of interface engineering,
"consisting of a systematic understanding of the interface,
controlling of the interface and tailoring of the
properties" are extensively discussed. It is concluded that,
there are a number of difficulties involved in the
postulation of the simple relationships between surface
interactions and the mechanical performances. A tentative
model is purposed to relate the interfacial strenghts to the
level of physical interactions, mainly to electron
donor-acceptor interactions at the interface. The mechanical
properties of polymer/polymer interfaces are shown to be
very sensitive to the detailed structure of the interface
and two major examples of this correlation presented are:

the role of chain ends and their spatial distribution in A/A healing as well as the role of entanglements in A/B fracture or in A/B slippage. The influence of interactions at polymer surfaces and interfaces on the properties of polymer systems, with emphasis on the acid-base interactions, are all reviewed in detail.

Several contributions review the methods of investigation of interfacial interactions and surfaces. The IGC method to evaluate the donor-acceptor interaction potential of components, as well as the classical techniques such as SEM, Auger, SIMMS, ion scattering and X-ray photoelectron spectroscopy. Various surface FT-IR techniques are extensively discussed and explained with applications. A new qualitative method (induction time approach) to study the trans crystal layers is also introduced.

Various strategies to control and modify surfaces and interfaces including physical, chemical and inherent modifications, in particular corona and cold plasma techniques are reviewed and discussed extensively.

Description of the mechanical properties of polymer composites are also made by considering the properties of particulate-long fiber and laminate composites through the different models generated in the literature. It is shown that, many advantages can be derived by use of liquid crystalline compounds as reinforcing fillers to produce blends with engineering thermoplastics.

During the meeting, there were also a number of presentations of students. Some of these are included in the book, too.

Finally, on behalf of the organizing committee and myself, I would like to thank to all lecturers for fulfilling their share in putting the parts together, to the participants for their active contributions and involvement as well as for creating the lively environment for the meeting. Our special thanks are due to Nato-ASI for making it all possible, and to the Basic Sciences and BAYG Groups of Turkish National and Scientific Research Council

(TUBITAK) for the additional financial assistances.

I am deeply grateful and would like to acknowledge to the members of the organizing committee for their guidance, kind cooperation and helps.

G. Akovali   Ankara,
                                                    Oct. 16. 1992

LIST OF PARTICIPANTS: (By Country in Alphabetical Order)

(a) Director

Prof. G. AKOVALI
Middle East Technical University
Orta Dogu Teknik Universitesi
06531-Ankara. **Turkiye**

(b) Lecturers

Prof. H. P. SCREIBER (Org. Comm. Member)
Genie Chimique-Ecole Polytechnique du Montreal
Montreal University. Case Postale 6079. Succursale A
Montreal. Quebec. H3C 3A7. **Canada**

Prof. M. R. WERTHEIMER
Dept. Genie Physique-Ecole Polytechnique
Case Postale 6079. Succursale A
Montreal. Quebec. H3C 3A7. **Canada**

Prof. P. G. de GENNES
Director, ESPCI-College de France
Physique de la Matiere Condensee
11. Place Marcelin-Berthelot
75231. Paris Cedex. 05. **France**

Prof. W WEISWEILER (Org. Comm. Member)
Universitat Karlsruhe-Inst. fur Chemische Technik
Kaiserstrasse. 12
7500-Karlsruhe. 1. **Germany**

Prof. L. NICOLAIS
Chairman, Dip. di Ingegneria Dei Materiali E Della
Produzione
Universita Degli Studi Di Napoli Federico II
Piazzale Tecchio. 80125. Napoli. **Italy**

Prof. M. NARDIN
CNRS- Centre de Recherches Sur La Physico-Chimie des
Surfaces Solides
24. Avenue du President Kennedy
68200 Mulhouse. **France**

Prof. F. J. MAURER
Chalmers Teknisca Hoskola-Inst. for Polymer Teknologi
S-41296. Goteborg. **Sweden**
(**Formerly** from DSM Research-Gleen. **The Netherlands**)

Prof. G. AKOVALI (Org. Comm. Member)
Middle East Technical University-Orta Dogu Teknik
Universitesi
06531-Ankara. **Turkiye**

Prof. E. Bayramli
Middle East Technical University-Orta Dogu Teknik Universitesi
06531-Ankara. **Turkiye**
Prof. U. YILMAZER
Middle East Technical University-Orta Dogu Teknik Universitesi- Dept. of Chemical Eng.
06531-Ankara. **Turkiye**

Prof. H. ISHIDA
Dept. of Macromolecular Science
Case Western Reserve University
10900 Ueclid Avenue. Olin 307
Cleveland. Ohio. 44106-7202. **USA**

Prof. E. M. LISTON
42 Peninsula Rd.
Belvedere. Ca.. 94820. **USA**

Dr. K. L. MITTAL
IBM-US Technical Education
500 Colombus Avenue
Thornwood. NY. 10594. **USA**

Dr. L. H. SHARPE
Editor in Chief, The J. of Adhesion
P. O. Box 3288
Hilton Head Island. SC. 29928. **USA**

Prof. J. P. WHIGHTMAN
Director, Centre for Adhesive and Sealant Science
Virginia Tech. Davidson Hall. Rm. 2. Blacsburg. VI. 24061-0102. **USA**

(c) **Other Participants:**
Prof. R. O. ABDULLAEV
370032 Baku-Achmetlinskoya Schosse 622/9
Kipa "Birlik" of Azerbaijan Academy of Sciences.
Azerbaijan

Prof. A. A. EFENDIEV
Director, Institute of Polymeric Materials
Azerbaijan Academy of Sciences
Sumgait. Samed Vurgun Street. 124
373200 Baku. **Azerbaijan**

Prof. S. FAKIROV
University of Sofia. Lab. on Structure and Prop. of Polymers
Bldg. A Ivanov . 1
1126. Sofia. **Bulgaria**

Dr. J. SAPIEHA
Ecole Polytechnique. Dept. Genie Physique.
Montreal University. Case Postale 6079. Succursale A
Montreal. Quebec. H3C 3A7 **Canada**

Dr. S. SAPIEHA
Ecole Polytechnique. Dept. Genie Physique.
Montreal University. Case Postale 6079. Succursale A
Montreal. Quebec. H3C 3A7 **Canada**

Dr. I. ULKEM
Ecole Polytechnique. Genie Chemie.
Montreal University. Case Postale 6079. Succursale A
Montreal. Quebec. H3C 3A7 **Canada**

Prof. F. BROCHARD-WYART
SRI, Universite Paris 6
11. Rue Pierre et Marie Curie
75005 Paris, **France**

Dr. T. NUGAY
Ecole Nationale Superiore de Chimie de Mulhouse
Rue Alfred Werner, 3-68093. Mulhose. **France**

Dr. N. NUGAY
Ecole Nationale Superiore de Chimie de Mulhouse
Rue Alfred Werner, 3-68093. Mulhose. **France**

Mr. M. FISCHER
Deutsche Kunstoff Institute, (DKI)
Schlossgarten Strasse 6
D. 6100. Darmstadt. **Germany**

Mr. E. EBERT
Univerisitat Karlsruhe. Inst. fur Chemische Technik
Kaiser Strasse. 12. 7500 Karlsruhe. **Germany**

Dr. M. T. SOUSA
FEUP- FAculdade de Engenhaira
Dept. Eng. Quimica
4099-Porto Cedex. **Portugal**

Dr. A. CUNHA
Universidade do Minho
DEP-ENgenhira de Polimeros
Largo do Papo. 4719. Braga Codex. **Portugal**

Prof. M. KISELEV
Inst. of Physical Chemistry-Dept. of Polymeric Materials
Russian Academy of Sciences
31. Leninskay av. Moscow. **Russia**. 117071

Mr. J. J. FERNANDEZ
CSIC-Instituto Nacional Del Carbon
La Cordoria. s/n. Apartado 73
E. 33080. Oviedo. **Spain**

Mr. E. FUENTE
CSIC-Instituto Nacional De. Carbon
La Corredoria. s/n. Apartado 73
E. 33080. Oviedo. **Spain**

Mrs. Z. BERDJANE
Corsejo Superior de Investigaciones Cientificas (CSIC)
Instituto de Estructura de la Materia. Rolasolano
Serrano. 119-123. 28006. Madrid. **Spain**

Prof. D. BALKOSE
Ege Universitesi Faculty of Engineering
Chem. Eng. Dept. 35100. Bornova. Izmir. **Turkiye**

Assoc. Prof. Dr. S. BASAN
Cumhuriyet Universitesi-Dept. of Chemistry
Sivas. 58140. **Turkiye**

Dr. F. YIGIT
Hacettepe University. Dept. of Chemistry
06532. Ankara. **Turkiye**

Assoc. Prof. Dr. M. TUNCAY
Istanbul University. Eng. Faculty. Chem. Dept.
Avcilar. Istanbul. **Turkiye**

Prof. G. GUNDUZ
Middle East Technical University
Orta Dogu Teknik Universitesi
Dept. of Chem. Eng. 06531. Ankara. **Turkiye**

Dr. Z. OKTEM
Middle East Technical University
Orta Dogu Teknik Universitesi
FEF-Dept. of Chem. 06531. Ankara. **Turkiye**

Dr. G. OKTEM
Middle East Technical University
Orta Dogu Teknik Universitesi
FEF-Dept. of Chem. 06531. Ankara. **Turkiye**

Miss. N. DILSIZ
Middle East Technical University
Orta Dogu Teknik Universitesi
FEF-Dept. of Chem. 06531. Ankara. **Turkiye**

Dr. H. CUKUROVA
Hacettepe University-Dept. of Chem. Eng.
06532. Beytepe. Ankara. **Turkiye**

Dr. A. AKMAN
Middle East Technical University
Orta Dogu Teknik Universitesi
Kimya Bolumu. 06531. Ankara. **Turkiye**

Dr. A. E. AKINAY
Middle East Technical University
Orta Dogu Teknik Universitesi
FEF Kimya Bl. 06531. Ankara. Türkiye

Dr. J. MAGUIRE
Southwest Research Inst. (SWRI)
Materials and Mechanics Dept.
6220 Culebra Rd. P.O. Drawer. 28510. San An·
78228-0520. USA

Prof. E. SANCAKTAR
Clarkson University,
Dept. of Mechanical and Aeronautical Eng.
Center for Advanced Materials Processing, (C.
Potsdam. NewYork. 13699-5725. USA

Dr. R. SIX
142 Schrenk Hall
University of Missouri at Rolla
Rolla. Missouri. 64501. USA

Prof. Z. RZAYEV
(Visiting Scientist From Azerbaijan)
His Adress for the 1992/93 is:
Middle East Technical University
Dept. of Chemistry. 06531. Ankara. Turkiye
His permanent adress afterwards is:
Vice Director, Inst. of Polymer Materials
Azerbaijan Academy of Sciences. Baku. Azerbaijan

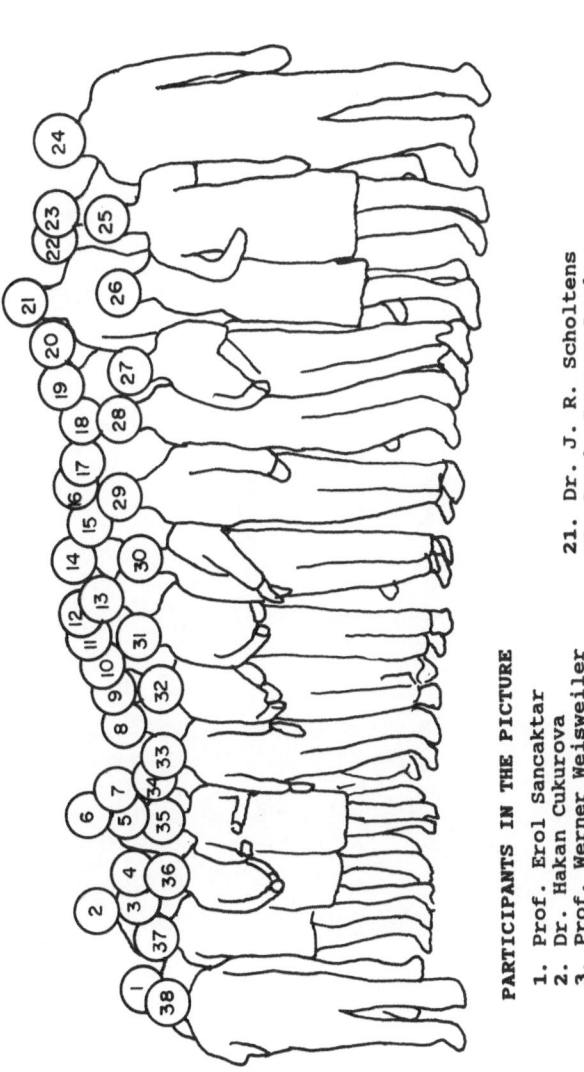

PARTICIPANTS IN THE PICTURE

1. Prof. Erol Sancaktar
2. Dr. Hakan Cukurova
3. Prof. Werner Weisweiler
4. Prof. Erdal Bayramli
5. Dr. Egon Ebert
6. Dr. Michael Fischer
7. Prof. Pierre G. de Gennes
8. Prof. F. Brochard-Wyart
9. A. Fuente
10. Prof. Z. Dobkowsky
11. Dr. Turgut Nugay
12. Prof. Michael R. Wertheimer
13. Ronald Six
14. Dr.Edward M. Liston
15. Prof. John Maguire
16. Mrs. E. M. Liston
17. Dr. A. Thiesen
18. Prof. Devrim Balkose
19. Prof. J. E. Klemberg-Sapieha
20. Dr. Ilhan Ulkem

21. Dr. J. R. Scholtens
22. Prof. Gungor Gunduz
23. Prof. Satilmix Basan
24. Dr. Zeki Oktem
25. Dr. Gulsu A. Oktem
26. Prof. Melda Tuncay
27. Sibel Aslan
28. Prof.J. Figueiredo
29. Prof. Hatsuo Ishida
30. Dr. Nihan Nugay
31. Ali Erkan Akinay
32. M. Teresa Sousa
33. Dr. Fatma Yigit
34. Dr. J. Sapieha
35. Nosrat Kojouwei
36. Nursel Dilsiz
37. Prof. Henry P. Schreiber
38. Prof. Guneri Akovali

INTERFACES, INTERPHASES AND "ADHESION": A Perspective

LOUIS H. SHARPE, PH.D.
Consultant and
Editor in Chief
The Journal of Adhesion
28 Red Maple Road
Hilton Head Island, SC 29928, U.S.A.

ABSTRACT. The terms "adhesion" and "interphase" are defined. The importance of interphases in the development of an understanding of the mechanical response of joint systems is discussed and examples of several, quite different, interphases are given. The adsorption, mechanical ("hooking"), electrostatic and diffusion theories of adhesion are briefly discussed and criticized. The rheological theory of adhesive joints is discussed and shown to be a basis for understanding the mechanical behavior of adhesive joints. The creation, by different processes, of several types of interphases in polyethylene is illustrated and discussed in light of their effects on the wettability and joinability of polyethylene. It is demonstrated that material in a joint may not be conserved during failure and that this may have an effect on conclusions about joint failure based on fractographic evidence. Finally, some of the difficulties inherent in the postulation of simple cause and effect relationships between surface interactions and the mechanical performance of joint systems are discussed.

1. Introduction

1.1 DEFINITION OF "ADHESION"

If one is going to talk about "adhesion", then it is necessary to define it. There are at least two common meanings.

1. In physical chemistry, the atomic or molecular attraction between a solid and a second, usually liquid, phase is called "adhesion" [1]. The magnitude of the so-called adhesive forces or "adhesion", in this case, is determined from equilibrium or quasi-equilibrium measurements of such quantities as contact angles of liquids on solids. This is "interfacial adhesion", meaning confined to the interface.

2. In technology, the term "adhesion" (in this case, labelled "practical adhesion" by Sharpe and Schonhorn [2]) is commonly used to mean the mechanical resistance to separation of a system of joined

1

G. Akovali (ed.), The Interfacial Interactions in Polymeric Composites, 1–20.
© 1993 *Kluwer Academic Publishers.*

materials (usually the breaking force or average breaking stress or work or fracture energy). This parameter is determined by a complex of many interacting factors, e.g., certain mechanical properties of the bulk and surface regions of the joined materials, the testing geometry and rate, the testing history and environment, and so on.

Clearly, the phenomena to which the two terms relate are not the same. The former has to do with interfacial measurements, mainly related to the creation of adhering systems which are at equilibrium or quasi-equilibrium. The latter, on the other hand, has to do with the processes related to the destruction of adhering systems and involves measurements which are clearly not obtained in systems at equilibrium. Failure in solid-solid systems is not, except under very unusual circumstances, an equilibrium, or reversible, process. Not recognizing the differences between these two usages of the same term, and the difference in the phenomena to which they relate, leads to a great deal of confusion and, sometimes, to incorrect concepts.

Viewed in a more conceptual way, the first is a qualitative matter relating to why, fundamentally, materials brought into contact may resist separation, without saying anything about the "goodness" or "poorness" of the resistance. The second, on the other hand, is a quantitative matter relating to level of resistance to separation of an adhering system.

1.2 DEFINITION OF "INTERPHASE"

Some time ago the author suggested (to ASTM) that the term "interphase" be defined as follows:

"A region intermediate to two (usually solid) phases in contact, the composition and/or structure and/or properties of which may be variable across the region and which also may differ from the composition and/or structure and/or properties of either of the two contacting phases."

One of the main purposes of this paper is to attempt to convince workers in adhesion science and technology that interphases do, in fact, exist and that they are of considerable importance in the performance of adhering systems of polymeric and other materials. In addition, that they influence and, in fact, may control, the mechanical and other behavior of macro-systems such as adhesive joints, coating/substrate systems and fiber- and particulate-reinforced composites. Finally, that we need to incorporate them into our models of these macro-systems, otherwise our models are flawed. We will return to this later in this paper.

1.3 IMPORTANCE OF INTERPHASES

In the past, and too often even now, models which are intended to describe the mechanical response (the "performance") of systems of adhering (or "joined") materials (more precisely objects which have, in addition to a certain composition and structure, a certain form) have been of two types. <u>Firstly,</u> there is the approach which emphasizes the role of the interface between the different elements of the joint as the determining factor in its response. It attempts to link changes in joint response directly (and solely) to changes in molecular structure or "bonding" (in the chemical sense) <u>at</u> the interface. Such a linkage presumes that it is possible to identify, to isolate, and to assign, simple cause and effect relationships between (two-dimensional) interfacial structure and mechanical response of a joint system. This is the area in which uncritical or imprecise usage of the term "adhesion" causes confusion. <u>Secondly,</u> there is the approach which completely ignores the role of the interface and attempts to understand system response in terms of the bulk response of the members of the system and a system geometry. The first view is that of the chemist--that is, consideration of molecular structure, interaction energies and bonding. The second is that of the mechanical engineer--that is, consideration of macroscopic response and fracture.

Neither of these approaches is capable of describing accurately the response of the systems with which we usually have to deal, because they both ignore the presence, in any real system, of boundary layers or "interphases" (as the author prefers to call them).

The surface region (i.e., its composition, structure, properties, scale, etc.) of a particular piece of material is influenced by the processing history of that particular piece of material. It will differ, in general, from the surface region of an originally identical piece of material processed in a different way. This can be illustrated by comparing, e.g., the variation in morphology of the surface oxide on a metal (e.g., copper) which results from surface treatment (Figure 1).

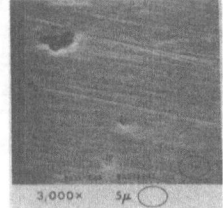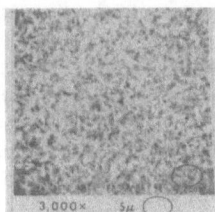

Figure 1. Surface morphology of copper: Left, polished copper; Right, copper subjected to a proprietary alkaline sodium chlorite treatment (Ebonol C^{TM}).

The situation in the case of polymers is rather more complicated due to the interplay of functional, compositional, structural and morphological factors in these materials. This complexity increases further if we are dealing (as we usually are) with formulated materials. Such materials contain fillers, plasticizers, extenders, mold release agents, etc., some of which are mobile and may also be surface active in the base polymer. The net result of these complications is that the surface regions of polymer systems are compositionally, structurally and morphologically quite complex, and are subject to variation with the details of processing. These surface regions, as well as those produced by various surface treatments (e.g., plasma, corona, oxidizing media), are interphases which have structure and properties, particularly mechanical properties, different from the bulk. Therefore, these interphases must be reckoned with in trying to understand the mechanical performance of adhering systems.

Polymers solidified from the melt, or polymerized from monomers or pre-polymers in contact with a solid may, while fluid, assume certain conformations at the surface of the solid (surface or interfacial structure) which are different from their conformations in the bulk solid, as well as those at the surface of the polymer in contact with a surrounding gaseous atmosphere. Such surface structure, and perhaps variations of it, may extend into the bulk perhaps tens to thousands of angstroms and may be preserved upon solidification, creating an interphase with properties different from the bulk.

2. Some Examples of Interphases

1. There are several early studies which illustrate the concept that a polymer adsorbed on a solid substrate has properties different from the bulk. For example, Kumins and Roteman [3] showed that the T_g of a PVAc/PVC copolymer was raised by several degrees in the presence of a TiO_2 filler. Kwei [4] proposed a model describing the effect of the filler on polymer segment mobility in filled polymer systems. Droste and DiBenedetto [5], studying a thermoplastic epoxy polymer filled with glass beads and attapulgite clay, found that the T_g of this polymer was also raised by incorporation of the filler. All of these authors attributed this effect to so-called "bound" polymer—polymer with reduced mobility due to adsorption on a solid. The work of Lipatov [6], over many years, fully supports the concept that polymer in the region of a polymer/solid interface has properties different from polymer in the bulk.

2. In a polymer system curing in contact with, e.g., a mineral surface such as aluminum oxide, or iron oxide, reactions can occur between the curing agent and the oxide. Salts may be formed, curing agent may be oxidized and, if a silane is present on the metal oxide, it may react with the curing agent to form products not present in the bulk polymer. These same processes can affect the crosslink density (or structure) of the polymer in the interfacial region or interphase, thus making it different from the bulk. An example of such a system comes from recent work of Boerio et al. [7]. They studied, by RAIR, ATR and XPS,

molecular structure in the interfacial region of aluminum/epoxy and steel/epoxy joints, primed with an aminosilane. They found that the interphase structure varied with the curing agent and the temperature and that it was different from the bulk.

3. A study by Comyn et al. [8] indicated that low (or no) cure took place in the interphase between an amine cured epoxy and aluminum because the amine was preferentially adsorbed onto the aluminum oxide on the aluminum. Garton et al. [9] showed that the acidic surface of a carbon fiber selectively adsorbed amine and catalyzed the reaction between the amine and an epoxy resin. Nigro and Ishida [10] found that homopolymerization of epoxy resin was catalyzed by a steel surface. Zukas et al. [11] discovered, in a model system of an amine cured epoxy resin and an activated aluminum oxide, a change in the relative rates of the reactions leading to crosslinking of the epoxy, so that the material in the interphase was structurally different from that in the bulk.

All of these studies illustrate that the course of polymerization reactions can be altered by the presence of certain solids in contact with the polymerizing material, leading to interphases with structures different from the bulk polymer.

4. The work of Schonhorn et al. on transcrystallinity (see below) illustrates an interphase which arises from mainly physical processes. Transcrystallinity arises in the interfacial region of crystallizable polymers when they are solidified from the melt in contact with a solid (nucleating) substrate. When crystallization is initiated, adjacent crystallites try to grow simultaneously. They interfere with each other's lateral growth and are forced to grow in columns. The rate of cooling from the melt essentially controls the depth of the transcrystalline region, which can vary from perhaps a few thousand angstroms to 100 micrometers or more. A transcrystalline region is another example of an interphase. It is one which is formed by the physical processes of nucleation and crystallization. Nevertheless, it is an interphase which has demonstrably different mechanical properties from the normal bulk solid. Kwei et al. (see below) found that the dynamic mechanical storage and loss modulus of a transcrystallized polyethylene and polypropylene, in the direction normal to the growth, were both considerably higher than in the normal bulk polymer. It follows that such an altered layer will have an effect on the mechanical response of a joint made with such a crystallizable polymer. (See Section 4.1.4 below for a discussion of the effects of transcrystallinity on joinability of polyethylene)

5. Several years ago, the author suggested [12] that it might be possible for roughness alone to create an interphase. The conceptual model was one in which the geometry of a roughened surface of a high-modulus material (say, a metal) was considered to vary more or less abruptly in every direction along the surface on some small scale, perhaps several hundred to several thousands of angstroms. If we apply

an external load to a macroscopic section of a joint made between such a roughened metal and a (lower-modulus) polymer, there will be produced, in the interfacial region, stresses and strains in the polymer which are characteristic of the <u>local</u> modes of loading induced by the <u>local</u> geometries. Since these geometries vary, the local stresses and strains will vary. The small scale of the constraints makes it possible for the stress fields in one local geometry to interact with fields in surrounding geometries, inducing the polymer to behave in a way quite different from its normal bulk behavior. The net result will be that the interfacial region of the polymer will deform and fail in a manner characteristic of the <u>local</u> geometries and the <u>local</u> constraints and not of the far-field (bulk) material. This would create <u>the effect</u> of a boundary layer, resulting solely from the geometric constraints of micro-roughness, which would cause the polymer <u>apparently</u> to behave mechanically in a manner different from the bulk simply due to the scale of the constraints.

There are many other examples which could be given for the existence, or probable existence, of interphases in systems of joined materials and of their probable effect on the mechanical response of such systems. However, it is believed that the evidence given here is quite convincing. The point to be made is that the interphase is a useful concept in attempting to understand the mechanical and other behavior of adhering systems, for the reason that interphases do, in fact, exist in real systems. Because of this, they must be reckoned with if we intend seriously to understand the performance, particularly the mechanical performance, of adhering systems.

3. Theories of "Adhesion"

What follows is a brief review of the theories of "adhesion", really "interfacial adhesion", although one of them, the diffusion theory, involves more than a two-dimensional view of the interface.

3.1 THE ADSORPTION THEORY

The origin of this theory is not clear. As generally accepted, it states that materials adhere by physi- or chemisorption and that:

1) The strength of joints is dermined mainly by interfacial forces

2) Strong joints result from primary valence bonds (chemisorption) across the interface or as a result of the presence of "polar groups"

3) Weak joints and failure "in adhesion" result from weak (van der Waals) forces across the interface

There are several bases on which on can criticize this theory. Among them are:

a) There is no direct and satisfying proof that chemical reaction (strong chemical bonds) <u>at</u> an interface contributes to the strength of a joint <u>system</u>

b) It is questionable whether such reaction is <u>ever</u> confined strictly to the interface--that is, if it is ever two-dimensional. In fact, there is good reason to believe that it is three-dimensional, i.e., it has depth (there is an <u>interphase</u>), particularly in the case of polymeric materials

c) A theory based strictly on interfacial interaction cannot be expected to explain or describe the mechanical response of a system of materials in which volume deformations are occurring.

3.2 THE MECHANICAL OR "HOOKING" THEORY

Hooking or interlocking is sometimes thought to be an essential feature in the interfacial adhesion of materials. It probably plays a direct role in the case of adhering systems of porous materials such as paper, cloth, wood or metallized plastics. However, because high-strength joints can be made with smooth adherends such as glass this theory cannot have general applicability. It is well known that roughness does have an effect on system strength. But that effect is not due, in general, to the simple, direct effect of hooking or locking or interference at the interface. It probably arises from the much more subtle effects of roughness in determining the microgeometry of the interfacial region. This microgeometry, in turn, affects the local microdeformation and failure in the interfacial <u>region</u> of a joint--thus influencing its macroscopic response.

3.3 THE ELECTROSTATIC THEORY

This theory, due to Deryagin [13] and his co-workers, treats the joint system as though it were a capacitor which is charged due to the contact of the two materials which make up the joint. The strength of the joint is presumed due to the existence of an electrical double layer at the interface. Apparently, the most important observation which led Deryagin to propose this theory was that electrical discharges and even electron emission can occur when one strips pressure-sensitive adhesive tape from a substrate. Presumably, this is due to separation of charge and development of a potential difference between the two halves of this capacitor which increases with separation until a discharge occurs.

This theory may be criticized on several grounds:

a) The electrical phenomena which are the basis of the theory occur only when the joint is broken. That is, the theory draws on a <u>result</u> of fracture--electrification--for explanation of a joining phenomenon

b) It is difficult to see how failure phenomena bear a direct one-to-one relationship to joining phenomena, because the rheological (and sometimes the chemical) state of at least one of the members of a joined system is different in the two instances (excepting pressure-sensitive adhesives)

c) There is no evidence to support the belief that the charged fracture surfaces are identically the same two (presumably) initially uncharged surfaces which were put together to form the joint system

d) Electrically conductive materials should not form joints, because they could not support separation of charge; however, they do

While electrostatic phenomena are of some interest and importance in the matter of adhering materials, there is no compelling evidence that they have a broad-based significance in relation to adhesion phenomena.

3.4 THE DIFFUSION THEORY

This theory, by Voyutskii [14], maintains that the extent of diffusion of polymers across the interface determines joint strength and that surface contact alone cannot be sufficient to create strong joints.

The major arguments for this theory are based on measurements of the breaking strengths of adhesive joints as a function of time of contact, temperature, polymer type, molecular weight, viscosity, and so forth. The author claimed that since the functional dependence of joint strength on these parameters is similar to what would be expected for a diffusion-controlled process, joint strength is determined by diffusion and mutual mixing of materials. By this view, then, the development of joint strength becomes a volume phenomenon; that is, where the materials join there must always be a diffusion-created layer.

One might level at least two criticisms at this theory:

a) It cannot have broad-based applicability, because it cannot explain development of joint strength in systems containing a hard solid, such as glass or a metal oxide. Diffusion of a polymer across a boundary or an interface could not occur in these cases at the temperatures or during the times used to make, e.g., adhesive joints

b) Diffusion-like behavior could simply be the result of "diffusion" (really local or short-range flow) of a polymer (or both polymers in certain polymer/polymer systems) to an interface to produce increasing area of contact

Despite these criticisms, however, we are not saying that diffusion-created interfaces or interphases do not exist or do not influence mechanical behavior of joints. They do. But Voyutskii's diffusion theory simply is not, and cannot be, the basis for a general understanding of the mechanical behavior of, e.g., adhesive joints or other adhering systems.

3.5 THE RHEOLOGICAL THEORY OF ADHESIVE JOINTS

This theory, due to Bikerman [15], is not a theory of interfacial adhesion. It states, in substance, that the strength (the breaking stress, the performance) of an adhesive joint is determined by the mechanical properties of the materials comprising the joint and the local stresses in the joint. It is not determined by interfacial forces, because clean failure "in adhesion" is a highly uncommon occurrence. Failure is essentially always cohesive, in the adherends and/or the adhesive or in some boundary layer.

It is a theory which attempts to answer the question..."What determines the strength or, more broadly, the mechanical response of adhesive joints?" It is not concerned directly with the fundamental question of what interfacial forces act across the interface. It is more directly concerned with the "real world" function of adhesive joints; that is to say their performance, and how it may be quantified and understood in some accessible fashion. As stated above, the theory gives little credence to the role of interfacial adhesion in performance. It postulates, on the basis of several arguments, that true interfacial failure rarely, if ever, occurs in the breaking of a joint by purely mechanical means. Therefore, we can neglect it as a contributing factor to the mechanical behavior of, e.g., an adhesive joint.

One can give several arguments for the improbability of "adhesional" or interfacial failure. Among them are:

a) The surface of a real adherend is generally a highly-irregular, three-dimensional contour, relative to atomic dimensions. On probability grounds, one should not expect failure to occur along this predetermined, highly irregular, three-dimensional path in response to some external loading

b) Because the surface is three-dimensionally irregular, a "simple" external mode of loading (tension, shear) is transformed into complex and locally varying modes of loading in the interfacial region

c) In many joints there is interpenetration (diffusion) of the materials. Therefore, no interface exists and true interfacial failure cannot occur

One should not think of this theory as applying only to the breaking strength of a joint. Rather, it should be thought of as applying more generally as a theory of the mechanical response of joints. One can then view joints as composite structures (structures comprised of differing materials) the mechanical behavior of which can be described and understood by application of the theories and methodologies of analytical mechanics and fracture. A very powerful point of view, indeed, when one remembers that it is the mechanical response in, e.g., an adhesive joint, that mainly concerns us because that is the major aspect of its performance. That is, are the joined parts going to stay together?

4. Interphases Illustrated by Studies of Poly(ethylene)

4.1 THE TREATMENT OF POLY(ETHYLENE) FOR JOINING

It is difficult to join the many varieties of poly(ethylene) (PE) with conventional adhesives (e.g., epoxies or polyesters) in a "structural" manner (i.e., fail the joint in the PE), without first treating the PE in some way. The reason usually given for this difficulty is that the material is "waxy" and non-wettable (correct) and that what one must do to improve its joinability is to improve its wettability (incorrect, as we shall see). To change its wettability, the material is sometimes flamed, or subjected to a corona discharge, or immersed in oxidizing media such as common laboratory glass-cleaning solution. All of these treatments change the chemical functionality of the surfaces as shown, e.g., by a reduced water contact angle. It is this improved wettability which is usually given as the reason for improved joinability, through oxidation and the generation of polar groups, thus, to stronger surface interactions with the adhesive, and so forth. In other words, the improvement is considered to be due strictly to surface effects.

4.1.1. *CASING.* There exists a body of work, all of it more twenty years old, which shows that wettability changes are probably secondary or side effects of the conventional "surface" treatments, and that the primary effect is a change in the mechanical properties of a thin surface layer in the PE, a <u>region,</u> of the order of perhaps hundreds to thousands of angstroms.

Schonhorn and Hansen [16] exposed PE to a radio-frequency excited glow discharge in several noble gases, a process which they called CASING (for Crosslinking by Activated Species of INert Gases). They found that PE subjected to CASING would retain its shape upon heating above its normal melting temperature, because the treatment produced a thin, tough, crosslinked "skin" on the PE. They obtained this skin for examination by extracting the soluble interior in a suitable solvent and found it to be quite thin. Figure 2 shows that the thickness of the skin (calculated from its dimensions and weight, assuming a density of 1.00) varied from about 3×10^2 angstroms at an exposure time of one second to about 10^4 at 10^4 seconds. They also found, Figure 3, that a 5-second exposure time was sufficient to maximize the joinability of PE with a conventional epoxy adhesive in an aluminum lap joint. It appears, therefore, that a layer of only about $5\text{-}10 \times 10^2$ angstroms governs this behavior. This is an example of an interphase that has been fabricated in a material, an interphase that markedly increases the strength of joints made with the material. This, despite the fact that the water contact angle of such treated PE was virtually unchanged from the unexposed material, provided that the exposure times were short, 5 seconds or less.

4.1.2. *Oxidation.* Morris [17] used an aluminum double lap joint to

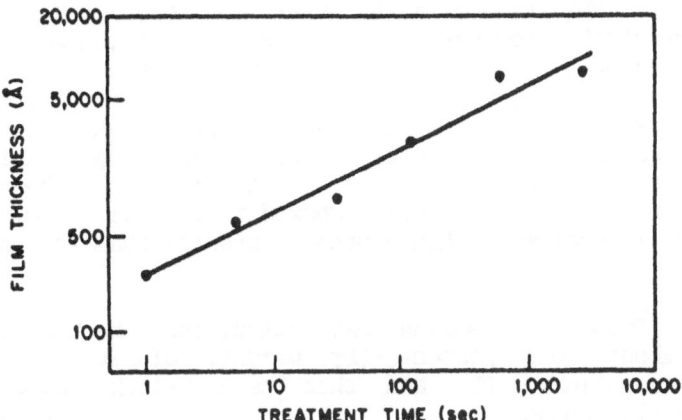

Figure 2. Thickness of crosslinked "skin" on PE as a function of CASING treatment time. The inert gas is helium. Adapted from Reference 16.

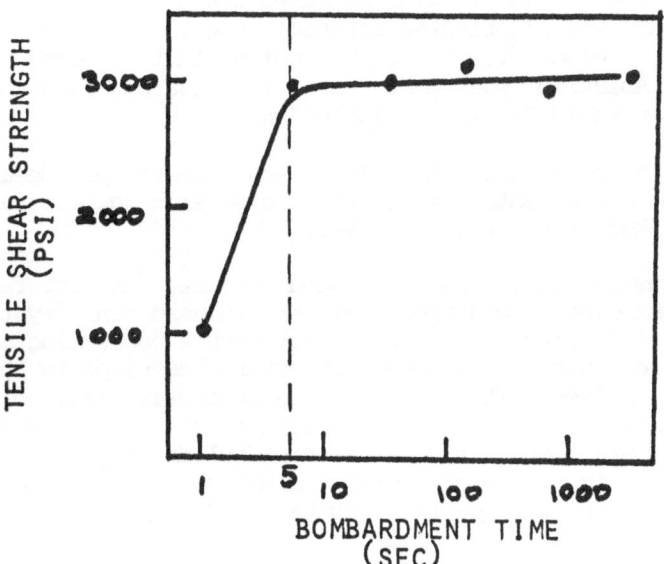

Figure 3. Tensile shear strength of Al/Epoxy adhesive/PE/Epoxy adhesive/Al joint as a function of CASING treatment. Adapted from Reference 16. See Fig. 4 for schematic of the joint.

assess the joinability of a PE that had been treated in aqueous ammonium persulfate for various times. She also found marked increases in joint strength with treatment time. Despite the fact that ammonium persulfate is a strong oxidizing agent, she found very little evidence of polymer oxidation either from IR spectra taken in the ATR mode (which samples a surface layer about 0.1 wavelengths thick) or from contact angle measurements (which effectively sample the surface). In all cases, the treated material showed a critical surface tension of wetting (CST) never more than 5 dyne/cm greater than for the untreated PE.

Morris also found, by solvent extraction, that the treated materials contained about one percent (by weight) of insoluble material, presumably crosslinked PE, and that the crosslinked material was the only substantial change produced in the material by the treatment. She concluded, as Schonhorn and Hansen did, that crosslinking of the surface region of the PE was the primary effect of the treatment.

4.1.3. *Flouorination*. In Figure 4 the strengths of the lap joint shown, which contained variously treated PE, are compared. The work is due to Schonhorn et al. and is a composite of the results of a number of studies. Joints made with PE treated with glass cleaning solution (highly oxidizing), by CASING, and even with elemental fluorine at ambient conditions, gave equivalent strengths although their CSTs are widely different. The fluorinated material, e.g., has a CST of about 20 dynes/cm [18], close to Teflon™, while that treated with glass-cleaning solution was more than 40 dynes/cm. Both exhibit, as does the CASING-treated PE, an insoluble "skin".

According to these studies the only common property produced by these widely-differing treatments, all of which give high joint strengths, is a thin, crosslinked or gel surface layer.

4.1.4. *Transcrystallinity*. As mentioned before, an interesting surface morphological condition may be induced in crystallizable polymers by solidifying them in contact with a nucleating substrate. One may observe that a columnar structure develops in the surface region of the polymer. This structure is obviously different, both in geometry and scale, from the larger spherulitic structure usually seen in the bulk. Figure 5 shows such a transcrystalline region in PE melted and solidified in contact with aluminum foil. The optical photomicrograph of the thin section was taken using polarized light.

Kwei et. al. [19] have measured the dynamic mechanical properties of the transcrystalline region in high-density PE normal to the direction of the column. His results are shown in Table 1. Both the storage modulus, E', and loss modulus, E'', of the transcrystalline region are considerably larger than those of the bulk material.

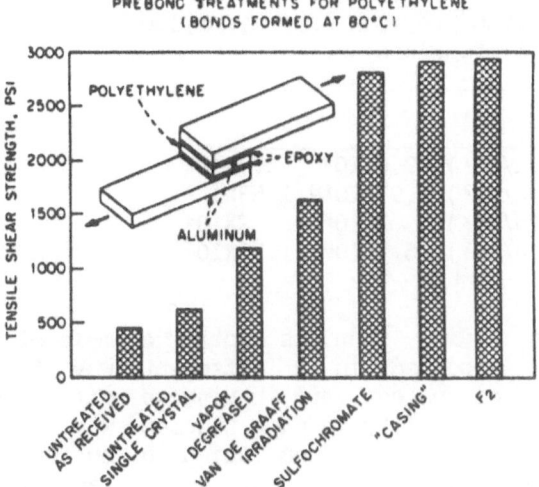

Figure 4. Breaking strength of joint shown for various treatments of polyethylene. ("sulfochromate" is common glass cleaning solution). Adapted from Reference 26.

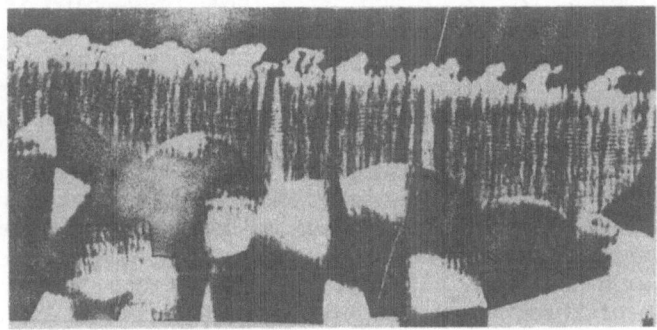

Figure 5. Transcrystalline growth in polyethylene. From Reference 20.

TABLE 1. Dynamic mechanical moduli of the bulk
and surface regions of polyethylene
and polypropylene

	PE	PP
$E_B{}'(\mathrm{dyn/cm^2})$	9.0×10^9	9.7×10^9
$E_S{}'(\mathrm{dyn/cm^2})$	1.97×10^{10}	1.53×10^9
$E_B{}''(\mathrm{dyn/cm^2})$	2×10^8	2×10^8
$E_S{}''(\mathrm{dyn/cm^2})$	6.7×10^8	1.20×10^9

The transcrystalline region, then, is another example of an interphase
in PE. This layer, produced in a different way from the other
treatments already discussed, and having a different structure,
nevertheless affects joinability in much the same way as the others do.
That is, joints made with PE from which a nucleating substrate
(aluminum) had been etched away in aqueous sodium hydroxide (presumably
without affecting either the PE or the transcrystalline structure),
gave approximately the same strengths as the same PE treated by CASING,
glass cleaning solution, or fluorine [20]. In addition, when the
nucleating substrate was gold, the transcrystalline PE showed much
higher CST values (70 dynes/cm) than for "normal" PE [21]. However,
heating it at 80°C for one hour in a nitrogen atmosphere caused
the CST to drop to its normal, lower value, without any sensible change
in either its transcrystallinity [22] or its joinability [23]. In a
similar way, the CST of FEP Teflon was also raised from 18 dynes/cm to
40 dynes/cm by nucleating and crystallizing it from the melt in contact
with gold [24].

The broad conclusion to be drawn from the preceeding is that the
mechanical response of a joint which contains PE joined to a more rigid
material (the epoxy adhesive) is governed primarily by the response of
the surface region, the interphase. It also may be appropriate at
this point to emphasize the following. Firstly, when one studies the
deformation and fracture of an adhesive joint one is studying, first
and foremost, a mechanical phenomenon. One is studying directly the
mechanical response of a composite structure to the application of a
load. Therefore, it makes a great deal of sense to try to understand
what that mechanical response means in mechanical terms, rather than
in strictly chemical terms. Secondly, in many instances interphases,
with properties different from the bulk materials, are involved in the
mechanical response of a composite structure such as an adhesive joint.
Therefore, one should try to characterize such interphases
compositionally and structurally, and their effects on the response of
the structure should also be considered and studied. If this is not
done, a most important, and many times controlling, aspect of the
mechanical performance of such structures has been neglected.

5. Material Conservation During Failure

It is common practice, using various techniques, to attempt to determine the locus of failure ("in adhesion", "in cohesion", "mixed", etc.) in failed systems, e.g., adhesive joints, and to use the results to try to assign a cause of failure and to correct it. So far as the author knows, it is always tacitly assumed in such studies that all material in the joint system is conserved during the failure process.

Conclusions which we reach about the locus of failure, and about the failure process itself, from fractographic examination may be flawed if we make the assumption that material in the joint is conserved during the failure process. That is, such conclusions may be flawed unless we do collateral studies to prove that material is not ejected from the joint during failure. The reason is that what is left behind to be "viewed" after failure may have been created by secondary processes which produced particulate ejecta, processes which may not have been related to those which were the proximate cause(s) of initiation and propagation of the failure. That is to say, the production of particulate ejecta may, in fact, result from an entirely different process or processes (e.g., reflection of a release wave from the traction points in a tension-induced failure) from that (or those) which resulted in initiation and propagation of failure. Therefore, such processes may alter or obscure the evidence associated with the primary failure process.

The work to be described which supports the concept of particulate ejecta production during failure was done by Logioco [25] many years ago but it was never published. Logioco made single-lap joints from transparent polycarbonate adherends using a simple UV-curable adhesive. He then loaded them in tension and photographed the initiation and propagation of failure with a movie camera at 30 frames/second, using a mirror to view the joint simultaneously from the front and side.

Figure 6 shows two photographs of such a joint, with the center 1/8-inch bonded, as it is about to fail (a), and as it finally separates (b). One can clearly see that considerable material, apparently particulate, is ejected from the joint at failure. Therefore, material was not conserved in this joint during failure.

If one observes the fracture surfaces of the adhesive on two adherends from a similar joint one concludes from its hackled, rough appearance that a large amount of energy was stored and then rapidly dissipated in the adhesive. The material almost "explodes" when the joint fails, explaining the relatively large amount of material ejected at failure.

The major point to be made is that the potential exists, even in joints which do not have relatively brittle or high-strength adhesives, for material ejection to occur, particularly if the joints are fractured at high strain rates. It is simply a matter of the degree to which this occurs and, therefore, the degree to which the interpretation of fractographic evidence is made more complicated and questionable. The author believes that this is an important matter

16

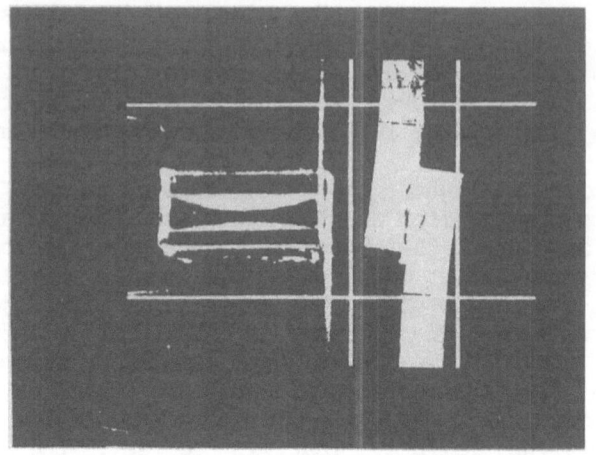

A

CONFIGURATION AT IMPENDING FAILURE

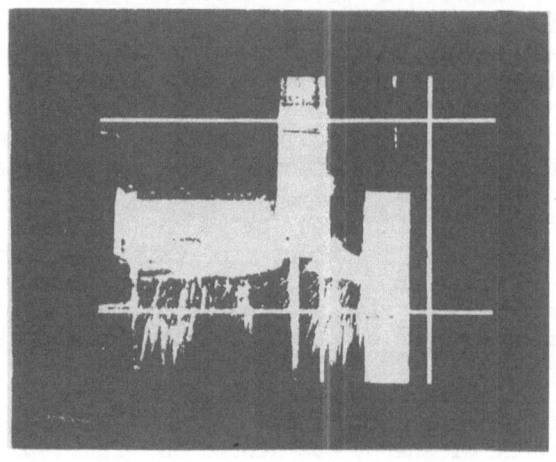

B

FINAL SEPARATION

Figure 6. Configuration of the joint described above at impending failure, A (top), and at the moment of final separation, B (bottom). Note the material ejected from the joint at failure.

which should be given some attention, because it raises the question as to whether or not interpretations of fractographic evidence are always sound and, in fact, whether conclusions drawn from the usual such evidence is always credible.

As Dickinson and co-workers [26] have shown, for fracture-induced material ejecta the aggregate surface area of the ejected material particles may be greater than the cross-sectional area of the fractured sample; therefore, "...ejecta should be considered in any description of fracture for most materials." They also point out that high strength materials yield the most finely-divided ejecta with high surface areas. This is, of course, because these are the materials which store the higher strain energies prior to fracture, therefore the ones which produce the more violent "explosions" at break.

6. Simple Cause and Effect Relationships

Even a "simple" adhesive joint, e.g. a lap joint, is a layer structure which exhibits highly complicated response to an external load. The nature of the response can vary greatly depending on the particular material combinations with which one is dealing. A (sometimes) minor change in a material property or its layer geometry, the nature of the loading or its rate, can induce different responses in the materials making up the joint and can thereby produce (sometimes large) changes in joint behavior. The points to be made are that an adhesive joint is a multi-layered composite structure, a system; that the response of this system is generally dependent on the response of its individual components; and that the mechanisms producing changes in system response may be many, sometimes interactive, and probably complex. Is it reasonable, then, in the face of this complexity, to draw cause and effect relationships between simple changes in chemical structure at an inter face and (macroscopic) system response?

In this connection, some of what the present author wrote about composite response, many years ago, may be worth repeating verbatim here [12]:

"It seems to me that we are failing to face the fact that a composite or composite structure is a system. The problem of response which faces us is a systems problem and it ought to be treated as such. That is, we need to be concerned with describing, explaining, and finally understanding, how the responses of the individual, not necessarily independent, parts of the system interact to determine response of the system as a whole.

It seems to me also that we need to be able to describe the system behavior phenomenologically, and be sure of that, before we can proceed to make sense of system response in a fundamental way. In fact, we may very well have to proceed through a hierarchy of levels of aggregation before we can finally reach the fundamental or molecular explanations of composite response which so many workers seem to be searching for.

What I am saying is that morphologies of one or more levels of scale
may intevene between the molecular level and the macroscopic level for
each of the materials in a composite. If this is so, then each of the
materials has to be viewed as being itself a composite material, the
response of which has to be described. But it is generally true that
the description of the response of these materials is no better than
phenomenological. Then how can it be reasonable to propose fundamental
explanations of composite response when one does not have a fundamental
understanding of the response of the elements of the composite?..."

"...If we cannot describe behavior on a macroscopic level, how can we
isolate, identify and assign causes and effects on a fundamental level?
For example, how can we say things in detail about the mechanism of
action of coupling agents in changing composite mechanical behavior,
when we do not even consider, much less use, a mechanical model of the
interface region in arriving at our conclusions? It seems to me that we
are going to have to work ourselves down the hierarchy of structure,
determining how each successive level determines response, rather than
to try to bypass the hierarchy and attempt to relate molecular
structure directly to mechanical response."

What is being pointed out here is that we do not, at present, have
well-developed mechanical models which connect the micro-cause(s)
(the molecular bases of the initiation--and propagation--of failure)
with the macro-effect (the measured ultimate mechanical performance
of a joint system). It is well known that changes in "surface"
properties produce changes in mechanical behavior of joint systems.
What we do not as yet fully appreciate is the mechanism (or mechanisms)
by means of which the influence of surfaces is transformed into changes
in mechanical behavior of joint systems. The concept of interphases
appears to be a basis for developing an understanding of this
connection. At the very least, the concept has the virtue of being
demonstrably close to reality.

7. Conclusions

Interphases exist. They influence (and may determine) certain
mechanical and other properties of systems in which they are present.
There needs to be a more general acceptance of their importance and
further studies need to be directed at answering the following
questions about interphases:

1. How are they created?

2. What is their composition?

3. What is their structure?

4. What are their properties?

5. How do they influence the (mechanical and other) performance of

systems of joined materials?

8. References

1. Adamson, Arthur W. (1982), Physical Chemistry of Surfaces, Wiley-Interscience, New York, p. 424 et seq.

2. Sharpe, Louis H., Schonhorn, Harold (1964), Surface Energetics, Adhesion and Adhesive Joints, Advances in Chemistry 43, 189.

3. Kumins, C. A., Roteman, J. (1963), J. Polym Sci, Part A, 1, 527.

4. Kwei, T. K. (1965), "Polymer-Filler Interaction. Thermodynamic Calculations and a Proposed Model", J. Polym. Sci. A-3, 3229.

5. Droste, O. H., DiBenedetto, A. T. (1969), J. Appl. Polym. Sci. 13, 2149.

6. See, for example, Lipatov, Yu. S. (1988), Colloid Chemistry of Polymers, Elsevier, New York, and references therein.

7. Dillingham, R. G., Boerio, F. J. (1987), "Interphase Composition in Aluminum/Epoxy Adhesive Joints", J. Adhesion 24, 313-335; Ondrus, D. J., Boerio, F. J., Grannen (1989), "Molecular Structure of Polymer/Metal Interphases", J. Adhesion 29, 27-42.

8. Comyn, J., Horley, C. C., Oxley, D. P., Pritchard, R. G., Legg, J. L. (1981), J. Adhesion 12, 171-188.

9. Garton, A., Stevenson, W. T. K., Wang, S. P. (1988), J. Polym. Sci., Chem. Ed. 26, 1377.

10. Nigro, J., Ishida, H. (1989), J. Appl. Polym. Sci. 38, 2191.

11. Zukas, Walter X., Craven, Kelly J., Wentworth, Stanley E. (1990), "Model Adherend Surface Effects on Epoxy Cure Reactions", J. Adhesion 33, 89-105.

12. Sharpe, Louis H. (1974), "Some Thoughts About the Mechanical Response of Composites", J. Adhesion 6, 15-21.

13. Deryagin, B. V., Smilga, V. P. (1960), Proc. 3rd Intern. Congress Surface Activity, Cologne, 1960, Universitatsdruckerei, Mainz, Vol. 2, p. 349. (Review)

14. Voyutskii, S. S. (1963), Autohesion and Adhesion of High Polymers, Polymer Reviews, Vol. 4, Interscience, New York.

15. Bikerman, J. J. (1968), The Science of Adhesion Adhesive Joints, 2nd edition, Academic Press, New York.

16. Schonhorn, H., Hansen, R. H. (1967), "Surface Treatment of Polymers for Adhesive Bonding", J. Appl. Polym. Science 11, 1461-1474.

17. Morris, C. E. M. (1970), J. Appl. Polym. Sci. 14, 2171.

18. Schonhorn, H., Hansen, R. H. (1968), "Surface Treatment of Polymers. II. Effectiveness of Fluorination as a Surface Treatment for Polyethylene", J. Appl. Polym. Sci. 12, 1231-1237.

19. Kwei, T. K., Schonhorn, H., Frisch, H. L. (1967), "Dynamic Mechanical Properties of the Transcrystalline Regions in Two Polyolefins", J. Appl. Polym. Sci. 38, 2512-2516.

20. Schonhorn, Harold, Ryan, Frank W. (1968), "Effect of Morphology in the Surface Region of Polymers on Adhesion and Adhesive Joint Strength", J. Polym. Sci., Part A-2, 6, 231-240.

21. Schonhorn, H. (1967), "Hetergeneous Nucleation of Polymer Melts on Surfaces. I. Influence of Substrates on Wettability", J. Polym. Sci. B5, 919-924.

22. Schonhorn, Harold (1968), "Heterogeneous Nucleation of Polymer Melts on High-Energy Surfaces. II. Effect of Substrate on Morphology and Wettability", Macromolecules 1, 145-151.

23. Schonhorn, H. Private communication.

24. Schonhorn, Harold, Ryan, Frank W. (1969), "Effect of Polymer Surface Morphology on Adhesion and Adhesive Joint Strength. II. FEP Teflon and Nylon 6", J. Polym. Sci. 7 (Part A-2), 105-111; (Hara, K., Schonhorn, H. (1970), "Effect on Wettability of FEP Teflon Surface Morphology", J. Adhesion 2, 100-105.

25. Logioco, J. W. (1974). Unpublished work.

26. Donaldson, E. E., Dickinson, J. T., Bhattacharya, S. K. (1988), "Production and Properties of Ejecta Released by Fracture of Materials", J. Adhesion 25, 281-302.

27. Sharpe, Louis H. (1972), "The Interphase in Adhesion", J. Adhesion 4, 51-64.

NOTE FROM THE AUTHOR: Some of the concepts presented above were discussed by the author in an earlier paper [27].

ASPECTS OF COMPONENT INTERACTIONS IN POLYMER SYSTEMS

H.P. SCHREIBER
Department of Chemical Engineering,
Ecole Polytechnique
P.O. Box 6079, Stn.A., Montreal (Que)
H3C 3A7, Canada

ABSTRACT. The paper examines the influence of interactions at polymer surfaces and interfaces on the properties of polymer systems, with emphasis on acid/base interactions. The method of inverse gas chromatography is used to evaluate the donor-acceptor interaction potential of components in polymer systems. The usefulness of the interaction parameters is established by their ability to rationalize diverse properties of polymer systems, including the adsorption of polymers on pigments, and the effectiveness of thermal stabilizers in pigmented polymers.   Various strategies for controlling surface and interfacial interactions in polymer systems are reviewed, with emphasis placed on the ability of polymers to adopt various surface orientations and compositions. These inherent surface modification effects are attributed to thermodynamic driving forces, and are shown to influence polymer adhesion, barrier and other properties dependent on surface and interfacial forces.

INTRODUCTION:

The use of polymers seems limitless in its variety; the technology associated with these uses rich in its sophistication. One of the main reasons for this is that polymers are virtually never used alone, but always in combination with other materials. These added materials may be stabilizers, plasticizers, reinforcing fibers, pigments or other polymers, formulated into multi-component systems with properties well suited for specified applications. Obviously, in multi-phase polymer systems, interfaces and interphases must exist. It seems obvious intuitively that the nature of these interfaces and interphases will affect the performance of the system as a whole. Inherent in that statement is the link between component interactions on the one hand, and the rheological, physico-chemical and mechanical properties of the system, on the other.

A concern for interactions requires the availability of methods able to describe them quantitatively. The first portion of this article examines approaches to the determination of component interactions, with emphasis on the technique of inverse gas chromatography (IGC), and on the use of acid/base concepts in that context. A corollary to a concern for interactions, is the ability to control them beneficially. The second portion of this article

21

G. Akovali (ed.), The Interfacial Interactions in Polymeric Composites, 21–59.
© 1993 *Kluwer Academic Publishers.*

briefly discusses techniques for the controlled modification of interfaces and of contact interactions.

PART 1. COMPONENT INTERACTIONS:CONCEPTS,MEASUREMENTS AND USES.

a. Solubility parameter.

The need to measure interactions in polymer systems was recognized early in the evolution of the polymer field. One widely practised approach is through the determination of "solubility" or "cohesion" parameters, δ. The parameter is, in effect, a cohesive energy density, as defined by Hildebrand[1] in

$$\delta = (\Delta H_v \ / \ V)^{1/2} \tag{1}$$

where ΔH_v is the molar vaporization energy of the substance and
V is its molar volume.

Originally intended for application to substances whose cohesion arose from dispersion forces, the parameter seemed to be of limited use with polymers, which generally decompose before vaporization enthalpies can be determined. The concept now has been greatly expanded. The overall δ can be divided into dispersion and polar contributions[2,3]. Often non-polar homomorphs of polar molecules can provide values of δ^d, and polar contributions, δ^p, can then be obtained from differences between δ and δ^d. Further refinements due to Hansen[3,4] have introduced a three-component solubility parameter, which separates non-dispersive contributions into polar and hydrogen bond components. This has been applied to organic liquids, and to some polymers. Calculations of δ for macromolecules also can be made from tabulated values of molar attraction constants[5], and extensive summaries of δ and of other cohesion parameters are readily available[6] to the potential user. Ultimately, however, the application of δ to polymer systems is impeded for the following reasons:

* No direct, experimental determinations of δ for polymers exist to corroborate the validity of calculations and inferences.

* Available solubility parameters generally apply to polymers as solutes at very high dilution. The concentration dependence of δ is difficult to assess.

* Data generally apply to room temperatures, and the evaluation of temperature dependence is problematic.

b. Interaction parameters from polymer solution theories.

Polymer solution thermodynamics, as developed first by Flory[7] and Huggins[8,9], expresses the interaction between a polymer and a liquid in terms of a dimensionless parameter, $\chi_{1,2}$. This can be written

$$\chi_{1,2} = (\mu_1 - \mu_1^0)/RT\varphi_1^2 - [\ln\varphi_1 + (1 - V_1/V_2)\varphi_2]\varphi_2 \qquad (2)$$

where subscripts 1 and 2 denote liquid (solvent) and polymer (solute), μ is the chemical potential, φ are volume fractions and V molar volumes.

$\chi_{1,2}$ expressed in this manner reflects intermolecular forces between the components of any polymer-liquid mixture, and therefore is not dependent on the choice of theory or theoretical model. Its usefulness in practise, however, once more is limited because it is usually determined by methods such as vapor pressure lowering, osmotic pressure, equilibrium swelling of polymers by liquids, light scattering, etc. In all of these cases, the interaction describes systems in which the polymer is at very high dilution. The temperature range over which the data may be collected is narrow and often far removed from conditions of interest. Moreover, these methods do not lend themselves readily to evaluations of what often are the most important interaction data - those between solid components of a polymer system.

c_i). **Inverse gas chromatography and some of its uses.**
IGC is a variation of coventional gas chromatography. Figure 1 shows a typical arrangement for IGC. In IGC a finely divided non-volatile material of interest (polymer, fiber, plasticizer, etc) is placed within a chromatographic column. It may be packed directly into the column, or coated onto a suitable support, or onto the walls of the column. A volatile "probe" of known characteristics is swept through the column by an inert mobile phase (eg. helium), and the output is monitored. The residence time of the probe and the shape of the output signal characterize the stationary phase and its interaction with the volatile phase.

The use of IGC has grown rapidly in recent years. The technique has been reviewed extensively[10,11] and the many uses for it have been summarized [12]. These uses include the relatively rapid, convenient evaluation of such polymer properties as the glass transition temperature, the degree of crystallinity and the surface energy. Adsorption isotherms for volatile gases on polymeric stationary phases are readily obtained from IGC [13] when significant quantities of the vapor molecule are introduced into the system. In most cases, however, the quantity of probe vapor injected into the carrier gas is extremely small. Thus, the retention datum relates to the thermodynamic interaction between polymer and vapor when the polymer is highly concentrated, as in most practical situations. Furthermore, IGC experiments may be carried out over appreciable temperature ranges, so that the temperature dependence of thermodynamic interactions is no longer indeterminate. The IGC technique has been extended[14,15] to allow for measurements of interations between mixed stationary phase components. Useful values of $\chi_{2,3}$ may be obtained for polymer blends or for mixtures of polymers with fibers, pigments, etc. Finally, recent developments have linked IGC data with acid/base concepts, and thus have provided a needed opportunity for clarifying the contribution of these specific interactions to the balance of properties in complex polymer systems.

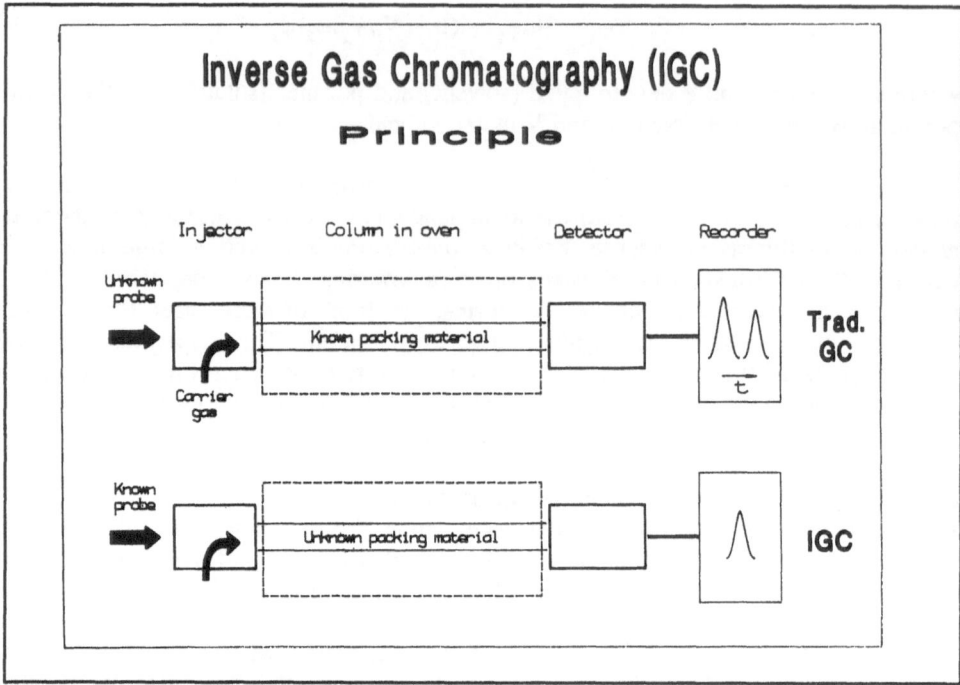

Figure 1   The principles of conventional and inverse gas chromatography.

The relationship between the basic datum of IGC, the specific retention volume, V_n, and the thermodynamic interaction datum $\chi_{1,2}$, may be written as

$$\chi_{1,2} = \ln (RTv_2/V_n\ p_1^\circ\ V_1) - (1 - V_1/V_2)\varphi 2 - p_1\ (B_{11} - V_1)/RT \qquad (3)$$

where v_2, V_2 and φ_2 are the specific volume, the molar volume and the volume fraction of the polymer, V_1 and p_1° refer to the molar volume and the saturation vapor pressure of the "probe", B_{11} corrects for non-ideality in the vapor (with values available from compilations such as those of Dreisbach[16]), R is the gas constant and T ($^\circ$ K) is the column temperature.

When $\chi_{1,2}$ values are determined for a given polymer or other non-volatile component of a polymer system and a series of vapors for which solubility parameter values (δ_1) are known, then the IGC method provides a unique way for the determination of δ for the polymer phase. The necessary relationship, introduced by DiPaola and Guillet[17], states that:

$$\delta_1^2/RT - \chi_{12}/V_1^\circ) = (2\delta_2/RT)\delta_1 - (\delta_2/RT + \chi S/V_1^\circ) \qquad (4)$$

A plot of the left hand side of eqn. 4 vs. δ_1 should lead to a straight line, with δ_2 obtained from its slope. The great advantage here is the ability to evaluate δ_2 for essentially pure polymers at temperatures relevant to identified applications. The generality and usefulness of the method has been demonstrated frequently, as for example in the work of Price[18].

Thermodynamic interactions between mixed stationary phases (2,3) may be obtained from IGC as first proposed in ref. 14. The approach calls for the experimental determination of $\chi_{1,2}$ and $\chi_{1,3}$, using common vapor probes to characterize any desired pure components 2 and 3. These solids may then be mixed to any suitable composition, a column constructed for IGC, and an overall interaction parameter, $\chi_{1(2,3)}$, evaluated. The latter is related to compositional variables through an extension of Scott's ternary solution theory[19], and the result is most frequently written as

$$\chi_{1(2,3)} = \left(\chi_{1,2}\right) \varphi_2 + \left(\chi_{1,3}\right) \varphi_3 - \left(\chi_{2,3}'\right) \varphi_2\varphi_3 \tag{5}$$

where $\chi_{2,3}' = \chi_{2,3} V_1/V_2$. The parameter for the interaction of mixed polymeric and/or non-polymeric solids is thereby normalized to the size of the vapor phase molecule. The attractiveness of this flexible and relatively easy experimental route to valuable information on the miscibility of system components is evident. Difficulties arise however, in that $\chi_{2,3}'$ is frequently found to vary with the selection of the vapor probe. This problem has been the subject of much discussion[20-23], and cannot be treated here in detail. Our view of the situation, recently published [24], states that the probe-dependence of $\chi_{2,3}$ is due to two major contributing factors. First, the surface composition of a mixed stationary phase will rarely, if ever, correspond to the composition of the bulk. Thermodynamic requirements to minimize the surface free energy of the stationary phase will favor the preferential concentration, at the surface, of the component with the lower (lowest) surface free energy γ_s: That aspect of the problem has been discussed by Lipatov among others[25], and is treated in more detail elsewhere in this volume (see Brochart-Wyart and deGennes). Thus values of φ_2 and φ_3, as defined by the bulk composition of mixtures, are inapplicable to eqn. (5). Instead a graphical method has been proposed[24] to evaluate the effective volume fraction and correct the problem. Second, since $\chi_{1,2}$ and $\chi_{1,3}$ will not usually be equal, it follows that the volatile phase will partition preferentially to the component with the lower pertinent $\chi_{1,x}$ value. Thus, the partitioning must vary with each probe, inevitably affecting the $\chi_{2,3}'$ datum. Far from invalidating the IGC route to thermodynamic information, these considerations shed new and valuable light on the nature of surface interactions in polymeric systems.

c_{ii}). IGC and acid/base interactions

Interactions at polymer surfaces and interfaces arise from two types of force. One of these, comparatively weak van der Waals or dispersion forces, are universal. Stronger interfacial forces arise from ionic, chemical or covalent bonds. These non-dispersion forces, of course, are not universally present. Fowkes [26] has proposed that non-dispersive interactions be represented quantitatively as (Lewis) acid/base, or electron acceptor/donor effects. Accordingly, the strength of an interface, as represented by the work of adhesion, W_a, can be written:

$$W_a = W_a + W_{ab} \tag{6}$$

Here ab represents acid/base effects, and the assumption is made that other non-dispersion (eg. dipole) forces may be neglected.

The above expression provides a convenient link to IGC. The free energy of adsorption per mole of "probe" molecules, $-\Delta G_{ads}$, is a function of the retention volume, V_n,

$$-\Delta G_{ads} = RT \ln V_n + \text{const} \tag{7}$$

It is also related to W_a via

$$-\Delta G_{ads} = N a W_a \tag{8}$$

where N is the Avogadro number, and a is the surface area of an adsorbed molecule. When only dispersion forces are involved at an interface, then according to Fowkes [26,27]

$$W_a = W_a^d = 2 (\gamma_s^d \cdot \gamma_l^d)^{1/2} \tag{9}$$

Thus the work of adhesion is given by dispersive contributions to the surface energies of the participating solid and liquid. It follows from the above that

$$RT \ln V_n = 2N (\gamma_s^d)^{1/2} a (\gamma_l^d)^{1/2} + \text{const.} \tag{10}$$

In cases where only dispersion forces are involved, the IGC experiment therefore yields γ_s^d from a plot of $RT \ln V_n$ vs. the product $a(\gamma_l^d)^{1/2}$. The principle has been demonstrated for a wide variety of solids[28-30] by using n-alkane vapors as the volatile probe. Figure 2 illustrates the case; here n-alkanes are used as solutes for a polycarbonate stationary phase at 110° C., the required a values being those given by Schultz and coworkers[29]. The calculated γ_s^d of 28.2 mJ/m2 is very reasonable for this polymer. The method, however, does not always yield reliable values of γ_s^d. The reason for this is in the uncertainty of a values. Molecular areas of adsorbed molecules may be distorted by forces exerted by the surface. Non-spherical molecules such as the alkanes may lie "flat" or "head-to-tail" in the interface. Consequently

IGC evaluations of $\gamma_s{}^d$ must be viewed with some caution. Other limitations to the reliable evaluation of $(\gamma_s)^d$ will be mentioned later in this article.

Equation 10 is a starting point* for the evaluation of acid/base interactions. This calls for the use of volatile probes which are themselves classified as Lewis acids and bases. In principle, when such vapors are used, then provided the solid phase is able to respond as either a donor or acceptor, the observed retention volume will exceed that expected from pure dispersion-force vapors (eg. the alkanes referred to above). The difference between an observed V_n datum and that given by a pure dispersion-force molecule with the same $\underline{a}$ value then identifies the acid/base interaction potential of the solid surface.

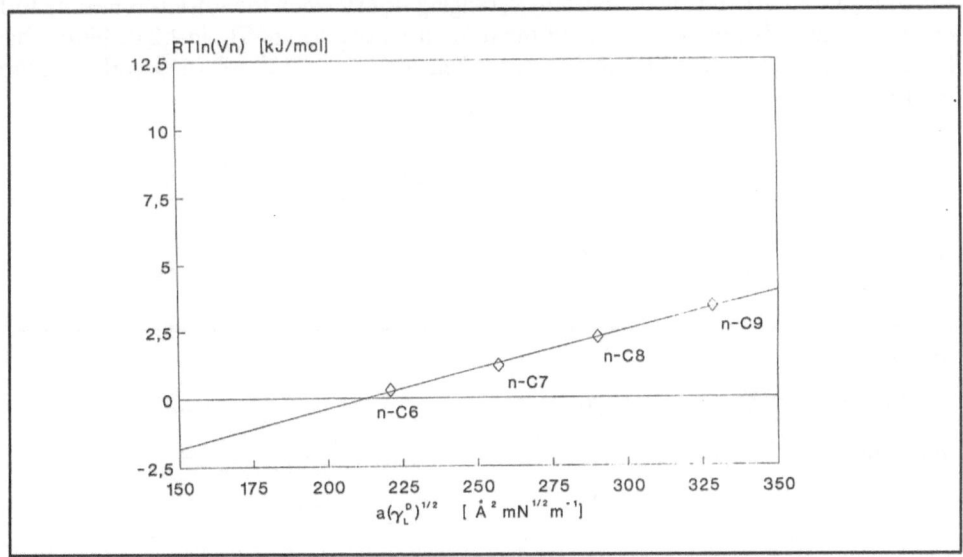

Figure 2   IGC data for retention of n-alkanes on polycarbonate at 110° C. Calculated $(\gamma_s)^d = 28.2$ mJ/m^2.

Application of the principle necessitates the selection of an appropriate theory of acidity/basicity. Currently, two are in vogue. One is that of Drago and coworkers[31], in which enthalpies of mixing an acid/base pair are equated to a set of four empirical parameters, C_a, E_a and C_b, E_b. The acidity and basicity of a substance thus is given by parameters expressing covalent (C) and electrostatic (E) contributions. The Drago approach has been much favored by Fowkes[26,27,32] ; using calorimetric data, C and E parameters have been obtained for a variety of organic and inorganic solids relevant to polymer systems. Recently Chen[33] has used IGC data to calculate Drago parameters for organic vapors and has used these in rationalizing the interaction of the molecules with polyolefin surfaces. The

---

*   But not the only one: Linear reference functions can also be generated by plotting RT In V_n against the saturation vapor pressure of the volatile probes, or against their normal boiling temperatures.

cumbersome need to obtain 4 empirical parameters, however, somewhat hinders the wide use of the Drago concept, and focusses attention on an alternative acid/base theory, namely that of Gutmann[34].

The Gutmann approach reduces to two the number of parameters needed to classify the acid and base character of organic substances. These are acceptor and donor numbers, AN and DN respectively. The simplicity is somewhat deceptive, however, because AN and DN have inconsistent units. Gutmann defines DN as the molar enthalpy of mixing a base with a reference acceptor, $SbCl_5$. In contrast AN is not an enthalpy parameter, but is defined as the relative ^{31}P-NMR shift in triethylphosphine oxide, when this substance is reacted with an acceptor solvent. AN is scaled arbitrarily[34], ranging from 0 when the solvent is hexane to 100 when the solvent is a dilute solution of the reference acceptor, $SbCl_5$, in 1,2 dichloroethane. Table 1 states relevant values for probes often used for acid/base (and surface energy) determinations.

Table 1. Properties of probes suited for acid-base characterization of solids

PROBE	$a(A^\circ)^2$	$\gamma_1^d \left(\dfrac{MJ}{m^2}\right)$	AN	DN	DN/AN	
n-hexane	51.5	18.4	---	---	---	
n-heptane	57.0	20.3	---	---	---	
n-octane	62.8	21.3	---	---	---	
n-nonane	68.7	22.7	---	---	---	
THF	45.0	22.5	8.0	20.0	2.5	BASES
D.E.Ether	47.0	15.0	3.9	19.2	4.9	
CHC13	44.0	25.0	23.1	0	0	ACIDS
benzene	46.0	26.7	8.2	0.1	0.01	
acetone	42.5	16.5	12.5	17.0	1.4	AMPHOTERIC

In spite of the noted problem, there has been much useful application of IGC data to the characterization of acid/base properties in solids,relevant to polymer systems. The approach of Papirer[28] has been followed with some success: This adopts the procedure noted above, and uses vapors with known values of $\underline{a}$, and with known (Gutmann) AN,DN (see for example those in Table 1). When the V_n of an acidic or basic probe is plotted according to eqn. 10, then the difference in $RTlnV_n$ between the probe and a reference probe of the same size capable only of dispersive interactions, is assumed to evaluate the acid/base contribution to ΔG.

$$-\Delta G_{ab} = RT \ln V_n/(V_n)_{ref}. \tag{11}$$

Since $\Delta G_{ab} = \Delta H_{ab} - T\Delta HS_{ab}$, then a plot of $\Delta G_{ab}/T$ vs $1/T$ explicitly evaluates the enthalpy term. As a first approximation, we write

$$\Delta H_{ab} = K_A \, DN + K_D \, AN.. \tag{12}$$

Rearranging the above, it follows that a straight line should be obtained when $\Delta H_{ab}/AN$ is plotted vs. DN/AN, the slope giving the acidity of the solid, K_A, and the intercept its basicity parameter K_D. The expectation has been confirmed in several cases[28,29]. It is illustrated for polyvinyl chloride (PVC) in Figure 3. The probes are those of Table 1. Clearly the polymer is distinctly acidic. Obtained in this manner, inherently the parameters represent acid and base interaction tendencies averaged over the temperature range of the experiments.

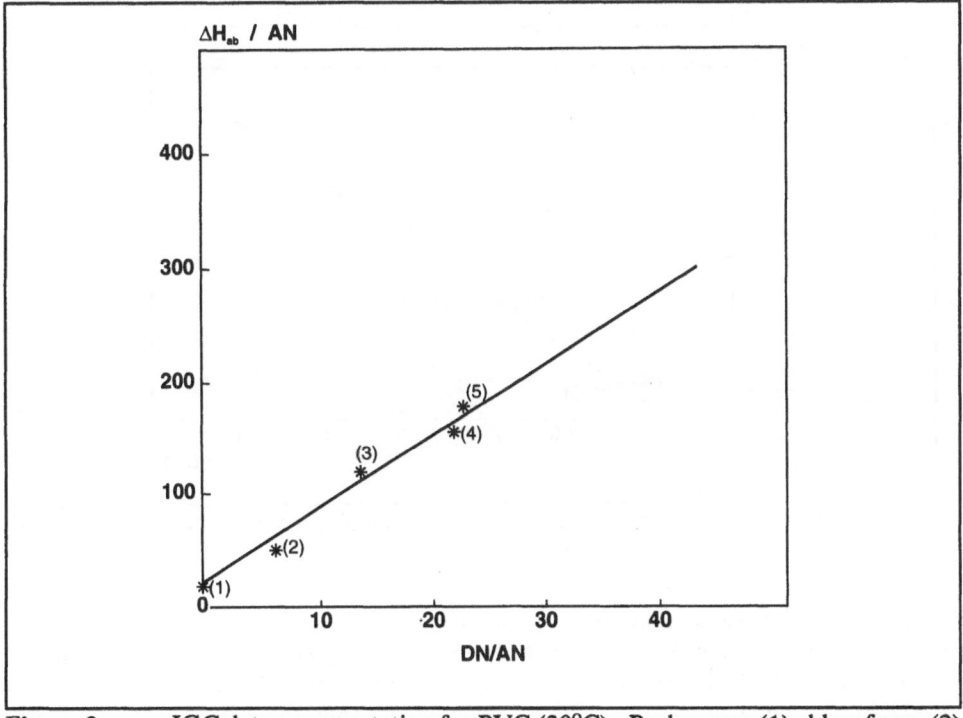

Figure 3   IGC data representation for PVC (30°C). Probes are: (1) chloroform; (2) acetone; (3) ethyl acetate; (4) diethyl ether; (5) THF.

An empirical way of assigning acid/base parameters to polymers and other non-volatiles has been practised in our laboratories[30,35]. Reference acid and base probes are used (generally chloroform and diethyl ether), along with the usual alkanes. The perpendicular distance identified as ΔG_{ab} in eqn 11 is here identified directly as the AN and DN number of the stationary phase. Figure 4A illustrates the situation for polycarbonate. The retention datum

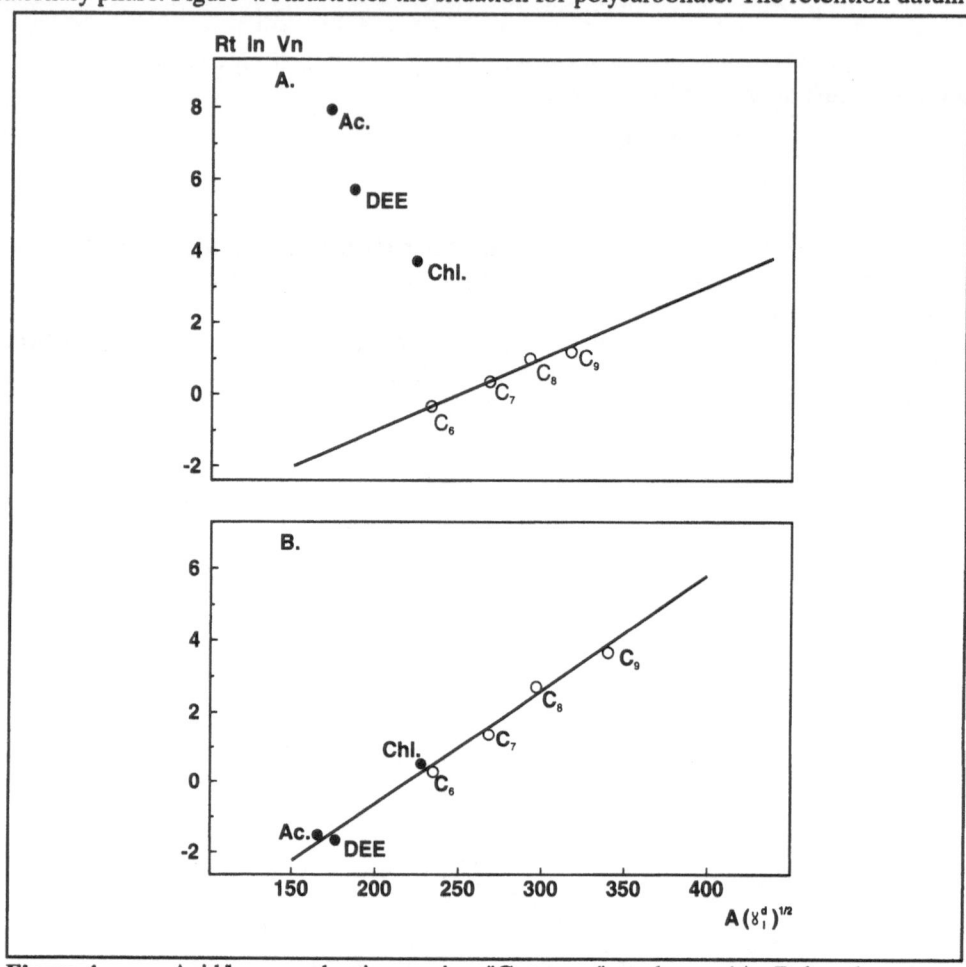

Figure 4   Acid/base evaluations using "Gutmann" probes.   4A. Polycarbonate at 70°C. 4B. Polypropylene at 30°C.

for chloroform, the acidic reference probe, defines DN for the polymer. The retention of the basic probe, DEE, evaluates the AN parameter of the polymer. The polymer AN and DN now have units of RT, although the non-uniform units of the reference probes represents an inherent problem to the use of these parameters. Thus the AN and DN of any **one** stationary phase do not rate its acidity and basicity on the same scale, but in a **series** of stationary phases useful comparisons can be made of relative acidities and basicities. Moreover, the interaction parameters are now evaluated at specific temperatures. Also shown in Fig. 4A is the datum for the amphoteric probe, acetone. The distance of this point from the alkane reference line is, within experimental uncertainty, the sum of the polymer AN + DN. The inference to be drawn is that the amphoteric vapor molecule can orient so as to interact with acidic as well as basic surface sites of the stationary phase. The conclusion also infers that both sets of sites are sterically accessible to vapor molecules with a near 40 A^2. Of course, in a polymer unable to act either as donor or acceptor of electrons, such as polypropylene, the points for chloroform, DEE and acetone would fall directly on the alkane line, so that the polymer's AN = DN = 0. This case is shown in Fig. 4B.

TABLE 2: ACID/BASE Interaction values for some polymers and components of polymer systems

(Data based on use of alkanes and reference probes chloroform and diethyl ether).

Substance	Conditioning	AN	DN	T($^{\circ}$C)	Type
LDPE	nil	0.4	0.1	35	neutral
LDPE	mild corona	2.7	1.5	35	acid
LLDPE	nil	0.2	0.1	35	"
Polyprop.	nil	0.4	0.1	35	"
Chlorinated PE	nil	7.2	1.7	30	acid
PMMA	nil	1.7	6.8	40	base
PVC	nil	10.3	2.8	30	acid
Polycarbonate	nil	7.7	8.8	40	amph/base
EVA (28%VA)	nil	4.7	8.3	35	base
Nylon 6	nil	2.9	7.7	40	base
Rayon	nil	23.0	14.6	30	amph/acid
Cotton	nil	29.9	9.8	30	acid
Cellulose	nil	31.1	24.2	35	amph/acid
Cellulose	AP silane	5.1	35.3	35	base
Cellulose	Chl. silane	43.4	3.2	35	acid
Epoxy	cured	9.6	7.7	50	amph/acid
Dioct.phthalate	nil	4.5	12.4	35	base
Sebacate	nil	3.7	12.2	35	base
Rutile	nil	11.2	9.7	30	amphoter.
Rutile	silicon	10.4	3.3	30	acid
Rutile	alumina	5.9	8.1	30	amph/base

The data in Table 2 provide some idea of the range of polymers, fibers, fillers etc., for which AN and DN values have been determined by this method. As expected, PVC again classifies as a distinctly acidic solid. It is interesting that non-volatile fluids such as phthalates, sebacates etc., are bases. These are known plasticizers for PVC, suggesting that the solubility of this polymer is dependent on acid/base coupling with its potential solvent. The AN and DN values for cellulose fibers show this to be a highly interactive, amphoteric solid. The interaction potential of the fiber can be influenced greatly by the use of surface modifiers, such as the silanes illustrated in the tabulation. Surface modification also plays a vital role in the interaction capabilities of certain solids, including pigments such as rutile TiO_2. The amphoteric surface properties of pure rutile are modified in commercial samples of the pigment toward acidity by surface coatings of SiO_2, or toward basicity by coatings of aluminium oxides. Clearly, in multi-component polymer systems using this pigment, one would expect to find very different states of interaction, depending on the specific choice of components. As an example, in the case of pigmented, plasticized PVC, significant differences in rheological and mechanical properties have been observed[30] when the interaction balance within the system was altered by substituting a basic rutile for one with acidic surface tendencies. Later in this article we will present further illustrations of the effects in polymer systems due to a shift in interaction balances.

TABLE 3. Temperature dependence of ACID/BASE interactions

T (°C):	30	40	50	60	70	80	90	100	120
1. PVC									
AN	10.3	10.4	10.0	9.6	8.3	7.2	5.5	4.6	---
DN	2.8	2.2	1.5	1.1	1.0	1.0	0.8	0.7	---
AN/DN	3.7	4.7	6.6	8.7	8.3	7.2	6.9	6.6	---
2. Polycarbonate:									
AN	7.7	7.4	6.5	6.1	5.7	5.0	4.1	3.3	2.6
DN	8.8	8.5	8.1	7.9	6.6	6.6	6.0	5.2	4.1
AN/DN	0.88	0.87	0.80	0.77	0.86	0.76	0.68	0.63	0.63

Another interesting use of empirical AN and DN indexes obtained by the method of refs.30,35 is exemplified in Table 3. This reports the temperature dependence of acid/base interactions in two commodity polymers, PVC and PC. The relevant AN and DN parameters for these polymers were evaluated by IGC at temperatures ranging from 30 - 130° C. As expected, at room temperature PVC again classifies as acid, and the relative acidity of the polymer surface increases steadily from 30 to about 70°C. Above 80 -90°C, however, the trend is reversed and acid/base interaction forces diminish. PC behaves as an amphoteric solid, its AN and DN finite and roughly equal at room temperatures. Above about 60°C however, the relative behavior of the polymer is that of a Lewis base. The results may be rationalized as follows:

In PVC, the binding energy of the basic probe with the surface is greater than that of the acidic chloroform probe. Consequently, with an increase in temperature, the relative change in retention characteristics is greater, initially, for the weakly-bonded acidic probe. The polymer surface therefore appears as relatively more acidic, and this is reflected by the ratio AN/DN. At sufficiently high temperatures the binding energy of acidic surface sites is gradually exceeded. The basic probe begins to desorb rapidly, the polymer surface now becomes weakly interactive and begins to resemble a dispersion-force solid. Whether or not the change in behavior is related to the glass transition temperature of the polymer, is not clear. The behavior of the PC specimen must lead to the conclusion that basic sites in this solid are more energetic than acidic ones. An increase in temperature therefore shifts the interaction balance to basicity, a trend that continues to about 90°C, whereupon again a reversal is observed.

Some practical consequences arise from the above observations. It follows that non-dispersive, acid/base interactions involving a polymer may be very different under use conditions, generally near ambient temperatures, than they are under processing conditions which generally take place at elevated temperatures. The appearance of miscibility in constituents of a complex system under one set of conditions is no guarantee, therefore, of miscibility under another, different set of conditions. Since in practise, polymer melts are usually cooled rapidly after processing, it is very likely that the solids are formed in unstable states of interaction. Frequently observed time-dependent changes in mechanical properties, in adhesive bond strengths and in other solid-state properties of multi-component polymer systems may often be triggered by the shift in thermodynamic interaction balances, as these attain equilibria dictated by their surroundings. This concept will reappear in the second part of this article.

Returning to the inconsistent units of Gutmann AN,DN values, the problem may be resolved by a procedure recently suggested by Riddle and Fowkes[36]. They have shown that the ^{31}P-NMR shift of triethylphosphine oxide,(Et$_3$PO), dissolved in acidic solvents,(the criterion used in Gutmann for the definitionof AN), can be divided into dispersive and true acid/base contributions. These authors also found that the dispersive contributions to the NMR shifts are directly proportional to calorimetric determinations of the enthalpies of dispersive interactions between Et$_3$PO and acidic liquids. A new acceptor number AN* is therefore obtained, and this has the same units as DN. Riddle and Fowkes [36] suggest that AN* be obtained from the simplified expression

$$AN^* = 0.288 (AN - AN^d) \tag{13}$$

where AN^d is the dispersion contribution reported by the authors.

The modified approach has been used by ourselves[37] in conjunction with IGC studies of various polymers. Although the determination of AN^d is somewhat cumbersome, the ability to define acid/base parameters in thermodynamically consistent units enhances the utility of the Gutmann model: It allows recognition of the fact that most macromolecules are amphoteric in character; that acid/base interaction potentials may be strongly T-dependent; that the balance of acid/base forces in any given substance may be represented by a unitless ratio number, such as AN*/DN. We expect considerable expansion in the near future of the available AN*, DN data base.

A final limitation to the interpretation of IGC results is to be noted. In order to favor the establishment of equilibrium conditions between the stationary and mobile phases in an IGC experiment, the quantities of vapor injected are extremely small (e.g. nanolitre range of concentration). As pointed out by Wesson and Allred[38], surfaces generally are not energetically uniform. The small amounts of available vapor therefore will tend to adsorb on the most energetic fraction of available sites. Surface energy and acid/base characteristics obtained from IGC will therefore describe the performance of these surface fractions, and not necessarily of the surface as a whole. More detailed descriptions of solid surfaces would necessitate the determination of adsorption isotherms. The suitability of IGC for this has been well documented[10,11]. In many instances, however, the dominant surface interactions of a solid will be those contributed by the high-energy sites. The time-consuming determination of isotherms may therefore be restricted to cases where surface characteristics must be known in great detail.

c_{iii}) Acid/base pair interactions and their use.

The ability to express quantitatively the acid/base interaction potential of individual polymers and of other constituents in polymer systems, raises the possibility of also obtaining the acid/base interaction between pairs of components. Among the first to consider the possibility were Papirer and coworkers[28], whose adoption of the Gutmann acid/base concept led to eqn. 12, and the definition of acid/base interaction constants K_A and K_D for a solid. Values of these parameters for two solids (1,2) could then be used to define a pair interaction value, A, as for example in ref 29:

$$A = (K_A)_1 (K_D)_2 + (K_A)_2 (K_D)_1 \tag{14}$$

Schultz and coworkers made use of the concept to show[29] that the adhesion of variously surface-modified carbon fibers to epoxy matrixes was closely correlated with the acid/base interaction between these materials. This clearly establishes the possibility of using IGC data as a guide to the optimization of material composites.

TABLE 4. Plateau adsorption for polymer/pigment pairs

(All adsorption at 30° C)				
Polymer:	P1	P2	P3	P4
$A_{max}(mg/m^2)$ on:				
Rutile R1	1.17	1.49	1.98	1.57
Rutile R2	0.93	1.45	2.17	1.74
Rutile R3	1.47	1.83	1.97	1.69
Quinto Magenta	0.53	0.73	0.94	0.80
Quinto Violet	0.62	0.82	0.99	0.85
Monastral Green	0.21	0.28	0.35	0.35

The empirical AN and DN indexes obtained by our approach[30,39] also lend themselves easily for calculations of pair interaction numbers. There is no formal theoretical guideline on how best to combine individual AN and DN numbers. Arithmetic, geometric and harmonic mean averaging may be used, with a decision as to preferred approach left to an empirical examination of results. One pair interaction number, I_{sp}, which has proven to be useful, is defined as follows:

$$I_{sp} = [(AN_1) \cdot (DN_2)]^{1/2} + [(AN_2) \cdot (DN_1)]^{1/2} \tag{15}$$

Some of the uses to which I_{sp} has been put may be illustrated. In one instance[40], interaction parameters were used to rationalize the dispersion behavior of pigments in solutions of polyesters used in automotive applications. The polymers were adsorbed onto inorganic and organic pigments. The resulting adsorption isotherms for given polymer/pigment pairs were found to be of the Langmuir type, the quantity of polymer adsorbed per unit area of pigment surface attaining a plateau value, A_{max}. Values of A_{max} for the adsorption of polyesters P1, P2, P3 and P4 from xylene solutions onto the pigments are reported in Table 4. The pigments included 3 rutiles with different surface finishes, and 3 organic colorants, identified in the table. Wide variations in the maximum adsorption values were observed. IGC data for the polymers and pigments were plotted as in Fig. 2, the retention of dispersive-force probes represented by a series of n-alkanes, with chloroform and diethyl ether used, respectively, as reference acid and base vapor. The resulting AN and DN parameters were then used with eqn 15 to obtain relevant I_{sp} numbers. These are given in Table 5. The acid/base pair interactions vary substantially, from very low values when the monastral green pigment was involved, to much higher ones when rutile was the adsorbent solid. It is again noteworthy that the three rutile surfaces are distinct.

TABLE 5. Pair interactions for Polymer/Pigment combinations

Polymer:	I_{sp} stated in kJ/mol			
	P1	P2	P3	P4
Adsorbent:				
Rutile R1	7.77	7.44	7.32	7.74
Rutile R2	7.40	7.15	7.04	7.44
Rutile R3	8.87	8.55	8.42	8.90
Q. Magenta	4.51	4.37	4.31	4.55
Q. Violet	6.23	6.02	5.93	6.27
M. Green	2.80	2.74	2.71	2.86

The relationship between A_{max} and I_{sp} is shown in Figure 5 for polymer P1 as the adsorbate. An excellent linear relationship is generated with an abscissa intercept at $I_{sp} \sim 2$. The correlation coefficient for the function of 0.96 is surprisingly good, considering that neither dispersion forces nor competing interactions between polymer and solvent are taken into consideration. In this instance, as in each of the four polymers, acid/base forces are dominant in the adsorption process. Indeed, the diagram predicts that unless these forces generate a pair interaction value in excess of about 1.5 - 2.0 kJ/mol., no significant adsorption will take place. Pigment dispersion in such a case would be difficult, and any dispersion made would be unstable.

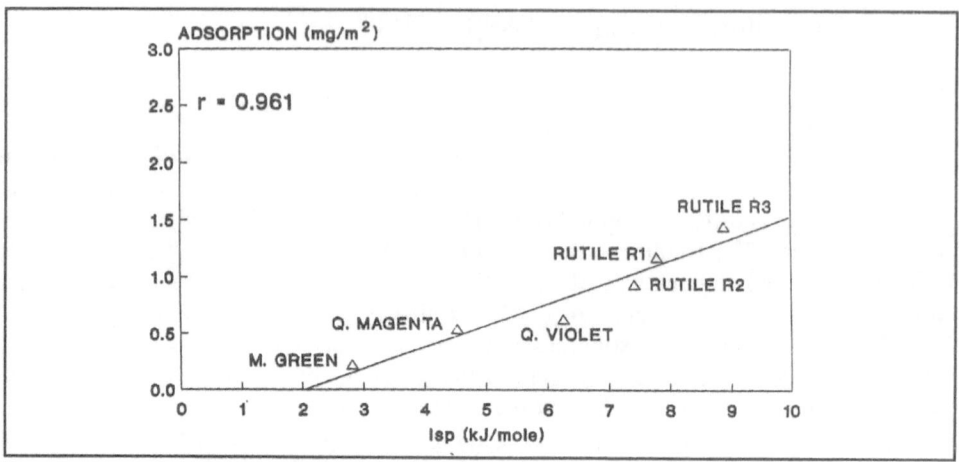

Figure 5   Correlating plateau adsorption of polymer P1 on pigment surfaces with relevant pair interaction numbers, I_{sp}.

The same polymer/pigment systems can be used to illustrate the value of acid/base interaction concepts for rationalizing the rheological behavior of the systems. For example, suspension viscosities of the various polymer solutions/pigment combinations may be evaluated with coaxial cylinder rheometers, as illustrated in Figure 6 for pigments at up to 25 volume-% concentrations in solutions of polymer P2. The curves fail to follow any of the widely accepted theories of suspension viscosity (eg. the relationship predicted by the theory of Maron-Pierce[41]). This is due to variable degrees of adsorption, and varying orientations of the adsorbed polymer molecule in the surface region of the pigments. An immobilized layer of polymer chains, of thickness ζ, is retained by the pigments, with the result that the effective pigment diameter increases, and with it also the effective pigment volume fraction. The change in pigment dimensions due to the adsorbed polymer layer may be represented by the ratio ζ/R, where R is the pigment radius, and plotted against I_{sp} for the various pigment / polymer pairs. This is done in Figure 7. Once again a strong correlation is evident, this time linking the thickness of the adsorbed layer with acid/base pair interactions. The largest change in ζ/R occurs for the very weakly interactive monastral green pigment. At first this may seem surprising; it reflects on the configuration of the adsorbed polymer. At large I_{sp}, the adsorbed molecule lies flat in the surface of the pigment, while an extended "head-to-tail" configuration is assumed to exist in cases such as the monastral green/P2. Once more, IGC data prove of considerable value in practical applications.

Another example showing the influence of pair interactions in complex polymer systems concerns the effectiveness of thermal stabilizers for polyolefins, when these are pigmented with a variety of rutiles. In the example, we chose polypropylene (PP) and linear low-density polyethylene (LLDPE) as hosts. These are materials without acid/base interaction potential. In order to stabilize them against thermally-initiated changes during processing, a variety of substituted phenolic substances is usually added. In the current example we have chosen two

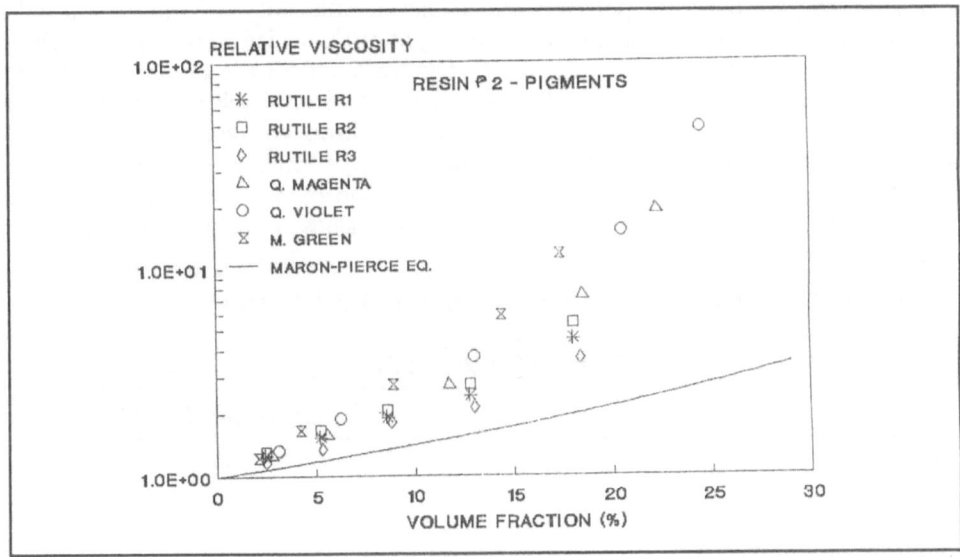

Figure 6   Relative viscosity vs. volume fraction for pigments dispersed in solutions of polymer P2. Note failure of theoretical curve to fit experimental data.

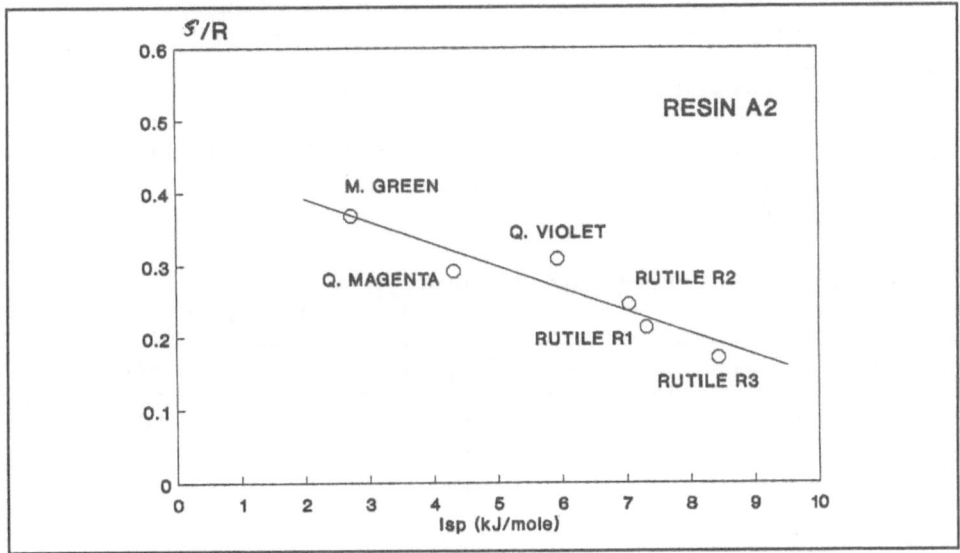

Figure 7   Relationship between acid/base interactions for P2/pigments and increase in effective pigment dimensions due to polymer adsorption.

commercial materials, coded SW and SX, and have added these at concentrations of 1.0 wt.% to the polymers. Also present is rutile at 1.0 wt%: in this study we used 3 surface coated pigments, coded R-1, R-2 and R-3. All were found to have similar surface areas at

$9 \ m^2/g$. When pure polymers, polymers with stabilizer, with rutile and with rutile/stabilizer combinations are examined by scanning calorimetry at 210^0C, an induction time,t, for thermal instability is readily detected. Table 6 shows that both SW and SX are effective in prolonging the stability of the polymers at 210^0C. Similarly, the rutiles individually contribute to thermal stabilization. However, when rutiles and thermal stabilizers are combined, the effects in thermal stability are not additive. Instead, varying degrees of "synergism" are produced. These may be expressed by the parameter t^*, which is the ratio of experimental induction times over the induction time calculated from the individual values in Table 6. The ratios t^* are entered in Table 7.

TABLE 6. Induction times for thermal instability of PP & LLDPE

(From DSC traces at 210° C)

	Induction time (min) for:
PP	22
PP + 1% SX	35
PP + 1% SW	47
PP + 1% R1	27
PP + 1% R2	33
PP + 1% R3	28
LLDPE	9
LLDPE + 1% SX	13
LLDPE + 1% SW	11

TABLE 7.   Relative induction times for thermal instability of PP & LLDPE; combined pigment/stabilizer effects

(Data refer to 210° C): Relative induction time $\ t* \ = \ \dfrac{t_{experiment}}{t_{calculated}}$

Host Polymer	Additive combination	t^*
PP	SX/R1	1.82
"	SX/R2	3.15
"	SX/R3	1.66
"	SW/R1	4.06
"	SW/R2	1.13
"	SW/R3	2.27
LLDPE	SX/R1	1.44
"	SX/R2	3.59
"	SX/R3	2.81

In order to account for the observations, IGC determinations of AN and DN were carried out. Here another simple, empirical approach was used to denote the interaction potential of each constituent in the compound. The parameter K_i was defined as

$$K_i = DN - AN ..$$ (17)

so that acidic materials report a negative value, and basic ones a positive K_i. The results, at 70^0 C, are shown in Figure 8. The host polymers, as expected, have K_i values near zero. The stabilizer SX and rutile R-2 are distinctly acidic, R-1 and SW are bases, while R-3 falls near the zero point. Unlike the polymers, however, this is an amphoteric solid with AN ~ DN , but with both of these indexes greater than zero. Assuming that at processing temperatures the materials retain their interaction characteristics (even though absolute AN and DN values would change), we may conclude that acid/base forces will promote the association of pairs such as R-2/SW and R-1/SX, but not of pairs such as R-1/SW or R-2/SX. The notion can be expressed by χK_i, the **sum** of individual K_i. χK_i is assigned a **positive** value whenever an acid and a base are involved, and a **negative** value for acid/acid, or base/base pairs (note, however, that this is not meant to imply repulsion of like pairs, but merely the failure of such pairs to associate preferentially!). The plot of Figure 9 clarifies the effects of acid/base forces. The thermal stability of the non-interactive host polymers is enhanced when pigment and stabilizer cannot interact through acid/base forces, but diminished when such interaction can take place. We may assume that decomposition products of SW and SX, needed for the stabilization of the polyolefins, bond to the appropriate rutile, becoming ineffective, or that the stabilizer itself adsorbs and its decomposition is hindered. Again, representations such as Fig. 9 are useful as guidelines to the formulation of polymer compounds with superior properties - in this instance enhanced tolerance to high temperature exposure. Other examples have been presented to support the importance of acid/base concepts in the balance of system properties. A note of caution is necessary, however. The importance of acid/base forces should not be exaggerated. Dispersion forces still exercise great influence on the behavior of multi-component systems, and their contribution is not to be neglected when interpreting property data, and when selecting materials for polymer compounds. Moreover, acid/base theories continue to be in a state of flux, and refinements to the Drago and Gutmann concepts favored at this time may alter our view of the interaction characteristics of commodity polymers and of additives frequently used with them.

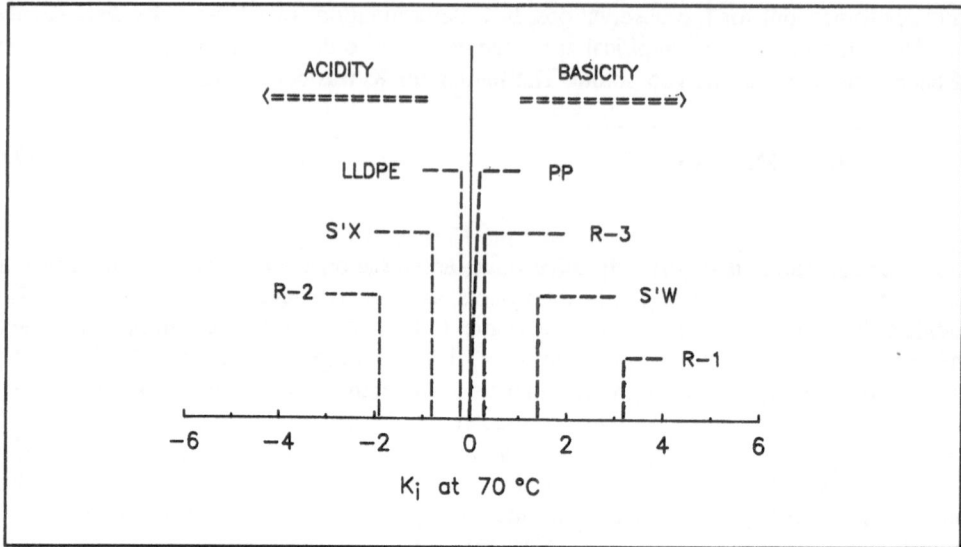

Figure 8   Acid/base interaction parameters for host polymers, pigments and thermal
              stabilizers in multi-component systems.

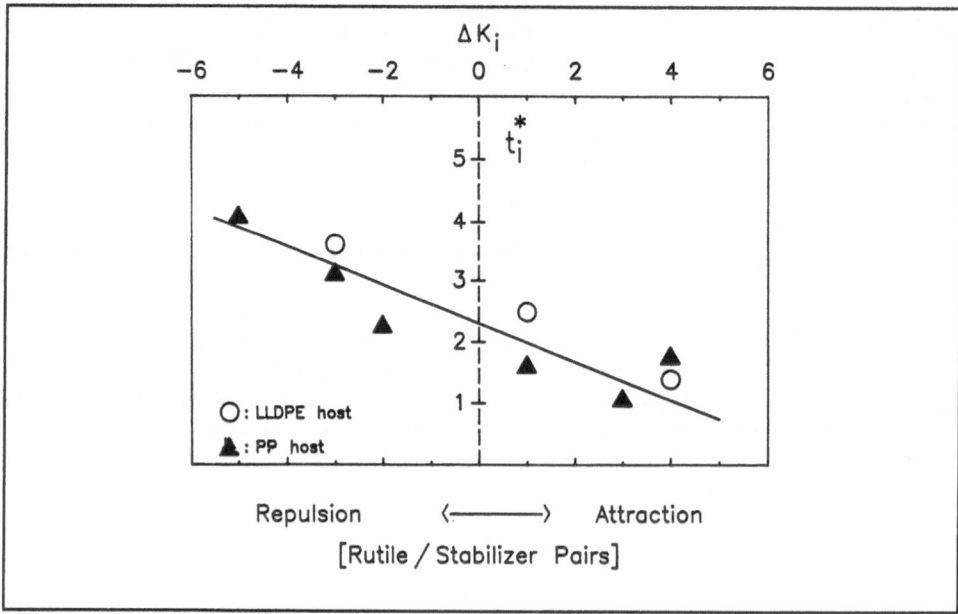

Figure 9   Showing dependence of synergism in thermal stability of polyolefins on
              acid/base interactions between stabilizer and pigment additives.

PART 2. CONTROL OF INTERFACIAL PROPERTIES.

A logical corrolary to the importance placed on interactions at polymer surfaces and interfaces is the ability to control these interactions through appropriate surface modifications. The desirability of controlling polymer surface properties has been evident for many years and various strategies have been developed toward that end. The strategies may be classified in the following, rather general way:

Mechanical modifications: Included here are simple washes and abrasion treatments of polymer surfaces. Thermal conditioning procedures also fall into this category.

Physical modifications: This relatively new and powerful category includes surface deposition and modification techniques using plasma and corona discharges.

Chemical modifications: This large category involves the use of modifying or coupling agents, notable among which are silanes.

Inherent modifications : Many, but not all polymers, are capable of changing surface properties by restructuring the surface region in response to the influence of a contacting medium.

The first three strategies are considered briefly in the present article. The physical modification category is discussed in considerable detail elsewhere in this compilation, with emphasis on plasma and corona procedures. Inherent modifications, which are an integral feature of the acid/base concepts advanced in preceding section, are discussed in greater detail.

a. Mechanical modifications:

Polymer surfaces, similar to those of metals or ceramics, are usually contaminated by extraneous materials which may interfere with attempts to wet, or bond to the polymer. From the earliest days of polymer use, simple surface preparation methods have become an integral part of processing technology. Often dust, or chemical contamination may be removed by simple washes using solvent or non-solvent fluids. A drying sequence generally follows the wash, that sequence being more exacting when the wash liquid is a solvent, so as to ensure its removal. The use of solvent washes has the added advantage of helping to remove asperities from the surface. The presence of these may impede the close contact between polymer surface and contacting medium needed to obtain adequate bonding. Asperities and other irregularities may also influence measurements of contact angles and of other data used for the characterization of polymer surfaces. Loosely bonded contaminants and surface irregularities may also be removed from polymer surfaces by simple abrasion treatments. Since the abrasive may itself become a contaminant, however, a wash procedure generally follows such a treatment.

The efficacy of these simple strategies for surface control may be illustrated by the bond strength of polyurethane/ABS joints, in which the PU surface was given a variety of

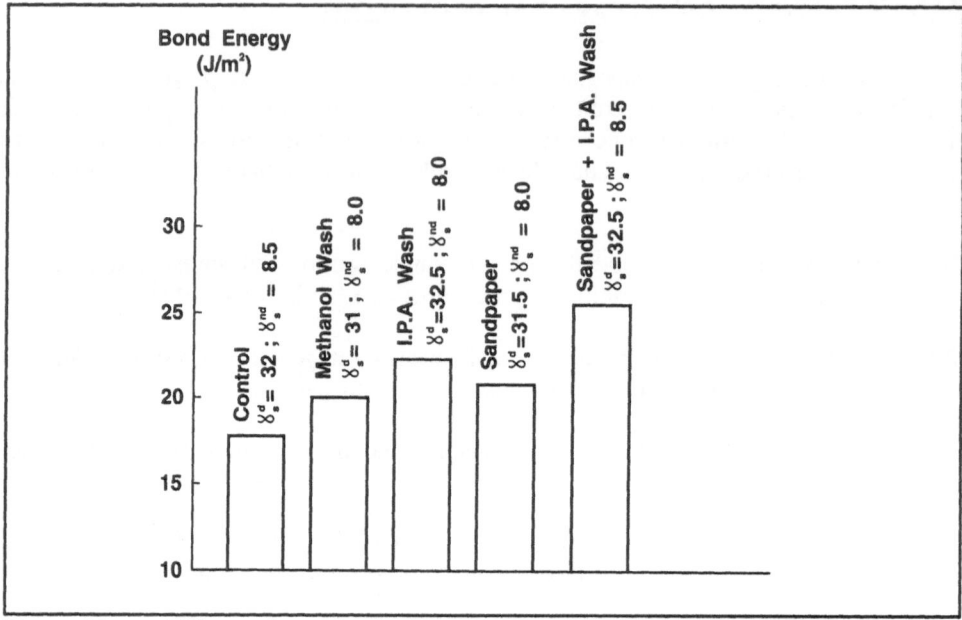

Figure 10   Effect of mechanical surface modifications on adhesive bonds of polyurethane /ABS joints.

mechanical surface treatments prior to being bonded to the washed ABS by a common compression molding method. The bar graph of Figure 10 shows the effects on bond strength of surface washes with methanol and isopropyl alcohol, of surface abrasion with sandpaper and the combination of abrasion/isopropyl alcohol wash. The dispersion and non-dispersion surface energies of the PU, stated in the Figure, are minimally affected by the surface treatments. The improved bond characteristics resulting from the surface treatments are attributed solely to the mechanical effects noted above.

Thermal conditioning can influence strongly the performance of some polymer surfaces. For example hydroscopic polymers, including polyesters and polyamides, must be dried before attempting to bond to them, in order to remove wholly or partially physi- and chemisorbed water. A particularly interesting case of thermal conditioning concerns polyolefins, notably low density and linear low-density polyethylene, LDPE and LLDPE. These polymers are known to have broad molecular weight distributions. Their surfaces are said to consist of mechanically weak, morphologically distinct "weak-boundary layers"[42,43]. These are composed of low molecular weight members of the distribution; their preferential concentration at the surface can be accounted for by thermodynamic arguments (see de Gennes, this book) and by kinetic considerations[44]. The presence of low molecular weight moieties strongly influences the cohesion of polyolefin surfaces, and fluids bonded to the surfaces (eg. printing inks) may be removed easily along with the weak boundary layers, giving the impression of apparent "adhesive" failures occuring at the interface. Of course, low molecular weight moieties must concentrate at surfaces of these polymers by diffusion mechanisms. There is ample time for this immediately following melt processing, if the hot, processed polymer is allowed to cool slowly to room temperature. When the hot polymer is quenched rapidly, however, solidification takes place before the low molecular weight species can accumulate and the cohesive strength of the polymer surface is substantially increased. The effect is temporary, because at room temperatures polymers like LDPE and LLDPE are well above their T_g values. Molecular diffusion is therefore slowed, but not stopped, by the high viscosity of the bulk polymer. Weak boundary layers eventually build, again affecting surface cohesion and the apparent adhesion of superposed materials.

TABLE 8: Effect of thermal history on peel strength of LLDPE/Aluminium bonds

(Peel strength determined at 30°C, draw rate = 50mm/mi on joints made at 200°C)

THERMAL CONDITIONING	PEEL STRENGTH (Kg/cm^2)
Slow cool (2°C/min.)	0.43
Slow Cool (7°C/min.)	0.48
Quench (> 200°C/min.)	2.88
Quench and aged @ R.T. for:	
1 hour	2.75
7 "	2.33
24 "	1.70
40 "	0.95
72 "	0.60
168 "	0.53
240 "	0.49

The temporary effects of thermal conditioning on adhesion to polymer surfaces have been reported in the literature[45]; work from our laboratories on LLDPE bonded to aluminium sheet is used to illustrate the effect in Table 8. Peel strength measurements (90°) on LLDPE bonded by compression molding at 200° C to degreased, dry Al sheet rise by almost an order of magnitude when the freshly prepared joint is quenched in running cold water, rather than being allowed to cool slowly to room temperature. After about 72 h. of storage, however, the effect is greatly diminished, and the peel strengths of "aged" joints now resemble those of joints allowed originally to cool slowly. Thermal conditioning procedures of this type (as opposed to prolonged exposures to high temperatures, during which permanent oxidative changes may occur in the polymer surface) have no detectable effect on the surface chemistry or energy of the polymer. Their usefulness is restricted, unless additional measures are taken to alter the polymer surface so as to block the re-establishment of weak boundary layers. Surface crosslinking by glow discharges in inert gases, referred to as CASING[46], is noted as one such measure. Its mention leads logically to a consideration of the next category.

b. Physical modifications:

$b_{i.)}$ Glow discharges at atmospheric and reduced pressures constitute the major items of this category. Discharges at atmospheric pressures, known as corona discharges, are perhaps the most widely used methods of surface conditioning. Corona treatment involves passing a polymeric solid through a gap, generally of the order of millimetres, between a grounded and a charged electrode. The latter may operate at potentials of up to 15 KV, and generate currents in the 10 - 50 μ amp. range. Under these conditions atmospheric gases, notably oxygen and nitrogen, are ionized and able to interact strongly with the substrate to be treated. Exposure times of the order of 1 - 30 seconds are usually sufficient to effect significant surface modification. The chemistry initiated by corona discharges is complex and depends not only on the variables of the corona , but on the substrate being treated[47].

By far the most common application of coronas is the surface treatment of polyolefins. These are significantly oxidized in corona discharges, and in addition, surface-localized chains undergo scission and a degree of surface crosslinking ensues. The latter effect tends to "weld" the inherent weak boundary layers to the polymer bulk and thereby increases the cohesive strength of the polymer surface. Polyolefins, for example LLDPE, have neither acidic nor basic interaction tendencies, their surface energies arising from dispersion forces only. Following corona treatment, however, finite values of γ^{nd}_s are recorded, an example being given in Figure 11. The figure also includes data for a polyurethane specimen, showing that the surface oxidation effect is not restricted to polyolefins. The corona-treated polyolefin, with a polar (acid/base) surface energy contribution of about 7 mJ/m^2, can now interact more strongly with polar wetting fluids, such as flexographic printing inks. Accordingly the adhesion of these fluids is much enhanced and thus corona treatment units have become a standard feature of most commercial polyolefin film production units.

As noted, surface modifications due to exposure to corona discharges are by no means limited to polyolefins but rather are generally applicable. The illustration in Fig. 11 for a polyurethane sample is indicative of that generality. Coronas may also play a vital role in the successful formulation of polymer composites. For example, there is growing interest in the use of cellulose fibers as reinforcing agents for polymer matrixes. The fiber surface is highly interactive and dominantly acidic , IGC measurements showing a hard-wood fiber to have values of AN = 23 and DN = 14 (see Section 1 and Table 2 for definitions of these parameters). Clearly, when the matrix polymer is an acid or is neutral, then only van der Waals forces will predominate at the matrix/fiber interface. The interface will be weak, and the mechanical properties of the composite will be correspondingly affected. Corona treatment of cellulose fibers, a subject of active study[48], strongly modifies the surface energy and leads to the formulation of composites with polyolefins in which mechanical and rheological properties are far superior to those formed with untreated components[49].

A final aspect of corona treatment worthy of note is activation of surfaces for subsequent chemical reactions. For example the anchoring of an ethylene-vinyl acetate copolymer (EVA) to $CaCO_3$ may be greatly strengthened by corona-activating the pigment surface immediately prior to contact with a solution of the copolymer. The data of Table 9 relate to this. When the polymer is adsorbed from xylene solution onto the untreated surface and the coated pigment is recovered and dried, the polymer can be quantitatively removed again from the

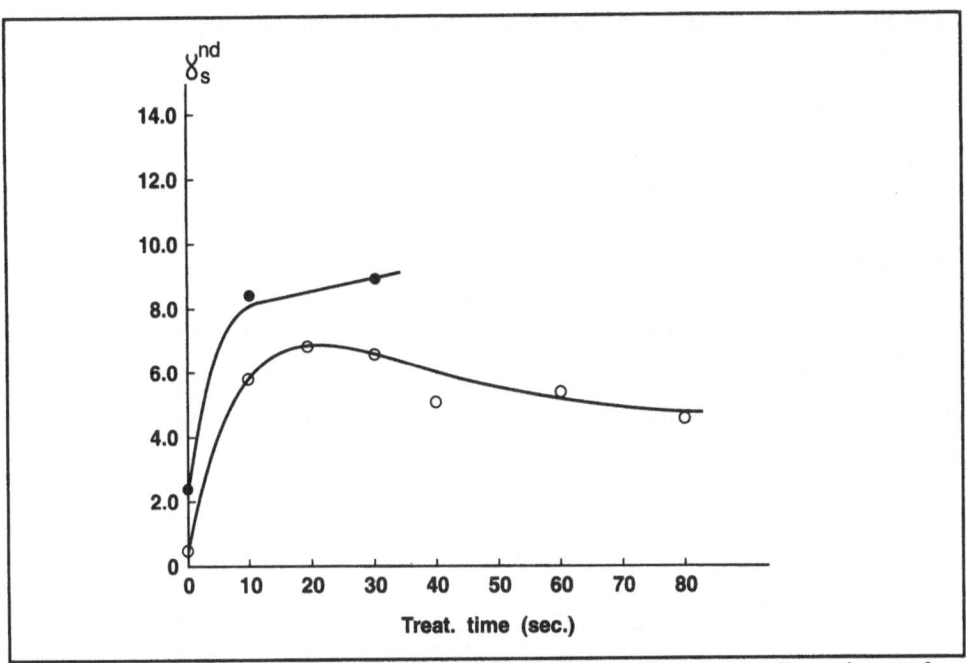

Figure 11   Effect of 15 KV, 20μamp, corona treatments on non-dispersion surface
energies of polymers.
○ LLDPE   ● Polyurethane

pigment by xylene extraction. Evidently, the bond strength at the pigment/polymer interface
is insufficient to resist the solvating power of this solvent. Following various degrees of
corona treatment, however, the quantity of permanently retained polymer increases. The
effect is also seen to be dependent on the time elapsed between the end of corona treatment
and contact with the polymer solution. Current evidence suggests that the polymer is
covalently bonded (grafted) to the corona-treated $CaCO_3$ surface, but that the efficacy of
grafting varies with the surface activity of the filler. That activity decays following corona
treatment. The ability to strongly anchor polymers to corona-activated surfaces again appears
to be invaluable to the economics and to the engineering aspects of composite formulation.

TABLE 9. Effect of corona treatment on anchoring of ethylene/vinyl acetate to $CaCO_3$.

(EVA adsorbed from xylene solution; recovered by Soxhlet extraction with xylene).

% EVA RECOVERED AFTER EXTRACTION PERIOD (h)

	1	5	12	24	48	72	144
A. Untreated $CaCo_3$	44	79	91	96	94	94	---
B. 12KV/60 sec. corona, immediate contact with EVA.	47	61	60	63	59	65	66
C. 15KV/30 sec. corona, immediate contact with EVA.	40	52	52	50	54	50	49
D. 15KV/30 sec. corona, contact with EVA after 5 min. in air.	43	57	62	66	66	64	---
E. As in D, but used after 24 h-in air.	41	64	69	73	70	75	75

$b_{ii.}$) Important surface modifications may be effected in discharges operating at reduced pressures and at radio (13.56 MHz) or microwave (2.45 GHz) frequencies. These are commonly referred to as cold plasma discharges. Their growing importance in both the fundamental and the applied fields of polymer science justifies the inclusion in this volume of chapters authored by Wertheimer and coworkers and by Liston, both devoted to the topic. The treatment here therefore can be cursory.

The subject of cold plasmas has been reviewed extensively in recent year[50]. They are usually applied to these major ends:

- Surface cleansing/ablation: Plasmas in air, oxygen, oxygen/fluorocarbon mixtures and in inert gases are frequently used to remove contaminants from surfaces, and in inert gas plasmas to crosslink the surface. The CASING technique, noted earlier[46], is an example of the latter application. One of the contributing mechanisms in plasma treatments is surface ablation, allowing for the removal of thin sections (generally sub-micron) from the surface. The preparation of ducts, or via-holes, through microelectronic wafers makes use of this technique. Surface oxidation produced by this method can result in important changes in the surface energy of the treated solid.

- Surface modification: Plasmas in inorganic vapors, such as nitrogen, ammonia, oxides of sulfur, nitrogen and carbon, are among many which modify surfaces by implanting chemical groups derived from the active species of the plasma discharge. These plasmas are capable of chemically modifying treated surfaces, without encapsulating them in "polymeric" layers. As in all plasma processes, the effects are localized in the surface region, so that bulk properties of plasma-treated materials remain unaffected.

- Plasma deposition or polymerization: Most organic vapors will oligomerize and finally form highly crosslinked, long chain molecules, frequently referred to as "plasma polymers"[51]. These are not truly polymeric species, however, since due to the complexity of plasma discharges no distinct pattern of repeat units is necessarily formed. The prefix plasma- is therefore fully justified in describing these products. An enormous range of plasma-polymers may be made, their chemistry depending on the selection of starting vapors. They are able to adhere tenaciously to the treated substrate and totally alter its surface chemistry and surface interactions, without affecting the bulk. A very wide range of applications is inherent in so powerful an approach to surface control. For example, surfaces may be rendered totally non-wetting by plasma deposits from fluorocarbon or organo-silicone starting vapors. They may be made compatible or incompatible with any given polymer matrix, again depending on the selection of starting vapor. Not surprisingly, the use of such plasmas to promote adhesion has become widespread [51,52].

Whilst plasma discharges are applied most frequently to polymeric substrates, other constituents of polymer systems may also be surface modified for specific, beneficial ends. As an example, the wettability of particulates by polymer fluids may be controlled by surface-modification. The ease of dispersing the solid in the polymer matrix is thereby also brought under control. We consider the case of $CaCO_3$ to be dispersed in PVC and PS melts. The acid/base interaction potentials of these materials, determined by the IGC techniques discussed above, are as follows:

	AN	DN
$CaCO_3$	3.7	9.6
PVC	7.9	4.0
PS	1.6	5.3

We may expect the basic particulate to be wetted strongly by the acidic PVC fluid, but not by the basic PS. These expectations are reflected in the dispersion behavior of the $CaCO_3$. When an intensive mixer (Brabender) is used to prepare dispersions of 5 wt.% solid in the polymers we find that a steady-state torque (viscosity) is attained in 5.5 min in PVC, but only after 9.5 min. of mixing with PS. In spite of the larger energy requirement in mixing with PS, microscopic examination of microtome sections taken from the filled solids show that in PVC the average $CaCO_3$ particle size attained is 6.5 μm, while in PS it is near 16 μm. Given that the inherent particle size for the particulate is 2.3 μm, this means that, on the average, triplet groups are formed during this particular dispersion process with PVC, while agglomerates of between 6 and 8 particles remain in PS. To cope with the situation, the $CaCO_3$ was surface treated in a microwave plasma at 300 watts power. One sample was treated in styrene vapor for 90 sec., at a vapor pressure of 0.8 torr. Another sample was treated in dichloroethylene for 60 sec. at 1.2 torr. These samples were re-examined by IGC and by XPS, and then used for dispersions in PVC and PS under identical conditions to those used with the unmodified particulate. The results were the following:

		AN	DN	Mix time (min)	Particle size (μm)
CaCO₃ (st.pl)		2.0	5.9		
CaCO₃ (dce. pl)		6.4	4.1		
CaCO₃ (st.pl)	in PVC			5.8	7.0
	in PS			6.2	9.5
CaCO₃ (dce.pl)	in PVC			9.0	14.5
	in PS			5.0	7.0

Styrene plasma treatment has formed a thin layer of plasma-polystyrene on the filler surface. The acid/base data for the surface are now similar to those of PS. Since the surface coverage is incomplete, the results do not match those of PS exactly. Treatment in dichloroethylene has created a chlorinated plasma-polymer, with AN and DN values which show the filler surface to be mildly acidic ! The dce-treated filler diperses rapidly and effectively in PS, in sharp contrast with the performance of the original particulate. In PVC the dispersion process is not aided by the new plasma coating; evidently the plasma-polymer is insufficiently alike PVC to be considered readily wetted by that fluid. Styrene treatment is beneficial in both cases: Acid/base forces still aid the dispersion in PVC, while the presence of a surface layer with the same general chemistry as the host polymer renders the acid/base question irrelevant to this case. The somewhat larger average particle sizes attained in the mixing experiment may be due to bridging of discrete particles by the plasma-polymer. Further examples of plasma treatments and their efficacy will be found in other chapters of this issue. The formation of a plasma-polymer layer or interphase at the surface of a substrate may be considered an alternative to similar structural modifications made by the use of more conventional chemical coupling agents. These are discussed briefly in the following section.

b$_{iii)}$: Chemical modifications.

The most widely used approach to surface modification is that of coupling or compatibilizing agents. A wide range of these chemicals is currently available; they include titanates, zircoaluminates, and similar bi- or multi-functional molecules, but by far the best known and the largest group is that of the silanes. Their structure and their ability to modify solid surfaces, thus acting as bridging or coupling agents between dissimilar materials, have been extensively reviewed[53,54]. The general structure of silanes is of the form $XSiY_3$, in which X represents a hydrolytically stable group, and Y is a hydrolizable functional group. Table 10, patterned after Pleuddemann and Walker [55,56], illustrates some of the commercially available variants. Any given case demanding stronger, more durable bonds, or enhanced mechanical properties of multi-component polymer systems, will best be served by one, or by a selected number of the many commercially available coupling agents. The present article cannot consider the technology of silane coupling agents and their appropriate selection in any comprehensive manner. The technical literature should be consulted before making a choice.

Table 10. Structure of some current silanes (after Plueddemann and Walker, refs. 55, 56).

Organofunctional group	Chemical structure
A. Vinyl	$CH_2=CHSi(OCH_3)_3$
B. Chloropropyl	$ClCH_2CH_2CH_2Si(OCH_3)_3$
C. Epoxy	$\overset{O}{\overset{/\backslash}{CH_2CHCH_2OCH_2CH_2CH_2Si(OCH_3)_3}}$
D. Methacrylate	$\overset{CH_3}{\overset{\mid}{CH_2=C-COOCH_2CH_2CH_2Si(OCH_3)_3}}$
E. Primary amine	$H_2NCH_2CH_2CH_2Si(OC_2H_5)_3$
F. Diamine	$H_2NCH_2CH_2NHCH_2CH_2CH_2Si(OCH_3)_3$
G. Mercapto	$HSCH_2CH_2CH_2CH_2Si(OCH_3)_3$
H. Cationic styryl	$CH_2=CHC_6H_4CH_2NHCH_2CH_2NH(CH_2)_3Si(OCH_3)_3HCl$
I. Cationic methacrylate	$\overset{CH_3 \quad\quad Cl-}{\overset{\mid \quad\quad\quad \oplus}{CH_2=C-COOCH_2CH_2-N(Me_2)CH_2CH_2CH_2Si(OCH_3)_3}}$

J. Chrome complex

K. Titanate	$\overset{CH_3}{\overset{\mid}{(CH_2=C-COO)_3TiOCH(CH_3)_2}}$
L. Crosslinker	$(CH_3O)_3SiCH_2CH_2Si(OH_3)_3$
M. Mixed silanes	$(C_6H_5Si(OCH_3)_3 + F$
N. Formulated	Melamine resin + C

It is clear from Table 10 that by varying the functional groups, for example from chloropropyl to diamino, a wide range of acid/base interaction potentials may be engineered into a surface, thus necessarily affecting the adhesive and cohesive strengths of interfaces and interphases in which the modified surfaces take part. Control of acid/base interactions is one of the mechanisms attributed to the functioning of coupling agents. The problem of mechanisms, however, is an exceedingly complex one, and is not devoid of controversy. A variety of contributing mechanisms have been proposed from time to time to account for the ability of silanes to improve adhesion, to control wettability and to resist the deterioration of bonds due to environmental exposure. The most important of these mechanisms appears to be the formation of covalent bonds between the silane and the surface (polymer, glass, mineral, etc) being modified[53]. The situation may be depicted as in the idealized example of Fig.12., again patterned after Plueddemann[55], in which a silane interphase couples a bulk polymer with a reinforcing glass fiber. However, the formation of covalent bonds is not a necessary condition: The adhesion between glass fibers and some thermoplastics may be strongly promoted by silane agents, even though no evidence for covalent bonding can be found [53].

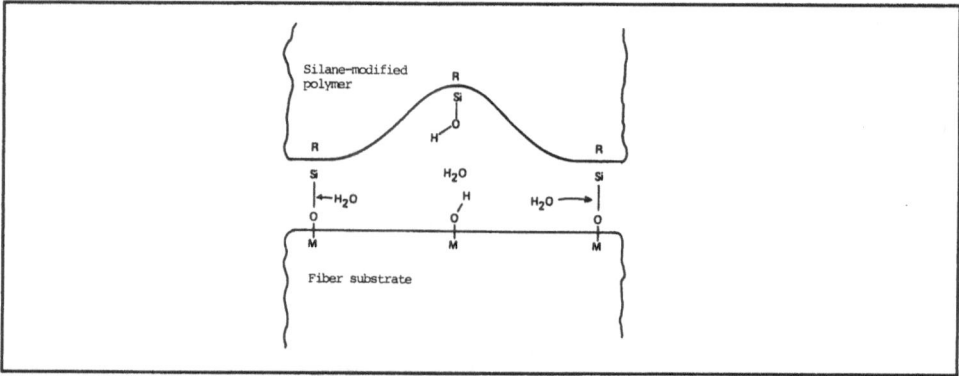

Figure 12   Proposed structure of silane-modified interphase between host polymer and reinforcing fiber.

The capability of silanes to improve the adhesion of joints which are difficult to bond is best illustrated by example. This is done with the aid of Figure 13. Here are compared the peel strengths of an epoxy/PP joint prepared without the use of a silane with one in which the epoxy was conditioned with γ-aminopropyltriethoxy silane (APS). In each case the PP was joined to the cured epoxy by compression molding at 190° C for 2 min., and the specimens were conditioned at 23° C, 50% RH for 48 hours prior to performing the peel test experiment. In that, the PP was peeled from the epoxy substrate at an angle of 90° and a speed of 50 mm/min. Two features are particularly noted in Fig. 13. One is the evident dissimilarity in peel strengths. The other is the marked difference in the regularity of the peel strength vs. extension diagram. Without APS, the peel strength averages near 130 J/m^2 , and there are major variations about that value. The failure was observed to be adhesive, with the PP peeling away from the substrate without tearing. However, the degree of wetting

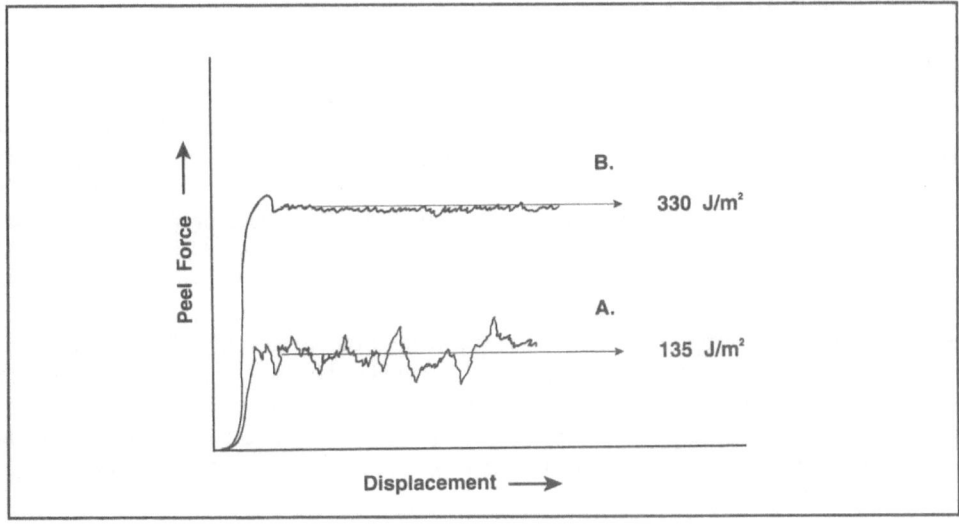

Figure 13   Peel force vs. displacement for PP/epoxy joints.
A. Without use of silane coupling agent.
B. Using APS to condition epoxy surface.

attained during the preparation step apparently was irregular, leading to sporadic contacts between the components and the consequent peaks and valleys of the trace shown in Fig. 13. The failure being adhesive, it seems likely that only dispersion forces acted across the interface in this instance. The presence of APS nearly triples the peel strength. Furthermore the failure now was observed to be cohesive, with traces of PP left on the epoxy surface. The relatively smooth trace in Fig. 13 is therefore more properly identified as a measure of the **cohesive** strength of PP. The bond across the epoxy/PP interface is stronger than the cohesion of PP and, clearly, created by much stronger forces than those of the control case.

The silanes listed above are "monomeric"; recently some advantages have been attributed to the use of polymeric silanes, that is, Si-O linkages associated with polymers such as polyethyleneimine[57]. These expand the range of situations in which strong, stable bonds can be forged between materials which may not normally bond adequately. A noteworthy limitation of silane-enhanced bonds is their inability to withstand prolonged exposure to temperatures in the range > 250° C. With the advent of high-performance, temperature-resistant polymers such as the polyimides, polyethers and polyether ketones, this limitation is becoming more keenly felt and motivates research aimed at remedying the problem. Promising routes, recently reported by Pape and Plueddemann[58], include blends of phenyl/aminosilanes, carboxysilane/zinc ion ionomer primer systems and vinylbenzylaminosilanes. Given the continuing evolution of this fertile area of physico-organic chemistry, it is not surprising to find a vast literature devoted to the subject. It is advisable periodically to consult reports of advances in order to derive maximum benefit from this chemical strategy to controlled surface modifications.

b$_{iv)}$. Inherent modifications. .

The final category of surface modifications to be considered has its origins in the thermodynamics of interfaces and interphases within multi-component polymers and is therefore an inherent feature of such systems. It is, in a sense, beyond the control of the scientist or engineer involved. Inherent modifications are noted here in some detail because they are apt to be neglected in practise, and yet represent an important "built-in" source of variations in the performance of bonds, barrier and mechanical properties of polymer systems.

A reasonable way to approach the occurrence of inherent surface modifications is through repetition of a caveat already stated in this article: Namely, that under equilibrium conditions, the surface and bulk compositions of polymers are never equal. Even in so (chemically) simple a polymer as polyethylene (PE), surface and bulk compositions must differ. This is due to the thermodynamic demand to minimize the surface free energy γ_s. As usual, that value is given by:

$$\gamma_s = H_s - TS_s \ ... \tag{22}$$

Here the entropy may be expressed as the sum of configurational (c) and non-configurational (nc) terms. Now the (ideal) PE chain is made up solely of CH_2 linkages, terminated by CH_3 groups. It is evident that $(S_{nc})_{CH3} > (S_{nc})_{CH2}$, so that to minimize γ_s, it follows that the polymer surface must be enriched in CH_3 groups. Polymers like PE have broad molecular weight distributions. The concentration of CH_3 groups varies inversely with the molecular weight, and the above condition is therefore satisfied by preferentially concentrating low molecular weight members of the ditribution in the region of the polymer surface. The presence of weak-boundary-layers (wbl) and of trans-crystallinity in polymers like PE undoubtedly may be traced to this thermodynamic origin. We have seen already that thermal treatments may counteract the effects of wbl temporarily. The presence and re-establishment of wbl structures, driven by the stated thermodynamic demands, may be viewed as one form of inherent surface modification, the negative effects of which require use of alternative modification strategies, for example corona or plasma treatment. Interestingly, wbl effects should be "frozen out" at sufficiently low temperatures, where the entropic contribution to γ_s fails to outweigh enthalpic contributions, which would favor the presence of -CH_2- links in the surface region. Unfortunately the kinetics associated with this kind of surface modification make the notion impractical.

If surface/bulk compositional differences occur in polyolefins, then it should not be surprising to find even more dramatic effects in chemically complex polymers. The case of poly(alkyl) methacrylates, for example (PMMA), is cited specifically[59]. This polymer may be considered to resemble a copolymer, in which methacrylic pendant chains are attached to a polypropylene backbone. Since the surface energy of the polyolefin backbone is lower than that of the pendant chain, then in order to minimize γ_s, the methacrylic chains should be oriented into the bulk of the polymer to the extent allowed by steric and kinetic factors. This would leave the surface enriched in the olefinic moiety. Supposing, as in ref.59, the polymer is allowed to dry against a variety of substrates, including liquid mercury, copper or gold foil. At the polymer/metal interface the thermodynamic drive is to minimize the interfacial surface energy, $\gamma_{1,2}$, this through the non-dispersive interactions favored for the particular combination of materials. In the case of PMMA the minimization of $\gamma_{1,2}$ requires the interfacial region to be enriched in the methacrylic chains. In other words, a thin film of PMMA, adhering to a high surface energy solid, is really a non-isotropic substance, in which the local compositions at the air interface, the substrate interface and the bulk all differ ! Quantitatively, the following is observed[59]:

The γ_s of a PMMA specimen dried in air is found to be 41.5 mJ/m^2, with a contribution of 3.0 mJ/m^2 from non-dispersion forces. When the thin film is formed against mercury, copper or gold, allowed to remain in contact with these metals for several days and then freed from the substrates (by acid etching in the case of Cu and Au) , the surface energies of the freshly exposed PMMA surfaces were found to be considerably higher:

Polymer film freshly removed from:	Hg	Cu	Au
γ_s (mJ/m2)	47.5	48.5	49.0
Non-dispersive contribution :	6.0	7.5	7.5

The minimization of interfacial tensions at the metal/polymer interface led to an increased concentration of polar groups in the polymer surface. The free films were non-isotropic, as anticipated, and consequently unstable. The instability was clearly evident in that the surfaces originally formed against the high surface energy metals relaxed over a period of hours, and re-attained the $(\gamma_s)^{nd}$ value dictated by the air environment.

The ability of polymers to display different surface compositions, depending on the medium in contact with their surfaces, involves a process referred to as molecular restructuring. The phenomenon has been reported for many homo-and copolymers[30,60,61]. It occurs frequently in biopolymers[62] and in biocompatible macromolecules including bicomponent polyurethanes. It is the major contributor to the "inherent surface modification" category and it can have profound effects on surface and interfacial properties of polymers, including adhesion, and barrier properties. The same considerations that trigger restructuring at "exposed" interfaces (eg. polymer/air) also apply to interfaces within a system. Restructuring events therefore must also take place at phase boundaries within multi-component systems. Thus, local differences in chain orientation and in composition will result in non-isotropic conditions within as well as at the external surfaces of polymeric compounds. Clearly, consequences on the properties of polymer composites must be anticipated. Of course, these conditions of non-isotropy are symptomatic of thermodynamic equilibria existing within such polymer compounds. The question now raised is whether equilibria are likely to be attained under customary polymer processing and use conditions.

The behavior of soft-segment polyurethanes (SPU) may be used to clarify the above question[63]. The polymers used in ref. 63 were bicomponent polymers with a polyurethane hard segment and an aliphatic polyether soft segment. The surface energy of the soft segment, at 36 mJ/m^2, is some 12 mJ lower than that of the hard segment. When the polymers are cast from solution against substrates (e.g. cleaned glass plates), are vacuum dried at 25°C for 24 h., and then conditioned at ambient temperatures and 0% RH for several days, their equilibrium surface energies against air may be obtained from conventional static contact angle measurements. These report values of 37 ± 1 mJ/m^2, with γ^{nd} of 2 - 3 mJ/m^2. In other words, the SPU surfaces are dominated by the soft, polyether segments, in keeping with thermodynamic demands. Massive molecular restructuring takes place when the polymers are immersed in orienting media, for example water, saline solutions, formamide, etc. Surface energies may be redetermined periodically, thus tracing the progress of the restructuring event. A typical illustration of this is given in Figure 14 for an SPU immersed in water. Over a period of some 10 - 18 days, depending on the temperature of the liquid, a new equilibrium surface state is attained, with γ^{nd} now 3-4 times higher than original. XPS analyses of N/O show[63] that the nitrogen content has been raised, so that restructuring has occured due to the diffusion of the hard segment into the interfacial region. The driving force is thermodynamic and the result is consistent with acid/base interactions between water and the various constituents of the SPU polymer. The time required to attain the new equilibrium is very considerable. Similar, long exposure times were needed to restructure the polymers used in previously cited references. From the temperature dependence of restructuring it is possible to obtain activation energies for the process. These fall[63] in the range of 5 - 10 Kcal/mol. Thus, even at polymer melt processing temperatures (e.g. 200° C), restructuring processes would require several tens of minutes to go to completion. Clearly, this indicates that in such industrial operations as film casting, coating from solutions, etc., where a restructurable polymer is contacted with an orienting medium (metal, ceramic, polar polymer), equilibria will not normally be attained. In the industrial product, generally used at ambient temperatures, the polymer surfaces and interfaces therefore will be in metastable states, from which they will deviate over long periods of time, as thermodynamic equilibria of orientation and composition are attained. During that period, adhesion, mechanical and other properties of the system must be expected to change: A physico-chemical origin of "aging" is thereby defined for polymer systems.

Further details on the nature of inherent surface modifications may be drawn from ref. 63. The rate of restructuring, and the magnitude of surface energy changes produced in a polymer, depend on the orienting strength of the contacting medium. That strength may be expressed in terms of the non-dispersive surface energy of the medium. The results shown in Table 11 document the point. The 3 SPU materials respond in various degrees to the orienting fluids, but the total changes in the polymer γ_s^{nd} increase systematically with the γ_l^{nd}. The times required to reach equilibrium also vary systematically, but inversely with the liquid's orienting strength. If the relationships in Table 11 are generally applicable - a matter yet to be established by broader experimentation - then it should be possible to predict the time needed to attain configurational and compositional equilibria at interfaces between any two contacting materials for which the non-dispersive surface energies are known or can be measured. Similarly, it should then be feasible to estimate the magnitude and duration of time-dependent changes in any of the properties of the system that depend on conditions of its surface, its interfaces and interphases. Adhesion, barrier and mechanical properties of composites are among those critical properties.

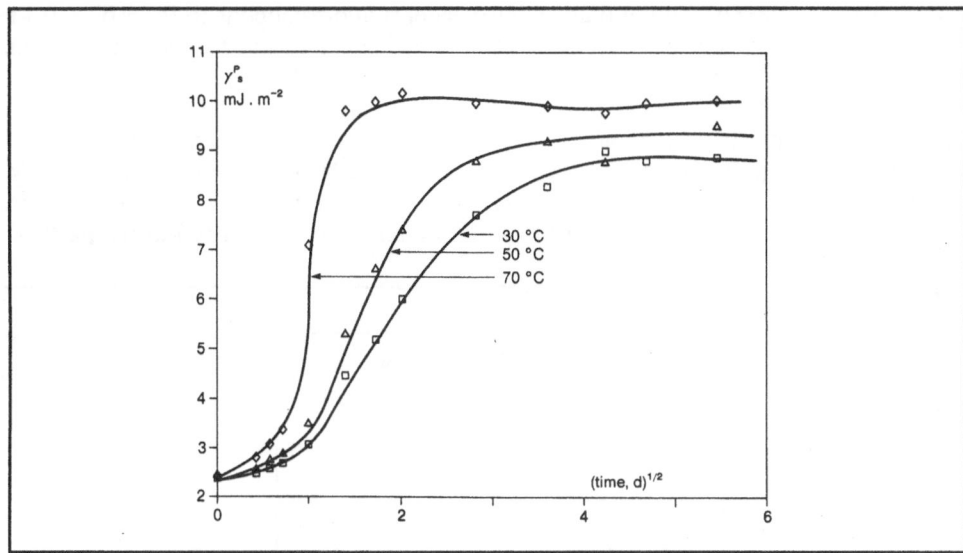

Figure 14   Restructing of SPU polymer on immersion in water.

TABLE 11. Initial, and equilibrium non-dispersion surface energies and equilibration times for immersion of SPU polymers.

Immersion Liquid:		Formamide	Ethylene Glycol	Water	5% salt solution
γ_l^{nd}(mJ/m^2		19.2	21.5	50.2	51.7
SPU - 1*:	$(\gamma^{nd})_i$	2.4	2.4	2.4	2.4
	$(\gamma^{nd})_{eq}$	6.5	7.3	9.5	10.7
	t_{eq} (h)	600	530	420	400
SPU - 2 :	$(\gamma^{nd})_i$	1.7	1.7	1.7	1.7
	$(\gamma^{nd})_{eq}$	5.4	6.3	8.5	9.3
	t_{eq} (h)	500	440	370	320
SPU - 3 :	$(\gamma^{nd})_i$	1.2	1.2	1.2	1.2
	$(\gamma_{nd})_{eq}$	4.6	5.0	6.8	7.5
	t_{eq} (h)	450	410	330	300

* All surface energies in mJ/m^2.

The importance of restructuring on adhesion properties of polymers is the subject of a final illustration. This involves the peel strength of bonds made between a polymeric adhesive tape (3M manufacture) and supported films of styrene/butyl acrylate (S/BA) copolymers with mole ratio compositions of 98/2 and 93/7.   Control specimens were made by casting the polymers from toluene solution onto stainless steel support coupons to form films 0.3 mm thick. These were vacuum dried and then conditioned at 0% RH and ambient temperatures for 72 h. prior to being bonded to the adhesive tape by the application of 0.8 tonnes pressure for 60 seconds. The adhesive tape was then peeled from the supported polymers under closely controlled conditions. The experiments were also performed on specimens in which the S/BA copolymers had first been preconditioned by immersion in water for varying periods. Following a given immersion cycle, the γ^{nd} of the polymers were obtained from contact angle measurements. The results of the exercise are summarized in the parts of Figure 15.

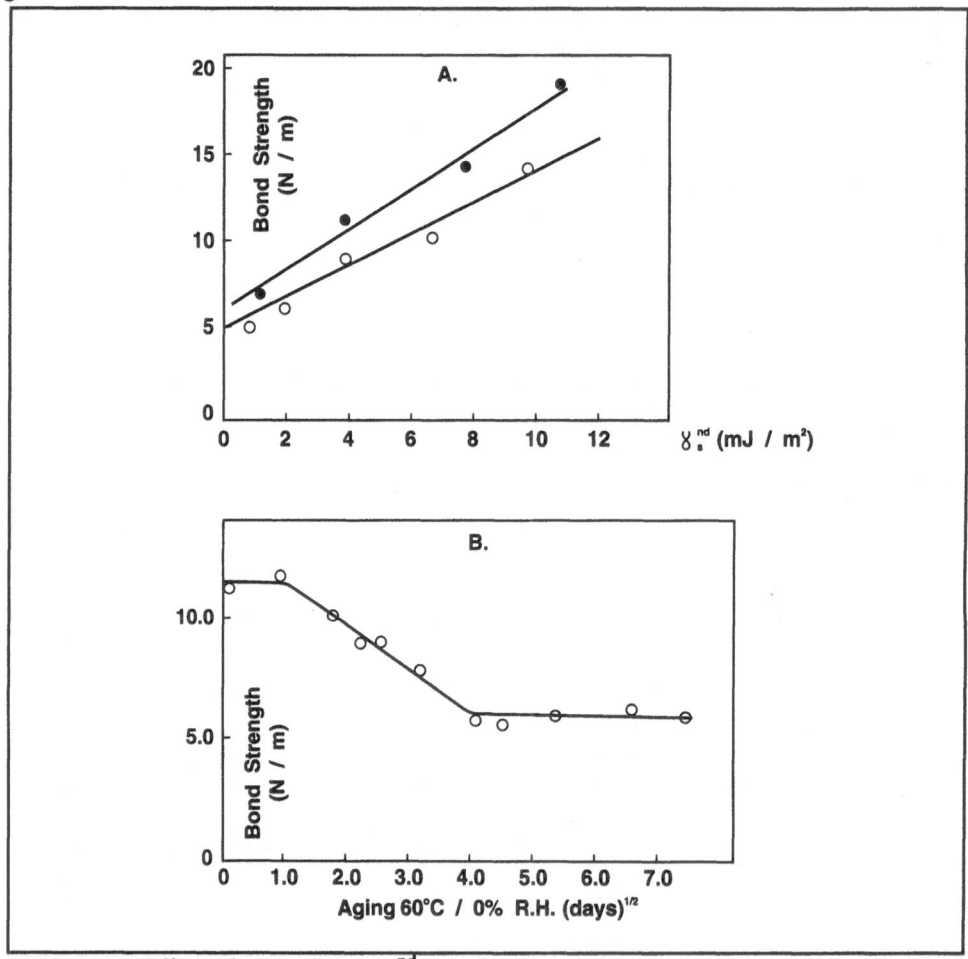

Figure 15A Effect of controlled $(\gamma_s)^{nd}$ on adhesion of S/BA copolymers to adhesive tape. Surface energy controlled by water immersion. ○ S/BA = 98/2 ● S/BA = 93/7

Figure 15B Decay of adhesion in joint with S/BA (98/2) originally at $(\gamma_s)^{nd} = 7.5$ mJ/m². Orienting force of tape medium estimated to be 4 mJ/m².

Analogous to the behavior of SPU, documented above, the S/BA copolymers also restructure in response to the orienting force of water, their non-dispersion surface energies rising from about 1 mJ/m^2 for the air-dried samples to greater than 10 mJ/m^2 for surfaces immersed for 10 days in water. As shown in Fig. 15A, the peel strength increases systematically with rising γ_s^{nd} of both copolymers. Evidently the contribution of non-dispersive (acid/base) forces to the peel strength can only take place when the polar BA moieties have had an opportunity to concentrate preferentially in the surface region of the copolymer. However, the most stable bond energies for the tape/polymer joint are not necessarily those obtained when the S/BA copolymer was preconditioned so as to attain the high (γ^{nd}) datum of about 10 mJ/m^2. This follows since the tape surface itself will exercise an orienting effect on the S/BA copolymer, and one that is not necessarily consistent with the surface structure created by lengthy exposure to water ! The results of Figure 15B relate to this point. When joints made with S/BA (98/2) that had been preconditioned in water to give a γ_s^{nd} of 7.5 mJ/m^2 are allowed to remain at 60° C, 0% RH for long periods, and re-tested periodically, it is evident that the initial high value of the peel strength diminishes to a new characteristic value, attained after 4 - 5 weeks of aging. By comparing the results in Fig. 15A and 15B, it may be concluded that the orienting strength of the adhesive tape necessitates the S/BA copolymer to adopt a surface composition and chain orientation consistent with a γ^{nd} value of about 4 mJ/m^2. The inherent surface modification principle must be heeded if optimum performance is to be realized from this combination of materials.

ACKNOWLEDGMENTS

Many of the results quoted here were obtained as a result of research supported by the Natural Sciences and Engineering Research Council, Canada. I thank the many graduate students and post-doctoral Fellows whose work has contributed to this article.

REFERENCES

1. Hildebrand,J.H. and Scott,R.L. (1964): Regular and Related Solutions, Van Nostrand - Reinhold, Princeton,N.J.
2. Blanks,R.F. and Prausnitz,J.M. (1964), Ind.Eng.Chem.Fundam.**3**,1.
3. Hansen,C.M. (1967), J. Paint Tech. **39**, 104.
4. Hansen, C.M. (1967), J. Paint Tech. **39**, 505.
5. Hoy, K.L. (1970), J. Paint Tech. **42**, 76.
6. Barton, A.F.L. (1983). Handbook of Solubility Parameters and Other Cohesion Parameters. CRC Press, Inc. Boca Raton, FL.
7. Flory, P.J. (1953). Principles of Polymer Chemistry. Cornell University Press, Ithaca, N.Y.
8. Huggins, Maurice L. (1941) J. Chem.Phys. **9**, 440.
9. Huggins, Maurice L. (1973) Polymer J. **4**, 502.
10. Braun,J.M. and Guillet,J.E. (1976) Advances in Polymer Science **21**, 108.
11. Gray,D.G. (1977). Prog. Polym. Sci. **5**, 1.
12. Lloyd, D.R., Ward,T.C. and Schreiber,H.P. (1989) Inverse Gas Chromatography. ACS Symposium Series **391**. Amer. Chem. Soc., Washington,D.C.
13. Dorris, G.M. and Gray,D.G. (1980) J. Colloid Interface Sci. **77**, 353.
14. Su,C.S., Patterson, D. and Schreiber,H.P. (1976). J. Appl. Polym.Sci. **20**, 1025.

15. El-Hibri,M.J., Cheng,Weizhuang, Hattam, Paul and Munk, Petr (1989). in Lloyd,D.R., Ward, T.C. and Schreiber, H.P. (1989). Inverse Gas Chromatography. ACS Symposium Series **391**,Ch. 10. Amer.Chem.Soc., Washington, D.C.

16. Dreisbach, R.R. (1961). Advances in Chemistry Series **29**, Amer. Chem. Soc., Washington, D.C.

17. DiPaola-Baranyi,G., and Guillet, J.E. (1978) Macromolecules **11**, 224

18. Price, Gareth J. (1989) in Lloyd, D.R., Ward, T.C. and Schreiber, H.P.: Inverse Gas Chromatography, ACS Symposium Series **391**, Ch. 5. Amer.Chem.Soc., Washington, D.C.

19. Scott,R.L. (1949). J. Chem. Phys.**17**,268.

20. DiPaola-Baranyi,G. and Degre,P. (1981). Macromolecules **14**, 1456.

21. Doube,C.P and Walsh,D.J. (1981) Eur. Polym. J. **17**, 63.

22. Sanchez, I. (1989) Polymer **30**, 471.

23. Chee,K.K. (1990) Polymer **31**, 1711.

24. Shi,Z.H. and Schreiber,H.P. (1991), Macromolecules **24**, 3522.

25. Lipatov, Yu.S. (1990) in Ishida, H.: Controlled Interphases in Composite Materials, Elsevier Publishers, New York, N.Y. p. 599.

26. Fowkes,F.M. (1980) in Lee,L.H. : Adhesion and Adsorption, Part A. Plenum Press, New York, N.Y. p. 43.

27. Fowkes F.M. (1964). Ind.Eng.Chem. **56**, 40.

28. Saint lour, C. and Papirer,E. (1983). J. Colloid Interface Sci. **91**, 63.

29. Schultz, J., Lavielle,L., and Martin,C. (1987) J. Adhesion **23**, 45.

30. Deng Zhuo and Schreiber, H.P. (1989) in Culbertson, Bill M.:Contemporary Topics in Polymer Science, Vol 6. Plenum Press, New York, N.Y. p. 385

31. Drago, R.S. and Wayland,B. (1965) J. Amer.Chem.Soc. **99**, 3571.

32. Fowkes, F.M. (1987) J. Adh. Sci. Tech. **1**, 7.

33. Chen, Fute (1988) Macromolecules **21**, 1640.

34. Gutmann, V. (1978) The Donor-Acceptor Approach to Molecular Interactions. Plenum Press, New York, N.Y.

35. Osmont,E. and Schreiber, H.P. (1989) in Lloyd, D.R., Ward, T.C. and Schreiber, H.P. :Inverse Gas Chromatography. ACS Symposium Series **391**. Amer.Chem.Soc. Washington, DC. Ch.17.

36. Riddle, F.L and Fowkes, F.M. (1990). J.Amer.Chem.Soc. **112**, 3259.

37. Panzer, Ulf and Schreiber,H.P. (1992). Macromolecules.In Press.

38. Wesson, Sheldon P. and Allred, Ronald E.,(1989), in Lloyd, T.C., Ward, T.C. and Schreiber, H.P.: Inverse Gas Chromatography. ACS Symposium Series, **391**, Ame. Chem. Soc., Washington, D.C. Ch. 15

39. Schreiber, H.P., Viau, J.-M., Fetoui,A. and Deng, Zhuo (1990). Polym.Eng.Sci. **30**, 263.

40. Lara,Javier A. and Schreiber, H.P. (1991). J.Coatings Tech. **63**, 81.

41. Maron, S.H. and Pierce, P.E. (1956). J. Colloid Sci. **11**,80.

42. Schonhorn, H. and Ryan, F.W. (1968). J. Polym. Sci. **A-26**,231

43. Sharpe, L. H. (1972). J. Adhesion **4**, 51.

44. Schreiber, H.P., Storey, S.H. and Bagley, E.B. (1966). Trans. Soc. Rheol. **10:1**, 275.

45. Tordella, J.P. (1970), J. Appl. Polym. Sci. **14**, 1627.

46. Schonhorn,H. and Hansen, R.H. (1967). J. Appl. Polym. Sci. **11**. 1461.

47. Wu, S. (1982): Surfaces and Interfaces. Marcel Dekker Inc., New York, N.Y.

48. Uehara, T. (1990). J. Appl. Polym. Sci. **41**, 1695.

49. Dong, S., Sapieha, S. and Schreiber, H.P. (1992). Polym. Eng. Sci. In Press.

50. Boenig, H.V. (1984).: Advances in Low-Temperature Plasma Chemistry, Technology, Applications. Technomic Publishing Co., Lancaster, PA.

51. Wertheimer, M.R., Klemberg-Sapieha,J.E., and Schreiber,H.P. (1984). Thin Solid Films **115**, 109.

52. Kaplan S.L. and Rose, F.W. (1991) Int. J. Adhesion and Adhesives **11**, 109.

53. Plueddemann, E.P. (1990).: Silane Coupling Agents. 2 edition. Plenum Press, New York, N.Y.

54. Plueddemann, E.P. (1974). Composite Materials, Vol.6. Academic Press, New York, N.Y.

55. Plueddemann, E.P. (1991). J. Adh. Sci.Tech. **5**, 261.

56. Walker, P. (1991). J. Adh. Sci. Tech. **5,**279.

57. Park, J.M. and Subramanian, R.V. (1991). J. Adh. Sci. Tech. **5**, 459.

58. Pape, P.G. and Plueddemann, E.P. (1991). J. Adh. Sci. Tech. **5**, 831.

59. Carre, A., Gamet, D., Schultz, J. and Schreiber, H.P. (1986). J. Macromol. Sci-Chem. **A23(1)**, 1.

60. Schultz, J. and Lavielle, L. (1989). in Lloyd, D.R., Ward, T.C. and Schreiber, H.P.: Inverse Gas Chromatography. ACS Symposium Series **391.** Amer. Chem. Soc., Washington, D.C. Ch. 14.

61. Chen, J.H. and Ruckenstein, E. (1990). J. Coll. Interface Sci. **135**, 496.

62. Andrade, J.D. (1988). Polymer Surface Dynamics, Plenum Press, New York, N.Y. Ch.1.

63. Deng, Zhuo and Schreiber, H.P. (1991). J. Adhesion **36**, 71.

Rheology at interfaces

P.G. de Gennes* and F. Brochard-Wyart**

*Collège de France, 11 place Marcelin-Berthelot, 75231 Paris Cedex 05, FRANCE
**PSI, Institut Curie, 11 rue P. et M. Curie, 75231 Paris Cedex 05, FRANCE

I. Introduction

II. Partial healing of an A / A interface

      A. Healing near the glass transition

      B. Fracture in the glassy state

III. A / B interfaces

      A. Interface structure for weakly incompatible pairs

      B. Toughness of A / B interfaces

      C. Tangential slip

      D. Suppression of slippage by block copolymers

IV. Concluding remarks

G. Akovali (ed.), The Interfacial Interactions in Polymeric Composites, 61–80.
© 1993 Kluwer Academic Publishers.

I. Introduction

Polymer / polymer interfaces play a dominant role in various mechanical features : coextrusion, adhesive properties, toughness of polymer blends, being typical examples. We begin to have precise experimental informations on interfacial structures using refined probes such as neutron reflectances. At the same moment, the toughness of bulk (glassy) polymers, which craze under tension, begins to be understood through an original idea of H. Brown [1]. It is thus tempting to extend the Brown ideas to various systems of "weak junctions". The junction may be (i) a partly healed contact between two identical polymer blocks A / A, as in the experiments of the Lausanne group [2] [3] (ii) a contact between two different polymers A and B. In all our discussion, we shall assume these junctions to be perfect -with full contact between the two partners, and no gaps : experimental arguments for the existence of these good contacts have been presented by Kausch and coworkers [2].

Our aim here is (a) to give a brief reminder on the theoretical description of the weak junctions (b) to show how some basic mechanical properties can be related to the structure. One of the major conclusions, for the A / A case, is that chain ends play a crucial role. Thus any attraction between a chain end and the free surface of one A block will react significantly on the A / A mechanical properties after welding.

This type of attraction was first suggested by systematic experiments on melts by D. Legrand and G. Gaines [4] [5], showing that the surface tension γ of oligomers was often lower than the surface tension γ_∞ of a high polymer, and that the correction

$$\gamma_\infty - \gamma (N) \sim N^{-x} \tag{1}$$

where N is the degree of polymerisation, and x an exponent of order 2/3. The fact that $x < 1$ shows that we are not dealing with a simple uniform dilution of chains ends (which would give a correction $\sim N^{-1}$). The most natural way of understanding the Legrand-Gaines result amounts to assume that the chain ends are attracted to the surface. The IBM group [6] has argued that a typical monomer along the chain suffers an entropy loss of order unity when it is located near the free surface, because the chain is "reflected" here, while the chain ends do not have this loss: thus one expects, on purely entropic grounds, a gain of free energy $\sim kT$ for each chain end brought to the surface. There are also enthalpic effects, which may increase or decrease the surface attraction. But, if, on the whole, the attraction is of order kT per chain end, we reach a simple regime [7], where all chains within one radius of gyration $R_0 = N^{1/2} a$ of the surface, put their ends on the surface - and the deeper chains are unperturbed. This leads to a surface fraction of chain ends ϕ_s of order $\dfrac{2}{N} \cdot \dfrac{R_0}{a}$ (where a is a monomer size), and thus :

$$\phi_s = \sim N^{-1/2}$$

(or x = 1/2) in this regime. We shall call this the <u>normal attractive regime</u>. The value x = 2/3 observed by Legrand and Gaines, may be the result of a cross over between 0 attraction and normal attraction.

This interpretation of the Gaines results is still controversial : Dee and Sauer [8] interpret γ (N) not by an effect of chain ends, but from the empirical N dependence of the (P, V, T) equation of state oligomers (the mean feature here being the change of the equilibrium density ρ (N)), plus a standard mean field analysis of the interfacial energy [9], as related to the equation of state and to the range of the intermolecular forces (the latter being assumed independent of N). Dee and Sauer get remarkable fits to the Gaines data -without involving any special localisation of chain ends ! Here, however, we shall keep in mind constantly the possiblility of chain end segregation near the free surface : indeed, we shall see that some of the neutron data on partial healing of A / A interfaces are more easily understood in the normal attractive regime than in 0 attraction.

II. Healing of an A / A interface

A) Healing near the glass transition. The basic healing experiment is idealized on fig. (1). We start with two blocks of the same polymer, which we call H and D [For certain experiments (D) may be a deuterated polymer, while (H) is the usual -proton carrying- species]. The two blocks are put into close contact under a mild pressure, at a temperature close to the glass point T_g, during a time t. The polymer chains from H and D begin to intertwine, and build up a diffuse profile for the D concentration ϕ_D (fig. 2). We are interested here primarily in this interdigitation process, at times t smaller than the reptation time of the chains T_{rep}. This corresponds to spatial widths of the profile e(t) which are smaller than the coil radius R_o.

Most experiments have been performed with polystyrene, and with H and D chains of comparable length : $N_{lt} \approx N_D = N$. The choice of N is non trivial :

a) We want $N \gg N_e$ (the distance between entanglements).

b) We want $\chi_{HD} < 1$, where χ_{HD} is the (small) Flory parameter describing a weak trend for segregation between the H and D species.

Typically N will be of order 2 000 – 6 000, while $N_e \sim 300$. The thickness e(t) of the partly healed zone is in the range of 100 Å - too small to be studied by forward scattering of charged particles. The main experimental tools used to measure the healing profile have been SIMS [10] and Neutron reflectance [11] [12]. Most data do show that the overall thickness e(t) grows like $t^{1/4}$:

$$e(t) \sim R_0 \left(\frac{t}{T_{rep}}\right)^{1/4} \qquad (t < T_{rep}) \qquad (2)$$

This is the natural law for spatial motions of one labeled monomer in an entangled melt [13] : after a time t , the chain carrying this monomer has moved along its own tube by a curvilinear length :

$$s(t) = (D_t t)^{1/2} \qquad (3)$$

where D_{tube} ($\sim N^{-1}$) is the tube diffusion coefficient. The corresponding distance as the crow flies is :

$$e(t) = (d\, s(t))^{1/2} \qquad (s > d) \qquad (4)$$

where $d = N_e^{1/2} a$ is the tube diameter, and (4) coincides with (2).

However, this simple agreement ignores an important fact, namely that near the contact surface, the chains were originally reflected, as is clear on fig. 1, and thus most of the tube motions do **not** give any intertwining. A first theoretical reflection on the problem was performed long ago by various authors [14] [15] [16], and will be summarized here.

a) For $N \gg N_e$, it is reasonable to assume that the "hairpin" processes of fig. 3a are negligible, the entropy of a hairpin on a lattice model is one half of the entropy of a free chain. More generally : hairpins are disfavored by a factor of order $\exp(- n / 2\, N_e)$, where n is the contour length of the hairpin.

b) Then, at the times of interest (where $e(t) > d$), all the interdigitation is due to the motion of chain ends ; one of them will start from some initial position (within $e(t)$ of the interface), and may cross (once or more) the interface. The number of monomers which it brings to the other side is a fraction of $s(t)$. Thus the total number v of monomers D going through the interface (per unit area) is of the form :

$$v \approx s(t) \int_{-e}^{0} \phi_e(z)\, dz \qquad (5)$$

where $\phi_e(z)$ is the initial distribution of chain ends. In the original discussions [14] [15] [16], it was assumed that $\phi_e(z)$ is uniform $\phi_e(z) = 2 / N$. But in our days, we know that chain ends may have been attracted to the original free surface of the block : as explained in the introduction, in normal attractive conditions, this will bring another (dominant) contribution to the integral [5], proportionnal to $\phi_{so} = N^{-1/2}$. Thus we have two cases :

$$v \sim s(t) \, e(t) \, N^{-1} \qquad \text{(no attraction)}$$

$$v \sim s(t) \, a \, N^{-1/2} \qquad \text{(normal attraction)}$$

$$\left.\begin{array}{c} \\ \\ \end{array}\right\} \qquad (6)$$

c)Because of the reflection of chains at the original interface, the profile is <u>discontinuous</u> (if we consider spatial scales larger than the tube diameter d). The general aspect is shown on fig. 2. Of major interest is the concentration ϕ_+ (t) of D monomers, on the H side, for $z \to 0$. We may write :

$$v \sim \phi_+ (t) \, e(t) \qquad (7)$$

Comparing (6) and (7), using the right normalisation factors, and inserting $\phi_{s0} \sim N^{-1/2}$, we then arrive at :

$$\phi_+ \sim \left(\frac{t}{T_{rep}}\right)^{1/2} \qquad \text{(no attraction)}$$

$$\phi_+ \sim \left(\frac{t}{T_{rep}}\right)^{1/4} \qquad \text{(normal attraction)}$$

$$\left.\begin{array}{c} \\ \\ \end{array}\right\} \qquad (8)$$

Thus, when chain ends were originally numerous at the surface, ϕ_+ (t) rises more rapidly.

On the experimental side, the most striking data on the profile comes from the neutron reflectance experiments of Reiter and Steiner [12]. They found that their profiles :
- could not be described by simple diffusion (giving an error function)
- could be described by the superposition of <u>two</u> errors functions E_{broad} and E_{sharp}

$$\phi_D (z) = 2\phi_+ (t) \, E_{broad}\left(\frac{z}{e(t)}\right) + (1 - 2\phi_+) \, E_{sharp}\left(\frac{z}{\sigma_c(t)}\right) \qquad (9)$$

where the errors functions E(z) are normalised by $E(0) = 1/2 \quad E(-\infty) = 1 \quad E(+\infty) = 0$.

For the "sharp" component (describing what we called the discontinuity), they found $\sigma_c(t)$ very weakly dependent of time -increasing from ~ 20 Å to 30 Å in the time interval $0 < T_{rep}$. For the "broad" component, the result is $e(t) \sim t^{0.17}$ -not too far from eq. (2).

But their most interesting result is related to $\phi_+(t)$. They found $\phi_+(t) \sim t^{0.22}$, very close to the prediction of eq. (8) for normal attraction between chain ends and the free surface. What is nice is that they obtained this without being biased by any theoretical prediction !

Thus the Reiter-Steiner experiment does suggest that (*in their conditions of sample preparation*) chain ends were originally attracted to the surface.

B) Mechanical toughness of partly healed A / A junctions. Kausch and coworkers [2] [3] have measured the fracture energy G_{1c} of partly healed A / A contacts : after healing over a time t, the sample is brought back to room temperature, where it is glassy, and then fractured along the junction. Experimentally, in most cases, the fracture energy G_{1c} increases with healing time : $G_{1c} \sim t^{1/2}$ $(t < T_{rep})$.

1) Bulk polymer toughness : the Brown model : To discuss this, let us first return to the case of bulk polymer fracture, in situations where crazing occurs, as shown on fig. (4). Over most of the crazed region, the stress σ (normal to the fracture plane), is nearly constant $\sigma = \sigma_y$, where σ_y is expected to describe plastic yield. However, near the crack tip, a stress concentration occurs, as emphasized by H. Brown [1]. At distance x from the crack tip, smaller than the length h_f of the ultimate fibrils, we would expect :

$$\sigma(x) \sim \sigma_y \left(\frac{h_f}{x}\right)^{1/2} \qquad (x \leq h_f) \tag{10}$$

Eq. (10) describes a square root singularity, as in a simple elastic medium. The coefficient is such that $\sigma \to \sigma_y$ for $x \sim h_f$ (beyond which we expect no stress concentration).

The singularity described above is cut off at the minimal value of $x : x \sim D$, where D is the interfibrillar distance. Thus the stress or the ultimate fiber is :

$$\sigma_1 \approx \sigma_y \left(\frac{h_f}{D}\right)^{1/2} \tag{11}$$

The length D is conditioned by capillary effects and cavitation instabilities at the point of birth I of the craze, and is expected to to be independent of molecular weight [17] [18]. Typically $D \sim 200$ Å.

Let us now restrict our attention to cases where the ultimate fibril breaks by <u>chemical scission</u> of its polymer chains. (This will be correct if the chains are long enough : N above a certain limiting value N*). Then the rupture condition is :

$$\sigma_i = \sigma_x = \frac{f_x}{a^2} \tag{12}$$

where f_x is the chemical force required to break a chain, a^2 is the chain cross section, and σ_x the corresponding stress. f_x is of the order U_{bond} / a, where U_{bond} is a covalent bonding energy, and is thus large (~ 1 nano Newton). The stress σ_x may be described as a material parameter of the polymer : however, we should keep in mind that chain scission will be sensitive to all dopants present in the polymer : catalysts, antioxydants,...

Inserting (12) into (11), we arrive at the Brown formula for the ultimate fibril length :

$$h_f \approx \left(\frac{\sigma_x}{\sigma_y}\right)^2 D \tag{13}$$

Because $\sigma_x \gg \sigma_y$, h_f can be of order several microns, as observed in materials like PS [18]. We can now turn to an estimate of the fracture energy :

$$G \approx \int_0^{h_f} \sigma(h) \, dk \approx \sigma_1 \, h_y \approx \frac{\sigma_x^2}{\sigma_y} D \tag{14}$$

Eq. (14) reflects the fact that most of the fiber pulling took place under the stress σ_y. The essential feature of this Brown formula is that $G \sim \sigma_x^2$, and is thus very large : eq. (14) is the basic explanation for the toughness of glassy plastics.

In fact, H. Brown was led to this type of formula by a series of systematic experiments on the toughness of an AB interface with A = p. styrene B = p. methyl metacrylate [19] [20]. Here, the cohesion between the two is established via block copolymers AB lying at the surface (fig. 5). In a number of cases the blocks ruptured very near their junction point (as shown by SIMS). It was found that the energy G_{ic} was proportionnal to the <u>square</u> of the number n of bridging chains per unit area : this fits with eq. (14) since, in the present case, σ_x is due to the copolymers only, and is thus proportional to n.

2) Transposition to partly healed interfaces : On fig. (3), the number of bridging chains per unit area is expected to be proportionnal to ϕ_+ (t) : any D monomer which has just crossed the border has a finite probability of being directly linked to the D side. Thus, to describe chemical rupture of the fibrils, we should perform the replacement :

$$\sigma_\chi \rightarrow 2 \, \phi_+ \, (t) \, \sigma_\chi \qquad (15)$$

where the factor (2) is fixed by the condition that σ_χ returns to its bulk value at $t > T_{rep}$ ($\phi_+ \rightarrow 1/2$). Eq. (14) then gives :

$$G_{1c}(t) = G_{1c}\big|_{bulk} \cdot 4 \, \phi_+^2 \, (t) \qquad (16)$$

If, and only if, the chain ends were originally numerous at the surface, we can then return to eq. (8) and write ϕ_+ (t) $\sim$ t $^{1/4}$, giving the experimental form $G_{1c} \sim$ t $^{1/2}$. Thus the Kausch law, combined with the Brown model, does suggest that a large number of chain ends were available at the interface when healing started.

A careful reader may be worried by the following point : in the Kausch experiments, the original blocks H and D were in fact obtained by rupture of one single sample. Could it be that chain ends were very numerous ($\phi_s > N^{-1/2}$, possibly $\phi_s \sim$ 1) on the interface at t = 0 ? We do not believe this to be the case,as explained on fig. (4). The separation of the two blocks took place via fibril rupture, but after this, the half fibrils on both sides have probably retracted to build again a compact layer of polymer on each lip of the fracture (around point F) : in this retraction process, chain ends may be buried in each layer. Most of the chains in this layer belonged to portions of the fibrils which were not disrupted chemically. Thus, if the retraction led to an equilibrium, we again expect $\phi_s \sim N^{-1/2}$.

III. A / B interfaces

A) Interface structure for weakly incompatible pairs. The qualitative aspect of an A / B interface is shown on fig. (6). A simple understanding of the structure can be obtained, starting from an abrupt interface, and allowing one A chain to protrude in the B side (fig. 7). If m monomers are exposed in this process, the enthalpy required is :

$$\Delta H_m \sim m \, \chi \, kT \qquad (17)$$

where χ is the Flory parameter [21] describing AB mixtures. The average value of m corresponds to $\Delta H_m \sim$ kT, and is thus :

$$\overline{m} = \chi^{-1} \qquad (\chi < 1) \tag{18}$$

(We constantly assume that $\overline{m}$ is much smaller than the overall chain length N).

Since the protruding chain is a random walk, the width e of the interface is the size of this random walk :

$$e \approx a\,\overline{m}^{1/2} = a\,\chi^{-1/2} \tag{19}$$

and e is much larger than a if χ is small : we shall constantly focus on this limit. Of course, the result (19) can be derived by more rigorous means, but the present approach is often illuminating.

The distribution of m values is given by a Boltzman exponential :

$$p_m = \frac{1}{\overline{m}}\,\exp\,(-\Delta H_m / kT) = \frac{1}{\overline{m}}\,\exp\,(-m/\overline{m}) \tag{20}$$

Of major interest for mechanical properties, is the probability that the protruding chain entangles with the surrounding matrix [2]. If we define an average chemical distance between entanglements N_e , we may write for the probability f of entanglements :

$$f = \sum_{N_e}^{\infty} p_m = \exp\,(-N_e\,\chi) \tag{21}$$

Of course, this formula is very approximate, because N_e needs not to be the same for the two partners A and B, and also not the same for the mixtures : a certain weighted average would then be required. But eq. (21) is still a reasonable starting point to discuss the mechanics of A / B contacts.

B) Toughness of A / B interfaces : Long ago a remarkable series of experiments was performed by Iyengar and Erickson [23]. They measured the adhesion energy G_{1c} of various polymers, on PET, by a 90° peeling test (at a fixed velocity 5 cm / sec.). The results were plotted as a function of the Hildebrand solubility parameter δ. They show a dramatic drop of G_{1c} as soon as the δ parameters of the two partners differed by more than one unit.

Can we establish contact between these data and eq. (21) for the probability of entanglements ? Let us assume that :

a) G_{1c} associated to the post craze fracture of a glassy A / B junction.

b) The entangled A or B chains in the junction must break.

Then we may apply Brown's eq. (14), provided that the chemical rupture stress σ_χ is suitably reduced : only the entangled chains at the junction contribute. This could give :

$$\sigma_\chi \rightarrow \sigma_\chi f \tag{22}$$

and :

$$G_{1c} = G_o f^2 = G_o \exp(-2 N_e \chi) \tag{23}$$

The result is an exponential drop in a $G_{1c}(\chi)$ plot : from the data, Iyengar and Erickson had proposed a different law : $G_{1c} \sim \exp[-k |\delta_A - \delta_B|]$ (Remember that, for simple Van der Waals interactions : $\chi \sim (\delta_A - \delta_B)^2$).

However, these differences are probably not very significative :

a) As already explained, N_e needs not be the same for all AB pairs : there need not be a universal plot $G_{1c}(\delta)$.

b) PET is always partly cristallised, and this complicates the picture.

The essential point is the rapid drop of G_{1c} when A and B become very different.

Of course, the best way of strengthening the AB interface, if A and B are strongly incompatible, amounts to bring an AB diblock copolymer at the interface [19] [20].

C) Tangential slip :

1) Long ago, we discussed the possibility of slip for a molten polymer against a solid surface [24]. One expects a significant slippage if :   a) the polymer does not bind to the surface   b) the polymer is not a glass (or a crystal) in the first few layers near the surface   c) the surface is not too rough.

Experimentally, the usual (no-slip) boundary conditions at the wall are often found to hold [25]. Certain observations on transparent extruders [26], or in plane-plane rheometry [27], do suggest a significant slip, but it is not clear whether this holds even in the linear regime (small shear stresses at the wall) which was considered in ref. [24].

2) Let us now consider an interface A / B between two molten polymers, with $1 > \chi \gg N^{-1}$. If χ is large, there will be no entanglements between A and B, and slippage may occur. This problem was first considered by Furukawa [28] -but without a full appreciation of the role of entanglements versus Rouse friction. A slightly improved (qualitative) discussion is given in ref. [29], and will be summarized here.

a) Consider first a case with no entanglements between A and B (fig. 8, 9). In the interfacial region, we have a steep velocity gradient [V] / 2e, and a weak Rouse viscosity $\eta_R(e)$. This viscosity is itself scale dependent , because e is smaller than the coil size R_o. As argued by F. Brochard (unpublished), we expect :

$$\eta_R(e) = \eta_l \frac{e^2}{a^2} \qquad (e < R_o) \tag{24}$$

Note that eq. 24 gives the right form for very short scales : $\eta = \eta_1$, and also for long scales : $\eta \sim \eta_1 N$ for $e \gtrsim R_0$.

Outside of the interfacial region, we have the strong viscosity of an entangled melt [30] :

$$\eta = \eta_l \frac{N^3}{N_e^2} \tag{25}$$

(where, for simplicity, we assume the same η_l and the same N for A and B). Writing that the stress $\sigma = \eta \frac{dv}{dz}$ is the same in both regions, we arrive at the extrapolation length b (as defined in fig. 9) :

$$b = e\left(\frac{\eta}{\eta_R(e)} - 1\right) \sim e\frac{\eta}{\eta_R(e)} \sim a\frac{N^3}{N_e^2} \chi^{1/2} \tag{26}$$

For instance, with $\chi = 0.1$, $N = 10^3$, $N_e = 10^2$ and a = 3 Å, we expect $b \sim 10$ microns.

b) Consider now a case where AB entanglement effects are dominant. Then a rough estimate of the inner viscosity is :

$$\eta_{in} = f\eta(e) \tag{27}$$

where $\eta(e)$ is the scale dependent viscosity for an entangled melt. From the argument in ref.[31] we expect :

$$\eta(e) = \eta_R(e)\left(\frac{N}{N_e}\right)^2 \qquad (e < R_0) \tag{28}$$

We are thus led to an extrapolation length b, in the entangled regime, of the form :

$$b \sim \frac{\eta}{f\eta(e)} - 1 \sim \frac{\eta}{f\eta(e)} \tag{29}$$

c) in intermediate regimes, we should add up the two types of friction (each proportionnal to $\frac{1}{b}$), and we arrive at :

$$\frac{e}{b} = \frac{\eta_R(e) + f\,\eta(e)}{\eta} = \frac{\eta_R(e)}{\eta}\left[1 + f\left(\frac{N}{N_e}\right)^2\right]$$

(30)

The cross-over is obtained when :

$$f = \exp\left(-N_e\,\chi\right) = \left(\frac{N_e}{N}\right)^2$$

(31)

(an estimate slightly different from that of ref. [29], but the difference occurs only in a log.). Typically, with $N = 10^3$ and $N_e = 10^2$, we require $N_e\,\chi \lesssim 5$ to be entangled.

There are a number of semi quantitative data on coextrusion of layered polymer systems, which indicate rather large values of b. In particular, we should mention the experiments of Miroshnikov and Andreeva [32], where a "jelly roll" of alternating A / B layers was extruded, with controled layer thickness of order 50 microns. The apparent viscosity of this composite melt structure was $\eta_{app} \sim \eta\,/\,20$ (where η is the average viscosity of A and B). This indicates that, for the particular system under study, the extrapolation length b was much larger than 50 microns.

D) Suppression of slippage by block copolymers [33].

Let us now assume that the number of mutual entanglements between A and B is exponentially small : if $N_e\chi \gg 1$, the two sides are decoupled and easily slide over one another. We now add (per unit area of the interface) a number ν of AB copolymers with degrees of polymerization Z_A, $Z_B \gg N_e$. Clearly the copolymers will connect the two blocks and increase the friction.

We assume that ν is small, i.e. $\nu\,Z_i\,a^2 < 1$. This places us in the "mushroom" regime, where two adjacents copolymers do not overlap. It will be seen that this is sufficient to block sliding.

1) Friction of a half block.

Consider now a polymer block (of Z_A monomers) which moves relative to the matrix at a velocity V which will be determined later. The semi-block cannot reptate because it is pinned at one end. In order to move, the matrix chains which traverse the "microgel" formed by Z_A monomers must disentangle. Their number is $p_A = Z_A{}^{1/2}$. For the microgel to advance an entanglement spacing $d = N_e{}^{1/2}\,a$, each of the p_A chains must reptate along it's own tube, by a tube length $L_t = N_A\,N_e{}^{-1/2}\,a$. Thus the curvilinear velocity of the chain is [34]

$$V_c \approx V\,\frac{L_t}{d} \approx V\,\frac{N_A}{N_e}$$

(32)

The dissipation per chain is $\zeta_1 V_c^2 N_A$. The total dissipation is :

$$T\dot{S} = \zeta_A V^2 = p_A \zeta_1 N_A \left(\frac{N_A}{N_e}\right)^2 \tag{33}$$

The friction coefficient ζ_A then has the Stokes form :

$$\zeta_A \approx \eta_A R(Z_A) \tag{34}$$

where $\eta_A = \zeta_1 a^{-1} N_A^3 / N_e^2$ is the matrix viscosity and $R(Z_A) = a Z_A^{1/2}$ is the mushroom size.

2) Interfacial friction.

Each copolymer moves at a velocity αV, such that the sum of the frictional forces acting upon it is zero :

$$\zeta_A \left(\frac{V}{2} - \alpha V\right) = \zeta_B \left(\frac{V}{2} + \alpha V\right) \tag{35}$$

This fixes α. The dissipation per unit interfacial area is then kV^2, with

$$k = v \frac{\zeta_A \zeta_B}{\zeta_A + \zeta_B} \tag{36}$$

3) Discussion.

Consider the symmetric case ($N_A = N_B = N$, etc). Then:

$$k \approx v \zeta_1 N^3 Z^{1/2} N_e^{-2} \tag{37}$$

$$b \approx \frac{\eta}{k} \approx [vR(Z)]^{-1} \tag{38}$$

Equating (41) and (34), we estimate the minimum concentration v^* in order to modify the sliding of the pure system :

$$v^* R^2 \approx N_e^2 N^{-3} Z^{1/2} \ll 1$$

In practice, trace amounts of copolymer is sufficient to prevent sliding.

The same effect is probably important to understand certain mechanical properties of the solid / polymer melt interface : We have seen that strong sliding is not generally observed. It is in fact sufficent for a few chains to be attached to special surface sites in order to supress sliding.

IV. Concluding remarks

The mechanical properties of polymer / polymer interfaces are clearly very sensitive to the detailed structure of the interface. We have seen here two major examples of this correlation   a) the role of chain ends, and of their spatial distribution in A / A healing   b) the role of entanglements in A / B fracture or in A / B slippage.

From a practical point of view, what can we do to modify the mechanical properties ? 1) For A / B systems, the most obvious additive is an A-B block copolymer. The main difficulty here is to bring the copolymer at the interface, since the kinetics of exchange between copolymer micelles and surfaces are very slow.   2) For A / A systems, we are facing an interesting chemical challenge : by suitable modifications of the chain ends, we may, or may not, encourage their segregation near the surface, and thus generate very different healing behaviors.

Acknowledgments : The authors have greatly benefited from discussions and other exchanges with H. Brown, A. Gent, H. Hervet, H. Kausch, E. Kramer, L. Léger, P. Pincus, G. Reiter, T. Russell, U. Steiner, and R. Wool.

REFERENCES :

(1)   H.R. Brown, *Macromolecules*, 1991, (to be published).

        "   *Annual Rev. Materials Sci.*, 1991, (to be published).

(2)   H. Jud, H. Kausch, J. Williams, *J. Material Sci.*, 16, 204 (1981).

(3)   H.H. Kausch, D. Petrovska Delacretaz, *Proc. IBM symposium on polymers*, Lech, 1990.

(4)   D. Legrand, G. Gaines, *J. Colloid Interface Sci.,* 31, 162 (1969).

(5)   "   "   "   "   42, 181 (1973).

(6)   A. Harihakan, S. Kumar, T. Russell, *Macromolecules,* 23, 3584 (1990).

(7)   P.G. de Gennes, *C.R. Acad. Sci. (Paris),* 307, 1841 (1988).

(8)   G. Dee, B. Sauer, *APS March meeting Cincinnati,* 1991, Abstract A 39-1.

(9)   C. Poser, I. Sanchez, *J. Colloid Interface Sci.,* 69, 539 (1979).

(10)   SIMS, S. G. Whitlow, R.P. Wool, *Macromolecules,* in print, 1991.

       "   "   "   paper presented at *Amer. Phys. Soc.,* 1989.

(11)   T.P. Russel, A. Karim, A. Mansour, G. Felcher, *Macromolecules,* 21, 1890 (1988).

       "   "   "   "   *Phys. Rev.,* B 42, 6846, (1990).

(12)   G. Reiter, U. Steiner, *J. Phys. (Paris),* in print.

       "   "   to be published in the proceedings of the Les Houches workshop on interfaces (D. Beysens and G. Forgacs editors, 1991).

(13)   P.G. de Gennes, *J. Chem. Phys.,* 55, 572 (1971).

(14)   "   *C.R. Acad. Sci. (Paris),* B 291, 219 (1980).

(15)   S. Prager, M. Tirrell, *J. Chem. Phys.,* 75, 5194 (1981).

(16)   R.P. Wool, K. O'Connor, *J. Appl. Phys.,* 52, 5953 (1981).

       R.P. Wool, B.L. Yuan, O. MC. Garel, *Polymer Engineering and Science,* 29, 1340 (1989).

(17)   H.R. Brown, *Materials Sci Reports,* 2, 315 (1987).

(18)   E. Kramer and L. Berger, *Adv. Polymer Sci.,* 91-92, p. 1 (1990).

(19)   H.R. Brown, V. Deline, P. Green, *Nature,* 341, 221 (1989).

(20)   K. Cho, H. Brown, D. Miller, *J. Pol. Sci. (Physics),* 28, 1699 (1990).

(21)   P. Flory, *Principles of Polymer Chemistry,* Cornell U. Press.

(22)   P.G. de Gennes, *C. R. Acad. Sci. (Paris),* 308 II, 1401 (1989).

(23)   Y. Iyengar, D. Erickson, *J. Appl. Polymer Sci,* 11, 2311 (1967).

(24)   P.G. de Gennes, *C.R. Acad. Sci. (Paris),* 288 B, 219 (1979).

(25)   J. Meissner, *Ann. Rev. Fluid Mech,* 17, 45 (1985).

(26)   J. Galt, B. Maxwell, *"Modern plastics",* (Mc Graw Hill ed.), Dec. 1964.

(27)   R. Burton, M. Folkes, N. Karm, A. Keller, *J. Materials Sci.,* 18, 315 (1983).

(28)   H. Furukawa, *Phys. Rev.,* A 40, 6403 (1989).

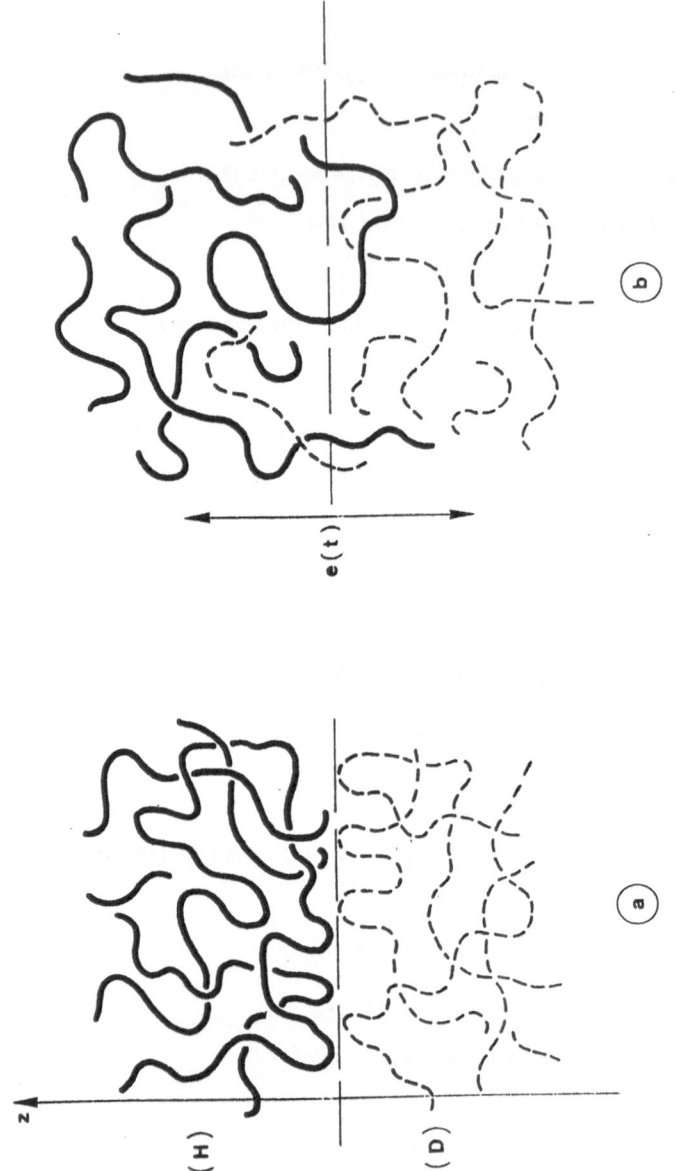

FIG: 1

(1)   Schematic representation of an interface between two chemically identical polymers H and D   a) before healing all chains are "reflected" on the contact plane   b) after partial healing (during a time t), the chains interdigitate. We are concerned here by the regime where the thickness e(t) of the mixing region is larger than a tube diameter (d), but smaller than the coil size (R_0).

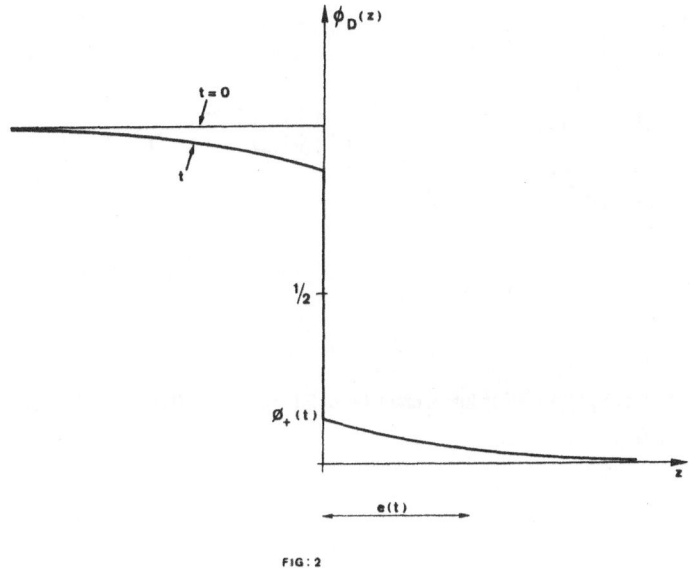

FIG: 2

(2) Concentration profile for the D species in the experiment of fig. 1. The tube diameter d is assumed to be much smaller than e(t) : then, the profile is discontinuous on the contact plane, because most chains are still reflected at this plane.

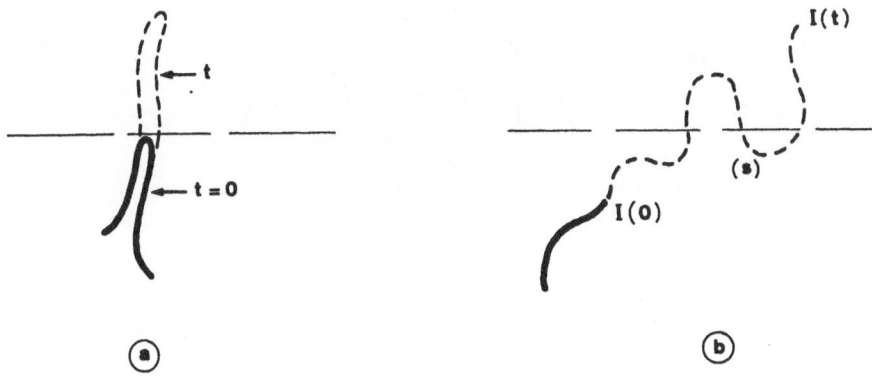

FIG : 3

(3) Two modes of interdigitation : (a) hairpins (b) chain ends crossing the contact plane. For strongly entangled chains ($N \gg N_e$), we expect process (b) to dominate.

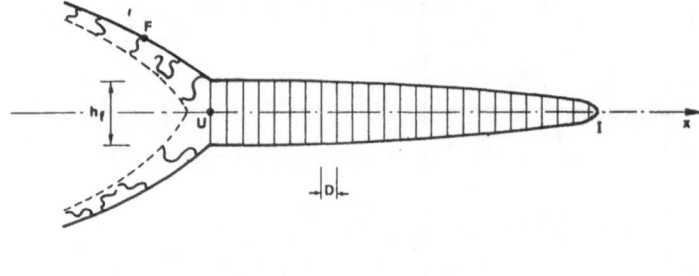

FIG:4

(4) A craze (initiated at point I), terminating into a crack (at point U). The fibrils break at U, and the resulting half fibrils retract around point F.

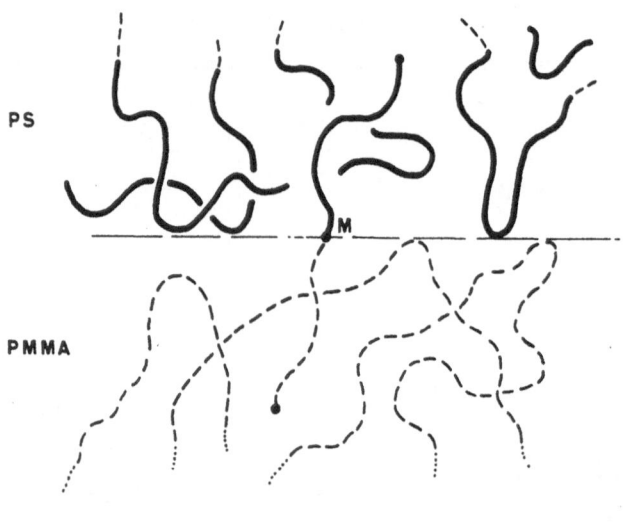

FIG:5

(5) An interface between two incompatible polymers, decorated by block copolymer molecules M.

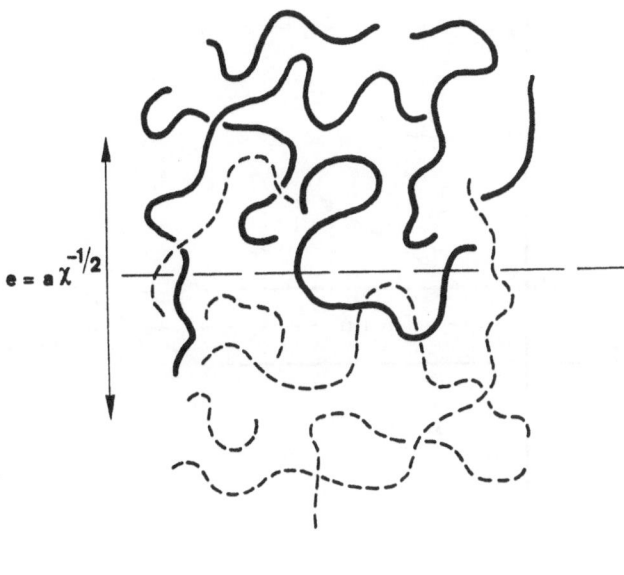

$$e = a\chi^{-1/2}$$

FIG : 6

(6)   An interface between weakly incompatible polymers ($\chi \ll 1$), with a width e.

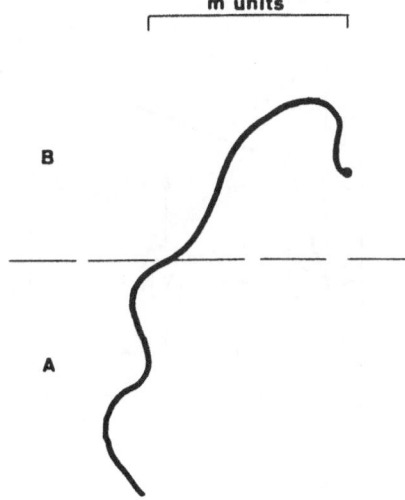

m units

B

A

FIG : 7

(7)   A simple approach to understand the structure of fig. (6) : the interface between the two polymers A and B is first assumed to be sharp. Then we allow one A chain to move partly into the B side, and discuss the resulting energy cost.

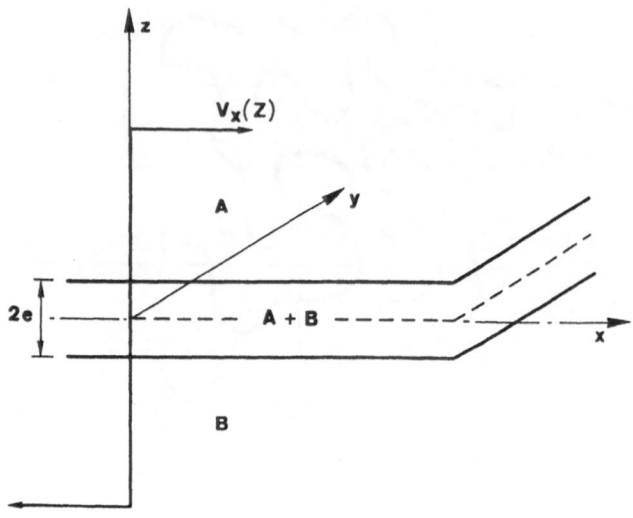

FIG : 8

(8)   Geometry of flow lines when two molten polymers A and B slip on each other.

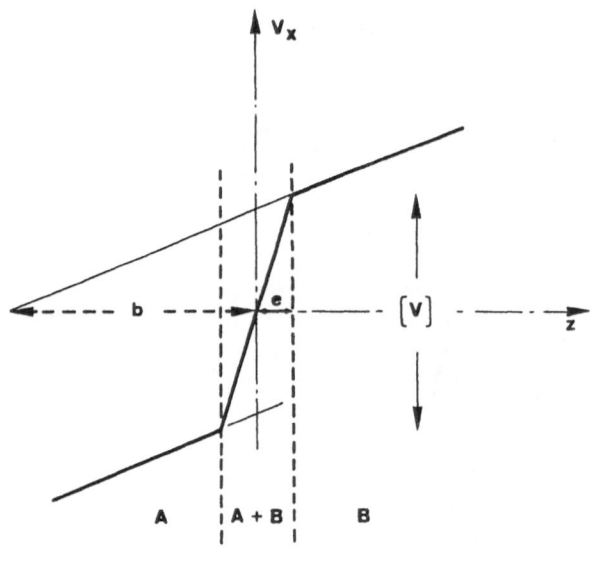

FIG : 9

(9)   Velocity profile in an A / B slippage process.

Interactions and Properties of Composites : a) Fibre-Matrix Adhesion Measurements.

Michel NARDIN and Jacques SCHULTZ

Centre de Recherches sur la Physico-Chimie des Surfaces Solides, C.N.R.S.,
24, avenue du Président Kennedy,
F-68200 Mulhouse, France.

1. Introduction

Great attention has been recently devoted to the development of fibre reinforced composites of high mechanical performance. Obviously, the final performances of a composite material are directly related to the properties of both basic constituents : the fibre and the matrix. The matrix gives materials its cohesion, while the fibres support most of the mechanical stresses. However, it is also well-known that the mechanism of load transfer at the fibre-matrix interface plays a major role in the mechanical and physical performances of composites and, consequently, the fibre-matrix interface can actually be considered as the third constituent of such materials.

For a good understanding of the fibre-matrix interface mechanics in relation to its physico-chemical properties, it is therefore necessary to determine the interface capacity to transfer the stresses from the matrix to the fibre. This is usually carried out by means of specific mechanical tests on real or model composites. The aim of the first part of this paper is to present and discuss briefly the most common micromechanical tests and theoretical approaches used nowadays for studying the mechanical behaviour of interfaces in composites. A second part will be concerned with the influence on the magnitude of the fibre-to-matrix stress transfer capacity, defined in terms of interfacial shear strength, of the interfacial adhesion between both constituents, on the one hand, and of the matrix alterations (interfacial layers) near the interface, on the other hand.

Mechanical tests that are the most commonly used for determining the strength of fibre-matrix interfaces in the field of composite materials are the followings :

a) **Tests on real composites** (usually unidirectional composites) with a high fibre-volume fraction (typically 60% by volume) :

G. Akovali (ed.), The Interfacial Interactions in Polymeric Composites, 81–93.
© 1993 *Kluwer Academic Publishers*.

- Measurement of the interlaminar shear strength (ILSS) by short beam shear test,
- Iosipescu's test,
- Transverse and off-axis tensile tests.

b) **Tests on model composites** (single fibre composites) constituted by a monofilament totally or partly embedded in the polymer matrix :
  - Pull-out and microdrop technique,
  - Microindentation test,
  - Fragmentation test,
  - Determination of fibre strain in single fibre composites by Raman spectrometry.

The usefulness and the validity of each micromechanical test as well as related theoretical approaches are now presented in turn. A comparison between results obtained on identical fibre-matrix systems by different micromechanical tests will be also analysed.

2. Tests on unidirectional composites

2.1. Interlaminar shear strength

This measurement is performed by a three-point flexure method on short beams, with the span-to-width ratio chosen to produce interlaminar shear failure [1,2]. The geometry of this test is described in Figure 1. Fibres are oriented in the direction perpendicular to the axes of the supports.

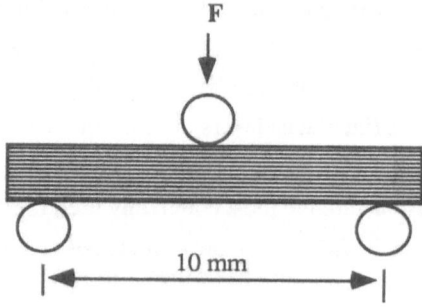

Figure 1 : Determination of the interlaminar shear strength

From elastic analysis of bending of beam, the apparent interlaminar shear strength τ_{ILSS} is given by :

(1) $\tau_{ILSS} = 0.75 \; F/A$

where F is the applied load at failure and A is the cross sectional area of the specimen. This method is often used, especially for industrial controls, since it is an easy way to rapidly determine the interlaminar properties in laminated unidirectional composites. However, shear and tensile failures generally simultaneously occur during measurement and, therefore, results have to be carefully analysed. Moreover, in particular in the case of thermoplastics based composites, the measurement of interlaminar shear strength can be limited by the intrinsic properties (yield strength for example) of the matrix. As a conclusion, this test is not well adapted to precisely analyse the mechanical behaviour of most of the fibre-matrix interfaces.

2.2. Iosipescu's test

In the method proposed in 1967 by Iosipescu [3], a state of uniform pure shear can be achieved within the test section of the specimen with the geometry defined in Figure 2. The value of the shear strength is simply equal to the ratio of the maximum force by the cross-sectional area between the two notch tips. Fibres can be aligned either in the direction or perpendicular to the direction of the applied force. Results obtained from this test are generally well reproducible whatever the nature of the fibres and the matrices.

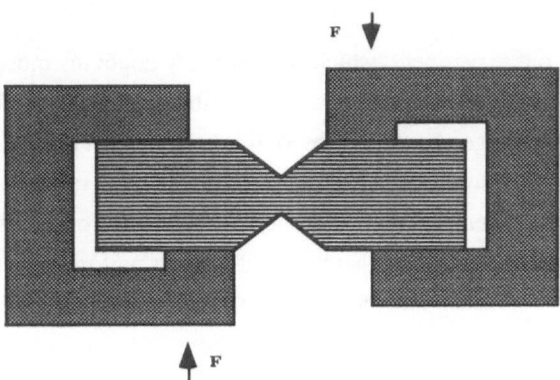

Figure 2 : Iosipescu' test

2.3. Off-axis and transverse tensile tests

A lot of data on mechanical behaviour of unidirectional composites, and in particular on the interfacial zone, can be obtained by subjecting these materials to off-axis tensile stress by varying the angle θ between the directions of the applied force and the fibres. Special analysis can be performed at $\theta = 90°$ corresponding to transverse traction. In that case, the interface is directly submitted to tensile load. Other experiments are generally made either at small values of θ (10 to 15°) or at $\pm 45°$ [4]. Due to the fact that the laminate is under a state of combined stress (existence of normal stress components) rather than pure shear, results on ultimate stress and strain have to be carefully analysed.

As a conclusion, it must be noticed that, in fact, all the tests performed on laminates are recommended only when the level of adhesion established between fibre and matrix is already known. The main problem involved in these types of tests is to control precisely the degree of alignment of fibres in the given direction. For example, a recent work [5] has shown that the mechanical responses of unidirectional laminates are very sensitive to the misorientation of fibres, even when this latter does not exceed 1 to 2%. Obviously, bridges or entanglements between fibres tend to increase greatly the measured mechanical properties and, thus, caution should be exercised when interpreting the results.

3. Tests on model composites

3.1. Pull-out and microdrop technique

The pull-out experiment, which was with any doubt the most widely used test during these last twenty years, consists of a single fibre partially or totally embedded in a block of resin of known geometry (Figure 3). It is believed that the results obtained from this test could be related to the fibre pull-out process which can occur during the failure of real composites. An increasing force is applied to the free end of the filament in order to pull it out of the matrix. Assuming that the shear stress is uniformly distributed along the immersion length, the mean value of the shear strength τ of the fibre-matrix interface is therefore given by :

(2) $$\tau = F_{max}/\pi d l_e$$

where d and l_e are the diameter and the embedded length of the fibre respectively, and F_{max} the maximum force corresponding to the pull-out process.

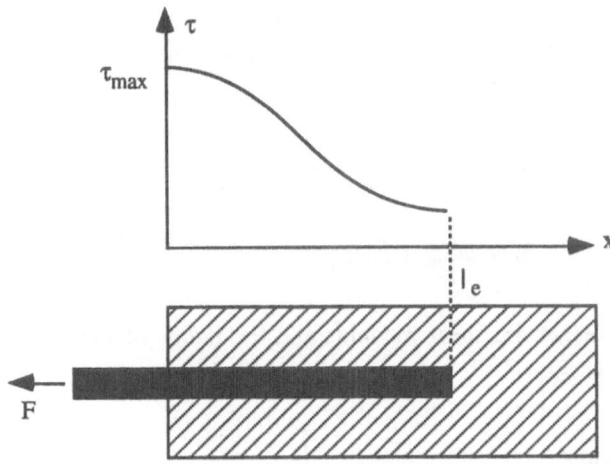

Figure 3 : Pull-out test

The shear lag analysis proposed by Greszczuk [6] shows that the shear strength τ_i is varied along the embedded length as described in Figure 3. This stress is given by :

$$\tau_i = \tau \alpha l_e (\frac{\cosh \alpha x}{\tanh \alpha l_e} - \sinh \alpha x)$$

(3)

with τ the mean shear strength given by equation (2) and α an elastic coefficient :

(4) $$\alpha^2 = 2G_i/(b_i r E_f)$$

where G_i and b_i are respectively the shear modulus and the thickness of an interfacial layer close to the fibre surface, E_f and r are the elastic modulus and the radius of the fibre respectively. At x = 0, the shear strength reaches its maximum value τ_{max} :

$$\tau_{max} = \frac{\tau.\alpha l_e}{\tanh \alpha l_e}$$

(5)

τ_{max} can be considered as a constant which characterizes the fibre-matrix adhesive strength. Therefore, according to equations (3) and (5), τ tends towards τ_{max} when l_e tends towards zero, whereas for large values of l_e, τ varies as $(l_e)^{-1}$. This analytical approach is generally well verified experimentally [7,8].

More recently, for the general case when significant deformations occur in both the fibre and the matrix, Chua and Piggott [9] have developped an energy approach by considering that the sum of the fibre strain energy U_f and shear energy U_m stored in the fibre and the matrix respectively equals the energy corresponding to the failure of the interface $U_t = 2\pi r l_e G_a$, where G_a is the interfacial fracture energy per unit surface area. Finally, these authors obtain the following expression for the pull-out force F :

(6) $$F^2 = 4\pi^2 r^2 (\beta r l_e E_f G_a).\tanh \beta l_e$$

(7) with $$\beta^2 = 2G_m/(r^2 E_f.\ln R/r)$$

where G_m and R are the shear modulus and the radius of the matrix respectively. Expression (6) reduces to :

•For long embedded length (tanh $\beta l_e \sim 1$) :

$$F \sim 2\pi r(\beta r l_e E_f G_a)^{1/2} \equiv l_e^{1/2}$$

•For short embedded length (tanh $\beta l_e \sim \beta l_e$) :

$$F \sim 2\pi r \beta l_e (r E_f G_a)^{1/2} \equiv l_e$$

These theoretical variations of F versus l_e are in rather good agreement with experimental results.

Finally, it must be noted that, very recently, Gent and Liu [10] have proposed an energy analysis when only friction forces are involved at the fibre-matrix interface during pull-out test.

When a high level of adhesion is established at the fibre-matrix interface, it appears that, beyond a given critical immersion length, the pull-out force becomes superior to the tensile force required to break the fibre. This problem is crucial when the fibre diameter is small (~ 10 μm for carbon or glass fibres), since it is necessary to reduce the embedded length to a hundred micrometers or less in order to determine the pull-out force experimentally. This can be achieved by putting directly a small axisymetrical drop of resin onto the fibre and thus performing the pull-out test as described in Figure 4. This type of geometry is usually called the microdrop technique and was first proposed by

Miller and co-workers [11]. The results are interpreted according to the theoretical approaches previously described.

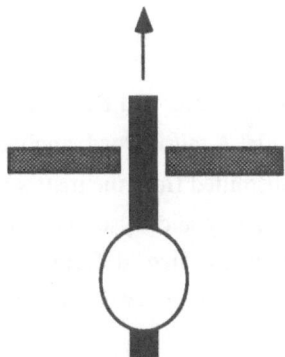

Figure 4 : Microdrop technique

3.2. Microindentation test

In the microindentation technique, first proposed by Mandell and co-workers [12] and presented in Figure 5, single fibres are compressively loaded to produce debonding at the interface or fibre slippage. This test can be applied directly to unidirectional composites insofar as it is possible to perform good cut and polishing of the surface perpendicular to the fibre direction. One of the advantages of such a procedure is to take into account the effect of the presence of the neighbouring fibres on the fibre-matrix interfacial behaviour.

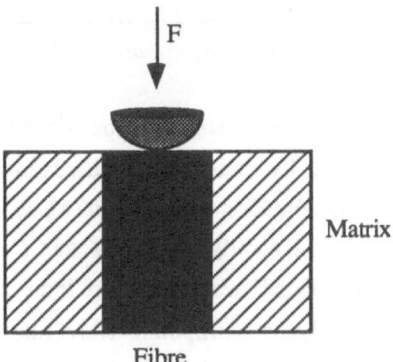

Figure 5 : Microindentation test

As for pull-out test, various shear lag or energy theoretical approaches and finite element techniques can be used for the mechanical analysis, in the case of strong level of interfacial adhesion as well as for fibre slippage. This test is mainly applied in the field of metal or ceramic matrix composites.

3.3. Fragmentation test

This method, proposed by Fraser and Di Benedetto [13], has been largely used and studied during the last decade. A tensile load, applied to a single fibre composite in the direction of the fibre, is transmitted from the matrix to the fibre across the interface and, as originally described by Kelly and Tyson [14], the fibre breaks into fragments (Figure 6) until a limiting fragment size, defining a critical length l_c, is reached. Therefore, the mean interfacial shear strength τ can be calculated from the expression :

$$(8) \qquad \tau = d.\sigma_f(l_c) / 2.l_c$$

where d is the diameter of the fibre and $\sigma_f(l_c)$ is the tensile strength of the fibre at a gaugelength equal to l_c.

Figure 6 : Fragmentation test

The measurement of the critical length l_c is performed directly by microscopical observations in the case of optically transparent matrices or by counting the acoustic events corresponding to the fragmentation process using an acoustic emission measuring device [15,16]. The value of l_c is generally small (less than 1 mm) and, consequently, it is impossible to carry out experimental measurements of individual fibre strength $\sigma_f(l_c)$ at these short lengths. Most analyses extrapolate fibre mean strength and strength distribution data obtained at higher gaugelengths, by means of a Weibull distribution analysis [17].

Expression (8) was obtained by a simple balance equation, or from the famous shear lag theory (fully elastic model) of Cox [18], assuming in particular that an ideal elastic stress transfer from the matrix to the fibre is performed during all the fragmentation process, even when the length of the fibre fragments reaches the critical length. The profiles of fibre tensile stress σ and matrix shear stress τ along the interface, as described by Cox' analysis, i.e. :

$$\sigma(x) = \varepsilon E_f\left(1 - \frac{\cosh \beta(\frac{l}{2} - x)}{\cosh \frac{\beta l}{2}}\right)$$

(9)

$$\tau(x) = \frac{r\varepsilon\beta E_f}{2}\left(\frac{\sinh \beta(\frac{l}{2} - x)}{\cosh \frac{\beta l}{2}}\right)$$

(10)

with β given by equation (7) and ε the applied strain, are schematically represented in Figure 7.

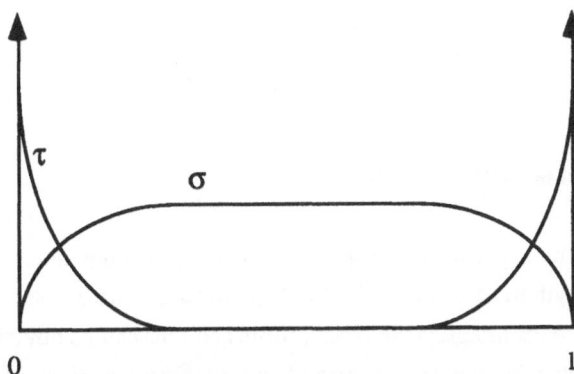

Figure 7 : Tensile and shear stress profiles along a fibre of length l.

However, all the purely elastic approaches cannot explain the stop of the fragmentation process for a given fragment length in single fibre composites. This phenomenon implies that inelastic deformations occur at a certain applied strain, leading to the saturation of the process at larger strains. Since the shear stress is maximum at the fibre ends, it is reasonable to consider that these inelastic deformations start from the fibre ends and are due to either an interfacial decohesion or the yielding of the matrix. For an interfacial failure, a friction stress acting on the unbound zone has to be taken into account. When matrix yielding occurs, the maximum shear stress near the fibre ends cannot exceed the yield strength of the matrix. Therefore, it is clear that equation (8), essentially based on the assumption of an elastic stress transfer, is no longer valid. Recent analytical investigations [19], which have not been confirmed experimentally yet, show that, in the case of interfacial failure as well as of matrix yielding, values of τ different from these predicted by equation (8) should be obtained, for a given value of the critical length l_c.

3.4. Determination of fibre strain in single fibre composites by Raman spectrometry

This method, proposed recently by Galiotis and co-workers [20], consists to apply a small deformation to a single fibre composite (see previous section) and to determine the fibre strain, due to stress transfer, by recording the Raman spectra of the fibre at different points x along this fibre. In fact, the location in wavenumber of some specific peaks of the fibre Raman spectrum is sensitive to the strain at which the fibre is subjected. Knowing the variation of the peak shift versus strain, previously determined on isolated fibres, it is thus easy to deduce the strain state of the fibre into the model composite. Furthermore, by a force balance equation on an element of the fibre, the shear stress τ_x can be calculated at each point x by the following expression :

$$\tau_x = \frac{E_f d}{4} (\frac{\delta e}{\delta x})$$

(11)

where e is the level of strain for a given x.

Most of the experimental results obtained confirm the stress or strain profile along the fibre length in agreement with Cox' approach (Figure 7). However, interfacial failure at the fibre ends modifies the strain profile and thus can be directly evidenced by Raman spectrometry. Moreover, the state of strain (strain mapping) of fibres into real composite due to residual thermal stresses can also be determined.

4. Conclusion

As shown by recent studies [21,22], it clearly appears that for a given fibre-matrix system (carbon fibre-epoxy resin systems particularly), large differences are observed between the values of the interfacial shear strength as measured by the different tests described in previous sections. Such a result indicates that the state of deformation and/or the interface failure mode, in particular for the tests involving model composites, are not identical for the different geometries. Moreover, as observed by Drzal and Herrera-Franco [21], none of the experimental techniques offers "a complete and unambiguous method for measuring the interfacial shear strength between fibre and matrix". However, the tests concerning model composites, i.e. single fibre composites, and in particular fragmentation test, should provide a clearer idea of both the adhesion strength and the failure mode at the fibre-matrix interface.

5. References

1. ASTM D. 2344-76.
2. C.A. Berg, J. Tirosh, M. Israeli, Composite Materials: Testing and Design (2nd conf.) ASTM STP 497, (1972), pp 206-218.
3. N. Iosipescu, "New accurate procedure for single shear testing of metals", J. of Materials 2(3), 537-566 (1967).
4. ASTM D 3518
5. P.J. Hine, B. Brew, R.A. Duckett and I.M. Ward, Composites Sci. Techn. 33, 35 (1988).
6. L.B. Greszczuk, Interfaces in Composites, ASTM STP 452, (1969), pp. 42-58.
7. G. Désarmot, M. Sanchez, Proc. 4èmes Journées Nationales sur les Composites, Paris 11-13 Sept. 1984, Ed. by G. Verchery (Ed. Pluralis), 449 ; G. Désarmot and J.P. Favre, Composites Sci. Techn. 42, 151 (1991).
8. M. Nardin and I.M. Ward, Mater. Sci. Techn. 3, 814 (1987) ; N.H. Ladizesky and I.M. Ward, J. Mater. Sci.24, 3763 (1989).
9. P.S. Chua and M.R. Piggott, Composites Sci. and Techn. 22, 33, 107, 185 and 245 (1985).
10. A.N. Gent and G.L. Liu, J. Mater. Sci. 26, 2467 (1991).
11. B. Miller, P. Muri, L. Rebenfeld, "A microbond method for determination of the shear strength of a fiber/resin interface", Composites Sci. and Techn. 28, 17-32 (1987).
12. J.F. Mandell, D.H. Grande, T.H. Tsiang and F.J. Mac Garry, Composite Materials: Testing and Design (7th conf.) ASTM STP 893, (1986), pp. 87-108 ; D.H. Grande, J.F. Mandell and K.C.C. Hong, J. Materials Sci. 23, 311 (1988).
13. W.A. Fraser, F.H. Ancker and A.T. Di Benedetto, 30th Ann. Techn. Conf., Reinf. Plastic/Composites Inst., Soc. Plastic Industry, 22-A, (1975) ; A.T. Di Benedetto and L. Nicolais, Plast 10 (5), 83-88 (1979).
14. A. Kelly and W.R. Tyson, J. Mech. Phys. Solids 13, 329 (1965).
15. D. Rouby and J.P. Favre, Proc. 16th European working Group on Acoustic Emission, London, 16-17 (Sept. 1987).
16. A.N. Netravili, L.T.T. Topoleski, W.H. Sachse and S.L. Phoenix, Composites Sci. Techn. 35, 13 (1989).

17. W. Weibull, The Royal Swedish Inst. for Engineering Research, Proc. n° 151, 4 (1939).

18. H.L. Cox, British J. Appl. Phys. **3**, 72 (1952).

19. I. Verpoest, M. Desaeger and R. Keunings, in "Controlled Interphases in Composite Materials", Ed. by H. Ishida, Elsevier, N.Y., 653 (1990).

20. C. Galiotis, R.J. Young, P.H.J. Yeung and D.N. Batchelder, J. Materials Sci. **19**, 3640 (1984) ; I.M. Robinson, R.J. Young, C. Galiotis and D.N. Batchelder, J. Materials Sci. **22**, 3642 (1987) ; C. Galiotis, N. Melanitis, D.N. Batchelder, I.M. Robinson and J.A. Peacock, Composites **19**, 321 (1988) ; C. Galiotis, Composites Sci. Techn. **42**, 125 (1991).

21 L.T. Drzal and P.J. Herrera-Franco, "Composite Fiber-Matrix Bond Tests", in "Engineered Materials Handbook, Vol. 3, Adhesives and Sealants", H.F. Brinson Technical Chairman, ASM International, 391 (1990).

22 Final Report on the "Round Robin on Interfacial Test Methods", Org. by M.J. Pitkethly and J.P. Favre, Royal Aircraft Establishment, Farnborough (UK), August 1991.

Some other references

a   G.S. Holister, C. Thomas, "Fibre reinforced materials", Elsevier, 1966.

b   D. Hull, "An introduction to composite materials", Cambridge University Press, 1981.

c   "Molecular characterization of composite materials", Ed. by H. Ishida and G. Kumar, Plenum Press, 1985.

d   "Interfaces in polymer, ceramic, and metal matrix composites", Ed. by H. Ishida, Elsevier, 1988.

e   M. Narkis, E.J.H. Chen, R.B. Pipes, "Review of methods for characterization of interfacial fiber-matrix interactions", Polymer Composites 9 (4), 245-251 (1988).

f   Proceedings of the first International Conference on "Interfacial Phenomena in Composite Materials - IPCM'89", Sheffield (UK), September 5-7, 1989.

g   "Controlled Interphases in Composite Materials", (Proceedings of the third International Conference on Composite Interfaces, Cleveland (Ohio, USA), May 21-24, 1990), Ed. by H. Ishida, Elsevier (N.Y., Amsterdam, London), 1990.

h   Composite Science and Technology, 42 (1-3), 1991.

i   "Interfacial Phenomena in Composite Materials '91", ed. by I. Verpoest & F. Jones, Butterworth Heinemann, Oxford, 1991.

Interactions and Properties of Composites : b) Adhesion-Composites Properties Relationships.

Michel NARDIN and Jacques SCHULTZ

Centre de Recherches sur la Physico-Chimie des Surfaces Solides, C.N.R.S.,
24, avenue du Président Kennedy,
F-68200 Mulhouse, France.

1. Introduction

It is now widely accepted that the adhesion between the reinforcing fibre and the polymer matrix is one of the most important parameters governing the performance of a composite material and that the interface could be considered as being the third constituent of the composite material.

However, high adhesive strength at the interface does not necessarily lead to optimum properties of the composite. It is therefore of great importance to clearly understand the role of the physico-chemical interactions at the fibre-matrix interface and to establish quantitative relationships between the nature and level of these interactions and the final mechanical behaviour of the composite materials.

More modestly, to a first approach, the aim of the present part is to analyse the influence of the interfacial adhesion W on the magnitude of the fibre-to-matrix stress transfer capacity. For all the study, these interfacial stress transfer capacity is defined in term of interfacial shear strength τ, measured by means of a fragmentation test on single fibre composites. According to the previous part of this study, τ is given by the following expression :

$$(1) \qquad \tau = d.\sigma_f(l_c) / 2.l_c$$

where d is the fibre diameter, l_c is the critical length of the fibre fragment at the end of the fragmentation process and $\sigma_f(l_c)$ is the fibre tensile strength at a gaugelength equal to l_c.

The methods for estimating the work of adhesion W at the fibre-matrix interface from the surface properties of both materials in contact are first presented. In each case,

95

G. Akovali (ed.), The Interfacial Interactions in Polymeric Composites, 95–105.
© 1993 Kluwer Academic Publishers.

quantitative correlations between W and τ are proposed. The limits of the theoretical and experimental approaches leading to the establishment of such relationships as well as their domain of validity are discussed. In particular, the effect of the existence of interfacial layers (interphases) exhibiting physical and mechanical properties different from the bulk, is examined. Two types of interphases are analysed as examples :

- a crystalline (maybe a transcrystalline) layer in the case of semi-crystalline matrices and, in particular for carbon fibre-poly(ether-ether-ketone) (PEEK) systems.
- a layer of polymer chains of reduced mobility near the fibre surface for carbon fibre-elastomer composites,

2. Estimation of the fibre-matrix adhesion

The thermodynamic model of adhesion, generally attributed to Sharpe and Schonhorn [1], is certainly the most widely used approach in adhesion science nowadays. This theory considers that the adhesive will adhere to the substrate because of interatomic and intermolecular forces established at the interface, provided that an intimate contact between both materials is achieved. The most common interfacial forces result from van der Waals (London, Debye and Keesom) and Lewis' acid-base interactions. The magnitude of these forces can generally be related to fundamental thermodynamic surface characteristics, such as surface free energies γ, of both materials in contact.

In agreement with Fowkes [2] and to a first approximation, the total energy of adhesion between the fibre and the matrix can be considered as the sum of three terms :

$$(2) \qquad W = W^D + W^P + W^{ab}$$

where W^D, W^P and W^{ab} are reversible adhesion energies corresponding respectively to dispersive (London), polar (dipole-dipole) and acid-base interactions between both materials. According to Fowkes [3], W^D is equal to twice the geometrical mean of dispersive component γ^D of the surface energy of fibre and matrix (subscripts f and m respectively) :

$$(3) \qquad W^D = 2(\gamma_f^D \cdot \gamma_m^D)^{1/2}$$

The validity of this equation is now well established. By analogy with the work of Fowkes, Owens and Wendt [4] and then Kaelble and Uy [5] have suggested that the non-dispersive part of interactions between materials can be expressed as the geometric mean of the non-dispersive components (usually called polar component) γ^P of their surface energy, although there is no theoretical reason to represent all the non-dispersive interactions by this type of expression :

(4)
$$W^P = 2(\gamma_f^P \cdot \gamma_m^P)^{1/2}$$

Hence, the work of adhesion W_1 becomes :

(5)
$$W_1 = W^D + W^P = 2(\gamma_f^D \cdot \gamma_m^D)^{1/2} + 2(\gamma_f^P \cdot \gamma_m^P)^{1/2}$$

More recently, it has been shown, in particular by Fowkes and co-workers [2,6,7], that electron acceptor and donor interactions, according to the generalized Lewis' acid-base concept, could be a major type of interfacial forces between two materials. This approach is able to take into account hydrogen bonds which are often involved in adhesive joints. Inverse gas chromatography at infinite dilution for example is a well adapted technique [8-10] for determining the acid-base characteristics of fibres and matrices. Retention data of probes of known properties, in particular their electron acceptor (AN) and donor (DN) numbers according to Gutmann's semi-empirical scale [11], allow the determination of acid-base parameters, K_A and K_D, of fibre and matrix surfaces. It becomes then possible to define a "specific interactions parameter" A at the fibre-matrix interface, as the cross-product of the coefficients K_A and K_D of both materials [10,11] :

(6)
$$A = K_A^f \cdot K_D^m + K_D^f \cdot K_A^m$$

superscripts f and m refering respectively to fibre and matrix.

In fact, since K_A and K_D coefficients are expressed in terms of enthalpy units, the parameter A can be considered equal to the variation of the enthalpy for acid-base interactions ($-\Delta H^{ab}$) at the fibre-matrix interface. Moreover, Fowkes and Mostafa [2] have suggested that the contribution of the polar interactions to the thermodynamic work of adhesion could be generally neglected compared to both dispersive and acid-base contributions. They have also considered that the acid-base component W^{ab} of the adhesion energy can be related to the variation of enthalpy ($-\Delta H^{ab}$) corresponding to the establishment of acid-base interactions at the interface, as follows :

(7)
$$W^{ab} = f.(-\Delta H^{ab}).n^{ab}$$

where f is a factor which converts enthalpy into free energy and is taken equal to unity, and n^{ab} is the number of acid-base bonds per unit interfacial area, close to 6 μmoles.m^{-2}. Therefore, from equations (3) and (7), the total work of adhesion W_2 becomes :

$$(8) \qquad W_2 = W^D + W^{ab} = 2.(\gamma_f^D.\gamma_m^D)^{1/2} + f.(-\Delta H^{ab}).n^{ab}$$

3. Relationships between τ and W

3.1. Relationships between τ and $W_1 = W^D + W^P$

For systems involving, on the one hand, poorly polar matrices, such as polyethylene, polyvinylchloride and polyurethane and, on the other hand, untreated and surface treated glass or carbon fibres, it has been shown in our laboratory [12,13] that a linear relationship can be established to a first approximation between τ and the reversible work of adhesion W_1 defined by equation (5). However for more polar matrices, like epoxy resin for example, such a relationship is no longer valid, since strong specific interactions are now established at the fibre-matrix interface.

3.2. Relationships between τ and A

More recent work [10] has shown that, for systems of untreated and surface treated carbon fibres in two types of epoxy resin, linear relationships (Figure 1) exist between the interfacial shear strength τ and the specific interaction parameter A, defined

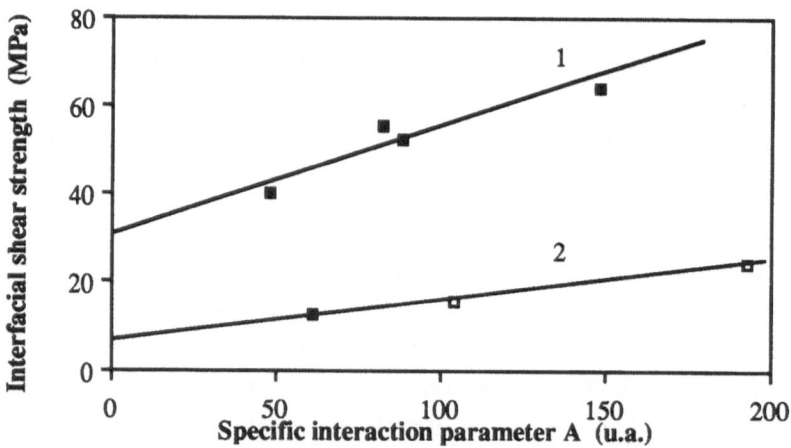

Figure 1 : τ versus A for carbon fibre-epoxy resin systems : 1) epoxy A ; 2) epoxy B

by equation (6) and which is roughly related to the level of the acid-base interactions at the fibre-matrix interface.

Such a strong correlation illustrates the importance of acid-base interactions for a better understanding of interfacial properties of composites. It is clear that the intercept at the origin should be different from zero, since dispersive interactions (W^D) still exists when A = 0. It also seems that the magnitude of this intercept depends on the mechanical properties of the matrix [14].

3.3. Relationships between τ and $W_2 = W^D + W^{ab}$

To reconcile both precedent approaches, a tentative correlation between τ and the work of adhesion W_2 defined as the sum of dispersive and acid-base interactions (equation (8)) has been recently proposed [15]. For all the different fibre-matrix systems, whatever the nature and properties of both the fibre and the matrix, Figure 2 shows that a linear relationship, passing through the origin, can be established between τ and W_2. Therefore, it is possible to write :

$$(9) \qquad \tau = k.W_2$$

where k is a coefficient of proportionality. Equation (9) indicates that physico-chemical interactions between the fibre and the matrix determine, to a large extent, the mechanical behaviour of the interface.

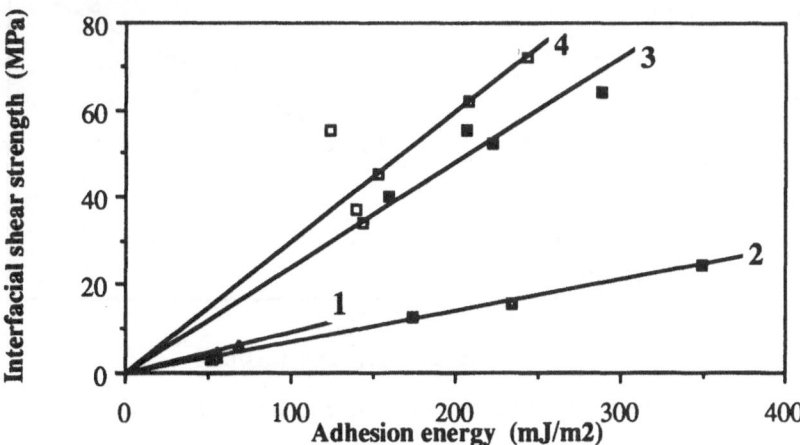

Figure 2 : τ versus W_2. 1) glass fibre-polyethylene ; 2) carbon fibre-epoxy B ; 3) carbon fibre-epoxy A ; 4) carbon fibre-PEEK.

100

However, the slope of the straight lines obtained are different for each system and, as already stated, could depend on the intrinsic properties of the fibre and the matrix, in particular their mechanical properties. In agreement with recent developments of the fully elastic model of interfacial stress transfer proposed by Cox [16], it has been shown that k is directly related to the elastic moduli E_f and E_m of the fibre and the matrix respectively, in order that τ can be expressed as :

(10)
$$\tau = \left(\frac{E_m}{E_f}\right)^{1/2} \frac{W_2}{\lambda}$$

where λ is a distance, independent on the system studied. The remarkable result is that this distance λ is equal to about 0.5 nm and, thus, corresponds to an equilibrium intermolecular centre-to-centre distance [17] involved in physical interactions, such as van der Waals and acid-base interactions. Moreover, rewriting equation (10) as follows :

(10')
$$\left\{ \left(\frac{E_f}{E_m}\right)^{1/2} \tau \right\} = \frac{W_2}{\lambda}$$

the left hand side of expression (10'), which corresponds to an interfacial shear strength "normalized" by the elastic properties of both the fibre and the matrix, can be considered as an adhesive pressure at the interface [17]. Figure 3 shows the master straight line relating this adhesive pressure to W_2.

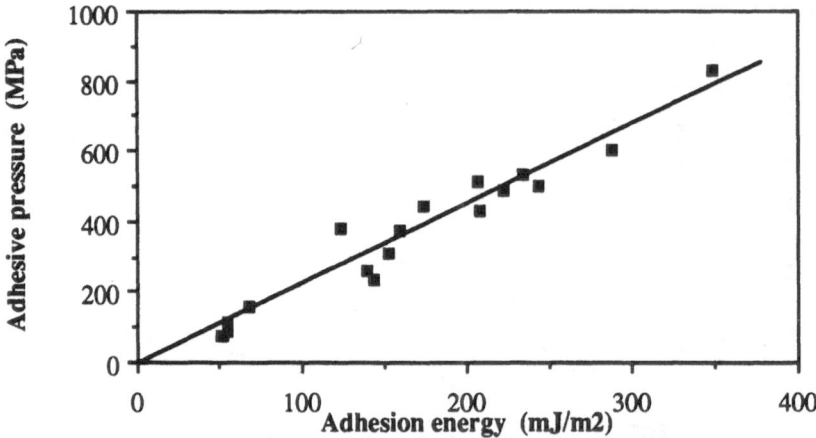

Figure 3 : Adhesive pressure versus adhesion energy W_2.

Finally, this new model clearly shows that both approaches studied : (i) on the one hand, the estimation of the free energy of adhesion, -thermodynamic quantity-, between the fibre and the matrix from the surface properties of both materials and, (ii) on the other hand, the measurement of the interfacial shear strength, -mechanical quantity- , by a fragmentation test, are consistent. Nevertheless, this model can be applied insofar as the following conditions are fulfilled :

- only physical interactions are involved at the interface, i.e. the existence of chemical bonds and/or mechanical interlocking for example are excluded,
- the mechanical behaviour of both the fibre and the matrix as well as the stress transfer mechanism are purely elastic,
- the bulk properties of matrices are not altered near the fibre surface or, in other words, no interfacial layers (interphases) exhibiting particular properties are formed.

The influence of the presence of two different kinds of interphase on the level of stress transfer at the fibre-matrix interface is briefly examined in the next section.

4. Effect of an interfacial layer

4.1. The case of a semi-crystalline matrix (PEEK) [18-21]

First, it is clear that using standard processing conditions for single fibre composites, i.e. a continuous cooling from the melt at a cooling rate of the order of 10 to 20 K/min, the previous model relating τ to W_2 is valid for carbon fibre-PEEK interfaces (Figure 2). However, different thermal treatments are also applied to model composites prior to fragmentation and, particularly, annealing treatments at different temperatures T_c comprized between the melting and the glass transition temperatures of PEEK. Such treatments lead to an isothermal crystallization of the matrix and it is observed that generally the values of τ are decreased, at least when a high level of adhesion (by using oxidized carbon fibres for example) is established at the fibre-matrix interface. More precisely, the results indicate that the crystallisation rate of the matrix at T_c is a major factor affecting τ (Figure 4). The greater the matrix crystallization rate, the lower the interfacial shear strength. This can be explained by the growth near the fibre surface of a more and more disorganized crystalline (and sometimes transcrystalline) structure when the crystallisation rate is increased.

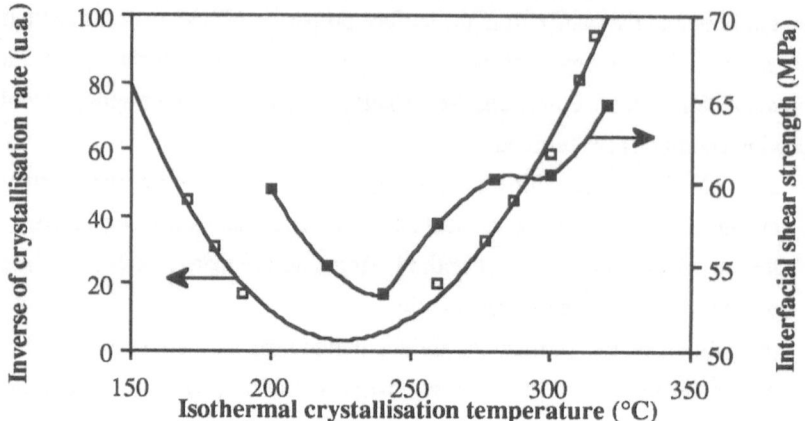

Figure 4 : Influence on τ of the PEEK crystallisation rate at different isothermal crystallization temperatures for systems involving oxidized carbon fibre (results concerning the crystallization rate come from ref [22])

Considering to a first approximation that a transcrystalline interphase is always formed near the fibre surface, it is clear that, during the fragmentation test, this interphase is subjected to a transverse tensile stress as refered to the direction of the crystallites (Figure 5). Moreover, assuming that our model is still valid insofar as the tranverse elastic modulus E_{it} of the transcrystalline interphase is taken into account in place of this of the bulk matrix in equation (10), an estimation of E_{it} can be made. It is found that the values of E_{it} are ranging from about 50 to 70% of the values of E_m, in good agreement with a simple mechanical analysis.

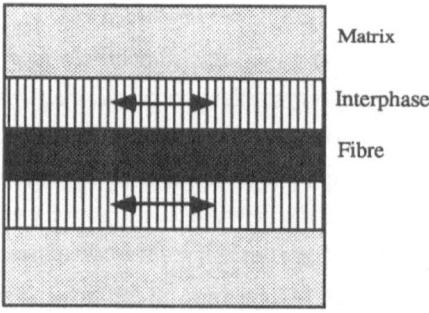

Figure 5 : Schematical representation of a transcrystalline interphase

However, it has been also shown that the amorphous phase of PEEK near the fibre surface plays a fundamental role in the stress transfer capacity of the interfacial region. In particular, the "constraint state" of this phase, which is mainly affected by the cooling procedure during the processing of composites, seems to be the major factor in this domain.

4.2. The case of an elastomeric matrix [23-25]

Whatever the nature of the elastomer used as matrix, i.e. polyurethane, poly(ethylene-vinylacetate) or styrene-butadiene rubber (SBR), it appears that the interfacial shear strength τ of carbon fibre-elastomer composites is much higher than theoretically expected from equation (10). Figure 6 illustrates the variation of τ versus W_2 in comparison with the prediction from our model in the case of carbon fibre-SBR systems. Any other theoretical approach is able to explain these high values of τ.

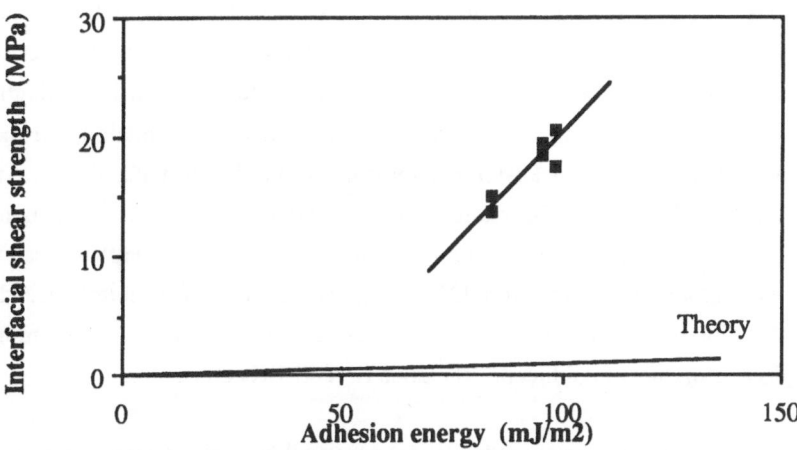

Figure 6 : τ versus adhesion energy for carbon fibre-SBR systems.

It was first evidenced that the hysteritic losses of bulk elastomer during the fragmentation test are not responsible for such a discrepancy between theory and experimental results. On the contrary, the high values of the interfacial shear strength are very well explained by the existence near the fibre surface of an interphase in which the polymer chain mobility is strongly reduced compared to the bulk matrix. Effectively, as stated in a recent work [25,26] and as observed by creep experiments at moderate applied load, this interfacial layer is able to age physically at temperatures far above the glass transition temperatures of the elastomer matrix and, consequently, exhibits a pseudo-glassy behaviour on the whole range of temperatures studied (about 100°C from glass transition temperature). It is clear that the mechanical properties of this layer are completely different from those of the bulk matrix and play a major role on the stress transfer phenomenon. Considering to a first approximation that our model remains valid in such conditions, the elastic modulus E_i of the interfacial layer can be estimated by taking it into account in place of E_m in equation (10). In each case, E_i is found equal to the elastic modulus of the elastomer in its glassy state whatever the temperature, in good agreement with the expected glassy behaviour of this layer. The variation of the interfacial shear strength versus temperature confirms this assumption.

5. Conclusion

A tentative model has been proposed to relate the interfacial shear strength at the fibre-matrix interface, measured by a fragmentation test on single fibre composites, to the level of adhesion between both materials. This last quantity has been estimated from the surface properties of both the fibre and the matrix and was defined as the sum of dispersive and acid-base interactions. This new model clearly indicates that the micromechanical properties of a composites are mainly determined by the level of physical interactions established at the fibre-matrix interface and, in particular , by electron acceptor-donor interactions. Moreover, to a first approximation, our model is able to explain the stress transfer phenomenon through interfacial layers, such as crystalline interphases in semi-crystalline matrices and interphases of reduced mobility in elastomeric matrices. An estimation of the elastic moduli of these interphases can also be proposed. Furthermore, recent work [21] has shown that the level of interfacial adhesion plays a major role on the final performances (tensile, transverse and compressive strengths and strains) of unidirectional carbon fibre-PEEK composites.

Finally, it can be concluded that the reversible work of adhesion at the fibre-matrix interface is an important parameter in determining the micromechanical behaviour of model composites as well as the ultimate properties of unidirectional laminates. Such a general approach could allow us to further improve the performances of advanced composites by controlling both the processing conditions and the level of interfacial interactions.

6. References

1.   L.H. Sharpe and H. Schonhorn, Chem. Eng. News **15**, 67 (1963).
2.   F.M. Fowkes and M.A. Mostafa, Ind. Eng. Chem. Prod. Res. Dev **17**, 3 (1978).
3.   F.M. Fowkes, Ind. Eng. Chem. **56**, 40 (1964).
4.   D.K Owens and R.C. Wendt, J. Appl. Polym. Sci. **13**, 1740 (1969).
5.   D.H. Kaelble and K.C. Uy, J. Adhesion **2**, 50 (1970).

6. F.M. Fowkes and S. Maruchi, Org. Coatings Plasics Chem. **37**, 605 (1977).

7. F.M. Fowkes, J. Adhesion Sci. Techn. **1**, 7 (1987).

8. C. Saint-Flour and E. Papirer, Ind. Eng. Chem. Prod. Res. Dev 21, 337 and 666 (1982).

9. E. Papirer, H. Balard and A. Vidal, Eur. Polym. J. **24**, 783 (1988).

10. J. Schultz, L. Lavielle and C. Martin, J. Adhesion **23**, 45 (1987).

11. V. Gutmann, "The Donor-Acceptor Approach to Molecular Interactions", Plenum Press, N.Y. &London, (1978).

12. H. Simon, PhD Thesis, Université de Haute-Alsace, Mulhouse, France, (1984).

13. J. Schultz, L. Lavielle and H. Simon, Proc. Intern. Symp. "Science and New Application of Carbon Fibers", University of Toyohashi (Japan), November 19-21, p. 125 (1984).

14. J. Schultz and L. Lavielle, in "Inverse Gas Chromatography", ACS Symp. Series, Ed. by D.R. Lloyd, T.C. Ward & P.H. Schreiber, Chap. 14, p. 185, (1989).

15. M. Nardin and J. Schultz, Comptes Rendus Acad. Sci. Paris **311**, Série II, 613 (1990).

16. H.L. Cox, British J. Appl. Phys. **3**, 72 (1952).

17. J.N. Israelachvili, "Intermolecular and Surface Forces", Academic Press, London, (1985).

18. M. Nardin, E.M. Asloun and J. Schultz, Polym. Adv. Techn. **2**, 109 (1991).

19. M. Nardin, E.M. Asloun and J. Schultz, Polym. Adv. Techn. **2**, 115 (1991).

20. M. Nardin, E.M. Asloun, F. Muller and J. Schultz, Polym. Adv. Techn. **2**, 161 (1991).

21. M. Nardin, E.M. Asloun, J. Schultz, J. Brandt and H. Richter, Polym. Adv. Techn. **2**, 171 (1991).

22. D.J. Blundell and B.N Osborn, Polymer **24**, 953 (1983).

23. E.M. Asloun, M. Nardin and J. Schultz, J. Mater. Sci. **24**, 1835 (1989).

24. M. Nardin, E.M. Asloun, M. Brogly and J. Schultz, in "Developments in the Science and Technology of Composite Materials", Ed. by A.R. Bunsell, P. Lamicq and A. Massiah, Elsevier Appl. Sci., N.Y. & London, p. 243 (1989).

25. M. Nardin, A. El Maliki and J. Schultz, J. Adhesion, submitted for publication (1992).

26. B. Haidar, Proc.Rubber Div. Meeting, ACS Symp., Las Vegas (NV,USA), May 29-June 1, (1990).

27. B. Haidar, in "Interfacial Phenomena in Composite Materials, IPCM'91", Ed. by I. Verpoest & F. Jones, Butterworth-Heinemann Ltd, Oxford, 145 (1990).

THE ROLE OF INTERFACE AT THE WALL IN FLOW OF CONCENTRATED COMPOSITES

Ülkü Yılmazer
Chemical Engineering Department
Middle East Technical University
06531, Ankara, Turkey

and

Dilhan M.Kalyon
Highly Filled Materials Institute
Chemistry and Chemical Engineering
Stevens Institute of Technology
Hoboken, New Jersey, 07030, USA

ABSTRACT. The rheological behavior of two highly concentrated composites (suspensions) were investigated in capillary and parallel disk torsional flows. Suspension I was bimodal in particle size distribution and contained 76.5% solids by volume. Suspension II was unimodal with 60% solids by volume. It was found that flow of the suspensions were strongly affected by slip at the walls of the instrument. The slip mechanism is thought to be caused by the formation of an "apparent slip layer" near the walls which mainly consists of the suspending medium. The slip velocity determined was found to be approximately linear with the shear stress at the wall in both types of flows. Suspension I exhibited Newtonian behavior at low shear stresses, but became pseudoplastic at high shear stresses. Suspension II was shear thinning at low shear stresses, but shear thickening at high shear stresses. The contribution of slip to the volumetric flow rate in capillary flows decreased with the shear stress at the wall in suspension I, whereas it increased with the shear stress at the wall in suspension II. This indicates that the contribution of slip to the total volumetric flow rate is more significant when the suspension exhibits higher viscosity. The flow was pluglike owing to slip at low shear stresses in suspension I and at high shear stresses in suspension II.

1. INTRODUCTION

The flow of highly concentrated composites (suspensions) is of interest in the processing of solid fuels, ceramics, detergents and plastics composites. The rheological properties of these suspensions depend on the rheological properties of the matrix, the particle size and its distribution, the particle content, the orientation of the particles in the flow field, the interactions between the particles and the interactions at the filler matrix interface (1-5). The flow of such suspensions is also strongly affected by slip at the walls of the instrument used in measuring the viscosity, and corrections need to be done in order to obtain the true viscosity of the suspensions (6-8). The flow of the suspensions may also exhibit instabilities in capillary flows, at low shear stress at the wall values, owing to the filtering of the matrix from the suspension at the entrance to the capillaries (9). This mechanism is called as "the mat formation" mechanism (9).

In this report methods of obtaining the slip velocity in capillary and torsional disk flows are reviewed, and the results on two different suspensions are compared.

G. Akovali (ed.), The Interfacial Interactions in Polymeric Composites, 107–123.
© 1993 *Kluwer Academic Publishers.*

2. MATERIALS

2.1. Material Preparation

Suspension I was prepared to simulate a solid rocket fuel and contained 76.5 percent by volume of solids in a hydroxyl terminated polybutadiene (HTPB) matrix in which di-octyl-adipate, DOA, was used as an antioxidant. The matrix was a Newtonian fluid with a shear viscosity of 5 Pa.s at 25°C and 1.3 Pa.s at 57.2°C. In this suspension aluminum and ammonium sulfate were used as fillers. Aluminum had an average particle size of 20 microns, and ninety percent of the particles were smaller than 29 microns. The ammonium sulfate was a bimodal blend of coarse and fine grades with average sizes of 200 and 20 microns respectively. The coarse fraction made up 75% of the total ammonium sulfate. The formulation of this suspension is given in Table 1.

TABLE 1. The Formulation of Suspension I.

Ingredients	Weight %	Density (kg/m³)	Volume %
HTPB	10.8	900	19.60
DOA	2.13	911.3	3.94
Aluminum	20.3	2710	12.61
Ammonium sulfate	67.09	1769	63.85

Suspension II was prepared by using a matrix of poly(butadiene acrylonitrile acrylic acid) terpolymer [PBAN] with 60% by volume of ammonium sulfate ground to an average particle size of 23 μm, exhibiting a standard deviation of 13 μm. The particles had low aspect ratios as observed with Scanning Electron Microscopy (SEM). The matrix in suspension II was also a Newtonian fluid with a viscosity of 37 Pa.s at 25°C.

Suspensions were prepared by compounding in a twin screw extruder and tested freshly. The details of the preparation methods are reported in References 6 and 10.

2.2. Rheological Characterization

Suspension I was prepared and tested at 57.2°C. Suspension II was prepared and characterized at 23°C. Parallel disk torsional tests were done with Rheometrics Mechanical Spectrometer, Model RMS 800 and System IV, by using parallel disk fixtures of 25 mm radius. The experiments were carried out by employing various gap heights in the 1 to 4 mm range. Fresh samples were used in each run. Multiple gap heights are needed to determine the slip velocity (6,11) in these flows.

The capillary experiments were done using an Instron Capillary Rheometer, Model TFD. For suspension I five capillaries were used with a constant L/D of 57.6 and diameters of 1.32, 1.59, 1.98, 2.52 and 3.05 mm. For suspension II four sets of capillaries with L/D ratios of 0, 19.2, 38.4 and 57.6 were used. In each set there were three capillaries with diameters of 1.32, 1.59 and 1.98 mm. Capillaries with constant L/D, but varying diameters are needed to determine the slip velocity at the wall (12), whereas capillaries with constant diameter, but varying lengths are needed to determine the exit and entrance effects (13).

3. BACKGROUND

3.1. Parallel Disk Torsional Flows

In torsional flows between two parallel disks, the apparent shear rate, $\dot{\gamma}_a$, (not corrected for slip effects) is a linear function of the radius, r, and is given by

$$\dot{\gamma}_a = \frac{\Omega r}{H}$$

(1)

where H is the gap height, r is the radial distance from the center of the disk and Ω is the angular velocity of the upper disk relative to the lower one.

The apparent shear rate, $\dot{\gamma}_a$, the true shear rate, $\dot{\gamma}$, and the slip velocity, u_s, are related by the following equation

$$\dot{\gamma}_a = \dot{\gamma}(\tau) + \frac{2u_s(\tau)}{H}$$

(2)

where $\dot{\gamma}$ and u_s are shown as functions of the shear stress. Here τ indicates $\tau_{z\theta}$ in which z is the axial and θ is the angular component in the cylindrical coordinate system. The shear stress at the outer edge of the desk, τ_R, is determined from

$$\tau_R = \frac{T}{2\pi R^3}\left[3 + \frac{d\ln T}{d\ln\dot{\gamma}_{aR}}\right]$$

(3)

where R is the radius of the disk.

In the last equation T is torque needed to rotate the upper disk and $\dot{\gamma}_{aR}$ is the apparent shear rate at the edge of the disk obtained by substituting r=R in Equation (1).

Yoshimura and Prud'homme (11) outlined a method in which two sets of experiments at two different gap heights are performed to determine the slip velocity. The following is a generalization of their method by using multiple gap heights:

Equation (2) also applies at r=R, thus

$$\dot{\gamma}_{aR} = \dot{\gamma}_R(\tau_R) + \frac{2u_s(\tau_R)}{H}$$

(4)

where $\dot{\gamma}_{aR}$ is the apparent shear rate and $\dot{\gamma}_R$ is the true shear rate at the edge of the disk.

This equation indicates that if plots of $\dot{\gamma}_{aR}$ versus 1/H are drawn at constant τ_R, then straight lines are obtained with intercepts of $\dot{\gamma}_R(\tau_R)$ and slopes of $2u_s(\tau_R)$. Thus the true shear rate, $\dot{\gamma}_R$, and the slip velocity, u_s, can be determined as functions of the shear stress at the edge of the disk.

Finally the apparent viscosity at the edge, η_a, can be calculated from

$$\eta_a = \frac{\tau_R}{\dot{\gamma}_{aR}}$$

(5)

and the true viscosity at the edge, η, can be found from

$$\eta = \frac{\tau_R}{\dot\gamma_R} \tag{6}$$

3.2. Capillary Flows

In order to perform the Bagley end correction (13) shear viscosity measurements are made at a constant diameter using different lengths. In cases where long capillaries are used the end corrections may be neglected.

To determine the slip velocity at the wall, capillaries with different diameters are used at constant L/D ratio and the slip velocity is determined (12) by

$$\frac{8V}{D} = \frac{4}{3} \int_{\tau_W}^{\tau_W} \tau^2 \dot\gamma \, d\tau + \frac{8u_s}{D} \tag{7}$$

Here D is the capillary diameter, V is the average velocity of the fluid across the cross-section of the capillary, τ_W is the shear stress at the wall, u_s is the slip velocity at the wall and $\dot\gamma$ is the true shear rate. Differentiating this equation with respect to 1/D at constant τ_W the slip velocity is obtained from (12)

$$\frac{(8V/D)}{(1/D)} \bigg|_{\tau_W} = 8u_s \tag{8}$$

Here 8V/D is equal to the apparent shear rate at the wall, $\dot\gamma_a$. The last equation shows that plots of 8V/D versus 1/D at constant τ_W give u_s as a function of τ_W.

The following analysis can be made to determine the true shear rate at the wall: The total volumetric flow rate, Q, is given by

$$Q = (\pi/4)D^2 V \tag{9}$$

and the contribution of slip to the total volumetric flow rate, Q_s, is found from

$$Q_s = (\pi/4)D^2 u_s \tag{10}$$

Using Equations 9 and 10 in Equation 7 and differentiating one obtains (7)

$$\dot\gamma_W = \frac{8(Q-Q_s)}{\pi D^3} \left[3 + \frac{d\ln(Q-Q_s)}{d\ln\tau_W} \right] \tag{11}$$

where $\dot\gamma_W$ is the true shear rate at the wall of the capillary. Thus the true shear viscosity, η, can be found from

$$\eta = \frac{\tau_W}{\dot\gamma_W} \tag{12}$$

4. RESULTS AND DISCUSSION

4.1. Parallel Disk Torsional Flows

In Figure 1 the shear stress at the edge, calculated from Equation 3, is shown versus the apparent shear rate, $\dot{\gamma}_{aR}$, in parallel disk torsional flow of suspension I.

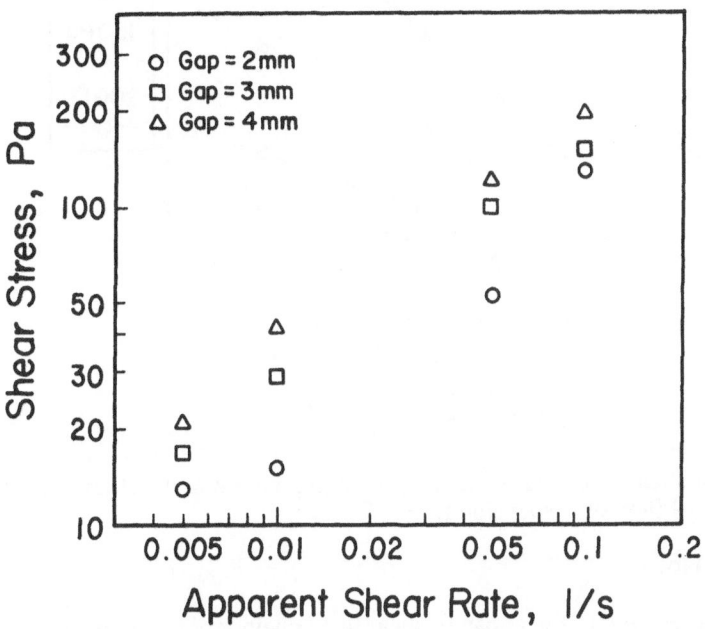

Figure 1. Shear stress versus the apparent shear rate in parallel disk torsional flow of suspension I (redrawn from Ref.10).

The data depend on the gap height indicating the possibility of slip at walls. The behavior of suspension II was qualitatively similar. The data shown in Figure 1 were then analyzed according to Equation 4. In Figure 2 plots of the apparent shear rate versus the reciprocal gap height are shown for suspension I at several selected values of the shear stress. The plots are linear as predicted by Equation 4 with slopes equal to $2u_S$. In suspension II two gap heights were used and the algebraic method outlined in References 8 and 11 were employed.

The slip velocities calculated from the slopes are shown in Figure 3 for suspension I and in Figure 4 for suspension II, along with the slip velocities calculated from capillary flow analysis.

The slip velocities were then fitted with a linear relationship between the slip velocity and the shear stress as

$$u_S = a\tau_R \qquad \text{or} \qquad u_S = a\tau_W \qquad (13)$$

the predictions of which are shown in Figures 3 and 4.The value of a was determined to be 7.4×10^{-4} mm/(Pa-s) in suspension I and 9.2×10^{-5} mm/(Pa-s) in suspension II.

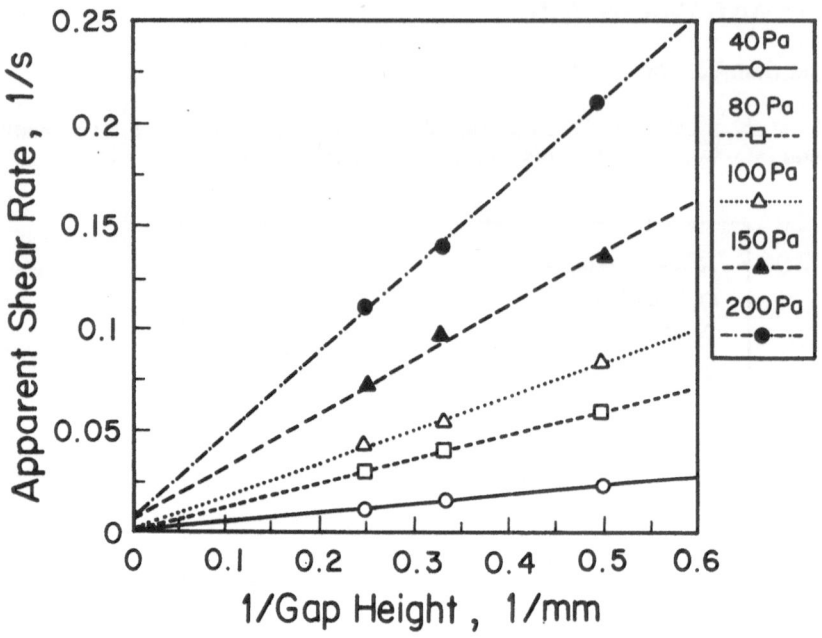

Figure 2. Apparent shear rate versus the reciprocal gap height at constant shear stress in parallel disk torsional flow of suspension I.

4.2. Capillary Flows

The wall shear stress versus the apparent shear rate behavior of suspension I in capillary flow is shown in Figure 5. For each capillary, flow instabilities were observed at low shear stress at the wall values. The ranges of the shear stresses at which the flow instabilities are observed in each capillary are shown in Figure 5.

The flow instabilities exhibited themselves as periodic oscillations of the pressure. The drops in the pressure were accompanied by squirting of the material followed by no-flow. This process is thought to be caused by time-periodic formation and break-up of mats of the solid filler at the entrance to the capillary (9,10,14). When the mat was formed, the liquid matrix was filtered out of the suspension. Thus the average of the oscillating pressure increased unboundedly giving rise to further instability. The mechanism of mat formation is discussed in detail in References 9 and 14. In the subsequent analysis only the stable region of the data in Figure 5 was considered. Suspension II gave similar results except in this case flow instabilities were not observed.

The separation of the data obtained with different capillary diameters suggests again the presence of slip. The data in Figure 5 were analyzed by using Equation 8 and plotting the apparent shear rate versus the reciprocal diameter at constant shear stress. The results are shown in Figure 6 for suspension I.

The plots in Figure 6 are substantially linear indicating slip and supporting the use of Equation 8. The slip velocities calculated by dividing the slopes of these lines by 8 are shown in Figure 3 for suspension I and in Figure 4 for suspension II.

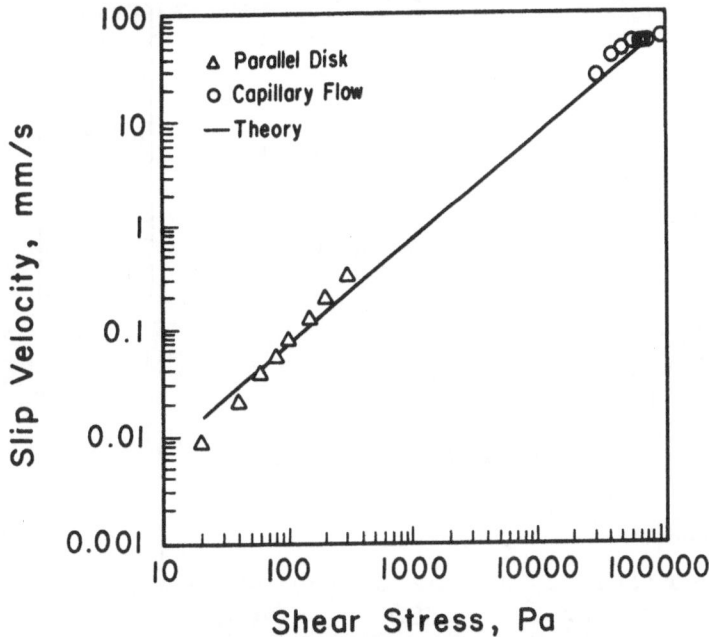

Figure 3. Slip velocity versus the shear stress in parallel disk torsional and capillary flows of suspension I (redrawn from Ref.10).

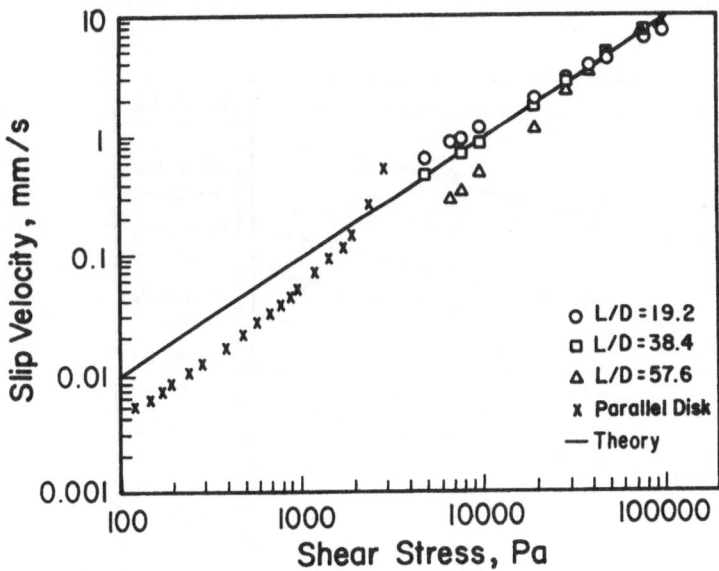

Figure 4. Slip velocity versus the shear stress in parallel disk torsional and capillary flows of suspension II (redrawn from Ref.6).

114

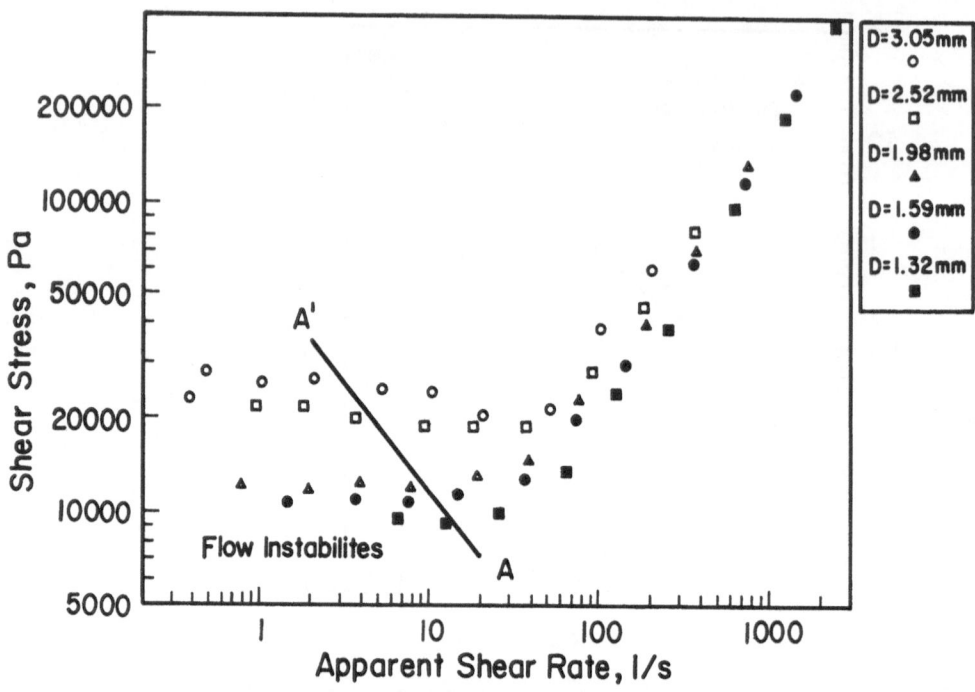

Figure 5. Shear stress versus the apparent shear rate in capillary flow of suspension I (redrawn from Ref.10).

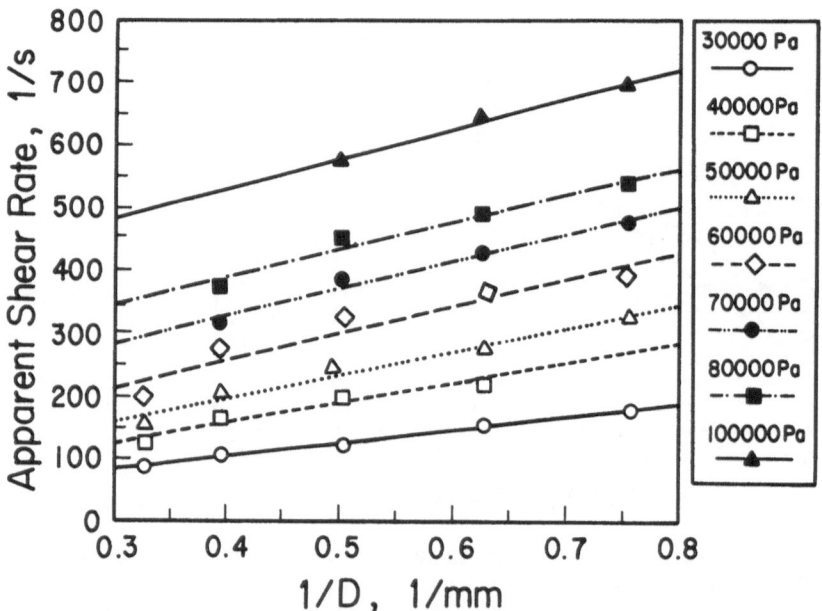

Figure 6. Apparent shear rate versus the reciprocal diameter at constant shear stress in capillary flow of suspension I.

4.3. Contribution of Slip to the Total Volumetric Flow Rate

The contribution of slip to the total volumetric flow defined as Q_S/Q is determined to be

$$\frac{Q_S}{Q} = \frac{u_S}{V} = \frac{8u_S}{D\dot\gamma_a}$$

(14)

The values of Q_S/Q are shown in Figure 7 as a function of τ_W by using the experimentally determined values of u_S and $\dot\gamma_a = 8V/D$ for each capillary. Theoretical curves are obtained from a constitutive equation developed later on.

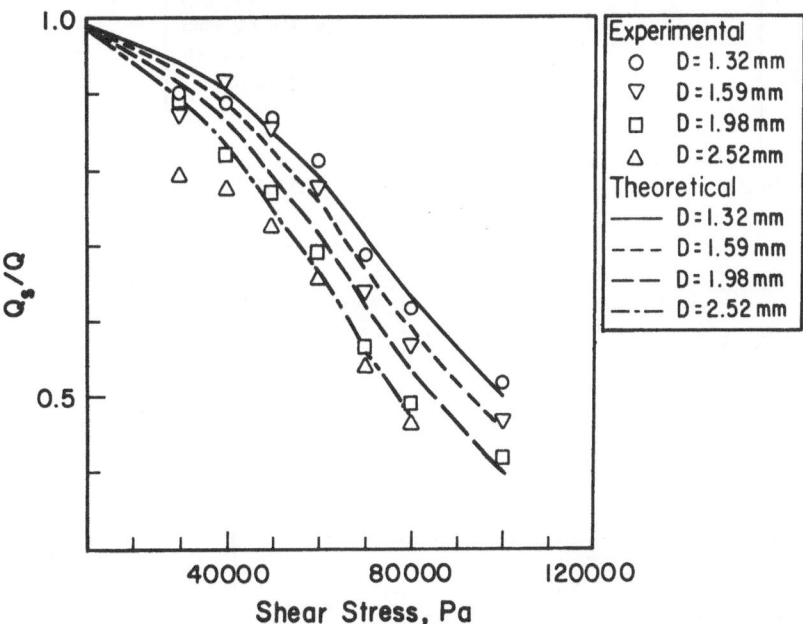

Figure 7. Q_S/Q values versus the shear stress at the wall, τ_W, in suspension I. The L/D is 57.6.

The results show that Q_S/Q decreases slightly with increasing capillary diameter and somewhat more strongly with the shear stress at the wall. In Figure 8, Q_S/Q values are shown versus the shear stress for suspension II. In this case again the Q_S/Q values decrease slightly with increasing capillary diameter except when Q_S/Q values are close to one. However in this suspension Q_S/Q increases with the shear stress at the wall and approaches one at high shear stresses.

The variation of Q_S/Q with the shear stress can be rationalized by observing the variation of viscosity with the shear stress in both suspensions. The true viscosity of the suspensions can be obtained by using Equations 3,4 and 6 in parallel disk torsional flows and Equations 8-12 in capillary flows.

The true viscosity vs. the shear stress behavior of the suspensions are shown in Figures 9 and 10 for suspensions I and II respectively.

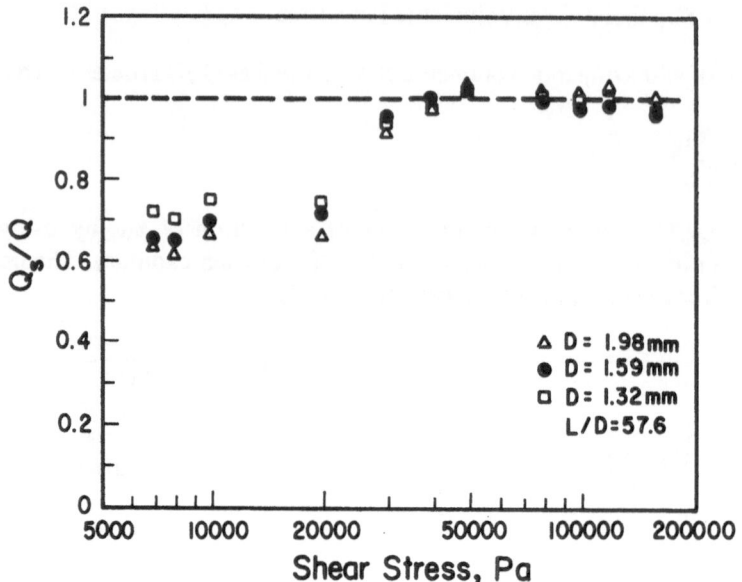

Figure 8. Q_S/Q values versus the shear stress at the wall, τ_W, in suspension II (redrawn from Ref.6).

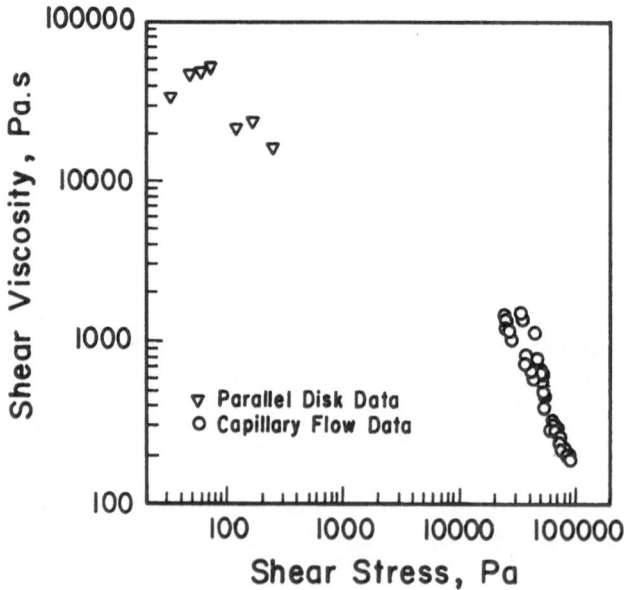

Figure 9. True viscosity versus the shear stress in suspension I.

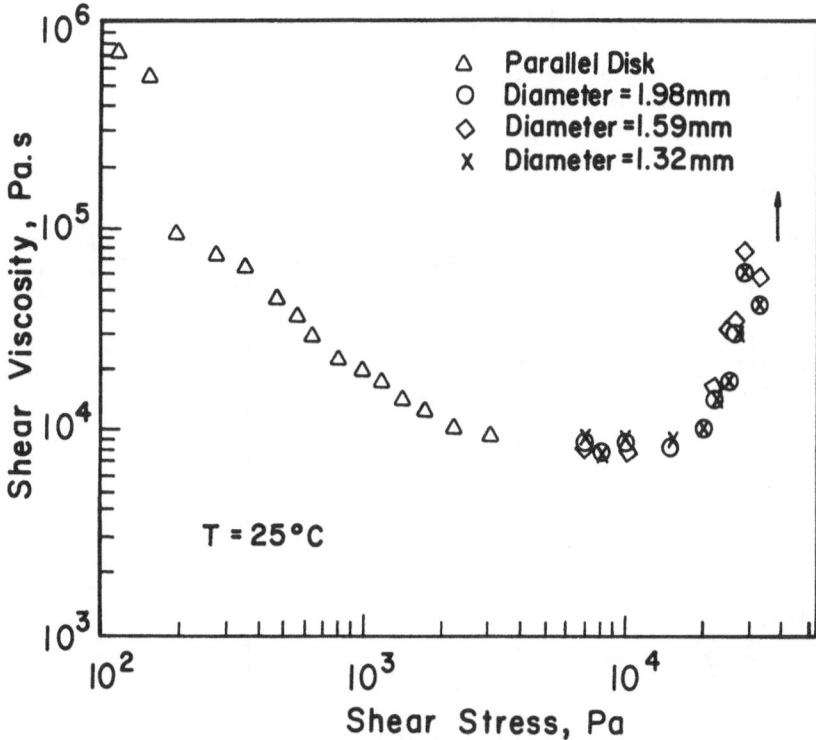

Figure 10. True viscosity versus the shear stress in suspension II (redrawn from Ref. 7).

Suspension I shows Newtonian-like behavior at low shear stresses, but it is shear thinning at high shear stresses. In suspension I, Q_s/Q is a decreasing function of the shear stress at the wall, and it approaches one at low shear stresses, implying plug flow at low shear stresses.

Suspension II exhibits shear thinning followed by shear thickening type of behavior. Comparing Figures 8 and 10, in suspension II, Q_s/Q (measured in capillary flows) and viscosity are both increasing functions of the shear stresses at high shear stresses. Specifically, the viscosity becomes unbounded at approximately a shear stress value of 40 000 Pa at which Q_s/Q becomes one. Above this value of stress the flow is plug-like aided by slip at the walls.

The relation between the shear sensitivity (pseudoplasticity vs. dilatancy) and Q_s/Q will be shown in the following discussion by developing a constitutive equation for suspension I and observing the predictions of this equation for Q_s/Q.

4.4. Bi-region Constitutive Equation for Suspension I.

The combined shear stress versus the true shear rate data in double logarithmic form are shown in Figure 11 and indicate the presence of a viscous Newtonian region at the low shear rates employed in the torsional flow experiments.

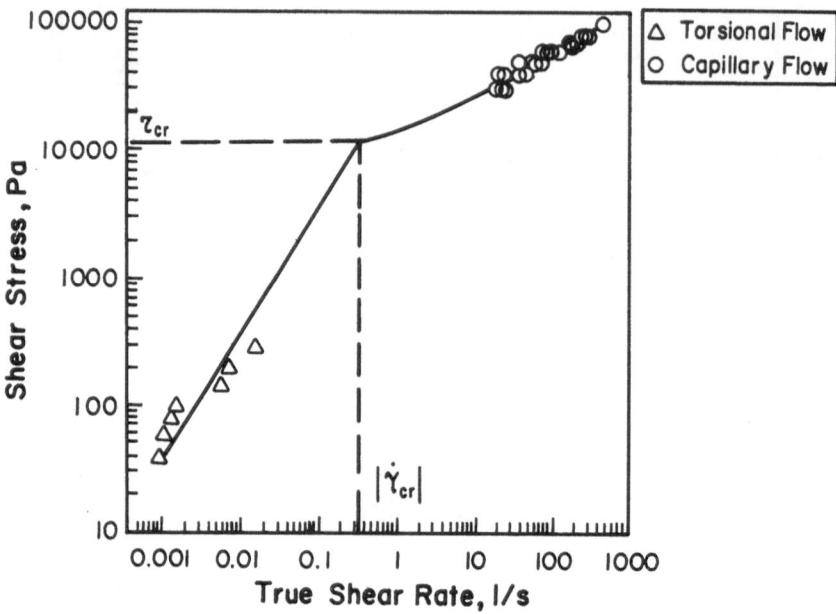

Figure 11. Shear stress versus the true shear rate at the wall from capillary and parallel disk torsional flow experiments on suspension I.

The existence of a viscous Newtonian region below a critical shear stress is also suggested by various experimental (15) and theoretical considerations (16-18). The experimental findings suggest the use of the following bi-region model to represent the shear stress, τ_{12}, versus deformation rate, $\dot{\gamma}$, behavior of the suspension (in one dimensional form):

$$\tau_{12} = -\mu\dot{\gamma} \qquad\qquad \text{for } |\tau_{12}| \leq \tau_{cr} \qquad\qquad (15)$$

and

$$\tau_{12} = \pm\tau'_{0} - m\, |\dot{\gamma}|^{n-1}\dot{\gamma} \qquad\qquad \text{for } |\tau_{12}| \geq \tau_{cr} \qquad\qquad (16)$$

Here m, μ, n, and τ'_{0} are material parameters. τ_{cr} represents the critical shear stress, below which the suspension behaves like a Newtonian fluid, with a relatively high viscosity, μ, as shown in Figure 11. The negative sign in Equation 16 is used for negative shear stresses.

The shear viscosity of the suspension exhibits shear rate sensitivity above this critical stress value, τ_{cr}. This model resembles in form to the purely viscous approximation of the Bingham Plastic fluid, suggested by Lipscomb and Denn (18). As shown in Figure 11, the curves represented by Equations 15 and 16 intersect at the critical shear rate, $|\dot{\gamma}_{cr}|$, which can be determined from:

$$\mu|\dot{\gamma}_{cr}| = \tau'_o + m|\dot{\gamma}_{cr}|^n \tag{17}$$

Then τ_{cr} is given by:

$$\tau_{cr} = \mu|\dot{\gamma}_{cr}| \tag{18}$$

The values of the material parameters, as well as the values of $|\dot{\gamma}_{cr}|$ and τ_{cr}, are shown in Table 2. The best fit to Equations 15 and 16 using these parameters is shown in Figure 11. The value of n is 0.4 and suggests shear thinning behavior at wall shear stress values which are greater than τ_{cr}.

TABLE 2. Best-fit values of the parameters used to describe Q_s/Q versus the wall shear stress behavior in capillary flow.

n = 0.4
s = 1/n = 2.5
μ = 35 000 Pa.s
m = 7 800 Pa.s$^{0.4}$
τ'_o = 6 660 Pa
τ_{cr} = 11 715 Pa
$\dot{\gamma}_{cr}$ = 0.3 s^{-1}

In capillary flows, the velocity profile and the resulting volumetric flow rate versus pressure drop behavior can be determined from Equations 15 and 16. There exist two regions of flow. Neglecting end effects and body forces, and assuming incompressibility, the isothermal solution can be obtained by using the boundary condition, $V_z(R) = U_s$. In the capillary, the velocity distribution for the region $r \geq r_{cr}$ and $r \leq r_{cr}$ are given by:

$$V_z = U_s + \frac{1}{\frac{dP}{dz}\frac{(s+1)}{2m}}\left\{\left(-\frac{r}{2m}\frac{dp}{dz}-\frac{\tau'_o}{m}\right)^{s+1} - \left(-\frac{R}{2m}\frac{dP}{dz}-\frac{\tau'_o}{m}\right)^{s+1}\right\} \quad \text{for } r \geq r_{cr} \tag{19}$$

and

$$V_z = U_s + \frac{1}{\frac{dP}{dz}\frac{(s+1)}{2m}}\left\{\left(-\frac{r_{cr}}{2m}\frac{dp}{dz}-\frac{\tau'_o}{m}\right)^{s+1}-\left(-\frac{R}{2m}\frac{dP}{dz}-\frac{\tau'_o}{m}\right)^{s+1}\right\}+\frac{1}{4\mu}\frac{dP}{dz}(r^2-r_{cr}^2)$$

$$\text{for } r \le r_{cr} \qquad (20)$$

Integrating the velocity profiles, given in Equations 19 and 20, the volumetric flow rate for $\tau_w \ge \tau_{cr}$ is determined as

$$Q = Q_s + \frac{\pi \tau_{cr}^4}{4\mu}\left(\frac{R}{\tau_w}\right)^3 + \pi R^3\left(\frac{\tau_w}{m}\right)^s\left\{\frac{(A^{s+3}-B^{s+3})}{s+3}+2\left(\frac{\tau'_o}{\tau_w}\right)\frac{A^{s+2}-B^{s+2}}{s+2}+\left(\frac{\tau'_o}{\tau_w}\right)2\frac{A^{s+1}-B^{s+1}}{s+1}\right\} (21)$$

where $\tau_w = -(dP/dz)(R/2)$, $\tau_{cr} = -(dP/dz)(r_{cr}/2)$, $A = (\tau_w - \tau'_o)/\tau_w$, $B = (\tau_{cr} - \tau'_o)/\tau_w$ and $Q_s = U_s \pi R^2$.

Equation 21 reduces to the expression for the volumetric flow rate of Herschel-Bulkley fluids in capillaries, as reported by Froishteter and Vinogradov (19), when $U_s = 0$, $\tau_{cr} = \tau_o$, $\tau'_o = \tau_o$ and $\mu = \infty$, where τ_o is the yield stress of Herschel-Bulkley fluid.

The validity of Equation 21 is shown for the capillary data with the values of the parameters determined previously. In Figure 12, $(Q-Q_s)$ values are shown in which Q_s is calculated using the U_s values from Figure 3, and the continuous curves are the predictions of Equation 21.

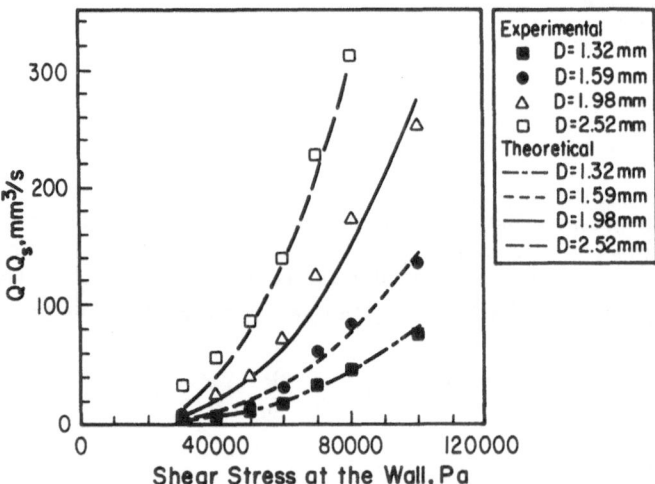

Figure 12. Q-Q$_s$ values versus the shear stress at the wall for various capillary diameters in suspension I.

Equation 21 is also used to predict the experimental values of the Q_s/Q ratio, shown in Figure 7 versus the wall shear stress, τ_w, for various capillary diameters. The experimental results and theoretical predictions indicate that the ratio Q_s/Q decreases with increasing capillary diameter in suspension I. With decreasing shear stress, Q_s/Q increases and approaches to one. This behavior of increasing Q_s/Q with decreasing shear stress is associated with the shear thinning nature of the suspension I at high-shear stresses. However, Q_s/Q values increase with increasing wall shear stress, τ_w, for suspension II which is shear thickening. The dependence of the Q_s/Q on the shear sensitivity parameter, n, can be shown qualitatively by simplifying Equation 21 by considering the special case where $\mu = \infty$, $\tau'_o = \tau_{cr} = 0$ and substituting $Q_s = \pi R^2 a \tau_w$ (as also considered for the Power Law fluid of Ostwald-de-Waele by Cohen (20)). The result is

$$\frac{Q_s}{Q} = \left[1 + \frac{R}{(s+3) \, am^s \tau_w^{1-s}} \right]^{-1} \tag{22}$$

Thus for shear thinning fluids (s>1) Equation 22 implies that, Q_s/Q may start from a value of one and decrease with the increasing wall shear stress, τ_w, as shown in Figure 7. However, for shear thickening fluids (s<1) Q_s/Q would increase with the shear stress as seen from Equation 22 and Figure 8. Finally, regardless of the shear thinning or thickening behavior, Equation 22 implies that Q_s/Q decreases with increasing capillary radius as observed in Figures 7 and 8. The radius dependence would disappear when Q_s/Q is close to one in the extreme case. This is observed in Figure 8 at shear stresses greater than 40 000 Pa.

The previous discussion clearly shows that the contribution of the slip to the total volumetric flow rate, Q_s/Q, increases when the viscosity increases, implying that in this study slip is the main mechanism of flow at extreme viscosities.

4.5. Slip Layer Thickness

In Reference 6 it was shown that the slip layer thickness, δ, is given by

$$\delta = \frac{u_s \eta_s}{\tau_w} \tag{23}$$

where η_s is the viscosity of suspending medium. In the derivation of Equation 23, a two phase Poiseuille flow with a Newtonian slip layer was assumed and isothermal, laminar, steady incompressible conditions with no end effects were considered. Substituting the linear relation between the slip velocity and the shear stress in Equation 23 the slip layer can be estimated from

$$\delta = a \eta_s \tag{24}$$

Using the slip coefficients "a" calculated in this report the slip layer thickness is estimated as 1 and 3.4 microns in suspensions I and II respectively. These values are a fraction of the average particle sizes used in these suspensions.

5. CONCLUSIONS

In the flow of highly concentrated suspensions studied here, slip at the wall was observed. The slip velocity was approximately linear with the shear stress at the wall, in both suspensions, in capillary as well as parallel disk torsional flows. The ratio of volumetric flow rate caused by slip to the total volumetric flow rate, Q_s/Q, decreased with increasing capillary radius. It increased with the shear stress in the shear thickening suspension (suspension II) and approached one as the viscosity became unbounded. However, it decreased with the shear stress in the shear thinning suspension (suspension I). The dependence of Q_s/Q on the diameter of the capillary and the shear stress was predicted by a bi-region constitutive equation developed. The slip layer thickness at the wall was estimated as a fraction of the average particle size. In this study, the suspension with bimodal particle size distribution exhibited pseudoplasticity, whereas the suspension with approximately uniform particles displayed shear thickening behavior at high shear stresses.

ACKNOWLEDGEMENTS

The experimental work was done at Highly Filled Materials Institute. It was supported by SDIO/IST as managed by ONR and by partial funding from APV Chemical Machinery. We thank Mr. J. Kowalczyk and Mr. J. Jones and APV for their suggestion and Dr. H. Gokturk, Dr. A. Lawal, Ms. B. Aral, Ms. P. Yaras, Messers. T. Fiske, J. Goradia and C. Jacob of Stevens for their contributions.

REFERENCES

1. Kamal, M.R. and Mutel, A. (1985), "Rheological Properties of Suspensions in Newtonian and Non-Newtonian Fluids", J.Polym.Eng., 5, 293-382.

2. Metzner, A.B. (1985), "Rheology of Suspensions in Polymeric Liquids", J.Rheol., 29, 739-775.

3. Khan, S.A. and Prud'homme, R.K. (1987), "Melt Rheology of Filled Thermoplastics", Rev.Chem.Eng., 3, 205-270.

4. Mewis, J. and Spaull, A.J.B. (1976), "Rheology of Concentrated Suspensions", Adv.Colloid and Interface Sci., 6, 173-200.

5. Vinogradov, G.V. and Malkin, A.Y. (1980), "Rheology of Polymers", Mir Publishers, Moscow, Chap.6.

6. Yılmazer, U. and Kalyon D.M. (1989), "Slip Effects in Capillary and Parallel Disk Torsional Flows of Highly Filled Suspensions", J.Rheol., 33(8), 1197-1212.

7. Yılmazer, U. and Kalyon D.M. (1991), "Dilatancy of Concentrated Suspensions with Newtonian Matrices", Polym.Comp., 12, 226-232.

8. Kalyon, D.M. and Yılmazer U. (1990), "Rheological Behavior of Highly Filled Suspensions which Exhibit Slip at the Wall", in A.Collyer and L.Utracki (eds.), Polymer Rheology and Processing, Elsevier Applied Science, London, pp.241-275.

9. Yılmazer, U., Gogos, C.G. and Kalyon D.M. (1989), "Mat Formation and Unstable Flows of Highly Filled Suspensions in Capillaries and Continuous Processors", Polym.Comp., 10, 242-248.

10. Kalyon, D.M., Yaras. P., Aral, B. and Yılmazer, U. (1992), "Rheological Behavior of a Concentrated Suspension. A Solid Rocket Fuel Simulant," Submitted to J.Rheol.

11. Yoshimura, A. and Prud'homme, R.K. (1988), "Wall Slip Corrections for Couette and Parallel Disk Viscometers", J.Rheol., 32, 53-67.

12. Mooney, M. (1931), "Explicit Formulas for Slip and Fluidity", J.Rheol., 2, 210-222.

13. Bagley, E.B. (1957), "End Corrections in the Capillary Flow of Polyethylene", J.Appl.Phys., 28, 624-627.

14. Yaras, P., Yılmazer, U. and Kalyon, D.M. (1992), "Flow Instabilities in Capillary Flow of Concentrated Suspensions", Submitted to Rheol.Acta.

15. Keentok, M. (1982), "The Measurement of the Yield Stress of Liquids", Rheol. Acta., 21, 325-332.

16. Donovan, E.J. and Tanner, R.I. (1984) "Numerical Study of the Bingham Squeeze Flow Problem", J. Non-Newtonian Fluid Mech.,15, 75-82.

17. Gartling, D.K. and Phan-Thien, N. (1984), "A Numerical Simulation of a Plastic Fluid in a Parallel-Plate Plastometer", J.Non-Newtonian Fluid Mech., 14, 347-360.

18. Lipscomb, G.G. and Denn, M.M. (1984), "Flow of Bingham Fluids in Complex Geometries", J. Non-Newtonian Fluid Mech.,15, 337-346.

19. Froishteter, G.B. and Vinogradov, G.V. (1980), "The Laminar Flow of Plastic Disperse Systems in Circular Tubes", Rheol. Acta, 19, 239-250.

20. Cohen. Y. (1981), Ph.D. Thesis, "The Behavior of Polymer Solutions in Non-Uniform Flows", University of Delawere.

APPLICATION OF SURFACE ANALYSIS TO HIGH PERFORMANCE POLYMERIC ADHESIVES AND COMPOSITES

J. P. WIGHTMAN
Chemistry Department
Center for Adhesive and Sealant Science
Center for Composite Materials and Structures
NSF Science and Technology Center
Virginia Institute for Material Systems
Virginia Polytechnic Institute and State University
Blacksburg, VA 24061 U.S.A.

ABSTRACT. The paper details the use of scanning electron microscopy, surface reflectance infrared spectroscopy, Auger electron spectroscopy, ion scattering spectroscopy, secondary ion mass spectroscopy, and x-ray photoelectron spectroscopy in the analysis of polymeric adhesives and composites. A brief review of the principle of each surface analytical technique will be followed by application of the technique to interfacial adhesion with an emphasis on polymer/metal, fiber/matrix, and composite/composite adhesion.

1. Introduction

Adhesion, as depicted in Figure 1, involves a detailed understanding of a number of broad areas including surfaces, polymers, and mechanics (1). The focus of the present review is on surfaces or more properly the interphase (2). The schematic diagram shown in Figure 2 serves to emphasize both the significance and the complex nature of the interphase (3). The importance of the interphase cannot be minimized in an understanding of the adhesion process. Indeed, poor mechanical properties of adhesively bonded structures are often a consequence of interphase properties. The interphase can be studied by microscopic, spectroscopic, thermodynamic, and kinetic techniques. Baun (4) has categorized fifty-four surface characterization methods according to which one of the six aspects of adhesion is being investigated. Smith (5) has given a detailed discussion of the role of surface energetics in adhesion.

The objective of this paper is to present a review of the results of some of the microscopic/spectroscopic techniques which have been used in the study of adhesion. The spectroscopic techniques to be discussed are listed in Table I adapted from Baun (4). A brief review of each technique will be followed by a discussion of results illustrating the application of the technique to polymer/metal, fiber/matrix, and composite/composite adhesion. A recent, more detailed review of the use of surface analytical techniques applied to polymer/metal adhesion has been published (6).

G. Akovali (ed.), The Interfacial Interactions in Polymeric Composites, 125–149.

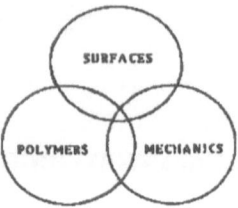

Figure 1. Major academic discipline areas in adhesion science.

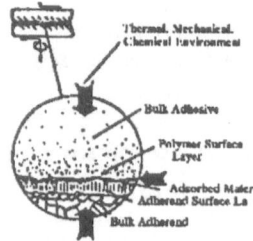

Figure 2. Schematic diagram of the interphase (3).

TABLE I SELECTED SURFACE CHARACTERIZATION METHODS (4)

ACRONYM	METHOD	ASPECT OF ADHESION
AES	Auger electron spectroscopy	A-C-D-E
ISS	Ion scattering spectroscopy	A-C-F
SEM	Scanning electron microscopy	B-D-F
SIMS	Secondary ion mass spectroscopy	A-C-E-F
SRS	Surface reflectance spectroscopy	B-C-D-E-F
STEM	Scanning transmission electron microscopy	B
XPS	X-ray photoelectron spectroscopy	A-C-E-F

A. Adherend chemistry

B. Adherend structure and morphology

C. Adhesive chemistry

D. Adhesive structure and morphology

E. Interaction of polymers with metals

F. Failure surfaces

2. Scanning Electron Microscopy (7)

Scanning electron microscopy (SEM) is a commonly used technique in the detailed study of adhesive bonding. The SEM technique has not only been used in post-failure analysis but also to characterize the adherend surface prior to priming/bonding. Evans and Packham (8) have examined the adhesion between polyethylene and copper, zinc, and steel. Anodizing treatments for copper and zinc and a hydrothermal oxidation for steel led to very rough surfaces as gauged by SEM. The peel strength of polyethylene to these substrates was high even when the polymer was stabilized with an anti-oxidant. The results listed in Table II are to be compared with a value of 0.14 N/mm for polyethylene with 1000 ppm antioxidant applied to etched steel. The significant point here is that high peel strength was attributed to a blade-like oxide coating formed on the steel adherend. The results suggests that high peel strengths are associated with thick residual polymer films observed with SEM following fracture.

A brief but comprehensive comparison of various microscopic techniques including SEM has been made by Ledbury et al. (9). SEM photomicrographs (10) of the titanium side of a failed Turco alkaline treated titanium wedge test panel show a thin layer of primer (BR127) covering the surface and the presence of iron particles in the interphase. SEM has proved useful in the analysis of adherend surfaces before bonding and after test failure. Post-failure analysis of titanium alloy lap shear specimens bonded with LaRC-13 polyimide adhesive with and without fluorosiloxane elastomer indicated that the failure mode shifted from cohesive to interfacial as determined from SEM photomicrographs (11). SEM has been used (12,13) to observe changes in surface topography of carbon fiber/polymer matrix composites after different surface pretreatments. For example, grit blasting was shown to remove the resin rich top surface layer and damage underlying carbon fibers.

TABLE II EFFECT OF CONCENTRATION OF ANTIOXIDANT IN POLYETHYLENE ON ADHESION TO STEEL TREATED TO GIVE A BLADE-LIKE SURFACE OXIDE (8)

P.p.m. antioxidant in polyethylene	Peel strength N/mm	No. of peels
0	2.73 ± 0.19	8
500	2.43 ± 0.09	8
1000	2.12 ± 0.23	4
2000	2.16 ± 0.18	4
5000	2.53 ± 0.29	4

(95% confidence limits are indicated)

3. Scanning Transmission Electron Microscopy (14)

The advent of high resolution scanning electron microscopy (STEM) represents a tremendous advancement in the microscopic examination of adherend and failure surfaces. Venables et al. (15) have demonstrated using STEM that the degradation mechanism causing bond failure of aluminum alloys in a humid environment is the conversion of the surface oxide with a cellular and whisker structure to a surface hydroxide with a "corn flake" type structure.

4. Surface Reflectance Infrared Spectroscopy (16)

There has been a resurgence of interest in the application of infrared spectroscopy, in particular, to surface and interphase analysis (17,18). The introduction of Fourier transform has given added impetus to such studies. The technique remains under-utilized however. A number of experimental techniques have been developed and are collectively termed surface reflectance spectroscopy (SRS). Schematic diagrams (19-21) of representative sampling attachments which can be used in infrared spectrophoto-meters are shown in Figure 3. An internal reflection element (IRE), say Ge or KRS-5 (TlBr + TlI), is used in an attenuated total reflectance attachment as shown in Fig. 3A. The sample is pressed firmly against both sides of the IRE. This technique works best with deformable solids. Specular reflection attachments are based on multiple, single, and grazing angle reflections where no IRE is used. A grazing angle specular reflection attachment is shown in Fig. 3B. A diffuse reflectance attachment, especially useful for powders, is illustrated in Fig. 3C. Interestingly, this attachment has been used to follow curing of a polyimide matrix in a composite (21). Photoacoustic attachments requiring minimal sample preparation are also available.

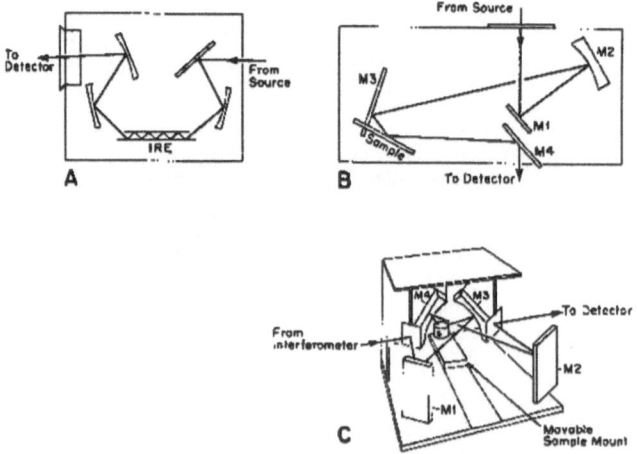

Figure 3. Attachments used in surface reflectance infrared spectroscopy (19-21).

Work has been reported on the use of specular reflectance in the analysis of failed adhesively bonded surfaces (22). For example, the spectrum shown in Figure 4 was obtained by a single reflection from a failure surface resulting from a lap shear test of a titanium alloy bonded with a polyimide. The absorption peaks in Fig. 4 match those for a neat polyimide film. Sung et al. (23) have used internal reflection techniques to show that A-1100 and A-151 silane films on aluminum surfaces are polysiloxane networks with incomplete crosslinking as evidenced by a marked decrease in the intensity of the CH$_3$ peak at 2970 cm^{-1} due to hydrolysis and polymerization. Boerio and Gosselin (24) have studied the interaction of γ -aminopropyltriethoxysilane (γ -APS) with aluminum and suggested adsorption of γ -APS as the hydrochloride from acidic solution onto aluminum followed by extensive hydrolysis of γ -APS, and, consequent formation of siloxane polymers. Differences in the bond durability of two different metal alkoxide coated titanium alloy adherends bonded with epoxy were attributed (25) to differing degrees of hydrolysis based on the intensity of the O-H stretch around 3500 cm^{-1}.

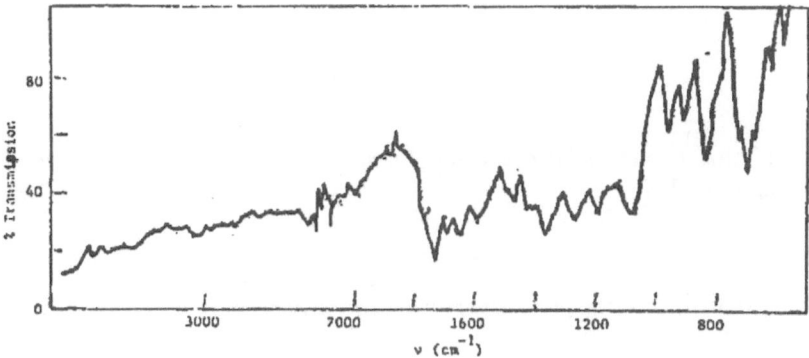

Figure 4. Reflectance infrared spectrum of a failed lap shear titanium alloy adherend bonded with a polyimide adhesive (22).

The infrared spectrum (13) of a carbon fiber/polyimide matrix composite obtained using a photoacoustic attachment is shown in Figure 5. Peak assignments for this spectrum are listed in Table III. However, it was not possible by this technique to detect changes in the surface composition of the composite following exposure to an oxygen plasma for up to 20 minutes. Recent spectra obtained in our laboratory (26) with grazing angle specular reflectance FT-IR are illustrated in Figure 6 and show the utility of this particular technique in obtaining good signal-to-noise infrared spectra on ultra-thin (7.5-15 nm) polymer films on reflective metal substrates. It is important to emphasize that infrared spectroscopy leads directly to the identification of groups such as methyl and methylene, carbonyl, ether, ester, hydroxyl, and amine in polymeric materials.

130

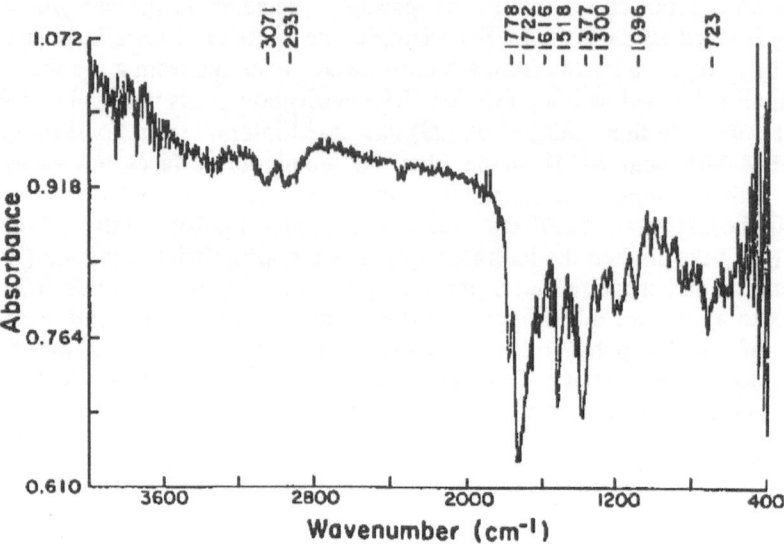

Figure 5. Photoacoustic FT-IR spectrum of a methanol washed carbon fiber/polyimide matrix composite (13).

TABLE III PAS-FTIR PEAK ASSIGNMENTS FOR METHANOL WASHED COMPOSITE (13)

Wavenumber (cm-1)	Assignment	
	Mode	Source
3071	C-H stretching	Aromatic
2931	C-H stretching	Aliphatic
1778	>C=O stretching - asym.	Ester or imide ring
1722	>C=O stretching - sym.	Ester or imide ring
1616	C=C stretching	Aromatic ring
1518	C=C stretching	Substituted aromatic ring
1377	Axial vibration	C-N-C of imide ring
1300	Axial vibration	C-N-C of imide ring
1096	Transverse vibration	C-N-C of imide ring
723	Out of plane vibration	C-N-C of imide ring

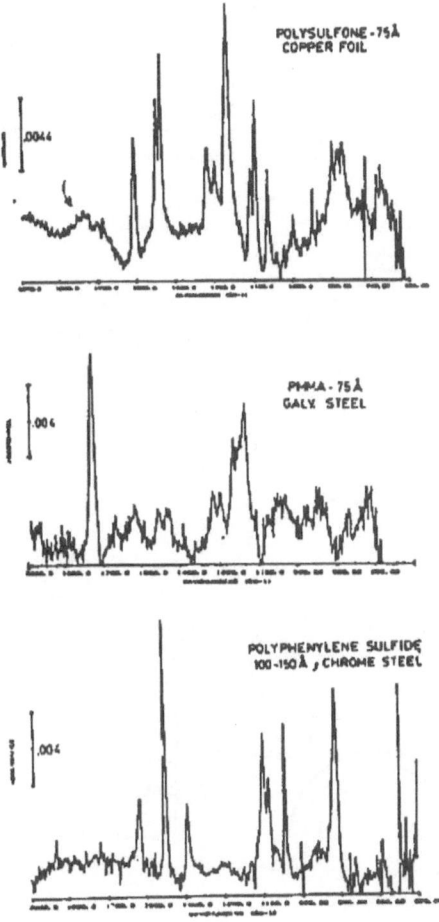

Figure 6. Specular reflection FT-IR spectra of ultra-thin polymer films obtained at grazing angle using parallel polarized infrared radiation (26).

5. Auger Electron Spectroscopy (27)

The principle of Auger electron spectroscopy (AES) is illustrated schematically in Figure 7. Here, a primary beam of electrons incident on a solid results in the emission of secondary (or Auger) electrons from the top 5 nm of the sample. The energy of the Auger electrons is characteristic of elements contained in the outer surface. Both elemental identification and atomic concentrations can be obtained from AES. The AES technique is not widely used in post-failure analysis of adhesively bonded

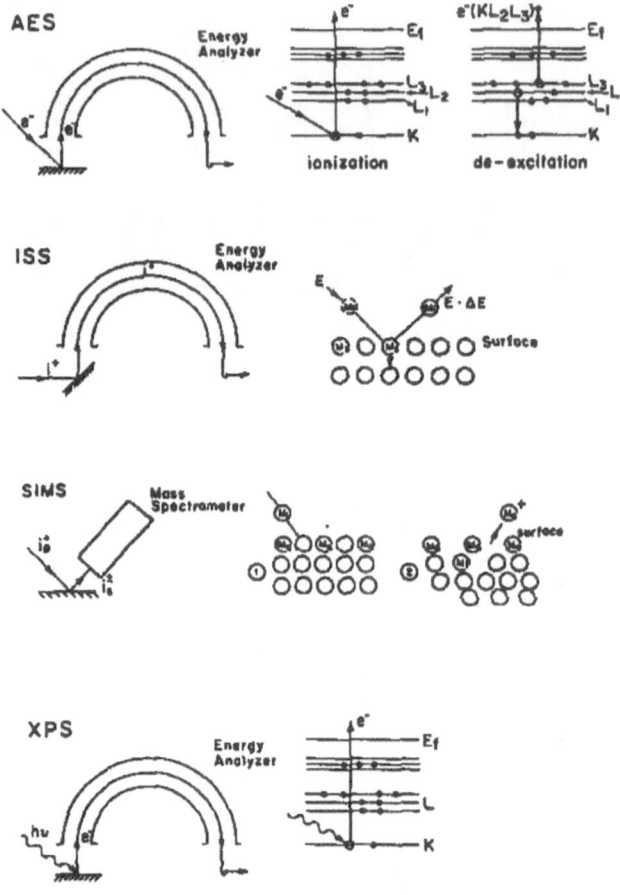

Figure 7. Schematic diagrams of four surface analytical techniques.

specimens perhaps due to possible beam damage of the adhesive. On the other hand, the use of AES in the analysis of inorganic adherend surfaces has been widespread. The Auger spectra of titanium alloy adherends pretreated by two different processes (10) are shown in Figure 8. Note the significant difference in the fluorine concentration following the two different pretreatments: MPF - modified phosphate/fluoride; CAA-chromic acid anodization. Depth profiling is also possible with AES (10) and a significantly thinner oxide is present on the titanium alloy adherend after a Dapco treatment (DA) than after chromic acid anodization (CAA) as indicated by the results in Figure 9. Sputtering time is proportional to oxide thickness so one compares the

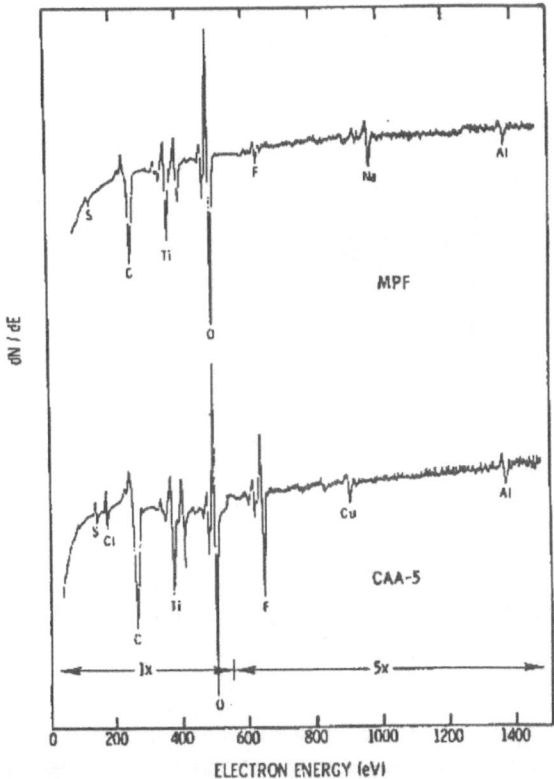

Figure 8. Auger spectra of MPF and CAA pretreated titanium alloy adherends (10).

sputtering time necessary to achieve equal concentration of titanium and oxygen for the two surface pretreatments.

6. Ion Scattering Spectroscopy (28)

The principle of ion scattering spectroscopy (ISS) is illustrated schematically in Figure 7. A primary ion beam of defined energy, typically helium or argon, undergoes inelastic collisions with a solid. The energy of the scattered beam relative to the incident beam is related to the masses of surface atoms with which the primary beam collided. An ISS spectrum then is used to identify atoms contained in the top surface layer of the sample. ISS is often stated to be the most sensitive of the surface analytical techniques. For example, A. Miller of Lehigh University states that if a monolayer of lead is deposited on a silicon substrate only lead will be detected by ISS. The ISS spectra (13) of a carbon fiber/polyimide matrix composite exposed to an oxygen plasma for increasing

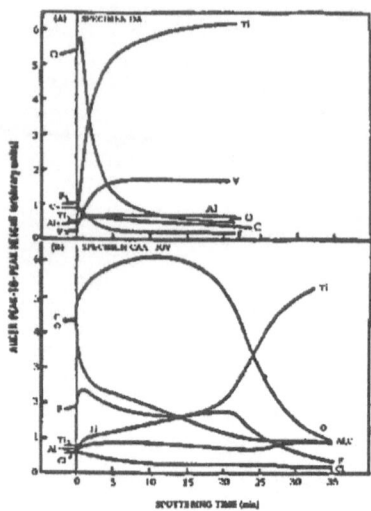

Figure 9. Auger depth profiles of (A) DA and (B) CAA pretreated titanium alloy adherends (10).

times are shown in Figure 10. The decrease in the fluorine signal with a concomitant increase in the oxygen signal at longer exposure times in an oxygen plasma is noted.

7. Secondary Ion Mass Spectroscopy (29)

The principle of secondary ion mass spectroscopy (SIMS) is illustrated schematically in Figure 7. A primary ion beam of sufficient energy focused on a solid causes production of simple and complex ions characteristic of the solid. The technique is then necessarily destructive; however, recent advances in so-called static SIMS minimizes sample damage and enables analysis of fractional monolayers. The advantage of SIMS in addition to high sensitivity is the ability to obtain information about the distribution of species across a solid surface. The SIMS technique is used in the chemical analysis of solid surfaces particularly in the fingerprinting of polymers by Briggs (30). Baun (28) has applied SIMS in conjunction with ion scattering spectroscopy to the analysis of bond failure surfaces in a double cantilever beam test. Chromium and aluminum signals increased on the adherend and adhesive sides, respectively. These results suggest a weak metal oxide surface layer rather than cohesive failure in the adhesive.

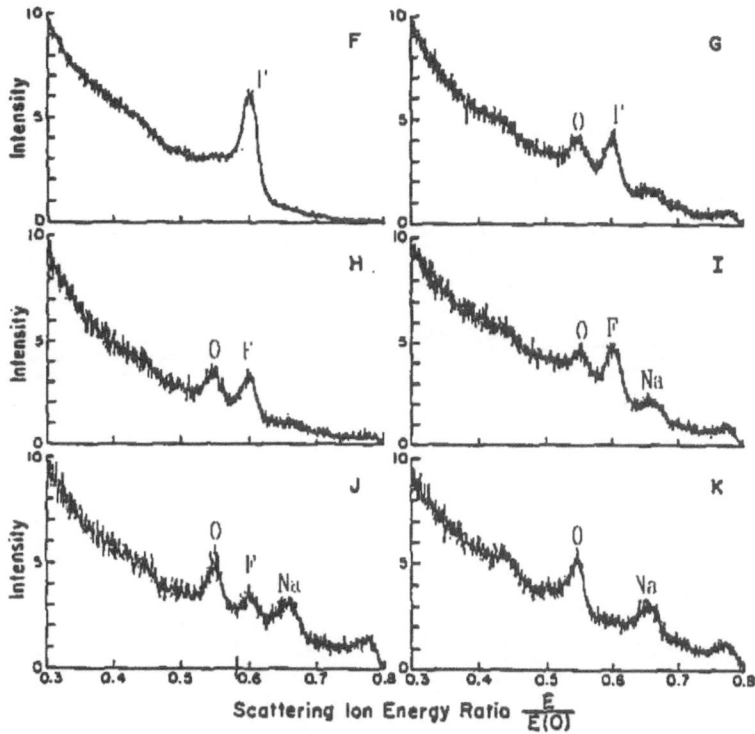

Figure 10. ISS survey spectra of carbon fiber/polyimide matrix composites exposed to an oxygen plasma for (F) 0.5 min (G) 1 min (H) 2 min (I) 5 min (J) 10 min and (K) 20 min (13).

8. X-Ray Photoelectron Spectroscopy (31-33)

X-ray photoelectron spectroscopy (XPS) or electron spectroscopy for chemical analysis (ESCA) involves electrostatic analysis at high resolution of electrons which are photoejected from the top 5 nm of a solid surface on irradiation with X-rays. Schematic diagrams of the technique are shown in Figures 7 and 11. The defining equation is written $h\nu = KE + BE + \phi$. The energy of the incident x-ray beam is $h\nu$, KE is the kinetic energy of the photoelectron, BE is the binding energy of the photoelectron, and ϕ is the work function of the spectrometer. The utility of XPS is the measurement of BE with high precision. The value of BE is characteristic of the element to which a given electron is bound. Further, and parenthetically the feature which has given XPS its impetus, the technique can detect small changes in BE which occur due to differences in chemical bonding between atoms - hence a "chemical shift". Chemical shifts for some

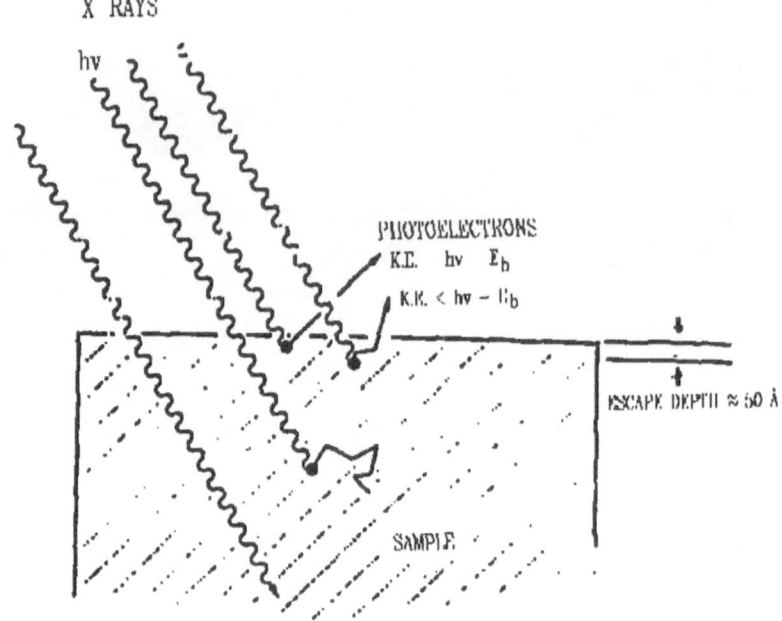

CHEMICAL SHIFTS OF OXIDATION STATES (eV)

Element	-2	-1	0	+1	+2	+3	+4	+5	+6	+7
					Oxidation State					
N 1s		0		+4.5		+5.1		+8.0		
S 1s	-2.0		0				+4.5		+5.8	
Cl 2p						+3.8		+7.1		+95
Cu 1s			0	+0.7	+4.4					

Figure 11. Schematic diagram of the XPS technique with representative binding energy shifts.

elements are also listed in Figure 11. The XPS technique is quite "sample friendly" and almost any solid can be analyzed conveniently. Thus, XPS is indeed a surface technique par excellence giving not only elemental identification but also bonding state inform-ation. It is a widely used surface analytical technique in adhesion studies. It should be noted that XPS and RAS spectra can be complementary. Group identification results from curve fitting of XPS photopeaks whereas the same groups are observed directly in RAS.

XPS analysis (34) of a chromic acid anodized titanium alloy surface is shown in Figure 12. The fluorine source is inorganic fluoride which is a residual resulting from the addition of hydrofluoric acid to the anodizing bath. The ratio of oxygen to titanium is approximately 2 as expected for a surface layer of titanium dioxide. The source of carbon is adsorbed organic compounds which parenthetically is the reason why the contact angle of water increases with time on freshly cleaned metal oxide substrates. An elegant XPS analysis (35) of a vulcanized rubber/brass interface which involved depth profiling showed that an interfacial film of reaction products is formed and becomes an integral part of the brass rubber band.

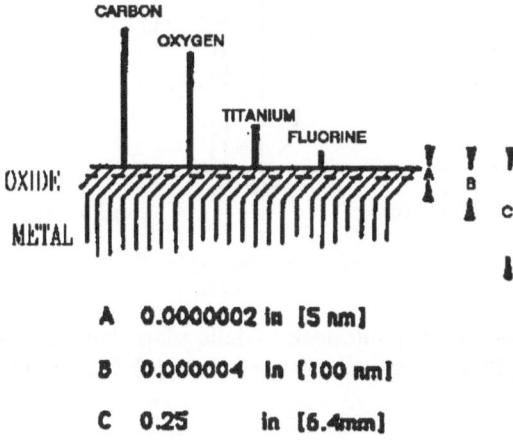

Figure 12. XPS atomic concentrations of a chromic acid anodized titanium alloy (34).

Scanning electron photomicrographs of failed lap shear specimens cannot be used to unambiguously establish whether interfacial failure has in fact occurred. For example, ultra-thin (<20 nm) polymer films remaining on adherend surfaces following single lap shear test may be undetected in an SEM photomicrograph. However, XPS analysis is ideally suited to detect residual polymer films with thicknesses < 5nm. For example, the XPS titanium 2p photopeak (22) from failure surfaces of two separate lap shear samples are shown in Figure 13. The adherend in both cases was a titanium alloy. The same polyimide adhesive before bonding was dissolved in diglyme in one case and in dimethylacetamide (DMAC) in the other. The intense titanium signal (see Fig. 13a) for the DMAC solvent system indicates interfacial rather than cohesive failure. The negligible titanium signal (see Fig. 13b) for the diglyme solvent system indicates residual polyimide adhesive remaining on the adherend surface. Since the XPS technique is limited to the top 5 nm, the average composition of the failure surface can be established by XPS at a level heretofore not possible. Other examples of failure surface analysis by XPS are cited below.

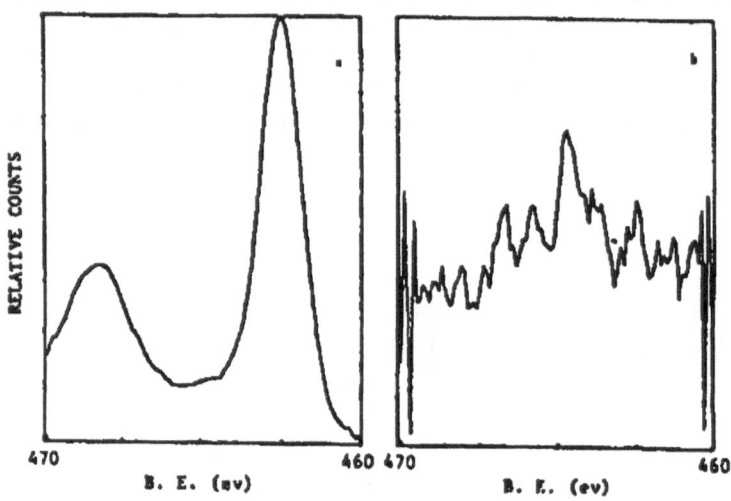

Figure 13. XPS titanium 2p photopeak of failed lap shear sample of titanium alloy bonded with polyimide adhesive solvent cast from (a) DMAC solution and from (b) diglyme solution (22).

The results of XPS analysis (36) of a failed T-peel sample of a titanium alloy bonded with epoxy are summarized in Table IV. Column 1 contains the surface composition of the chromic acid anodized adherend before bonding. Columns 2 and 3 are the corresponding results for the two failure surfaces following the T-peel test. The surface composition for all three surfaces is similar. Thus, the immediate conclusion is that failure did not occur cohesively, that is, within the epoxy; nor did failure occur interfacially, that is, at the epoxy/metal oxide interface. Instead, failure occurred within the titanium oxide layer. The fact that a small but significant amount of elemental titanium [Ti(0)] was detected in addition to Ti(IV) on one surface (see column 2) means that the average failure plane passed very near the oxide/metal interface. A schematic diagram of the failure surfaces based on the XPS results is given at the bottom of Table IV. One begins to appreciate the power of the XPS technique to make unequivocal

TABLE IV ATOMIC COMPOSITION OF ANODIZED TITANIUM ALLOY SAMPLES. [#1 - ANODIZED SAMPLE BEFORE BONDING; #2, #3 - FAILURE SAMPLES.] (36)

ELEMENT	ATOM PERCENT		
	#1	#2	#3
F 1s	1.2	0.4	2.4
O 1s	13.2	23.6	16.9
V 2p$_{3/2}$	0.1	0.1	NSP*
Ti (IV) 2p$_{3/2}$	6.9	7.8	7.1
Ti (O) 2p$_{3/2}$	NSP	0.4	NSP
N 1s	0.6	0.7	0.9
C 1s	76.6	65.5	71.3
Cl 2P	0.4	0.3	NSP
Al 2s	1.0	1.3	1.4

* NSP - No significant peak

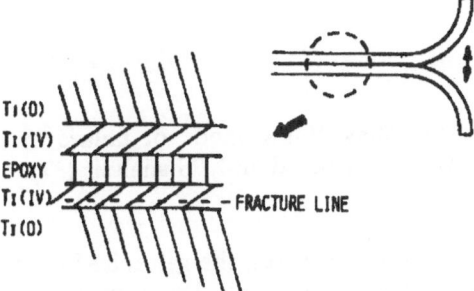

assignments of the composition of the failure surfaces and in turn the assignment of the failure mode (cohesive or interfacial or other as in the above case).

The failure surfaces produced following tests of lap shear or wedge samples of titanium alloy bonded with epoxy depended on the surface pretreatment (34). Simple acid etching of the adherend produced primarily interfacial failure between the oxide and epoxy whereas chromic acid anodization of the adherend resulted in failure within the oxide layer as in the case discussed above.

Bonding of oily, galvanized steel with epoxy produced some surprising failure mode results depending on the type of galvanized steel (37). A schematic diagram of three different failure modes for three different galvanized steels is shown in Figure 14. The galvanized steel labeled 'A527' failed interfacially, that is, between the epoxy and the surface oxide layer. The galvanized steel labeled 'MS' showed cohesive failure, that is,

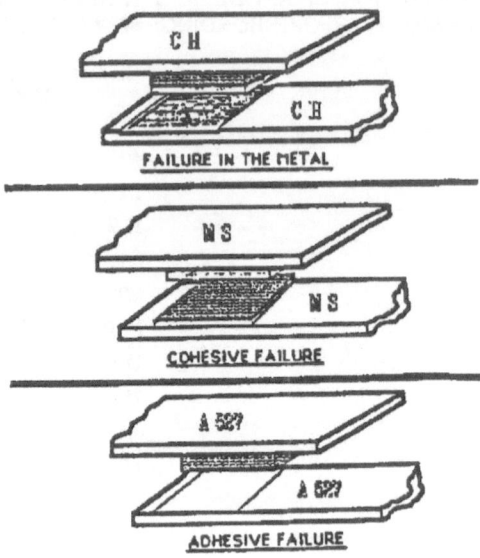

Figure 14. Schematic diagrams of the mode of failure of lap shear samples of galvanized steel boned with epoxy based on XPS analysis (37).

within the epoxy adhesive itself. However, the galvanized steel labeled 'GH' failed at the steel substrate/galvanized layer interface. Depth profiling using AES was used to complete the failure mode assignment in this last case.

The often used FPL etch of an aluminum-lithium alloy bonded with polysulfone leads to interfacial (at the metal oxide/polymer interface) failure (38) which is a surprisingly uncommon type of failure. The results leading to this assignment are shown as XPS C 1s and O 1s narrow scan spectra in Figure 15. This definitive assignment of failure mode is based on the fact that one failure surface has an oxygen photopeak similar to the pretreated adherend before bonding and the other failure surface has an oxygen photopeak similar to the adhesive.

In a study of fiber/matrix adhesion, the effect of surface pretreatment of carbon fibers in gas plasmas on adhesion have been described (39). The atomic concentrations of Hercules IM7 carbon fiber as determined by XPS before and after plasma exposure are listed in Table V. The nitrogen content of carbon fibers is increased in an ammonia plasma whereas the oxygen content is increased in an air plasma. No effect of the air plasma treatment of carbon fibers on adhesion to polyethersulfone is seen as noted in

Table VI. However, the debonding force is greater following exposure of carbon fiber to an ammonia plasma, and quenching (Q) of the sample.

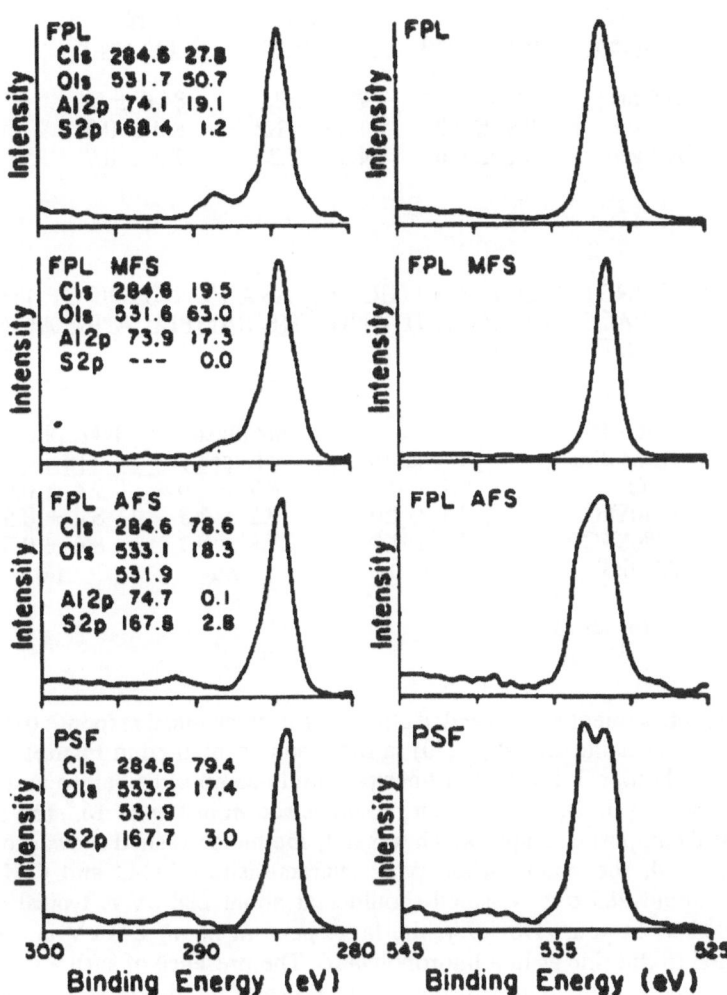

Figure 15. XPS carbon 1s and oxygen 1s photopeaks for pretreated aluminum alloy after acid etch and before bonding (FPL); two failure surfaces following bonding with polysulfone and testing in the wedge configuration (FPL MFS and FPL AFS); and, neat polysulfone (PSF) (38).

TABLE V. XPS ANALYSIS OF IM7 CARBON FIBERS BEFORE AND AFTER PLASMA TREATMENTS. ATOMIC CONCENTRATIONS EXPRESSED IN PERCENT (39).

Plasma Treatment	as-received	air plasma (15 sec.)	NH$_3$ plasma (15 sec.)
Carbon	85.0 ± 1.3	76.0 ± 2.2	84.8 ± 0.7
Oxygen	9.9 ± 0.8	19.8 ± 1.2	8.0 ± 0.3
Nitrogen	5.1 ± 0.6	4.1 ± 1.3	7.2 ± 0.7

TABLE VI. DEBONDING LOAD (IN GRAMS) AS A FUNCTION OF THE FIBER SURFACE TREATMENT AND THE SAMPLE ANNEALING CONDITIONS (39).

Surface Treat.-> Annealing	as-received	air plasma (15 sec.)	NH$_3$ plasma (15 sec.)
Q	8.1 ± 0.8	8.7 ± 1.0	9.4 ± 0.8
A240SFC	7.1 ± 1.0	7.2 ± 1.3	8.4 ± 0.5
A280SFC	9.2 ± 1.1	7.8 ± 0.7	8.6 ± 0.5
A280Q'	8.2 ± 0.6	nd	nd

nd: not determined

That surface pretreatment has a decided effect on the mechanical response of adhesively bonded composites is illustrated (13,40) in the results contained in Figures 16 and 17. The XPS C 1s photopeak for carbon fiber/polyimide composites before and following exposure for varying times to an oxygen plasma is shown in Figure 16. The surface of the as-received composite sample which was only methanol washed shows a complex C 1s photopeak with the main carbon peak characteristic of C-C and C-H bonding appearing at about 285 eV. A small shoulder at about 288 eV is typical of carbon doubly bonded to oxygen. However, the large peak at about 293 eV is evidence of carbon bonded to fluorine as in a fluoropolymer. The presence of such a group on the composite surface is consistent with the expected use of mold release materials in the fabrication of the composite. It is obvious that increasing exposure times to an oxygen plasma results in the disappearance of the fluorocarbon contaminant with a concomitant increase and leveling off of the carbon-oxygen contribution to total carbon 1s photopeak. The presence of this fluoropolymer on the composite surface has a demonstrable effect on bondability. The average crack length of the as-received sample bonded with polyimide adhesive in a wedge configuration aged for two hours at 204°C is shown in Figure 17. Note the higher crack length and greater error bars for the as-received sample. Contrast these results for those of the 5 minute oxygen plasma characterized by both a smaller crack length and smaller error bars.

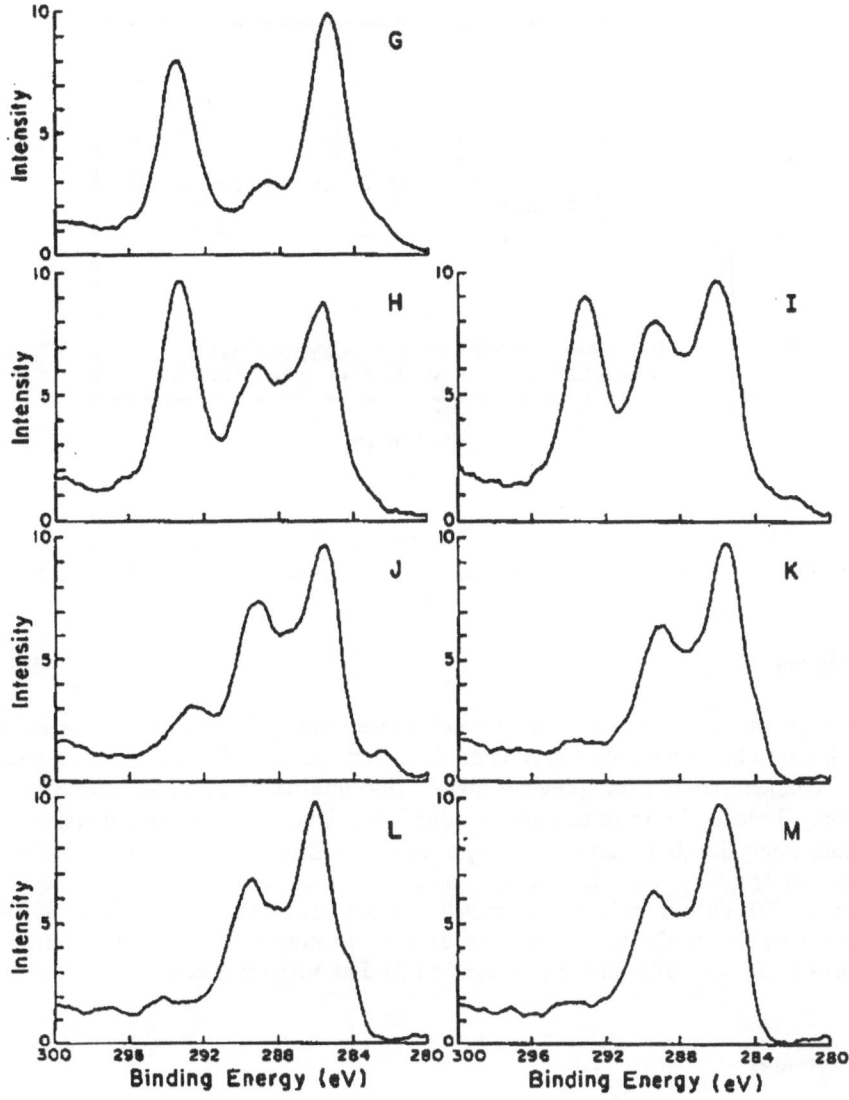

Figure 16. XPS carbon 1s photopeak for carbon fiber/polyimide matrix composites methanol washed (G) and exposed to an oxygen plasma for (H) 0.5 min (I) 1 min (J) 2 min (K) 5 min (L) 10 min (M) 20 min (13).

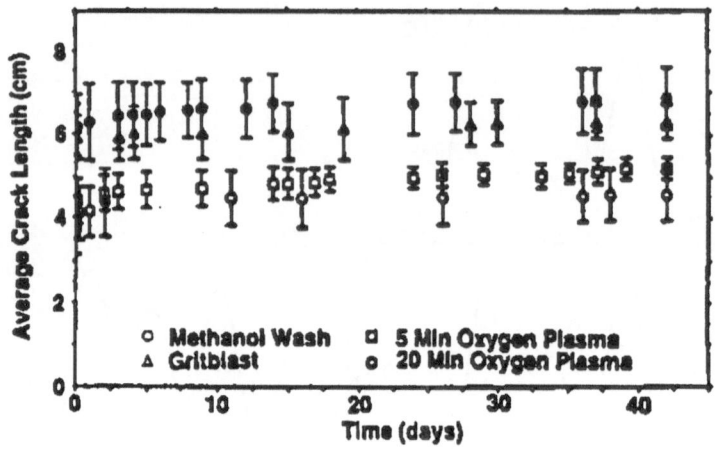

Figure 17. Effect of pretreatment on crack propagation of carbon fiber/polyimide matrix composites bonded in the wedge configuration and aged 1000 hours at 204°C (40).

9. Summary

In summary, there are a number of microscopic/spectroscopic experimental techniques, of which only a limited number have been discussed above, which are readily adaptable to the characterization of polymer/metal, fiber/matrix, and composite/composite adhesion. Indeed, basic questions in adhesion science such as the mechanism of adhesion, bond durability and the composition of failure surfaces, can be addressed experimentally today with increasing confidence due to the availability of these techniques. The author and his associates have summarized (6, 11-13,22,25,26,34,36-62) the results of the application of surface analysis to polymer/metal, fiber/matrix, and composite/composite adhesion on systems studied at Virginia Tech.

10. References

1.   H. F. Brinson, "Mechanics Applied to Adhesion Science," Applied Mechanics Review, 38, pp. ii-iv (1985).

2.   L. H. Sharpe, "Some Thoughts About the Mechanical Response of Composites," J. Adhesion, 6, 15-21 (1974).

3.   L. T. Drzal, M. J. Rich and P. F. Lloyd, "Adhesion of Graphite Fibers to Epoxy Matrices. I. The Role of Fiber Surface Treatment," J. Adhesion, 6, 1-30 (1982).

4. W. L. Baun, "Applications of Surface Analysis Techniques to Studies of Adhesion," Applications of Surface Science, 4, 291-306 (1980).

5. T. Smith, "Surface Energetics and Adhesion," J. Adhesion, 11, 243-256 (1980).

6. J. A. Filbey and J. P. Wightman, "Surface Characterization in Polymer/Metal Adhesion," in Adhesive Bonding, L. H. Lee, ed., pp. 175-202, Plenum, New York (1991).

7. O. Johari and A. V. Samudra, "Scanning Electron Microscopy" in Characterization of Solid Surfaces, P. F. Kane and G. P. Larrabee, eds., pp. 107-131, Plenum, New York (1974).

8. J. R. G. Evans and P. E. Packham, "Adhesion of Polyethylene to Metals: The Role of Surface Topography," J. Adhesion, 10, 177-191 (1979).

9. E. A. Ledbury et al., "Microstructural Characterization of Adhesively Bonded Joints," Proc. 12th Natl. SAMPE Techn. Conf., pp. 935-950, SAMPE, Azuza, CA (1980).

10. B. M. Ditchek, K. R. Breen, and J. D. Venables, "Bondability of Ti Adherends," Martin Marietta Labr. Report, MML TR-80-17c, Baltimore, MD (1980).

11. W. Chen and J. P. Wightman, "A Fundamental Approach to Adhesion: Synthesis, Surface Analysis, Thermodynamics and Mechanics," NASA Report NSG-1124, NASA-LaRC, Hampton, VA (Jan., 1979).

12. T. A. DeVilbiss, D. L. Messick, D. J. Progar, and J. P. Wightman, "SEM/XPS Analysis of Fractured Adhesively Bonded Graphite Fibre-Reinforced Polyimide Composites," Composites, 16, 207-219 (1985).

13. D. J. D. Moyer and J. P. Wightman, "Characterization of Surface Pretreatments Carbon Fiber/Polyimide Matrix Composites," Surface and Interface Analysis, 14, 496-504 (1989).

14. J. D. Venables et al., "Oxide Morphologies on Aluminum Prepared for Adhesive Bonding," Appl. Surface Sci., 3, 88-98 (1970).

15. J. D. Venables et al., "Effect of Moisture on Adhesively Bonded Aluminum Structures," Proc. 12th Natl. SAMPE Techn. Conf., pp. 909-923, SAMPE, Azusa, CA (1980).

16. N. J. Harrick, Internal Reflection Spectroscopy, Interscience, New York (1967).

17. R. G. Nuzzo, L. H. Dubois and D. L. Allara, "Fundamental Studies of Microscopic Wetting on Organic Surfaces. I. Formation and Structural Characterization of a Self-Consistent Series of Polyfunctional Organic Monolayers," J. Am. Chem. Soc., 112, 558-569 (1990).

18. H. Ishida, "Quantitative Surface FT-IR Spectroscopic Analysis of Polymers," Rubber Chem. Technol., 60, 497-554 (1987).

19. R. H. Honeycutt, M.S. Thesis, Virginia Polytechnic Institute and State University, Blacksburg, VA, 1973.

20. H. F. Webster, M.S. Thesis, Virginia Polytechnic Institute and State University, Blacksburg, VA, 1985.

21. P. R. Young and A. C. Chang, Proc. SAMPE Techn. Conf., pp. 136ff, SAMPE, Azusa, CA (1984).

22. T. A. Bush, M. E. Counts and J. P. Wightman, "The Use of SEM, ESCA, and Specular Reflectance IR in the Analysis of Fracture Surfaces in Several Polyimide/Titanium 6-4 Systems," in Adhesion Science and Technology, Part A, L. H. Lee, Ed., pp. 365-394, Plenum, New York (1975).

23. C. S. P. Sung, S. H. Lee and N. H. Sung, "Role of Coupling Agents in Metal Polymer Adhesion. I. The Structure of the Silane Film at Metal-Polymer Interface" in Adhesion and Adsorption of Polymers, Part B, L. H. Lee, ed., pp. 757-773, Plenum, New York (1980).

24. F. J. Boerio and C. A. Gosselin, "Structure-Property Relationships of Silane Films on Aluminum Substrates," Proc. 36th Ann. Tech. Conf., SPI Reinforced Plastics Composites Inst., Sec. 2.G pp. 1-4 (1981).

25. J. A. Filbey and J. P. Wightman, "Metal Alkoxide Primers in Titanium/Epoxy Bonds," J. Adhesion, 28, 23-29 (1989).

26. H. F. Webster and J. P. Wightman, "Effects of Oxygen and Ammonia Plasma Treatment on Polyphenylene Sulfide Thin Films and Their Interaction with Epoxy Adhesive," J. Adhesion Sci. Technol., 5, 93-106 (1991).

27. T. A. Carlson, Photoelectron and Auger Spectroscopy, Plenum, New York (1975).

28. W. L. Baun, "Study of Adhesive Bonding and Bond Failure Surface Using ISS-SIMS," in Characterization of Metal and Polymer Surfaces, Vol. I, L. H. Lee, ed., pp. 375-390, Academic Press, New York (1977).

29. A. Benninghoven, "Developments in Secondary Ion Mass Spectroscopy and Applications to Surface Studies," Surf. Sci., 53, 596-625 (1975).

30. D. Briggs, "Recent Advances in Secondary Ion Mass Spectrometry (SIMS) for Polymer Surface Analysis," Br. Polymer J., 21, 3-15 (1989).

31. K. Siegbahn et al., ESCA-Atomic, Molecular and Solid State Structures Studied by Means of Electron Spectroscopy, Almquist and Wiksells, Uppsala (1967).

32. D. M. Hercules, Anal. Chem., 44, 106R (1972).

33. D. M. Hercules and J. C. Carver, Anal. Chem., 46, 133R (1974).

34. J. A. Filbey and J. P. Wightman, "Factors Affecting the Durability of Ti-6Al-4V/Epoxy Bonds," J. Adhesion, 28, 1-22 (1989).

35. W. J. van Ooij, A. Kleinhesselink and S. R. Leyenaar, "Industrial Applications of XPS: Study of Polymer-to-Metal Adhesion Failure," Surface Science, 89, 165-173 (1979).

36. J. P. Wightman, "The Application of Surface Analysis to Polymer/Metal Adhesion," SAMPE Quarterly, 13, 1 (1981).

37. P. Commercon and J. P. Wightman, "The Application of Surface Analysis Techniques in the Adhesive Bonding of Oily Automotive Steel," J. Adhesion, 22, 13-21 (1987).

38. C. U. Ko and J. P. Wightman, "Characterization of Surface Pretreatments of Al/Li Alloy and Related Mechanical Properties of Polysulfone Adhesive Bonds," J. Adhesion, 24, 93-107 (1987).

39. P. Commercon and J. P. Wightman, "Surface Characterization of Plasma Treated Carbon Fibers and Adhesion to a Thermoplastic Polymer," J. Adhesion, 36 000 (1992).

40. D. J. D. Moyer and J. P. Wightman, "The Effect of Surface Pretreatment on Carbon Fiber-Polyimide Matrix Composite Bonding," Surface and Interface Analysis, 17, 457-464 (1991).

41. R. V. Siriwardane and J. P. Wightman, "Surface Characterization of Ti and Ti (6% Al - 4% V) Metal Powders and Interaction with Primer Solutions," J. Adhesion, 15, 225-239 (1983).

42. D. L. Messick, D. J. Progar and J. P. Wightman, "Surface Analysis of Graphite Reinforced Polyimide Composites," NASA Techn. Memo. 85700, NASA-LARC, Hampton, VA (1983).

43. S. Dias and J. P. Wightman, "The Application of Thermodynamic and Spectroscopic Techniques to Adhesion in the Polyimide/Ti 6-4 and Polyphenylquinoxaline/Ti 6-4 Systems," in Adhesive Chemistry, L. H. Lee (ed.), Plenum, New York, pp. 481-488 (1984).

44. T. A. DeVilbiss and J. P. Wightman, "Surface Characteristics of Carbon Fibers," in Composite Interfaces, H. Ishida and J. L. Koenig, eds., Elsevier, New York, pp. 307-316 (1986).

45. J. A. Filbey, J. P. Wightman and D. J. Progar, "Sodium Hydroxide Anodization of Ti-6Al-4V Adherends," J. Adhesion, 20, 283-291 (1987).

46. J. P. Wightman, "Surface Analysis Examines Fundamental Adhesion Questions," Adhesives Age, pp. 30-32, August, 1987.

47. T. A. DeVilbiss, D. J. Progar and J. P. Wightman, "SEM/XPS Analysis of Fractured Adhesively Bonded Graphite Fibre Surface Resin-rich/Graphite Fibre Composites," Composites, 19, 67-71 (1988).

48. C. U. Ko and J. P. Wightman, "Experimental Analysis of Moisture Intrusion into the Al/Li - Polysulfone Interface," J. Adhesion, 25, 23-29 (1988).

49. J. A. Filbey and J. P. Wightman, "Factors Affecting the Durability of Titanium/Epoxy Bonds," in Adhesion 12, K. W. Allen, ed., pp. 17-32, Elsevier, London (1988).

50. J. A. Filbey and J. P. Wightman, "Factors Affecting the Durability of Titanium/Epoxy Bonds," in Adhesion Science Review 1, H. F. Brinson, J. P. Wightman and T. C. Ward, eds., pp. 1-15, CAS, Blacksburg (1988).

51. J. P. Wightman and J. A. Skiles, "Analysis of Chromic Acid Anodized Ti-6Al-4V Adherends with High Temperature Structural Adhesives," in Proc. 33rd SAMPE Symposium, pp. 473-483, SAMPE, Azusa, CA (1988).

52. J. A. Skiles and J. P. Wightman, "Analysis of Chromic Acid Anodized Ti-6Al-4V Adherends with High Temperature Structural Adhesives, SAMPE J., 24, No. 4, 21-24 (1988).

53. J. A. Skiles and J. P. Wightman, "Heat Resistant Thermoplastic Chromic Acid Anodized Ti-6Al-4V Single Lap Bond Evaluation," Int. J. Adhesion & Adhesives, 8, No. 4, 201-206 (1988).

54. J. A. Skiles and J. P. Wightman, "The Influence of Ti-6Al-4V Chromic Acid Anodization Conditions Upon Oxide Thickness and Topography," J. Adhesion, 26, 301-314 (1988).

55. J. A. Skiles and J. P. Wightman, "Influence of Chromic Acid Anodization Conditions Upon Anodized Ti-6Al-4V/Polysulfone Single Lap Bond Strengths," SAMPE J., 25, No. 1, 41-45 (1989).

56. C. U. Ko, E. Balcells, T. C. Ward and J. P. Wightman, "Effect of Surface Topography on the Relaxation Behavior of Thin Polysulfone Coatings on Pretreated Aluminum Substrates," J. Adhesion, 28, 247-260 (1989).

57. D. J. D. Moyer and J. P. Wightman, "Surface Characterization and High Temperature Adhesive Bonding of Carbon Fiber-Polyimide Matrix Composites," in Proc. Am. Soc. Composites - 4th Techn. Conf., pp. 367-376, Technomic Publishing Co., Lancaster, PA (1989).

58. L. L. Smith, J. G. Dillard, L. S. Horning, J. D. Rancourt, L. T. Taylor and J. P. Wightman, "Cobalt Doping of a Polyimide Adhesive," Int. J. Adhesion and Adhesives, 11, 80-86 (1991).

59. B. Menon, R. A. Pike and J. P. Wightman, "Metal Alkoxide Primers in the Adhesive Bonding of Mild Steel," J. Adhesion Sci. Technol., 5, 883-893 (1991).

60. J. W. Chin and J. P. Wightman, "Adhesion to Plasma-Modified LaRC-TPI I. Surface Characterization," J. Adhesion, 36, 25-37 (1991).

61. J. B. Hollenhead and J. P. Wightman, "The Adhesive Bonding of Steel with Polysulfone," J. Adhesion, 37, 121-130 (1992).

62. J. W. Chin and J. P. Wightman, "Surface Characterization and Adhesion of Oxygen Plasma-Modified LaRC-TPI," SAMPE Q., 23(2), 2-10 (1992).

11. Acknowledgements

The author thanks Ms. Dorothy E. Tastet and Dr. Jennifer A. Filbey for their help in researching the literature for this paper. A number of the results contained in this paper have been taken from the papers and graduate theses/dissertations of the following persons whose research accomplishments are gratefully acknowledged: Ms. Diane Allen, Mr. Therman Bush, Mr. Wen Chen, Ms. Joannie Chin, Dr. Pascal Commercon, Dr. Mary Ellen Counts, Ms. Surani Dias, Dr. David W. Dwight, Dr. Jennifer A. Filbey-Arney, Mr. James Hollenhead, Mr. Hugh Honeycutt, Dr. James S. Jen, Dr. Chan U. Ko, Mr. Tim Lin, Ms. Beena Menon, Mr. Donald Messick, Dr. Denise J. D. Moyer, Ms. Kelly A. Sanderson, Dr. Ranjani Siriwardane, Dr. Jean Ann Skiles, Ms. Laura Smith, Ms. Ai-Ling Tsou, and Dr. H. Francis Webster. Our surface analysis results as related to adhesion would not have been obtained without the financial support of the following sponsors: The Adhesive and Sealant Council, Inc., Alcoa, Elf Aquitaine, Ford Motor, Johnson Wax, NASA-Langley Research Center, National Science Foundation, Office of Naval Research, Phillips Petroleum, and the Virginia Institute for Material Systems.

SOME EXPERIMENTAL METHODS OF CHARACTERIZING SURFACES

E.BAYRAMLI
Department of Chemistry, Middle East Technical University
06531 Ankara, TURKEY

ABSTRACT. The methods of surface characterization can be divided into two broad classes as being spectroscopical and thermodynamical. Among the first group all types of analytical techniques of surface analysis, which aims at functional group and elemental analysis at various depth profiles and lateral analysis areas, can be included. The thermodynamical methods are mostly based on titration, adsorption,calorimetry and contact angle determination. The contact angle methods are unique in the sense that they give information on the work of adhesion which is directly relevant to the composite physical properties. The London-dispersion or van der Waals(LW) interactions at an interface can be estimated from the contact angle data which employ probe liquids with only LW interactions. For polymeric composites the other interaction term is the Lewis acid-base type component specific interactions which are more difficult to determine experimentally. The experimentally determined contact angle itself is an ill-defined parameter that requires careful measurements and fulfill a set of conditions to be meaningful from a thermodynamical point of view. Within these limitations ,wetting measurements are used to augment the spectroscopical techniques to characterize surfaces.

1. INTRODUCTION

Understanding the interfacial characteristics of the polymer composites is a requirement for optimizing the filler or the fiber to resin compatibility. A good adhesion between the fiber or the filler and the structural polymer will exploit the full potential of the load transfer property of the reinforcing material. The degree of adhesion is determined by the surface chemical affinity of the components in the composite which can be a chemical bond or secondary attractions such as van der Waals type or hydrogen bonding. For the continous fiber composites , in particular, poor adhesion may be due to insufficient physical contact between the resin and the fiber as a result of incomplete wetting . In the secondary type attractions and the impregnation process an important experimental parameter is the contact angle. The equilibrium contact angle is defined by the well known Young-Dupre relationship for a homogeneous and smooth surface. In practice very few solid surfaces meet the above criteria. Nonetheless, the contact angle data is widely used to predict the composite properties basicly because it can provide direct information on the work of adhesion with relatively simple instrumentation.

In the discussion which follows a critical review of the contact angle data will be made emphasizing the hysteresis effects and the measurements on low and high energy solid surfaces. The use of

151

G. Akovali (ed.), The Interfacial Interactions in Polymeric Composites, 151–168.
© 1993 *Kluwer Academic Publishers.*

contact angles to obtain surface free energy and interaction terms will be summarized. The theory proposed by Pimentel[1] and later extended to surface interactions by Fowkes[2,3] will be applied to polymer composite surfaces and interfaces. Finally, an interpretation of surface functional groups data obtained by Electron Spectroscopy for Chemical Analysis(ESCA) will be correlated with the surface energy values calculated from contact angle measurements.

2. THE NATURE OF CONTACT ANGLE

The contact angle formed between three phases at the three phase contact line(TPL) is defined by the Young-Dupre equation as a function of surface and interfacial energies of the respective phases when one of the phases is a solid. The assumptions made in the derivation is that the solid surface is nondeformable and homogeneous in the chemical and the physical sense(smooth)[4], furthermore, there should be no formation of a film on the solid surface extending from the TPL onto the solid surface. The definition of solid surface heterogeneity has not yet been solved satisfactorily. The questions about the scales of physical roughness and chemical heterogeneity that are significant in creating a hysteresis effect in the measured contact angles may be considered as unsolved problems[5]. From an experimental point of view ,a lack of hysteresis between the so called the advancing and the receding contact angles can most probably be accepted as an indicator of a true equilibrium contact angle. In practice very few solid surfaces exhibit no hysteresis in carefully made measurements. When the TPL is made to move at velocities less than 1μm per second what in most cases observed is that the contact line pivots around a fixed location until a critical value is reached in both the advancing and the receding mode[6]. The two limiting values may relax slightly to reduce hysteresis when the TPL motion is stopped but, a unique contact angle value in most cases is never reached.

An investigation of the contact angle phenomenon may best be done by separate studies on low and high energy surfaces. Most of the polymeric surfaces can be characterized as low energy and the filler surfaces as high energy except the carbon fiber surface. There are specific problems associated to these cases which is discussed in the following sections together with some numerical simulations of advancing and receding TPL motion on model rough surfaces. The theoretical experiments described below are done in a way that simulates the tensiometric experiments performed on low and high energy surfaces. The similarities in the form of hysteresis observed and the jumps of the contact line between equilibrium positions exist in both cases.

2.1. Simulated Wetting Experiments

Huh and Scriven[7] have studied liquid meniscus shapes around vertical cylinderical rods by numerically solving the Young-Laplace equation with specific boundary conditions, presenting the solutions in extensive numerical tables. When the cylinder radius is small the approximate solution of James[8] can be used to calculate the meniscus shape without significant error. We have calculated for various axisymmetric rod type geometries the variation of the contact angle as the TPL is allowed to travel both ways on these surfaces[9]. For sinusoidal circumferential grooves on a cylinder a large hysteresis is observed with non-equilibrium jumps between the stable meniscus configurations during advancing and the receding (fig.1).

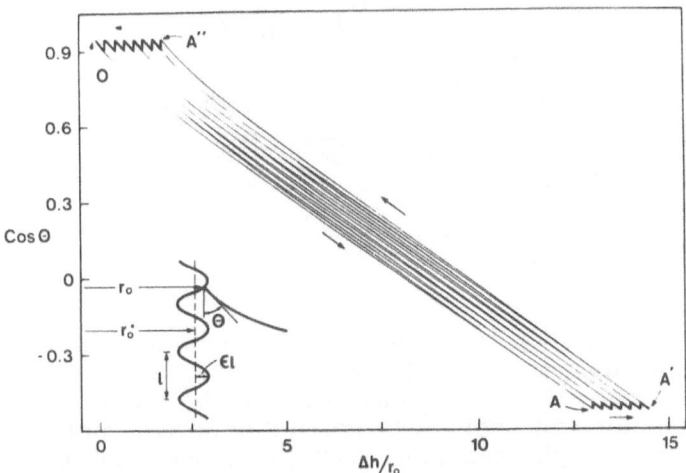

Figure 1. Cosine of the apparent contact angle, Ø, vs. vertical displacement of the rod ,Δh, for sinusoidal circumferential grooves on a cylinder with average radius of 0.025 cm and E=0.2.; starting with O a hysteresis loop is obtained by vertically immersing the cylinder into the liquid phase from O to A' and withdrawing from A' to O. The intrinsic Ø is 70°[9].

Randomly rough surfaces without axial symmetry are difficult to treat mathematically, not only because the capillarity equations become very complicated to solve[10], but also because the exact location ot the contact line is not known a priori . We can investigate the more tractable problem of wetting of a surface with one dimensional roughness. A unidirectional randomness on a cylinder surface is generated by connecting the randomly generated points with polynomials. A sample of such a surface is given in Figure 2.

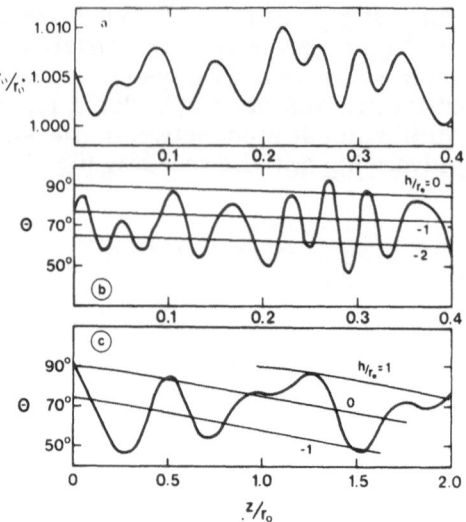

Figure 2. (a) A section of the numerically generated surface profile. (b) The possible apparent contact angles($\emptyset=70°$) for system in a and in (c) for a different surface. The lines indicate the meniscus height at a constant immersion depth with varying apparent contact angle; the intersections are the available equilibrium positions for the contact line[9].

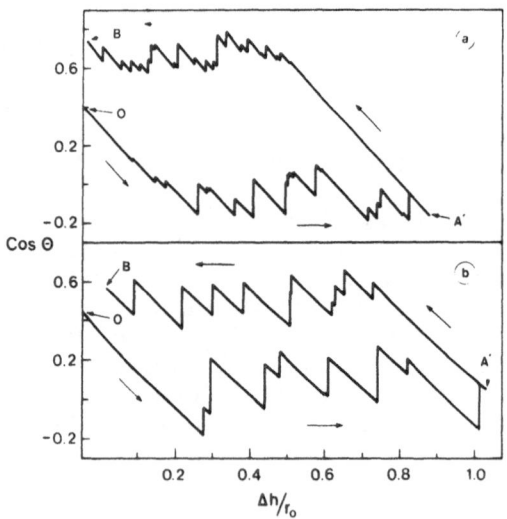

Figure 3. Hysteresis loops obtained on randomly rough surfaces . (a) smaller roughness scale, (b) larger roughness scale.

Using Fig. 2 one can construct theoretical wetting hysteresis loops by gradually increasing and decreasing the immersion from depth, say, h=0. Figure 3a and 3b are for smaller and larger roughness scales respectively. Conceptually it is easier to form Fig. 3 using Fig. 2, but in numerical calculations it is much more practical to calculate the equilibrium meniscus heights at regular intervals and search for the locations of non-equilibrium jumps to the nearest equilibrium positions as h continously increases or decreases. At smaller roughness dimensions we obtain more frequent TPL jumps together with large non-equilibrium jumps. Also the contact angle hysteresis is larger at smaller dimensions. The number of rugosities skipped over during jumps increases with decreasing roughness dimensions. It may be speculated that as the dimensions of roughness are progressively reduced, one will eventually observe a smooth motion of the contact line where, at a microscopic level, the contact line advances or recedes with very small jumps from one equilibrium site to the next.

The mean advancing contact angle and the receding contact angle are shown in Fig. 4 as a function of the roughness size of the solid surface when the standard deviation of the slope distribution is equal to 11.65. The bars indicate the standard deviation from the mean apparent contact angle. For smaller roughness dimensions only the extreme surface slope values contribute to the hysteresis; at larger roughness dimensions the hysteresis asymptotically approaches a constant value.

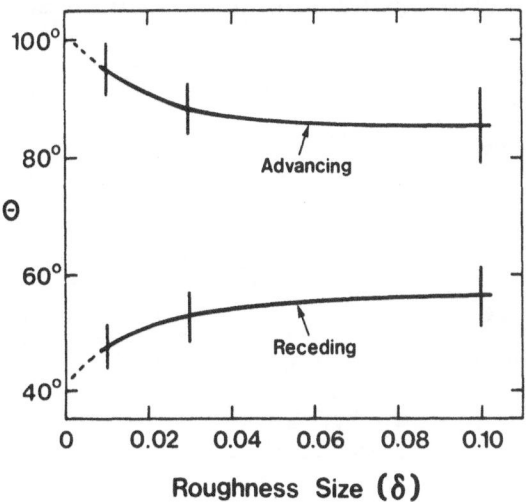

Figure 4. Effect of the roughness size on apparent contact angle for Ø=70° for a typical randomly rough surface investigated[9].

The question of the minimum roughness dimension which can be considered as a cause of wetting hysteresis seems to be a basic one. It has been claimed that surface roughnesses below 0.1μm does not cause hysteresis[11], whereas the work of Oliver et al.[12] indicates that 0.05 μm steps on a mica surface inhibit spreading. It seems that the nature of the roughness (surface texture)[13] rather then its absolute size is a determining factor as well[14]. Also the formation of composite surfaces during wetting due to constraints arising from the solid surface geometry is a cause of irreversibility during wetting. The irreversible stick-jump phenomena which is experimentally observed in certain systems makes treatments of wetting based on equilibrium thermodynamics somewhat suspect. It should also be mentioned that from a mathematical point of view there is no difference between a chemically and a physically rough surface. The analysis regarding the apparent and the intrinsic contact angles is the same.

2. 2. Low Energy Surfaces

The surface energies of 25 to about 100 mN/m can be considered as low energy. The generally accepted assumption is that in the absence of a film pressure on these surfaces , the measured solid surface energy is representative of the surface itself. The examples cited here are for a polar polymer Nylon-6 water, nonpolar polytetrafluoroethane (Teflon) decane and dimethyldichlorosilane(DDS) coated glass water systems[6]. The fibers are suspended from one of the arms of an electrobalance and they are immersed into the liquid by means of a hydraulicly driven stage capable of very slow reproducible velocities. The capillary force around the fiber is measured by the balance and if the surface tension of the liquid is known the apparent contact angle can be calculated easily. In Fig. 5 The buoyancy corrected tensiograms of the above systems is given. All systems exhibit significant amount of hysteresis which do not relax to an equilibrium values when the stage is stopped.

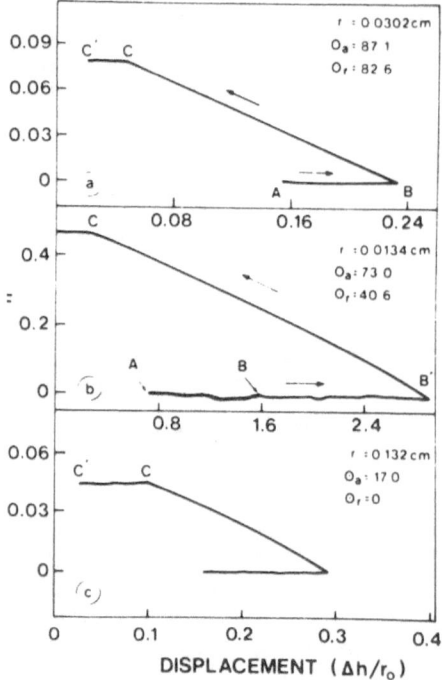

Figure 5.   Tensiograms of low energy surfaces. (a) DDS coated glass with stage velocity 0.2µm/s where F is the nondimensional buoyancy corrected capillary force. (b) Nylon-6 water; stage velocity 0.017 µm/s from A to B and 0.2 µm/s from B to B' ; note the more detailed trace at lower TPL velocity. (c) Teflon decane with 0.2 µm/s stage velocity. When

the stage velocity is reversed the meniscus pivots at a fixed position until the receding contact angle value at C is reached and then starts to move[6].

When the contact line is moved at a higher velocity than the experiments given in Fig. 5 a relaxation effect is also observed. In Figure 6 for the DDS coated glass water system at 0.34 µm/s velocity a relaxation of the contact angle is observed when the stage is stopped[15]. Relating this phenomenon to the numerical studies one can speculate that at higher velocities the liquid can only reside at extreme physical or chemical rugosities resulting in a higher hysteresis value.

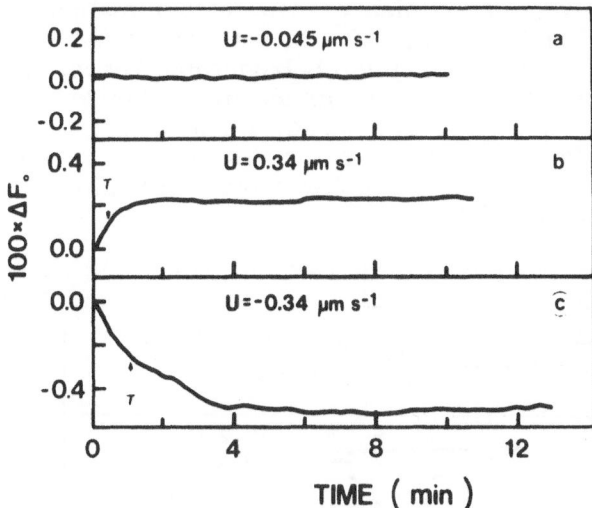

Figure 6. The relaxation of the tension relative to the total capillary force, ΔF, in time : (a) for a receding TPL at 0.045 µm/s with no relaxation ; (b) for an advancing TPL ; (c) for a receding TPL. The relaxation half time is denoted by T ,and the stage velocity by u.

For more viscous liquids such as epoxy resins and polymer melts the contact angle may never relax to equilibrium advancing and receding values in practical time scales and the absolute value of the dynamic contact angle is most probably much larger than observed for water.

2. 3. High Energy Surfaces

An observation which makes wetting of high energy surfaces controversial is that the reported contact angles of water on most of them are much larger than zero. Fox et al.[16] attributed this to the adsorption of liquid molecules with polar and non-polar end groups which form monomolecular films on the solid surface. Contact angles on these films are larger than zero. An alternative explanation is provided by Shrader's[17,18] experimental work in which complete wetting of gold, silver and copper surfaces is only possible in ultra high vacuum, because it was claimed that on these surfaces it is impossible to avoid surface contamination of some kind or other in experiments performed at standard temperature and pressure. The unavoidable presence of minute amounts of surface active impurities(SAI) alters the nature of the surface.

The surfaces investigated here is a relatively low high energy surface quartz with 200-260 mN/m and platinum with 1700-1800 mN/m that has an extremely high surface energy. The wetting loop of a freshly cleaned quartz fiber with water as the wetting liquid showed complete reversibility with zero contact angle at 1 μm stage velocity(Fig. 7a). The slope of the trace is due to the buoyancy force. When the fiber-liquid system is allowed to rest several hours before forming a meniscus , we observe a hysteresis which eventually diminished during the advance of the TPL (Fig. 7b). After a waiting period of 12 h the hysteresis loop showed a stick-jump behavior(Fig. 7c) with an average jump length of about 75μm. The contact angle change during these jumps was, on the average 12° for the advancing line. The magnitude of these jumps tended to decrease as the TPL advanced. Apparently a quartz surface has no intrinsic structure which could give rise to jumps over such large distances[6].

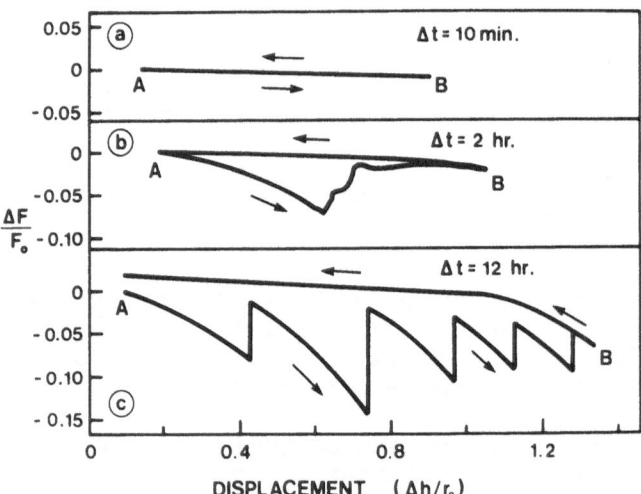

Figure 7. Tensiogram of a quartz fiber with water where Δt denotes the time elapsed before the formation of the meniscus and the start of the TPL motion. In all cases the stage velocity is 0.1 μm/s and the direction of the TPL motion is reversed at B.

The effect of the liquid surface contamination on the observed tensiogram is demonstrated by touching the water surface by the tip of a steel needle upon which the system which showed no hysteresis before now exhibited a hysteresis and stick-jump behavior which diminished as the contact line advanced. The results of a platinum-water system presented below indicate a mechanism similar to that operating in the wetting of quarts surfaces. The regular stick-jump in Fig. 7c is found to be absent when the stage velocity is reduced to 0.045 μm/s, similar to the case of platinum fibers (see below).

The tensiogram of a platinum fiber-water system is shown in Fig. 8. The advancing TPL at 0.3 μm/s exhibited stick-jump phenomena similar to those observed for numerical experiments for circumferential axisymmetric regular grooves. The TPL advanced from A to B with an average 67° contact angle. The contact line hinged at B, when the motion of the instrument stage was reversed, until the receding contact angle value (29.4°) was reached at C and stayed constant from C to C'. The inset in Fig.8 shows that the TPL motion was velocity dependent. When the velocity was reduced from 0.3 to 0.045 μm/s the stick-jump behavior was replaced by a smoother TPL motion. This effect was completely reversible ; increasing the velocity back to 0.3 μm/s re-established the regular stick-jump behavior[6].

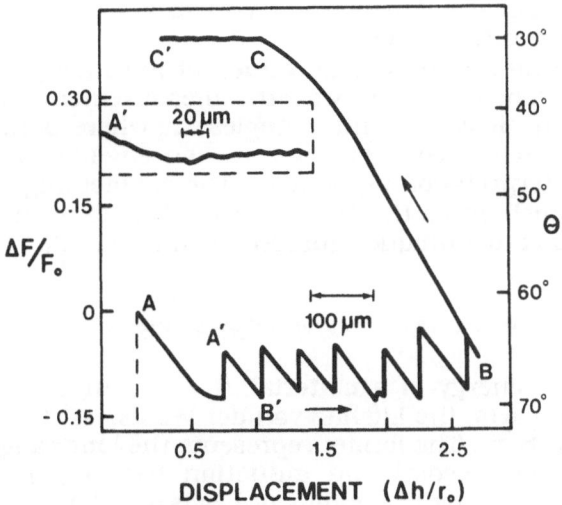

DISPLACEMENT ($\Delta h/r_o$)

Figure 8. Tensiogram of a platinum fiber with radius 0.220 mm in water, showing the stick-jump behavior while the stage was advancing with a velocity of 0.3 μm/s. The effect of velocity on the advancing TPL is shown in the inset which shows that at 0.045 μm/s stage velocity no regular jumps occured.

The physical phenomena occuring in Figs 7(c) and 8 can be explained by assuming the formation of a very thin ring of low surface energy material on the solid near the TPL which can act as a barrier to the advancing TPL. This is followed by an increase in the observed contact angle until the occurence of the next jump. Because the fiber diameter is small and the TPL region extremely thin , even trace amounts of contaminants will suffice for the formation of such a ring. The recent work of Cazabat et al.[19] observe the similar phenomena on high energy surfaces. In their review article Mysels and Florence[20] convincingly discuss the sources of impurities on water surfaces and attribute many "often-observed" phenomena to contamination. Very small amounts of SAI which will have no measurable effect on the liquid surface tension may significantly alter the wetting of high energy surfaces.

One can actually calculate a diffusion coefficient for the diffusion of SAI on the platinum surface from the value of the critical velocity above which the irreversible jumps take place. From the analysis of Fig. 8 we find that the TPL moves with a velocity of 0.09μm/s in between the jumps which should be very close to the critical velocity. Forcing the TPL to move faster than the critical velocity causes the observed behavior. Immediately after a jump the TPL can move over the surface with an advancing contact angle characteristic of a clean fiber-water interface. However, almost instantaneously a ring a SAI is formed and as a result the TPL advances over the solid surface with a velocity close to the critical velocity while the meniscus pivots till the characteristic contact angle is reached. We would expect an increase in the displacement between consecutive jumps as the stage speed approaches the critical velocity. Indeed, when it is decreased to 0.15 μm/s an increase in the displacement is observed without a change in the the the TPL velocity between the jumps.

It is important to note that values of advancing and receding contact angles depend on the way the measurements are done. In principle one can measure contact angles anywhere between the two limiting values. In the contact angle measurements utmost care is needed in the interpretation of the data. The contact angle depends on how and with which speed the TPL is moved. Due to hysteresis contact angles do not relax to a unique equilibrium value[21,22].

3. INTERACTIONS IN COMPOSITE INTERFACES

The total surface energy of a material, γ_i , can be considered as composed of two parts, the Lifshitz-van der Waals, γ_i^{LW} ,and the acid-base component γ_i^{AB} . The former represents the long range dispersion forces, orientation(Keesom) and induction (Debye), and the latter represents the short range H-bonding or Lewis acid-base interactions. It is given as the sum of of two components ,say, for a solid S and a liquid L

$$\gamma_S = \gamma_S^{LW} + \gamma_S^{AB} \tag{1a}$$

$$\gamma_L = \gamma_L^{LW} + \gamma_L^{AB} \tag{1b}$$

where the acid-base term is,

$$\gamma_S^{AB} = 2 (\gamma_S^+ \gamma_S^-)^{1/2} \tag{2}$$

The superscripts plus and minus denote the surface energy contributions due to Lewis acid and Lewis base respectively. Similarly for the liquid phase, we have

$$\gamma_L^{AB} = 2(\gamma_L^+ \gamma_L^-)^{1/2} \tag{3}$$

The adhesion between a solid and a liquid is given as

$$(1 + \cos \Theta) \gamma_L = 2 [(\gamma_S^{LW} \gamma_L^{LW})^{1/2}] + \gamma_{SL}^{AB} \tag{4}$$

$$\gamma_{SL}^{AB} = 2 [(\gamma_S^+ \gamma_L^-)^{1/2} + (\gamma_S^- \gamma_L^+)^{1/2}] \tag{5}$$

where γ_{SL}^{AB} represents the short range Lewis acid-base interactions denotes the contact angle of the liquid on the solid surface. In principle one needs to know γ_S^+, γ_S^-, γ_L^+ and γ_L^- to calculate the total adhesion theoretically provided γ_S^{LW} is determined by a non-polar liquid such as methylene iodide. Experimental determination of the total amount of adhesion is possible from Eq. 5 together with the acid-base contribution for a given liquid solid system. The experimental difficulty is to measure a meaningful contact angle when a polymeric liquid resin is employed. Ideally one would like to characterize solid resin(polymer) and the filler(fiber) surface energies in terms of LW , acid and base components with suitable probe liquids and calculate the work of adhesion for any given resin-solid system. In practice , γ_L^+ and γ_L^- of the probe liquid are not known. However, the relative polarity of the probe liquid i, with respect to water , δ_{iw} , can be measured . For liquid i, the relative polarity of the Lewis acid, δ_{iw}^+ , and lewis base, δ_{iw}^- , component is defined as

$$\delta_{iw}^+ = \left[\frac{\gamma_i^+}{\gamma_w^+} \right]^{1/2} \quad \text{and} \quad \delta_{iw}^- = \left[\frac{\gamma_i^-}{\gamma_w^-} \right]^{1/2} \tag{6}$$

From the above definitions we can obtain

$$\delta_{iw}^+ \, \delta_{iw}^- = \frac{2(\gamma_i^+ \gamma_i^-)^{1/2}}{\gamma_{iw}^{AB}} \tag{7}$$

where $\gamma_w^{AB} = 2(\gamma_w^+ \gamma_w^-)^{1/2}$ is the Lewis acid-base component of the water surface tension. Rearranging the above equation it is possible to show that

$$\frac{\delta_{iw}^+}{\delta_{iw}^-} = \left(\frac{\gamma_i^+}{\gamma_i^-} \right)^{1/2} \left(\frac{\gamma_w^-}{\gamma_w^+} \right)^{1/2} \tag{8}$$

Since ($\gamma_w^-/\gamma_w^+)^{1/2}$ is a constant, relative surface acidity or basicity of two liquids can be compared if their γ_i^{AB} are almost identical.

Applying the data from reference 23 to ethylene glycol and formamide it can be shown that $\delta^+_{iw}/\delta^-_{iw}$ for these two liquids are 0.316 and 0.240 respectively. From this comparison it is clear that ethylene glycol is more acidic than formamide . Both liquids have predominantly basic characteristcs. Based upon these considerations ethylene glycol can be used as the base probe and formamide as the acid probe. Since total acid-base components of these liquids are approximately equal, 19.0 mN/m and 18.6 mN/m , respectively, it is possible to analyze the solid surface energies by measuring the contact angles with the forementioned probe liquids.

3.1. Carbon fiber Composites Surface Energies

Contact angles on single carbon fibers are measured tensiometrically using the set-up given in Fig.9 . In some runs a hysteresis of about 5° is noticed where the advancing angle is taken to be the the thermodynamically significant angle.

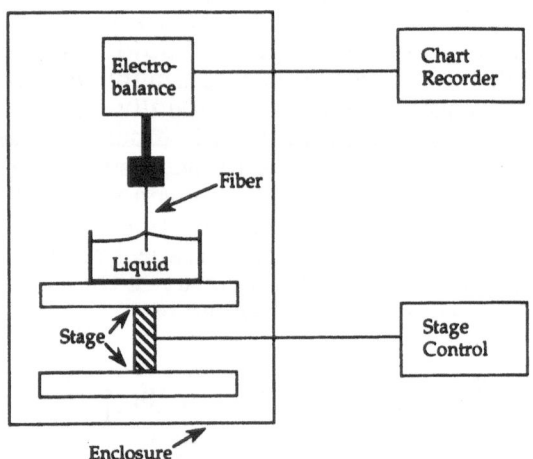

Figure 9. Experimental set-up for contact angle measurement.

A sample run is shown in Fig. 10 where the formation and the rupture of meniscus around the carbon fiber is indicated. The stage motion is stopped at A after advancing and at B the stage is lowered for a receding contact line. In this case there is no hysteresis. The increase in force before rupture is the end effect where the liquid is suspended at the edge and the force goes through a maximum.

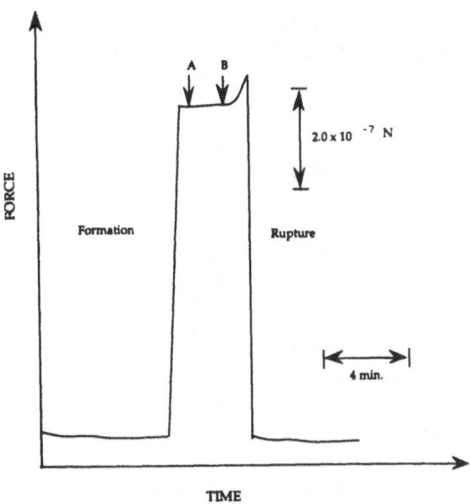

Figure 10. A typical tensiogram of IM7 carbon fiber with formamide as the wetting liquid(fiber perimeter= 23.0 µm).

The results of surface free energy measurements for four types of carbon fibers are summarized in Table.1[24]. Although the δ_L^{AB} for ethylene glycol and formamide are almost equal, their interaction with the surface produces very different δ_{SL}^{AB} values which can be explained by a difference in their electron donating capabilities.The highest average value of δ_{SL}^{AB} is obtained for IM7, therefore its surface characteristic in terms of acid-base reactivity is the highest with a slight acidic tendency. The IM7 fiber can best be characterized as amphotheric whereas the T650-42 fiber is clearly a basic fiber and G40-800 fiber shows acidic surface characteristics. T40 fiber shows the least amount of short range , H-bonding interactions and it is ranked lowest in terms of total functionality among fibers investigated.

TABLE 1. Components of the surface energies for carbon fibers, LW of the solid, ethylene glycol/ carbon fiber   and formamide/ carbon fiber (all in mN/m) acid-base interaction terms.

Fiber	δ_S^{LW}	δ_{SL}^{AB}(EG)	δ_{SL}^{AB}(F)	Rank
IM7	31.4	30.1	32.5	1 (amphoteric)
T650-42	30.8	31.7	27.2	2 (basic)
G40-800	31.5	24.1	34.8	3 (acidic)
T40	33.2	23.8	23.2	4 (amphotheric)

3 .2. ESCA Results on Carbon Fibers

After establishing the elemental compositions of the carbon fiber surfaces by ESCA, wet chemical derivatizations were carried out to determine the number of -COOH, -COH and -C=O functional groups on the very outer fiber surface[24]. It is posible to use pentafluorophenylhydrazine for detecting carbonyl ,chloro silanes for alcohol and carboxylic acid groups by pentafluorobromotoluene(fig. 11).

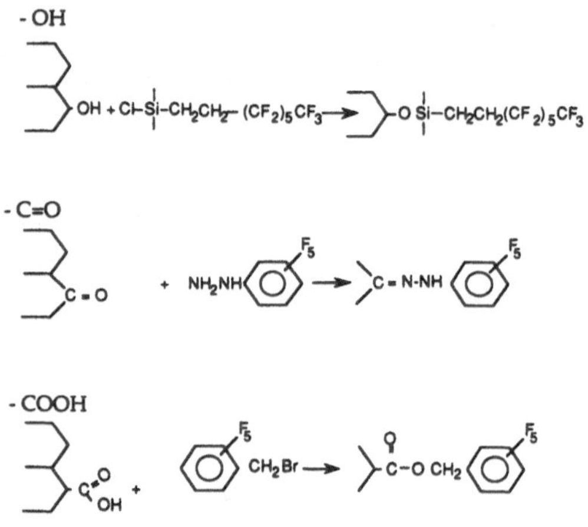

Figure 11. Selective derivatization of various functional groups on the carbon fiber surface.

Forming the derivative groups result in good spectral resolution due to signal enhancement from the incorporation of several fluorine atoms. In addition quantitive information on the functionality of these oxygen containing groups can be obtained , that is, the functionality per hundred surface atoms. Knowing the specific functionality on the fiber surface, it is possible to correlate ESCA results with those obtained from contact angle measurements.

The elemental composition of the approximately top 100 A° of the tested fibers is given in Table 2.

TABLE 2 . The elemental composition of the underivatized carbon fiber surfaces(top 100 A° in percentages)

Fiber	C	O	N	F	Na	Si	S
IM7	81	14	5.2	-	0.7	-	-
T650-42	85	11	2.0	-	-	0.2	0.2
G40-800	78	15	7.2	-	-	-	-
T40	95	2.7	1.9	-	-	0.7	-

What is striking in Table 2 is the extremely carbonized structure of T40. The functional group analysis of the derivatized carbon fiber surfaces is given in Table 3.

TABLE 3 . ESCA data using selective derivatization   (values are in number of functional/ 100 surface atoms)

Fiber	OH	C=O	COOH	Sum	Rank
IM7	0.7	1.1	1.8	3.6	1
T650-42	0.5	1.4	1.6	3.5	2
G40-800	0.7	1.3	1.2	3.2	3
T40	0.2	1.8	0.6	2.6	4

The combination of the ESCA results based on surface derivatives with the results obtained from contact angles give potentially meaningful information. For example, by examining the oxygen atomic percent of the IM7 and G40-800 fibers' surfaces which have high levels of oxygen may be acidic. The values in Table 1 clearly shows that G40-800 is clearly acidic whereas IM7 is amphotheric, perhaps slighly acidic. We find that the sum of the functional groups from ESCA is the highest for IM7 which correlates well with the surface energy data. If one to compare the data on T40 which is the most graphitic of the tested four fibers ,a balanced acid-base character is revealed by the surface energy data with the lowest values. T40 has the lowest number of functional groups on its surface. For the fibers T650-52 and G40-800 it is not possible to assess the acidity or basicity of their surfaces, but, from the surface energy data we can deduce that T650-42 is clearly acidic and G40-800 has a basic surface characteristic.

3 . 3 . Surface Energy of Bismaleimide-Polyimide Polymer Blend

Bismaleimide(BMI) polyimide(PI) blend exhibits exceptionally good thermal and mechanical properties, but, as in most imides wetting and a good adhesion to the carbon fiber as well as good mixing is not as easy as in the high temperature epoxy systems. A knowledge of the acidic , basic nature of the structural polymers is essential to find a suitable carbon fiber and to establish the correct blending ratio without comprimising the adhesive properties. Tensiometric tests on these surfaces are carried out on polymer coated glass Wilhelmy plates[25]. As discussed for carbon fiber surfaces methylene iodide, formamide and ethylene glycol is used to characterize surface energies at different BMI to PI ratios(Table 4). We obtain a larger acid-base interaction with foramide which signifies that the blend surface is acidic. The amount of interaction increases with the PI percentage. In terms of crosslinking density high PI will be detremental to the overall performance of the material. Measurements over 30% is not reported because of the swelling tendency of the resin with the probe liquid.

TABLE 4 . The Lifshitz-van der Waals(LW) and acid-base components of BMI-PI resin interaction with methylene iodide, formamide and ethylene glycol (in mN/m)

Composition (% BMI)			
100	18.4	16.0	30.7
90	23.0	16.0	33.3
80	31.1	14.4	31.3
70	31.4	17.6	35.4
60	36.4	-	36.0
0	36.6	-	35.0

An interesting minimum is observed around 20% PI value for both probe liquids in Table 4. It suggests that it is not an experimental artifact,since, separate coated plates are used in each case. Certain morphological changes or phase separations at the surface region can give rise to a chemically heterogeneous surface which will increase the recorded contact angle to give a low adhesion tension.

4. CONCLUDING REMARKS

In this article we have discussed the problems associated with the experimentally measured contact angle. The observations are the presence of a significant contact angle hysteresis with most liquid-solid systems, the relaxation of contact angle to lower values for advancing

TPL,and, to higher values for receding TPL and the presence of multitude of metastable contact angles inbetween the two limiting values. The conclusion is that the formation of the meniscus for a contact angle measurement is as important as the recorded or calculated value of the contact angle in a given experiment. A cited value of a contact angle should also mention how the measurement is carried out.

The use of Fowkes' method of acid-base interactions for problems of adhesion at an interface gives valueable information about the nature of short range attractions in a composite system. The notion that two adjoining surfaces require reciprocal Lewis acid-base components for effective interaction makes the choice of components more of a calculated guess. A surface energy characterization done as summarized in this study also paves the way for a better surface modification routine to obtain high performance polymeric composites. The use of spectroscopical surface analysis methods in conjugtion with surface energy methods can provide a better understanding of the composite properties.

5 . REFERENCES

1 . Pimentel, G.C. and McClellan , A.L. (1960), The Hydrogen Bond, Freeman Press, San Francisco .

2 . Fowkes, F. M. (1983), Physico-Chemical Aspects of Polymer Surfaces , in K.L. Mittal (ed.), Plenum Press, New York, Vol.2, p.583.

3 . Fowkes, F. M. (1987), 'Bonding and adhesion in Polymer Interfaces', J. Adhesion Sci. Techn. 1, 7.

4 . Gibbs, J. W. (1926), ' The Collected Works of J. W. Gibbs, Vol. 1, Thermodynamics', Yale University Press, p. 314.

5 . Israelachvili, J. N. and Gee, M. L. (1989), 'Contact Angles on Chemically Hetergeneous Surfaces', Langmuir 5, p. 288.

6 . Bayramlı, E., van de Ven, T.G.M. and Mason, S.G.(1981) ,'Tensiometric Studies on Wetting lll. Low and High Energy Surfaces' , J. Colloid Interface Sci. 3, p. 131.

7 . Huh, C. and Scriven, L.E.(1969) 'Shapes of axisymmetric Fluid Interfaces of Unbounded Extend', J. colloid Int. Sci. 30, p.323.

8 . James, D. H.(1974),'The Meniscus on the Outside of a Small Cylinder', J. of Fluid Mech. 63, p.657.

9.Bayramlı,E., van de Ven T.G.M. and Mason,S.G.(1981), 'Tensiometric Studies on Wetting I. Some Effects of Surface Roughness (theoretical), Can. J. of Chem. 59, p.1954.

10. Neumann, A.W., Renzow, D. and Richter, I.E. (1971),'Temperature Dependence of the Wetting of n-hexatriacontane by Water', Fortschrittsber, Kolloide Polym. 55, p.49.

11. Dettre, R. H. and Johnson, R.E.(1977) 'Dynamic Contact Angles and Contact Angle Hysteresis' J. Colloid Int. Sci. 62, p.205.

12. Oliver, J.F., Huh, C. and Mason, S.G.(1977), 'Resistance to Spreading by Sharp Edges', J. Colloid Int. Sci. 59, p.568.

13 . Huh, C. and Mason, S.G.(1977),'Effects of Surface Roughness on Wetting "Theoretical" ', J. Colloid Int. Sci. 60, p.11.

14 . Mason, S.G. (1978), 'Wetting, Spreading and Adhesion in J.F. Padday(ed.)', Academic Press, New York, p.348.

15 . Bayramlı, E., van de Ven, T.G.M. and Mason, S.G. (1981), 'Tensiometric Studies on Wetting IV. Contact Angle and Surface Pressure Relaxation', Colloids and Surfaces 3, p.279.

16 . Fox, H.W., Hane, E.F. and Zisman(1934), W.A., J. Phys. Chem. 59, p.1097.

17 . Schrader, M.E. (1970),'Ultrahigh Vacuum Techniques in the Measurement of Contact Angles ll. Water on Gold', J. Phys. Chem. 74, p.2313.

18. Schrader, M.E. (1974), 'Ultrahigh Vacuum Techniques in the measurement of Contact Angles. lll. Water on Copper and Silver', J. Phys. Chem. 78, p. 87.

19 . Cohen, M.A.C. and Cazabat, A.M. (1987),' A Moving Contact Line: Further Studies of "Haines' Jumps', Progr. Colloid Polymer Sci. 74, p.64.

20 . Mysels, K.J. and Florence, A.T. (1970),' Clean Surfaces in G. goldfinger(ed.)' , Marcel Dekker, New York, p.219.

21 . Johnson, R.E. and Dettre, R.H. (1964),'Contact Angle Hysteresis (1)Idealized Rough Surface', J. Phys. Chem. 68, p.2744.

22 . Ibid. (1965),' Contact Angle Hysteresis (IV) Contact Angle Measurements on Heterogeneous Surface', J. Phys. Chem. 69, p.1507.

23. van Oss, C.J., Chaudhury, M.K. and Good, R.J. (1989),'Estimation of the Polar Parameters of the Surface Tension of Liquids by Contact Angle Meassurements on Gels', J. Colloid Int. Sci. 128, p.313.

24 . Chan, D., Hozbor,M.A., Bayramlı, E. and Powell, R.E.(1991),' Surface Characterization of Intermediate Modulus Graphite Fibers via Surface Energy Measurements and ESCA', Carbon 29(8), p.1091.

25 . Bayramlı, E., Accepted,'Surface Energies of Bismaleimide-Polyimide Blends', METU J. Pure and Applied Sci.

CONTROLLED INTERPHASES IN GLASS FIBER AND PARTICULATE REINFORCED POLYMERS: STRUCTURE OF SILANE COUPLING AGENTS IN SOLUTIONS AND ON SUBSTRATES

HATSUO ISHIDA
Department of Macromolecular Science
Case Western Reserve University
Cleveland, Ohio 44106-7202

ABSTRACT

The structure of silane coupling agents in solution and on solid substrate is reviewed with special emphasis on the fundamentals of structural development. Factors affecting the molecular weight, adsorption behavior, and chemical bond formation are discussed. Molecular aspects of the reinforcement mechanisms are discussed in relation to the interfacial bond formation. Effects of surface treatment on the rheological and hydrothermal properties of filled systems and composites are described.

G. Akovali (ed.), The Interfacial Interactions in Polymeric Composites, 169–199.

TABLE OF CONTENTS

1. Introduction to Composite Interfaces and Interphases

It is now well-recognized that the interface and interphase play an important role in the mechanical and physical properties of composite materials (1-9). The number of studies dealing with the characterization of interfaces and interphases and their influence on mechanical properties has increased dramatically over the past decade. Among many areas of composite interface studies, the fundamental understanding of the application of silane coupling agents has advanced substantially. This paper intends to describe the structure and role of silane coupling agents on glass fibers and particulate fillers. A special emphasis will be placed on the systematic description of the understanding and controlling of the structure of silanes.

The terms interface and interphase have been sometimes used interchangeably. However, an *interface* is a hypothetical molecular or atomic plane separating two dissimilar phases and is seldom associated with an actual material unless there is a bond passing through the separating plane and the structure differs from either of the phases. On the other hand, *interphase* is defined as an "interfacial region" whose properties are similar to, but distinguishably different from, the bulk properties. Thus, the interphase is associated with a finite thickness which depends upon the cause of the structural difference from the bulk. The thickness of the measured interphase also depends strongly on the characterization methods used despite studying the same interphase. Accordingly, one must be careful to mention the characterization methods whenever interphase thickness measurements are discussed. The definitions of the interface and interphase are schematically illustrated in Figure 1.

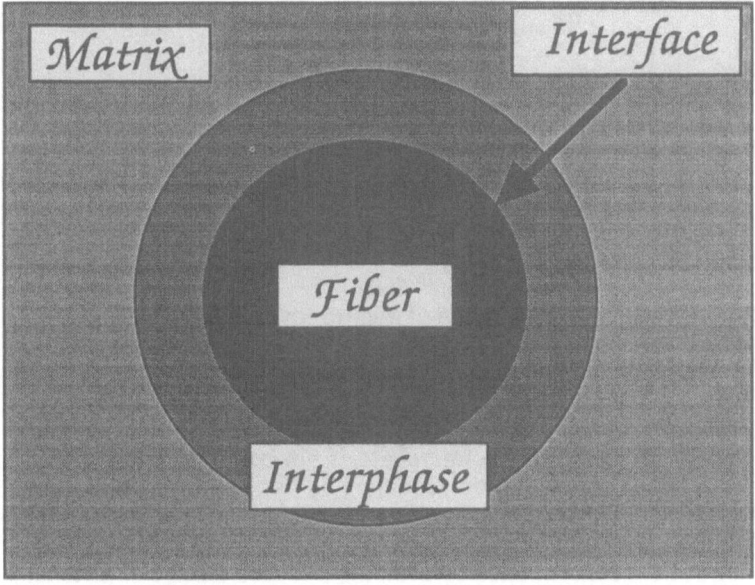

Figure 1. Conceptual drawing of interface and interphase in fiber reinforced composites.

The nature and thickness of the interphase are influenced by the substrate surface and its specific interaction with the matrix resin. Representative effects of the substrate surface are illustrated in Table 1.

Table 1. Effect of Substrate Surface on the Structure of Matrix Interphase

- **Chemical Effects**

  ◊ Reaction with matrix
  ◊ Catalytic activity
      - Acceleration of polymerization
      - Degradation of matrix
  ◊ Catalyst poisoning
  ◊ Selective reaction with a certain component

- **Physicochemical Effects**

  ◊ Preferential adsorption of a catalyst
      - Inhibition of cure
  ◊ Diffusion of a surface treating agent
      - Interpenetrating networks
  ◊ Topochemical reaction
      - Restriction of reactivity

- **Physical Effects**

  ◊ Nucleation of crystallites
      -transcrystallization
  ◊ Chain packing
      -epitaxial crystallization
  ◊ Restriction of mobility
      -molecular orientation
  ◊ Preferential adsorption of non-reactive component

An interphase may possess a sharp or graded boundary, examples of which are listed in Table 2. For a sharp boundary, recognizing the filler/matrix interface or matrix interphase/matrix bulk is relatively straightforward. For a graded boundary, determining the thickness of the interphase is not possible unless either an average quantity or relative concentration reduction is defined to be the boundary of the interphase. Determination of the concentration profile thus becomes an important task to understand the structure of a graded interphase. A graded interphase is often useful for diffusing a stress concentration by avoiding a sudden change of material properties.

The concept of an interphase is well-established. Yet, it can only be defined through a collective description of many complex phenomena, and thus at the present time, no single explanation or approach exists which enables the complete understanding of the nature of any particular interphase. Therefore, it is essential to characterize the interphase using many characterization methods prior to attempting its structural control. If the interphase structure can be controlled, tailoring the mechanical and physical properties of a composite through the knowledge of the structure/property relationship becomes feasible. This systematic approach may be termed *interphase engineering*. It is the purpose of this article to discuss the principles of silane structure formation in solutions and on substrates rather than to present a summary of all published work in this field. Therefore, papers published from our laboratory will be reviewed whenever applicable, although there may be many other papers on the subject.

Table 2. Structure of Matrix Interphase

• **Sharp Boundary**

◊ Epitaxial crystal
◊ Transcrystal
◊ Interfacial reaction with matrix
◊ Totally incompatible filler/matrix combination
      - No surface treatment

• **Graded Interphase Structure**

◊ Diffusion of coupling agent
      - Interpenetrating networks (IPN)
◊ Preferential adsorption
◊ Restricted mobility

2. Introduction to Silane Coupling Agents

2.1. Structure of Silanes

Silane coupling agents usually possess a dual functionality of the form, $YSiX_3$, where Y is the organofunctional group and X the hydrolyzable group. X is typically an alkoxy group and sometimes Cl. Some representative silanes are listed in Table 3. The structure of an organofunctional group is chosen so as to copolymerize with a thermosetting resin or to be compatible with a thermoplastic resin.

2.2. Hydrolysis Reactions of Silanes

Upon contact with water, the following reactions take place.

$$RSi(OR')_3 \xrightarrow{H_2O} RSi(OH)_3 + 3R'OH$$

The hydrolysis of the trialkoxy groups takes place in a stepwise fashion. As the first alkoxy group is hydrolyzed, the steric hindrance to the other two alkoxy groups decreases. It is expected that the final alkoxy group is easier to hydrolyze than the first alkoxy group. The hydrolysis of silanes must be carefully controlled to be effective. Ordinary neutral silanes form oily droplets upon mixing with water. Thus, the dissolution of the silane molecules into water is the kinetic limiting step. The formation of the first silanol group increases the solubility of the silane into water. However, it is not known whether the unhydrolyzed trialkoxysilane first dissolves into water with the alkoxy groups hydrolyzed in a stepwise fashion or if the formation of the first silanol group is the driving force for the silane to dissolve into water.

Table 3.   Typical industrially used silane coupling agents

Organofunctional Group	Chemical Formula
Cationic styryl	$CH_2=CHC_6H_4CH_2NH(CH_2)_2NH(CH_2)_3Si(OCH_3)_3 \cdot HCl$
Chloropropyl	$Cl(CH_2)_3Si(OCH_3)_3$
Cycloapliphatic epoxide	$\overset{O}{\diagup}\!\!\!\diagdown\!\!\!-(CH_2)_2Si(OCH_3)_3$
Diamine	$H_2N(CH_2)_2NH(CH_2)_3Si(OCH_3)_3$
Epoxy	$\overset{O}{\overset{\diagup\diagdown}{CH_2CHCH_2O}}(CH_2)_3Si(OCH_3)_3$
Mercapto	$HS(CH_2)_3Si(OCH_3)_3$
Methacrylate	$\overset{CH_3}{\underset{\vert}{CH_2}}=C\text{-}COO(CH_2)_3Si(OCH_3)_3$
Primary amine	$H_2N(CH_2)_3Si(OCH_3)_3$
Vinyl	$CH_2=CHSi(OCH_3)_3$

Usually, acidic water with a pH ranging from 3.0 to 4.0 is used to catalyze the hydrolysis reaction. However, a glycidoxy-functional (epoxy-functional) silane readily opens its oxane ring in this pH range. For this silane, pH 5-6 may be needed to prevent the ring opening reactions. These neutral silanes may need to be hydrolyzed for tens of minutes at room temperature for complete hydrolysis, provided that the concentration of the silane is relatively low.

For an aqueous alcohol solution and other organic solution, the situation changes due to homogeneous reactions. Because of this, a high concentration of silane in water changes its behavior after certain degree of hydrolysis due to the increased alcohol content produced as a byproduct of the reaction. Amino-functional silanes readily dissolve into water and their alkalinity catalyzes the hydrolysis. For this reason, aminosilanes utilize neutral water for hydrolysis. Hence, upon contact with water, the hydrolysis will immediately take place in a matter of seconds. Since silanols are unstable in an alkaline media, immediate condensation also follows. For a few percent by weight of neutral silane solution, nearly all of silane molecules are silanetriol. In contrast, the same concentration of an aminosilane yields a solution with nearly one hundred percent oligomeric silanes. The content of the monomeric aminosilanetriol dramatically increases if the silane concentration in water is a fraction of a percent (10).

2.3. Condensation Reactions of Silanes

Upon drying, or with time in solution, the silanol groups condense to form siloxane groups, SiOSi:

$$2RSi(OH)_3 \longrightarrow \begin{array}{cc} R & R \\ HO\text{-}Si\text{-}O\text{-}Si\text{-}OH \\ O & O \\ H & H \end{array}$$

- - - - - - - - - - -

$$RSi(OH)_3 \longrightarrow \begin{array}{ccccc} R & R & R & R & R \\ HO\text{-}Si\text{-}O\text{-}Si\text{-}O\text{-}Si\text{-}O\text{-}Si\text{-}O\text{-}Si\text{-}OH \\ O & O & O & O & O \\ H & H & | & H & H \\ & & HOSiR \\ & & O \\ & & H \end{array}$$

The simplified structures shown above are only to demonstrate that the molecular size increases by condensing with silanetriols and oligomeric silanes. The actual structures may be far more complex since the trifunctionality of the silane can lead to many possible structures. A similar condensation reaction takes place with the surface silanol and other hydroxyl groups of a filler and fiber, thus allowing the silane to bind chemically to the surface. This is a typical reaction path for a treatment using a prehydrolyzed silane solution. The rate of silanol condensation to siloxane depends strongly on the pH, concentration and temperature of the solution, but weakly on the structure of the organofunctional group.

The molecular weight of the condensed silane is also influenced by the same factors. Another factor that is important for the silane coupling effect is the molecular architecture of the siloxane networks. Caged structures and ladder structures are the extreme cases of the different siloxane structures (11,12). The solution pH, topochemical effects of the solid surface, the structure of the organofunctional group, and the concentration of the silane are all believed to influence siloxane structure, although no detailed study has yet been reported.

An indirect observation is shown in Figure 2 where γ-methacryloxypropyl-trimethoxysilane (γ-MPS) is hydrolyzed and condensed in an aqueous alcohol solution in the absence of a filler (13). The solvents were evaporated at room temperature to simulate the typical treatment of a filler. The molecular weight was then measured by size exclusion chromatography as a function of time. When the solution was adjusted below pH 4.0, Figure 2 resulted showing a discrete increase of the molecular size until the column no longer separated individual species. On the other hand, the solution near neutral pH, as well as alkaline pH's, exhibited a quite different behavior. A monomeric species existed until the solvents were mostly evaporated and suddenly a large molecular weight species appeared as shown in Figure 3, unlike stepwise increase of the molecular weight in an acidic media. This may be an indication of a ladder-like structure or more open structure than the cage-like structure.

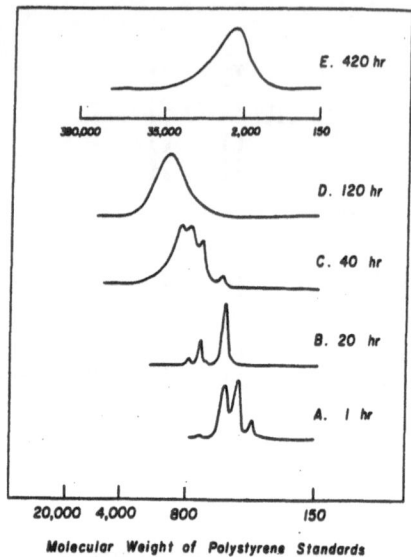

Figure 2   SEC chromatograms of the hydrolyzate of γ-MPS after the solution pH was adjusted to 3.6. Time from pH adjustment: (A) 1 h in solution, (B) 20 h during drying, (C) 40 h, dry, (D) 120 h, dry, (E) 420 h, dry.

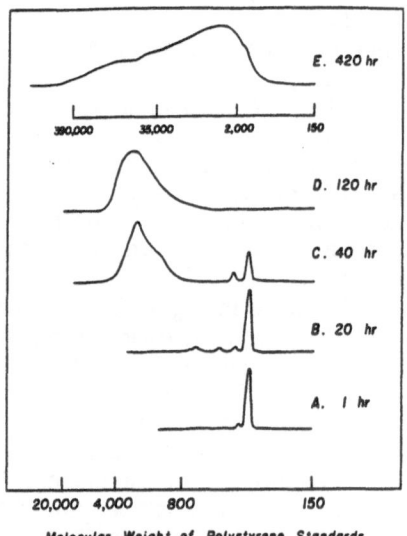

Figure 3   SEC chromatograms of the hydrolyzate of γ-MPS after the solution pH was adjusted to 6.9. Time from pH adjustment: (A) 1 h in solution, (B) 20 h during drying, (C) 40 h, dry, (D) 120 h, dry, (E) 420 h, dry.

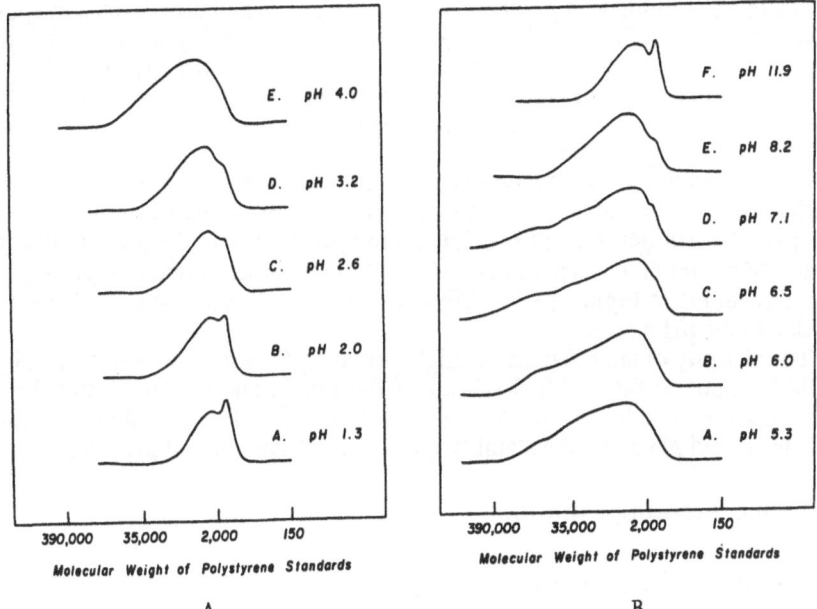

Figure 4 SEC chromatograms of the pH-adjusted hydrolyzate of γ-MPS.
Figure 4A: (A) pH 1.3, (B) pH 2.0, (C) pH 2.6, (C) pH 3.2, (D) pH 4.0.
Figure 4B: (A) pH 5.3, (B) pH 6.0, (C) pH 6.5, (D) pH 7.1, (E) pH 8.2, (F)
pH 11.9.

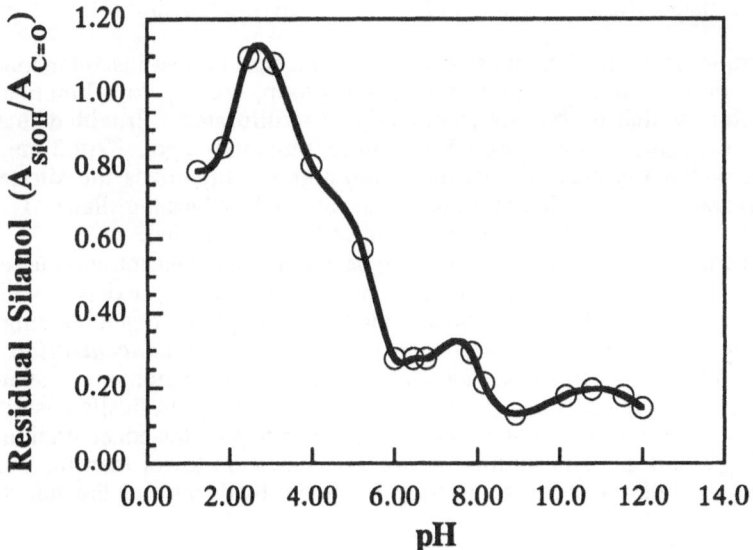

Figure 5. Stability of silanol as a function of the pH of the original solution
from which the precipitates were prepared

A similar solution was prepared with varying pH's and the solvents evaporated. The molecular weight of the precipitated silane is illustrated in Figure 4 as a function of the solution pH. The average molecular weight of the silane strongly depended on the initial pH of the solution. In both acidic and alkaline pH's, the molecular weight was equivalent to a few thousand of the polystyrene standard. On the contrary, the neutral pH's yielded quite large molecules.

In summarizing the above two experiments, it is possible to suggest a change in the condensation mechanism around pH 4 to 5. The mechanisms in the neutral and alkaline pH's are similar, except that the silanol group in an alkaline media forms a silanolate, SiO^-, which is resistant to condensation. Thus as shown in Figure 4, small oligomers resulted at higher pH's. Similar structures can be observed on a solid surface due to the pH effect.

The stability of silanol depends highly on the pH of the environment. Figure 5 shows the amount of the residual silanol of the precipitated silane molecules from aqueous solutions which are adjusted to respective pH's. It shows that the silanol is most stable around pH 3 and the stability quickly decreases around pH 4-5 (13).

2.4. Silane Treating Solutions

2.4.1. Aqueous Treatment

Ordinarily, silanes are applied to a substrate as monomeric, hydrolyzed species. The effectiveness of a silane is better when applied as monomeric silane rather than condensed oligomers. These silane films are very thin with a thickness on the order of several nanometers and, thus the strength of the film is rather insignificant in terms of the overall mechanical properties of a composite material. However, a silane primer is applied as a rather concentrated, condensed species, forming a film of a few hundred nanometers. Thus, the mechanical properties of the film itself can contribute to the composite properties. For this type of silane primer as well as the structure of silane on a solid surface, consideration of a kinetic effect and a thermodynamic effect is very important.

For example, an aminofunctional silane can condense almost instantaneously after hydrolysis due to the high pH of the amine group, leading to a kinetically controlled structure which is thermodynamically unequilibrated. Provided that a sufficiently high concentration was used, the solution turns into a gel. With time, in the presence of water, the catalytic action of the amine reorganizes the siloxane structure by rehydration and condensation reactions, and redissolves the silane. Thus, the time factor is also important when silane structures are to be studied.

When an aqueous solution of a silane is to be applied, the concentration used is on the order of a fraction of a percent. In this range, a concentration exists at which isolated silane monomers can form associated molecules through hydrogen bonding of the silanol groups. This concentration may be termed *onset concentration of association* (14). The associated silanes adsorb differently from the monomeric silanes as described in the following section. Furthermore, the amount of physically adsorbed silanes, also discussed in the following section, vary depending on the concentration of the silane treating solution. The solution with a concentration lower than the onset concentration of association yields less physisorbed silanes by decreasing the defect of the adsorbed layers.

2.4.2. Integral Blends

In addition to being used as a substrate pretreatment prior to mixing with a matrix resin, silanes can also be used as *integral blends*. This method eliminates the steps of pretreatment, drying, and repulverization by incorporating the unhydrolyzed silane directly into an uncured resin. However, the integral blending method usually requires a greater amount of silane than direct treatment due to its lower efficiency of silane utilization. The integral technique relies on the preferential adsorption of the silane onto the substrate. Immediately following blending, many of the reactive groups may remain completely or partially unhydrolyzed. Hydrolysis may proceed with time if sufficient water is available from either the resin or the substrate surface. An additional source of water is the adsorbed water on the substrate surface. Direct chemical reaction with the substrate can also proceed via the following exchange reaction.

$$
\begin{array}{c}
\text{SiOH} \\
\text{SiOH} + (CH_3O)_3SiR \longrightarrow \quad \text{Si-O-SiR} \quad + CH_3OH \\
\text{SiOH}
\end{array}
$$

Substrate

Catalysts such as amines or heat treatment helps accelerate the reaction since the reactivity of the above reaction is lower than the condensation reaction of silanol groups. Because there are few silanols available, the silane in this treatment method tends to be near a monomolecular equivalent. A similar situation is also observed when a filler is heated in a silane solution of a hydrophobic solvent such as toluene.

3. Structure of Silane on Solid Surfaces

3.1. Adsorption and Deposition of Silanes

The adsorption process of a silane molecule onto a substrate surface is influenced by the number of functional groups which are capable of strongly interacting with the surface via hydrogen or ionic bonding. Although it is difficult to distinguish it from the adsorption phenomenon, deposition of silane molecules will occur when the solvent is a poor solvent for the silane monomer or oligomers. It is important to note that the solubility changes as a function of the silane molecular weight. For example, water dissolves the silanetriol of the methacryl-functional silane (γ-MPS). However, the oligomers greater than the trimer precipitate from the solution. *Adsorption* is related to the active participation of the adsorbent toward the substrate usually with a special mechanism to anchor the molecule to the surface. Studies dealing with the adsorption phenomenon of a silane coupling agent have been reported by Nishiyama et al. (15,16).

On the other hand, *deposition* is the process in which the adsorbent molecule is forced out of the solution and onto the substrate surface. Deposition and orientation of methacryl-functional silanes have also been studied as a function of the length of the aliphatic spacer group (17). When the solubility of the silane is poor or has become poor by the structural or concentration changes due to evaporation of the solvent, deposition may take place. Although both of these processes lead to the accumulation of silane layers on the substrate surface, it is important to recognize that the molecular architecture of the condensed silane is different for each situation.

The orientation of a silane molecule influences the molecular architecture of the condensed silane and the chemical reactivity of the silane with the substrate surface as well as the matrix. Silanes usually possess multiple sites with quite different electronegativity or hydrophilicity on one molecule. The pH of the solution influences the state of charge density on the functional group as well as the substrate surface. Hence, the pH of the solution strongly influences the orientation of the silane either attracting or repelling a certain portion of the silane molecule. Under ordinary treatment conditions, silanols are believed to adsorb onto the hydroxyl-covered surface. However, if the silane molecule possesses another ionic portion that can compete with this process, it is possible for the molecule to adsorb upside-down (3).

3.2. Mode of Adsorption

Silanes adsorb onto solid surfaces differently depending upon the roughness of the surface with respect to the dimension of the silane molecules. For example, a high surface area silica typically adsorbs a few monomolecular layer equivalents whereas a smooth surface such as glass fibers adsorb many layers. The thickness of the silane on the glass fiber increases monotonously until the onset concentration of association at which the hydrolyzed silane molecules start forming hydrogen bonded species (14). Thus, the orientation of the associated silanes is disturbed and the thickness buildup slows. The disturbance of the packing is schematically shown in Figure 6.

The silane molecules tend to adsorb head-to-head in an extreme case where relatively rigid organofunctional groups such as cyclohexyl and phenyl groups are on the silane molecule. The crystal structure of cyclohexylsilanetriol exemplifies this extreme case (18). It is these silanes that form relatively stable silanetriol crystals due to good packing. In these crystals, silanols are hydrogen bonded in a bifurcated manner as shown below, meaning that the crystal possesses a rather disorganized hydrogen bonding structure.

$$\equiv Si - O \overset{\displaystyle H}{\underset{\displaystyle H}{\diamond}} O - Si \equiv$$

The packing of the silane molecules in the coupling agent layers strongly depends on the structure of the organofunctionality. This trend of organized packing relaxes as the flexibility of the organofunctional group increases, resulting in an increased free volume. An important ramification of this statement is that the degree of penetration of the matrix molecules into the coupling agent layers is highly influenced by the molecular packing. It is interesting to note that many useful coupling agents belong to this flexible organofunctional type.

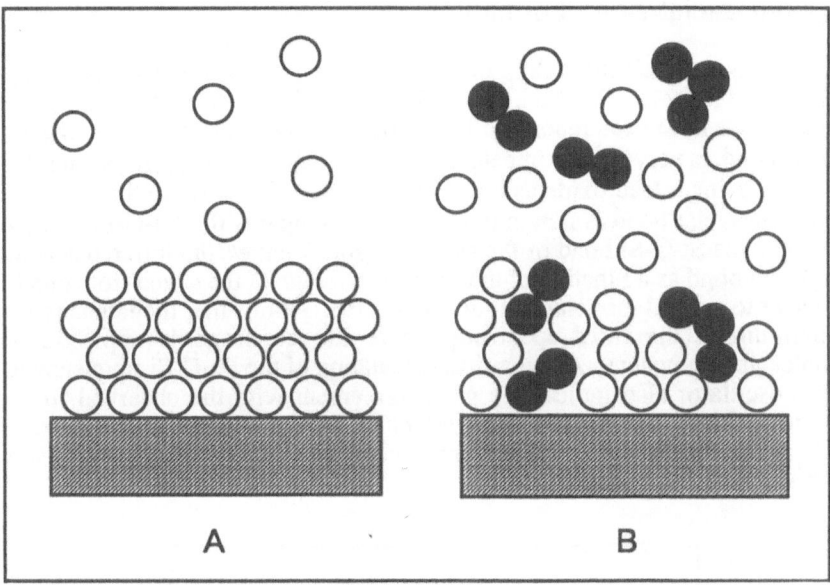

Figure 6. Schematic representation of the adsorption of silane coupling agent molecules onto glass fiber surface. A) The silane solution is a low concentration below the onset of association, and B) the silane solution is a high concentration above the onset of association. Open circle indicates isolated silanes whereas the closed circle denotes hydrogen bonded or oligomeric silanes. Note the disturbance in packing is created at a high concentration.

3.3.   Physisorbed and Chemisorbed Silanes

3.3.1.   Chemisorbed Silanes

Silanes can form many structures due to their tri-functional condensation mechanism. Monomeric silanetriol can directly react with the surface of a filler and fiber. Silane oligomers can also react with the surface as some unreacted silanol groups remain in the oligomers. The notion of physisorbed and chemisorbed silanes is used based on the reaction with the substrate surface. *Chemisorbed silanes* are the ones that possess at least one covalent bond with the substrate. Within the chemisorbed silanes, further distinction may be made between *tightly chemisorbed silanes* and *loosely chemisorbed silanes*. Tightly chemisorbed silanes use many silanol groups for covalent bonding with the substrate surface whereas the loosely chemisorbed silanes use only a few silanols to bond to the surface. An extreme case of this loosely chemisorbed silane is the oligomers or polymers with one covalent bond to anchor the entire molecule to the surface. Both silanes do not desorb when washed with an anhydrous organic solvent.

182

3.3.2. Interfacial Bond Formation

Although chemical bonding of a silane with the substrate surface is expected from the knowledge of the traditional chemistry, few direct observations of the interfacial bonds have been reported. This is because the interfacial bond, $Si_{substrate}$-O-Si_{silane} bond, is very similar in nature to the $Si_{substrate}$-O-$Si_{substrate}$ and Si_{silane}-O-Si_{silane}. Using a lead oxide as a model filler, this difficulty can be circumvented since the interfacial bond is now a unique Pb-O-Si against the Pb-O-Pb bond of the substrate and the Si-O-Si bond of the silane. Figure 7 shows the infrared spectrum of the interfacial bond as a function of the surface coverage of the silane from one fifth of a monolayer to several monolayer equivalents (19). Notice that the intensity reached the maximum at the third spectrum where the coverage is approximately a monomolecular equivalent. A theoretical calculation of the Pb-O-Si frequency using a harmonic oscillator also indicates a good agreement with the observed frequency. Furthermore, this band was not observed either in the silane condensate or in the substrate as the absorbance contribution of the organic functionality and the substrate have been subtracted.

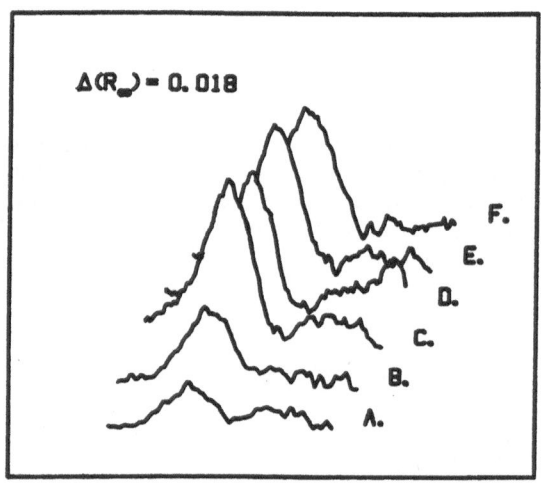

Figure 7. Diffuse reflectance infrared difference spectra from 100 to 910 cm^{-1} of only the plumbosiloxane contribution near 965 cm^{-1}. Spectra were obtained using the digital subtraction method and are not scale expanded. The adsorbate concentrations and measured peak to baseline intensity differences in Kubelka-Munk units are respectively, (A) 0.40 mg/m^2 and 0.005, (B) 0.80 mg/m^2 and 0.009, (C) 1.50 mg/m^2 and 0.018, (D) 2.00 mg/m^2 and 0.015, (E) 4.00 mg/m^2 and 0.016, (F) 6.00 mg/m^2 and 0.014.

In addition to the interfacial bonds, a quantitative observation of the inter-silane bonds can be made. Figure 8 exhibits the intensity of the inter-silane bonds as a function of the surface coverage (20). Two observations are especially noteworthy. First, the intensity of the inter-silane bonds was near-zero until certain surface coverage was reached despite the presence of the silane could be observed. This is an indication that the silane did not adsorb as an island but rather adsorbed separately in a random fashion so that the inter-silane distance is too far to form a siloxane bond. As the surface coverage increases, the distance becomes sufficiently short to form the siloxane bond. The second observation is the presence of a plateau near the surface coverage of the 0.5 mg silane/g substrate. If the silane molecules are forming the second and higher layers randomly, instead of first completing the bottom layer and then the second layer, the intensity increase is expected to be monotonous. It seems that during the plateau, the same process as the first layer is repeated. The adsorption is random in position thus inter-silane bond formation is retarded. As the coverage of the second layer becomes high, the inter-silane bonds again started forming. This process may be repeated for the third layer and beyond. As the number of layers increases, the registration of the orientation becomes more disarrayed, and it becomes more difficult to observe the plateau discussed above. However, a much smoother surface like mica showed the same phenomenon to the third and possibly the fourth layers (21). Understanding the mode of surface coverage is important because the interaction of the treated filler with the matrix is strongly influenced by the way the silane covers the surface.

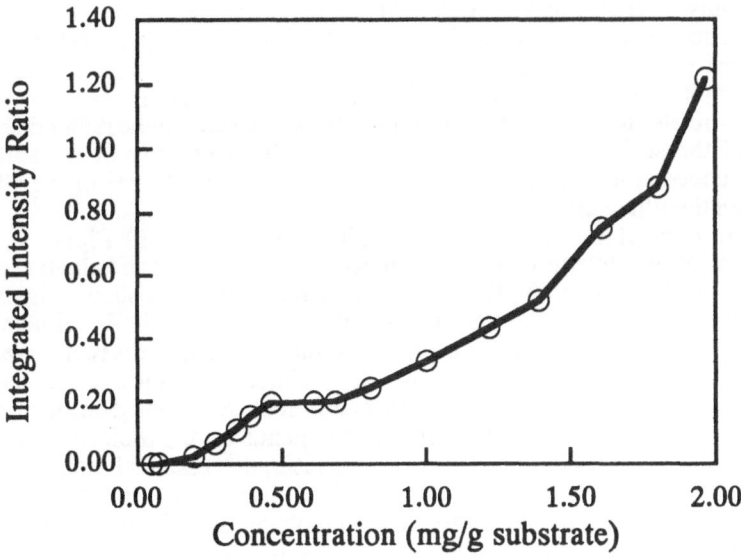

Figure 8. Intensities of the infrared band due to the intersilane bonds as a function of the amount of silane on a substrate

3.3.3. Physisorbed Silanes

Those silanes which do not possess any covalent bonds with the substrate surface are termed *physisorbed silanes*. The physisorbed silanes may be small oligomers or large polymers. The unreacted silanol groups on the molecules must be sterically hindered such that they are not readily available for condensation with the substrate surface or with the chemisorbed silane. Quite often, oligomers take the form of cyclic silanes where concentration of the unreacted silanol group in the molecule is low despite the small molecular size. A perfect cubical octamer for example does not have any silanol groups although the molecular weight is less than a few thousand.

These physisorbed oligomers can be easily distinguished by washing with an organic solvent. However, high molecular weight physisorbed silanes are not as easily distinguished as the small oligomers since the rate of desorption is much slower than for the small oligomers. Many silanol groups, though they may be sterically hindered for covalent bond formation, can form hydrogen bonds with the substrate surface or chemisorbed silanes. These hydrogen bonded silanols are in a dynamic equilibrium in terms of their interaction with the substrate. Thus, when a driving force to desorb exists, the hydrogen bonds are gradually broken and, with time, even a large physisorbed molecule desorbs. If a small amount of water exists in the solvent used, water can hydrolyze the siloxane bonds that were used to chemisorb the silane molecule to the substrate. The time scale for desorption and hydrolysis of the siloxane bonds might be comparable for a large, loosely chemisorbed molecules. For small oligomers, it is not important since the time necessary to hydrolyze the siloxane bonds is much longer than the extraction time.

The structure of the physisorbed silane molecules tends to be cyclic oligomers, and their molecular weight strongly depends on the pH of the filler (22). The pH of the filler is conveniently measured by the slurry pH method where a certain amount of filler is boiled in water for a specified time and the pH of the liquid measured. This method provides composite information about both the solubility of the selective surface components as well as the concentration of the surface acid/alkaline centers. This slurry pH directly relates to the molecular weight of the condensed silane rather than the concentration of the surface acid centers alone, since the solubility of the surface species and the concentration of acid centers both contribute to the pH of the silane solution that is on the filler surface.

As in the case of the aforementioned simulation shown in Figure 4, the molecular weights of the physisorbed silanes are small at both acidic and alkaline pH's whereas they are large near neutral pH's (23). In an acidic media, the silane oligomers are small because of the tendency to form cage-like molecules while in an alkaline media the silane oligomers are small due to the termination of the molecular weight growth by the silanolate group. Thus, the molecular architecture of the siloxane chains in these oligomers with similar molecular weight is different. Since the diffusion of the physisorbed silane molecules into the matrix resin depends on the molecular weight when a silane treated filler or fiber is mixed with the resin, it is essential to understand the nature of the physisorbed silane. The physisorbed silanes often contain residual silanol groups and these groups bridge physisorbed molecules via hydrogen bonding. Thus a desorption process is dependent not only on the molecular weight but also on the molecular architecture of the physisorbed silane.

Once the basic molecular architectural trend is predestined by the effect of the substrate surface, a heat treatment only accelerates the completion of the trend rather than creating a new class of structure (22). Figure 9 and 10 shows SEC chromatograms of γ-MPS hydrolyzates collected from the surface of a filler and from the precipitate of a silane solution without a filler. Both chromatograms of the initial samples are similar in the molecular weight and its distribution. However, after a mild heat treatment at 80 °C for 10 hr, the silane from the filler surface stays almost the same size whereas the silane from the precipitate polymerized to a size such that it can no longer dissolve in the tetrahydrofuran (THF) carrier liquid. The peak near the molecular weight of 300 may be an additive present in the initial silane.

Consequently, if the molecular architecture of the siloxane networks needs to be changed, it is necessary to change the solution pH of the silane, treat the filler by an acid or alkaline material to control the slurry pH, or add another silane which has different tendency toward forming siloxane networks. For example, a silane with a flexible organofunctional group such as γ-MPS tends to form a cyclic structure wherein a rigid functional group leads to a more open structure. Hence, if a larger molecular weight γ-MPS condensate is desired, it can be mixed with a rigid silane like phenyltrimethoxysilane or vinyltrimethoxysilane. The molecular weight changes for mixed silane systems are plotted in Figure 11 where 10 mol % of the second silane has been added (22). It seems that dilution of the organofunctional group by the second silane does not grossly affect the effectiveness of the first silane so long as the composition of the second silane does not dominate. Localization of silane near the substrate surface can be achieved either by improving chemical bonding of the silane to the substrate or by reducing the solubility to the matrix by increased molecular weight.

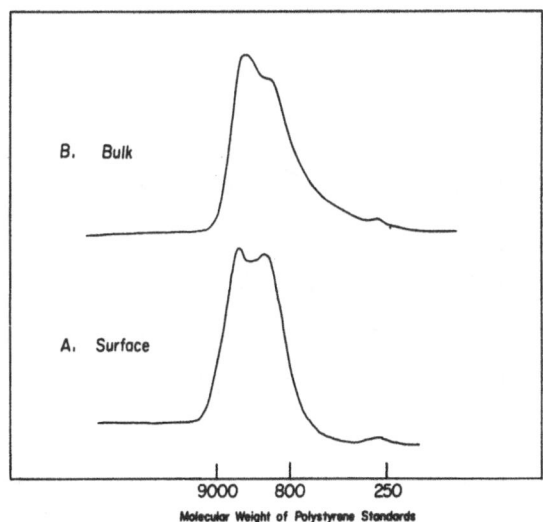

Figure 9. SEC chromatograms of the condensates of γ-MPS: (A) extracted from the surface of particulate clay, (B) dried in bulk.

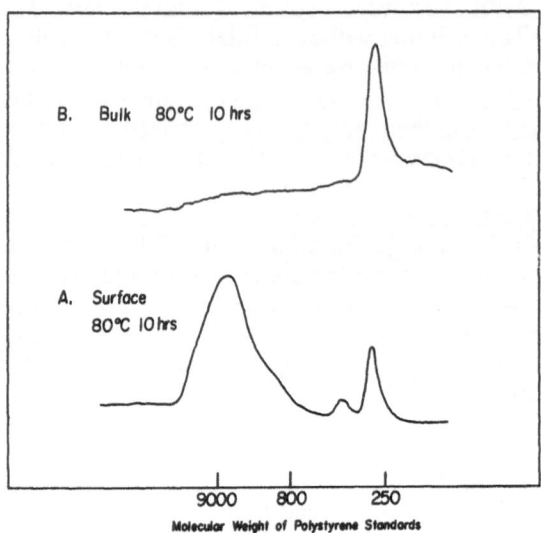

Figure 10. SEC chromatograms of the silane shown in Figure 9 except that the condensates were heat treated at 80° C for 10 h.

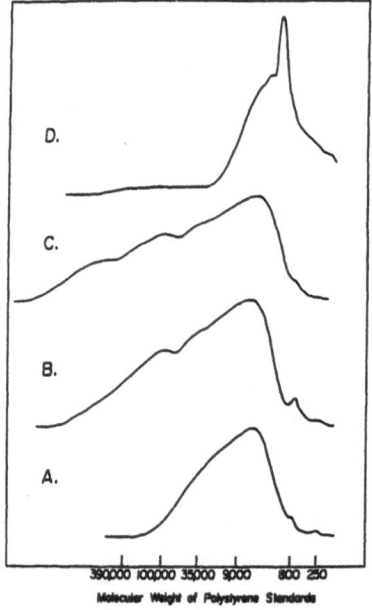

Figure 11. SEC chromatograms of of the bulk hydrolyzate of γ-MPS containing 10 mol % of a second species and heated at 80o C for 85 h: (A) 100 mol % γ-MPS, (B) with 10 mol % tetramethoxysilane, (C) with 10 mol % vinyltrimethoxysilane, (D) with 10 mol % isopropyltridioctylpyrophosphato-titanate.

3.3.4. Mechanical Properties of the Silane Interphase

At the present time, no *in-situ* measurement of the mechanical properties across an interphase containing physisorbed silane has been made. However, a simulation of the mechanical properties can be made by preparing copolymers of precipitated silane oligomers with a matrix resin. Figures 12 and 13 show the flexural modulus and the flexural strength of the copolymers of γ-MPS and unsaturated isophthalic type polyester resin as a function of the silane concentration (24). Beyond 50 % silane content, phase separation was observed and thus excluded from the study since a valid comparison could not be made. However, in reality, regions with such compositions do exist and indeed the phase separated area might be observed. Phase separation is also observed with the copolymers of γ-aminopropyltriethoxysilane (γ-APS) and an epoxy resin (diglycidylether of bisphenol-A) (25). The flexural modulus of the γ-MPS precipitate/polyester copolymers slightly increased until 20 % by weight silane and decreased beyond this composition.   On the other hand, the flexural strength monotonously decreases at all silane compositions. This is a typical behavior of anti-plasticization. It is interesting to note that for the γ-MPS/unsaturated polyester, any region with physisorbed silane is mechanically weaker than the bulk matrix.

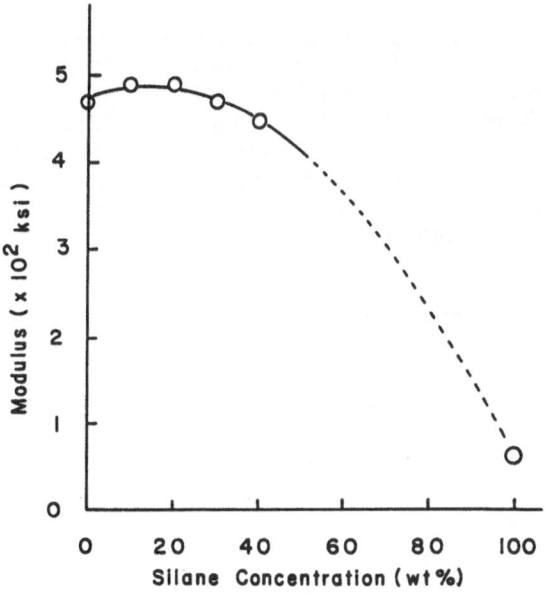

Figure 12. Flexural modulus of the copolymers of the precipitates of γ-MPS hydrolyzates and unsaturated polyester resin.

188

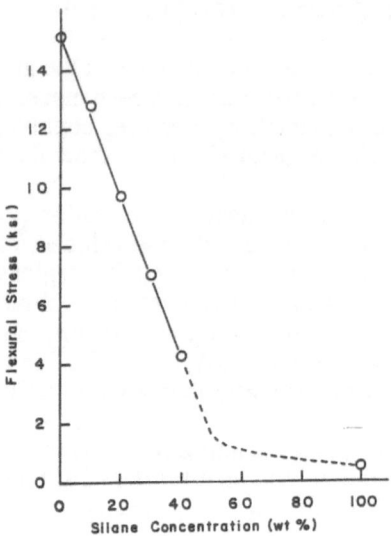

Figure 13. Flexural strength of the copolymers of the precipitates of γ-MPS hydrolyzates and unsaturated polyester resin.

. Provided that it is overly simplistic, a conceptual mechanical property profile across the interphase is schematically illustrated in Figure 14 using the expected diffusion of the silane and the observed mechanical properties of the materials with various compositions. A mechanically weak portion in the diffusion region of the physisorbed silane is illustrated and the gradual increase of the strength over distance to the bulk matrix is also incorporated. Thus, if the physisorbed silanes are removed from the treated reinforcements, stronger composite materials can be manufactured. Indeed, such observations have been reported (26,27).

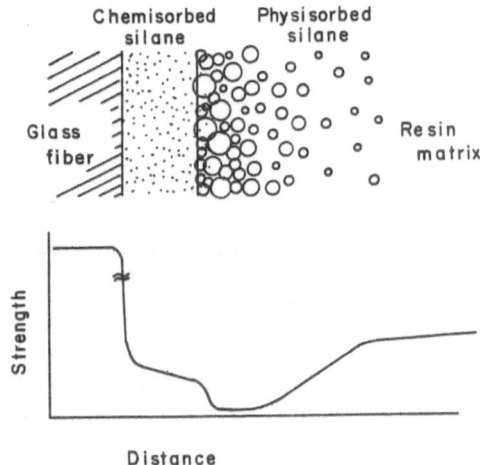

Figure 14. Conceptual mechanical property profile across the glass fiber/matrix interface.

4. Structure of the Silane/Matrix Interphase

4.1. Copolymer Formation with the Matrix Resin

The chemical reaction of the organofunctional group and the matrix is, in principle, the same as any other organic chemistry. However, the availability of the organofunctional group may be hindered if the silanes are tightly packed. Restricted molecular mobility, including the preferential interaction with the substrate surface, reduces the reactivity of the organofunctional groups. Due to the wide distribution of possible structures of the silane molecules in the silane layers, a systematic study of the silane/matrix reactivity is very difficult, although some attempts as follows have been made. Figures 15 and 16 show NMR $T_{1\rho}$ relaxation measurements of the aminosilane/epoxy copolymers at different silane contents (25). The different slopes of each carbon relaxation process, and thus different $T_{1\rho}$ values, indicate that the system is phase separated. However, the composition shown in figure 16 shows the same $T_{1\rho}$ values for all carbon atoms, suggesting that the system is homogeneous. The heterogeneity detected by the NMR method is said to be smaller than 1 nm in size.

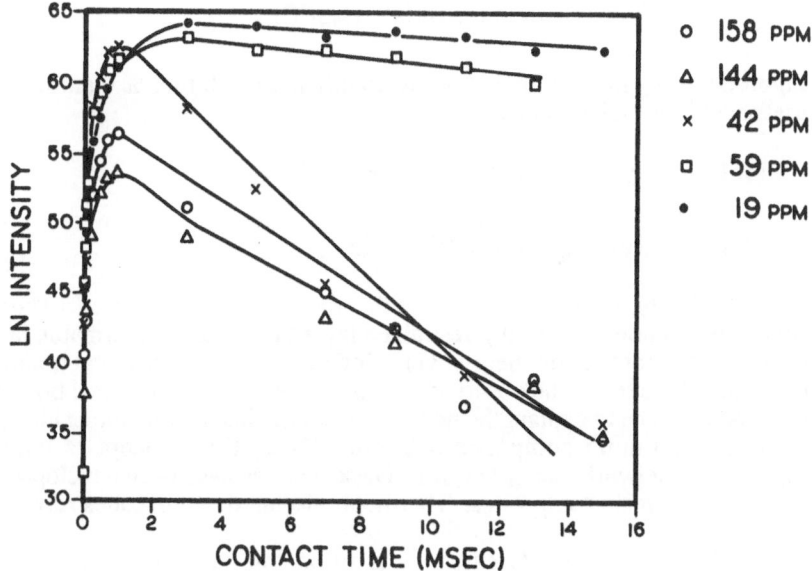

Figure 14. $T_{1\rho}$ measurements of the copolymer of 60 mol % γ-aminopropyl-triethoxysilane and 40 mol % epoxy.

In spite of the altered reactivities of surface species, the general trend of copolymerization still holds. Thus, it would be helpful to know the copolymerization parameters, r and Q factors, of the organofunctional group and the matrix functional group. Depending on the r and Q values, the silane and the matrix interphase may form an alternating copolymer or terminal reaction, leading to a quite different interphase.

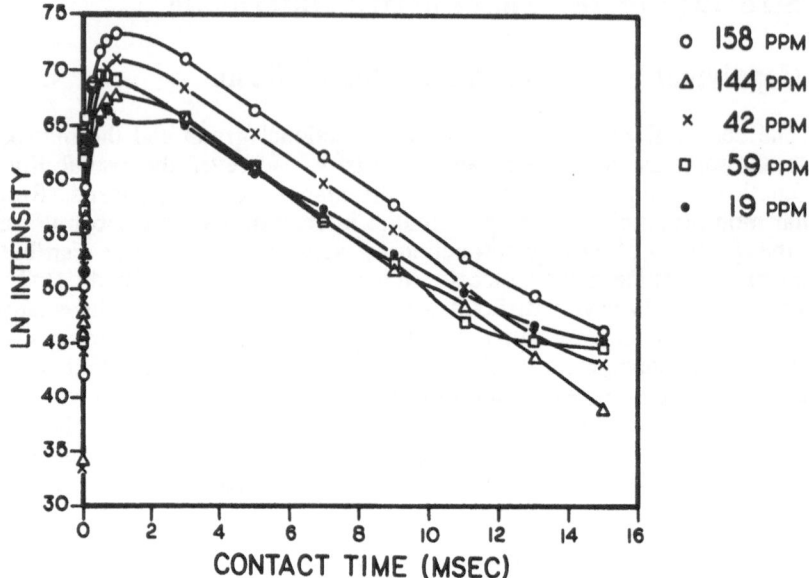

Figure 15. $T_{1\rho}$ measurements of the copolymer of 30 mol % γ-aminopropyl-triethoxysilane and 70 mol % epoxy.

5. Reinforcement Mechanisms

5.1. Various Reinforcement Theories

Historically, many reinforcement theories have been proposed. Those include the chemical bonding theory (28), restrained layer theory (29), deformable layer theory (30), and coefficient friction theory (31). However, only the chemical bonding theory could sufficiently explain the observed results. However, the chemical bonding theory alone is not adequate to explain the necessity of more than a monomolecular equivalent of silane for optimum composite strength. Thus, this concept is coupled with interpenetrating network theory (31,32). These theories have been developed primarily for thermosetting resin composites. Thermoplastic-matrix composites rely on different mechanisms.

If no chemical bonds exist, as is often the case in thermoplastic-matrix composites, it is helpful to have a hydrophobic matrix or semicrystalline matrix to reduce the number of water molecules diffusing to the fiber/matrix interface. It is believed that isolated water molecules are not as harmful in causing permanent damage to the composite properties as the condensed water. Even isolated water molecules can act as a plasticizer and reduce the strength of the matrix, although such property changes are often recoverable upon drying. This is in contrast to the interfacial damage caused by the condensed water where an enormous osmotic pressure causes debonding at the fiber/matrix interface.

5.2. Concentration of Interfacial Bonds

It is interesting to consider the minimum number of interfacial bonds which is necessary to provide sufficient composite strength. Very few papers are available on this subject. Figure 17 shows the strength of a composite with coupling agents specially designed to evaluate the number of interfacial bonds (33). A methacryl-functional silane is copolymerized with styrene in different concentrations in order to vary the average molecular weight, Mn, of the inter-pendant groups. The flexural strengths of the dry and wet composites are plotted as a function of the Mn. Provided that this is not exact evaluation of the interfacial bond density, qualitative information can still be obtained. Both the dry and wet flexural strengths of the composite do not markedly change until the Mn reaches approximately 1000. The dry strength gradually reduces as the Mn increases implying that compatibility alone can provide substantial dry strength. On the other hand, the wet strength decreases suddenly beyond Mn = 1000. Thus, chemical bonds are essential to provide permanence of the bond in a humid environment.

Chemical bonds can maintain bonding even in the presence of small hydrophilic molecules such as water and alcohol which compete for the hydrophilic reinforcement surface with the silane. Physical bonds including van der Waals' force, ionic bonds, and acid-base interactions are quite adequate to provide good dry strength but these effects can yield misleading results if the wet strength is also of interest.

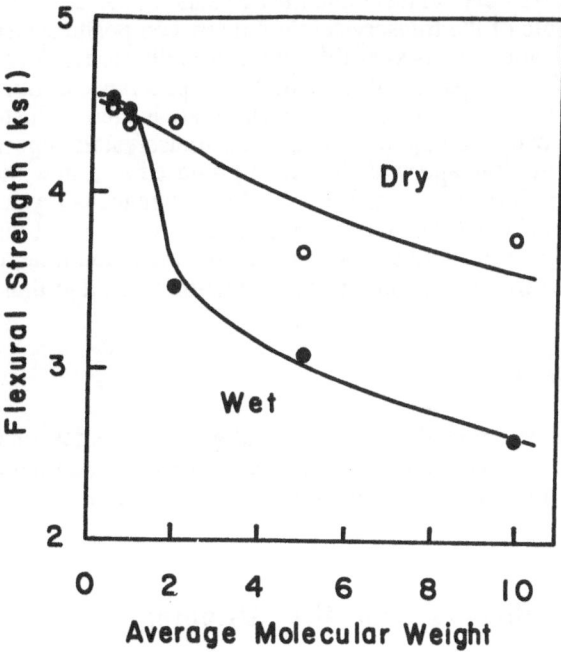

Figure 17. Flexural strength of glass cloth reinforced polystyrene with polymeric silane-treated glass: the silane to styrene ratio was varied so that the molecular weight of the polystyrene between two silane molecules become from 500 to 10,000 as shown as average molecular weight.

5.3. Adhesion of Thermoplastics to Substrates

Another adhesion mechanism for thermoplastic-matrix composites is the formation of transcrystalline layers around the reinforcement fibers. A transcrystal is a collection of columnar crystals growing transverse to the fiber axis. When a massive nucleation at the fiber surface occurs, the initial spherulitic crystals soon touch one another leaving only the direction transverse to the fiber axis to grow. This form of morphology is considered useful for the adhesion of a thermoplastic resin to the reinforcement fibers. However, the improved adhesion might be offset by the residual interfacial stress caused by the crystallization shrinkage. At the present time, the overall benefit of the transcrystalline morphology to composite properties is therefore poorly understood, although there are some papers showing the benefits of this type of morphology to the mechanical properties of composite materials.

Transcrystals can be formed by many methods as long as massive nucleation at the fiber surface is promoted. Some of those methods include an applied stress, epitaxial crystallization, coating of the fiber by a nucleating agent or liquid crystalline material. All these variations complicate the study. It is thus important to consider the thermodynamics of the interphase. Until recently, few quantitative studies of transcrystallization have been reported due to the difficulty in applying the heterogeneous nucleation theory. Heterogeneous nucleation theory allows one to study the interfacial energy difference function, $\Delta\sigma$, of the fiber/matrix interface by measuring the growth rate of the transcrystal layer and the nucleation rate. It is straightforward to measure the growth rate of the transcrystal with many composite systems. However, determination of nucleation rate is very difficult, especially for reinforcements such as carbon fibers. Even for transparent fibers, a massive appearance of nuclei makes such measurements tedious. Furthermore, the nucleation rate is thermal history dependent. For these reasons, few of the papers dealing with transcrystals reported $\Delta\sigma$ values. Recently, the induction time approach has been proposed as a new method to study transcrystallization (34,35). When an isothermal heterogeneous crystal growth takes place, a threshold period exists when the formation and remelting of the precursors of nuclei takes place. This threshold period is called the induction time, t_i. The rate of nucleation, I, is inversely proportional to the induction time, and thus the following relationship exists.

$$I \cdot t_i = C \text{ (constant)}$$

By using t_i, which is determined from the crystal growth kinetics, one can eliminate the use of I, an aforementioned difficult quantity to measure. At the present time, no quantitative study of the effect of silane structures on the transcrystal formation has been reported.

6.   Effect of Silanes on the Rheological Properties of Composites

Although the rheology of silane-treated filled systems has been reported (36-38), few reports deal with the silane structure/rheology correlation.

The viscosity of a filled system is usually much higher than a non-filled system mainly due to the formation of filler aggregates. The aggregated filler shows a higher apparent filler fraction, ϕ_a, than the true filler fraction, ϕ, because the interstices of the fillers are not easily accessible space for the resin to flow. The viscosity is a nonlinear function of the filler fraction, and the viscosity increases much more rapidly at higher filler fractions, especially near the closest packing fraction, ϕ_m. A drastic viscosity reduction of a highly filled system can be achieved by breaking up the aggregates of the filler. This can be achieved by applying a shear or extensional force which mechanically separate the aggregates. The efficiency of the separation depends on the deformation rate. An alternative method is to change the surface energetics of the filler so that the filler/matrix interaction force overcomes the aggregation force.

The aggregation force might be due to the bridging effect of water molecules, direct hydrogen bonding of the surface hydroxyl groups, or ionic interactions. All of these effects can be reduced by treating the filler surface with silane coupling agents. Silanes can compete for, and react with, the surface hydroxyl groups and cover the filler with organo-functional groups. If the organo-functional group has a hydrophilic group, it tends to bend toward the hydrophilic filler surface and hence to expose the hydrophobic portion of the group. Since the van der Waals' force between the organo-functional groups is similar to that between the silane and resin, a small shearing force can easily break up the agglomerated filler particles.

In addition to the effects caused by the chemically bound silanes, a physisorbed silane can further reduce the viscosity. A silicon fluid is a good lubricant or a release agent. It has been previously discussed that a filler surface tends to produce cyclic oligomers of varying molecular sizes. An acidic or alkaline filler especially produces silane oligomers with a molecular weight of a few thousand. The amount of physisorbed silane varies depending on the concentration of the silane treating solution. In general, the higher the concentration of the silane in the treating solution, the more the physisorbed silane. This is especially true in the treatment method used for particulate fillers. Fillers tend to form a cake if a slurry of the filler/silane solution was dried. In order to avoid this additional processing, a treatment method termed an integral blend is adopted where a rather concentrated alcohol aqueous solution of a silane is sprayed onto a tumbling filer. Thus, the filler does not completely wet and immerse itself in the silane solution.

A mixture of a silane-pretreated filler and a polymer exhibits complex viscosity behavior. In addition to the breakup of agglomerates, silanes can interact physically and chemically with the polymer matrix. A reactive silane, such as an azide -functional silane, can form chemical bonds with aliphatic polymers and crosslink the polymer chain forming a matrix interphase around the filler. Polymer melts are usually kept at elevated temperatures and thus ordinarily unreactive silanes can form bonds with the polymer. Polymers also cleave, forming reactive polymeric free radical chain ends. These chain ends can graft onto silanes. Thus, the viscosity increases, counteracting the viscosity reduction by the break-up of the agglomerates. Another influencial factor is the acid-base interaction of the organo-functional group with the functional group of the polymer chains. Strong acid-base interactions can act as physical crosslinks modifying the rheological properties of the interphase as the silanes are multifunctional by the oligomeric or polymeric structures. Physisorbed silanes can diffuse out of the filler/matrix interface and the thickness of the physically crosslinked matrix interphase increases. As a result, the viscosity would also increases.

When the integral blending method was used to treat a filler surface and the amount of added solution was varied to change the amount of the silane coverage, peculiar results were observed. As the amount of the added solution was increased, the quantity of the deposited silane monotonously increased as expected. However, this increase in the deposited silane halted suddenly for the quantity of silane solution where the filler was completely wetted by the solution as shown in Figure 18. It is likely that the silane molecules may have coevaporated with the alcohol molecules. Accordingly, there is an upper limit of silane that can be deposited by the integral blending method. It is not clear wether this phenomenon applies to a compound such as an amino-functional silane which tends to form oligomers quickly by the catalytic action of the amine group.

A highly filled system is of great technological significance in the ceramic, artificial marble, and other industries. Due to the very high filler contents, the viscosity of the filled system is usually extremely high to the extent that the mixture of initial ingredients is dough-like in consistency, even if a silica filler is mixed with a such a low viscosity material as methyl methacrylate. The initial viscosity of the mixture can be as high as above 100 Pa.s (100,000 cp). If a silane is added and the mixture is ball-milled, the viscosity drops in time almost three orders of magnitude to below 0.1 Pa.s (100 cp), and the mixture can be readily poured (39). The viscosity reduction process is a function of the surface coverage of the filler by the silane. The complete coverage of the filler surface is necessary to observe low viscosity. Since this type of treatment is done in an organic solvent without any water added, the rate of hydrolysis is very slow. The consumption of the silane alkoxy group relies mostly on the water adsorbed on the filler surface and the aforementioned surface silanol/alkoxy group exchange mechanism. Thus, the addition of an alkaline catalyst greatly accelerates the chemical reaction of the silane with the filler.

Ordinarily, in the aqueous treatment of fillers and fibers, silane molecules are almost completely hydrolyzed in the solution prior to mixing with the filler and fiber. Thus, the silanol groups of the silane can be a driving force for adsorption onto the substrate surface. In the case of an anhydrous organic system, this type of adsorption mechanism may be modified. Due to the limited amount of water molecules, only one of three hydrolyzable alkoxy groups may be hydrolyzed and the incompletely hydrolyzed silane may be adsorbed. These incompletely hydrolyzed groups may be hydrolyzed with time on the filler surface. In the other mechanism, when an unhydrolyzed silane molecule encounter the surface silanol groups, the aforementioned exchange mechanism shown below takes place, releasing an alcohol molecule and the silane becomes part of the solid (40).

$$\blacksquare\ SiOH\ +\ (CH_3O)_3SiR\ \longrightarrow\ \blacksquare\ Si\text{-}O\text{-}Si(CH_3)_2R\ +\ CH_3OH$$

In either process, the rate of adsorption is expected to be extremely slow. Hence, the addition of an alkaline catalyst greatly aids in accelerating these adsorption processes. Heat and/or time can achieve the same viscosity reduction effect, because both increased temperature and prolonged time lead to more adsorbed silanes. Very few silane oligomers are produced in this method since the hydrolyzed silane molecules do not exist in quantity at any time of treatment. Unlike the mixing of a silane pretreated filler, it is essential to consider the kinetics of silane adsorption in this type of treatment method.

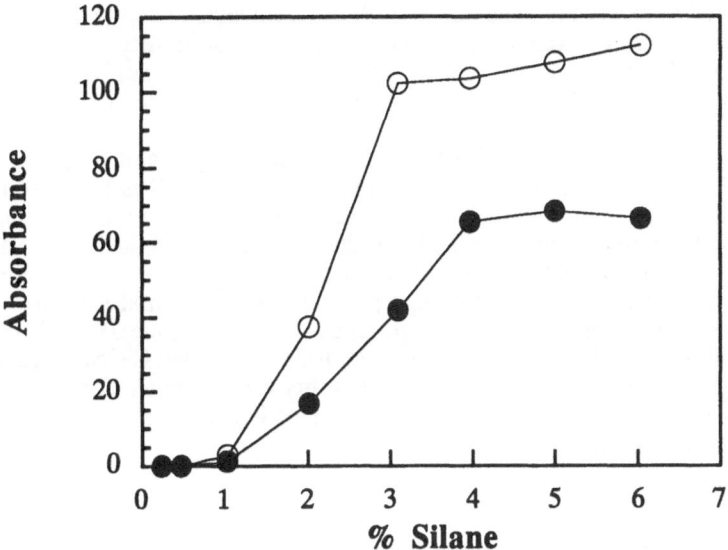

Figure 18. Amount of total and chemisorbed silane on a filler as a function of the initial silane concentration with respect to the filler

7.   Hydrothermal Stability of Silane Coupling Agent

Siloxane and silanol groups establish an equilibrium depending upon the humidity level as shown below.

$$\equiv SiOSi\equiv \xrightleftharpoons{H_2O} \equiv SiOH + \equiv SiOH$$

Silanol groups are very unstable and the equilibrium is toward forming the siloxane group. Acidic and alkaline pH's catalyze the hydrolysis and condensation reactions. The bond energy of the siloxane is very high and thus the thermal stability of siloxane bonds is excellent, as exemplified by the excellent properties of polydimethylsiloxane at elevated temperatures. However, in the presence of benzoic acid catalyst, the activation energy of the hydrolysis becomes as low as 5 kcal/mol as compared to the bond energy of the siloxane, 89.3 kcal/mol. In spite of this low activation energy of hydrolysis in the presence of a catalyst, a silane coupling agent is effective even in a humid environment. This is probably because of the highly skewed equilibrium toward forming siloxane bonds.

The hydrothermal stability of the siloxane networks is not strongly dependent upon the structure of the organofunctional group. The hydrothermal stability of a silane is studied by desorption of the silane from the substrate surface. In spite of the

196

similar time required to cleave siloxane bonds from different silane coupling agent, the desorption kinetics differs markedly due to the difference in solubility of the newly formed oligomers. If an additional mechanism exists, such as surface induced polymerization of the organofunctional groups, further difficulty for desorption is encountered.

The hydrothermal stability of the SiOR bonds where R represents C, Si, Al and other elements depends strongly on the element in the R group. It is believed that electronegativity difference between the silicon atom and the element in the R group is one of the important factors determining the hydrothermal stability of the bond. Some evaluations of the various oxane bonds have been reported (41). Pure silica, alumina, and titania were treated by a silane coupling agent and hydrothermal stability was studied by monitoring the rate of silane desorption as shown in Figure 19. It is clear from the figure that the stability of the bonds is in the following order: SiOSi > SiOTi > SiOAl. It is intriguing to speculate that a similar stability difference also exists on the surface of the glass fiber, as it consists of various surface hydroxyl groups as well as alkaline and alkaline earth elements.

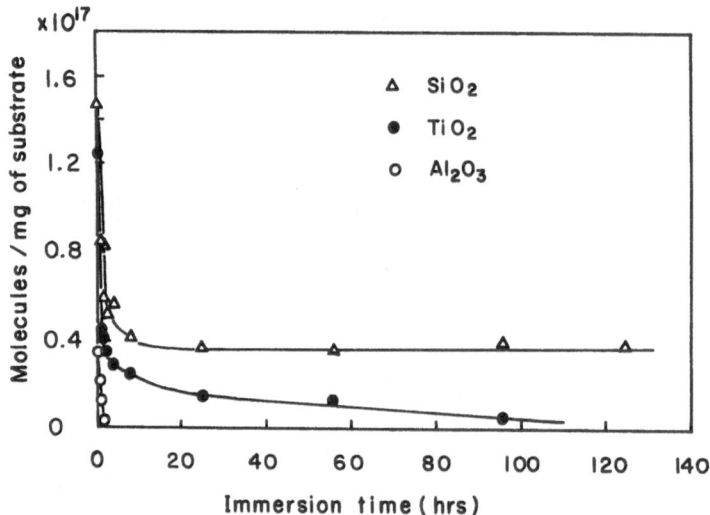

Figure 19. The desorption kinetics of γ-aminopropyldimethylethoxysilane from various substrate surfaces: (△) silica, (●) titania, (○) alumina.

When water reaches the glass/silane interface, water molecules might exist in a separate gas-like state or condensed liquid-like state. It is believed that this liquid-like water damages the interface more than the gas-like molecules. A recent dielectric study by Steeman and Maurer shows a rare insight on this subject. They measured the dielectric constant of glass bead filled polyethylene as a function of the silane coverage on the glass beads surface and the relative humidity. When the water molecules are less than a monomolecular equivalent, the effect of the water on the dielectric constant was small. As the coverage reaches several layers equivalent, the hydrolytic damage dramatically increased.

A simple concentration consideration alone is difficult to explain the above phenomenon. It is possible that the associated, liquid-like water can build a strong osmotic pressure at the interface that will stress the oxane bonds. The oxane bonds at the interface will be subjected to stress. When a tensile stress is applied to the bond, the mechanochemical effect lowers the apparent activation energy of the hydrolysis of the oxane bonds, making it easier to be hydrolytically cleaved. The aforementioned differences in the hydrothermal stability of various oxane bonds may become especially important under such circumstances. Easily cleaved, non-siloxane bonds may become the source of water collection and exert pressure on the surrounding siloxane bonds, which will be cleaved easier than the siloxane bonds with pure silica surface.

It is also necessary to consider the dissolution of the alkaline and alkaline earth elements into the water. If the water is acidic, the dissolution of the surface elements readily takes place. When large atoms near the glass fiber surface are leached out, the siloxane and other networks will be placed under tension, making the glass/silane bonds more vulnerable to hydrolytic attack. All of these phenomena described above do not exist in pure silica's such as quartz powder or fumed silica. It is known that pure silica filled polymers exhibit good hydrothermal stability. It is possible that the lack of these mechanisms to weaken the interfacial bonds may significantly contribute to the stability of the composite. While there are no rigorous studies to prove the aforementioned phenomena nor are effort correlating these subjects with the hydrothermal stability of a composite, these phenomena are well-accepted in the literature for other materials.

8.　Summary

The molecular structure of silanes in solution and on substrates has been reviewed. Major advancements over the past decade have helped to elucidate the structure of silane coupling agents. With an improved understanding of the silane structures, it is now possible to control some structures of silane on a substrate. Once structure/property correlation is understood, a specific property of a composite material can be tailored by controlling the interfacial structure. Our current knowledge of silane coupling agents allows us to attempt this systematic approach that has been defined in this paper as interphase engineering.

9.　Acknowledgment

This work was in part supported by the grant from The Office of Naval Research.

198

10. References

1. E.P. Plueddemann, Ed., "Interfaces in Polymer Matrix Composites," Academic Press, New York (1974).

2. D.E. Leyden, and W. Collins, Eds., "Silylated Surfaces," Gordon and Breach Science, New York (1980).

3. E.P. Plueddemann,"Silane Coupling Agents," Plenum Press, New York (1982).

4. H. Ishida and G. Kumar, Eds., "Molecular Characterization of Composite Interfaces," Plenum Press, New York (198).

5. H. Ishida, and J.L. Koenig, Eds.,"Composite Interfaces," Elsevier Science Publishing, New York (1986).

6. D.E. Leyden, Ed.,"Silanes, Surfaces, and Interfaces," Gordon and Breach Science, New York (1986).

7. H. Ishida, Ed., "Interfaces in Polymer, Ceramic, and Metal Matrix Composites," Elsevier Science, New York (1988).

8. F.R. Jones, Ed., "Interfacial Phenomena in Composite Materials," Butterworths, London (1989).

9. H. Ishida, Ed.," Controlled Interphases in Composite Materials," Elsevier Science, New York (1990).

10. H. Ishida, S. Naviroj, S.K. Tripathy, J.J. Fitzgerald, and J.L. Koenig, *J. Polym. Sci. -Phys.*, **20**, 701 (1982).

11. J.F. Brown, Jr. and L.H. Vogt, *J. Am. Chem. Soc.*, **84**, 4313 (1965).

12. J.F. Brown, Jr., *J. Am. Chem. Soc.*, **84**, 4317 (1965).

13. J.D. Miller and H. Ishida, *Polym. Composites*, **5**, 18 (1984).

14. H. Ishida, S. Naviroj and J.L. Koenig, in "Physicochemical Aspects of Polymer Surfaces," K.L. Mittal, Ed., Plenum, New York (1983) p.91.

15. N. Nishiyama, K. Horie, R. Shick and H. Ishida, *Polymer*, **31**, 380 (1990).

16. N. Nishiyama, *J. Dent. Mat.*, **5**, 519 (1986).

17. J. Jang, H. Ishida and E.P. Plueddemann, Proc. 42nd Ann. Tech. Conf., Composites Inst., SPI, Sect. 21-F (1987).

18. H. Ishida, J.L. Koenig and K.Gardner, *J. Chem. Phys.*, **77**, 5748 (1982).

19. J.D. Miller and H. Ishida, *J. Chem. Phys.*, **86**, 1593 (1987).

20. J.D. Miller and H. Ishida, *Langmuir*, **2**, 127 (1986).

21. B.D. Favis, L.P. Blanchard, J. Leonard and R.E. Prud'homme, *Polym. Comp.*, **5**, 11 (1984).

22. J.D. Miller and H. Ishida, *J. Polym. Sci. -Phys.*, **23**, 2227 (1985).

23. J.D. Miller and H. Ishida, *Macromolecules*, **17**, 1659 (1984).

24. H. Ishida, unpublished results.

25. K. Hoh, H. Ishida and J.L. Koenig, *Polym. Comp.*, **9**, 15 (1988).

26. R.T. Graf, J.L. Koenig and H. Ishida, *J. Adhesion*, **16**, 97 (1983).

27. M. Kokubpo, H. Inagawa, M. Kawahara, D. Terunuma and H. Nohira, *Kobunshi Ronbunshu*, **38**, 201 (1981).

28. J. Bjorksten and L.L. Yaeger, *Mod. Plast.*, **29**, 124 (1952).

29. C.A. Kumins and J. Roteman, *J. Polym. Sci.*, **1A**, 527 (1963).

30. R.C. Hooper, Proc. 11th Ann. Tech. Conf./Reinforced Plast. Div., SPI, Sect. 8-B (1956).

31. E.P. Plueddemann, "Silane Coupling Agents," Plenum Press, New York (1982) p.134.

32. H. Ishida and J.L. Koenig, *J. Polym. Sci. -Phys.*, **17**, 615 (1979).

33. E.P. Plueddemann, *J. Paint. Technol.*, **10**, 1 (1968).

34. H. Ishida and P. Bussi, *Macromolecules*, **24**, 3569 (1991).

35. H. Ishida and P. Bussi, *J. Mat. Sci.*, **26**, 6373 (1991).

36. C.D. Han, C. Sanford and H.J. Yoo, *Polym. Eng. Sci.*, **18**, 849 (1978).

37. J.D. Miller, H. Ishida and F.H.J. Maurer, *Rheol. Acta*, **27**, 397 (1988).

38. C. Scott, H. Ishida and F.H.J. Maurer, *J. Mat. Sci.*, **26**, 5708 (1991).

39. H. Ishida and T. Gayet, unpublished results.

40. P. Dreyfuss, *Macromolecules*, **11**, 1031 (1978).

41. S. Naviroj, J.L. Koenig and H. Ishida, *J. Adhesion*, **18**, 93 (1985).

CONTROL AND MODIFICATION OF SURFACES AND INTERFACES BY CORONA AND LOW PRESSURE PLASMA

J.E. KLEMBERG-SAPIEHA, L. MARTINU, S. SAPIEHA*,
AND M.R.WERTHEIMER
Department of Engineering Physics, École Polytechnique
Box 6079, Station "A"
Montréal, Québec H3C 3A7, Canada
*PAPRICAN and Department of Chemical Engineering

ABSTRACT. Physical and chemical effects in reactive gases, assisted by corona or by "cold" plasma at low pressure, allow one to modify the surface and interfacial properties of various materials or their combinations, for example of polymers. This, in turn, can greatly improve the materials' performance in specific applications, such as adhesion.

We begin with a succinct review of plasma principles, and then describe various aspects of plasma-surface interactions with the above-named objectives in mind. This will be illustrated, in particular, by our own experimental results obtained with the use of corona, on the one hand, and of low pressure plasma on the other. We distinguish between the effects of reagent gas type, of ion bombardment, and of ultraviolet radiation upon the chemical structures of polymeric surfaces and upon the resulting physical property changes, principally their adhesion characteristics and aging phenomena. The fundamental aspects are illustrated by numerous examples based on property evaluations of diverse polymeric composite systems.

1.   Introduction

In recent years, we have witnessed a remarkable growth in the use of synthetic organic polymers in technology, both for "high-tech" and for consumer-product applications. Polymers have been able to replace more traditional engineering materials, for example metals, on account of their many desirable physical and chemical characteristics (high strength-to-weight ratio, resistance to corrosion...) and their relatively low cost. However, fundamental differences between polymers and other engineering solids have also created numerous important technical challenges, for instance during manufacturing operations. An important example is the characteristic low surface energy of polymers and their resulting intrinsically poor adhesion [1-3]. Since adhesion is fundamentally a surface property, governed by a layer of molecular dimensions, it is possible to modify this surface-near region without affecting the materials' (desirable) bulk properties.

Over the years, several methods have been developed to modify polymer surfaces for improved adhesion, wettability, printability, dye-uptake, etc. These include mechanical or wet-chemical treatments, and exposure to flames, corona discharges, and glow discharge plasmas. A basic objective of any such treatment is to remove loosely bonded surface contamination and provide intimate contact between the two interacting materials, on a molecular scale. The reason is that molecular energies across an interface decrease drastically with increasing intermolecular distance r: typically the attractive energies associated with both the non-polar London dispersion force and the Keesom dipolar interaction decrease with r^{-6} [1]. The simplest method one can envisage is to mechanically roughen a surface, thereby enhancing the total contact area and mechanical interlocking, one of four basic mechanisms which have been proposed to explain

G. Akovali (ed.), The Interfacial Interactions in Polymeric Composites, 201–222.
© 1993 *Kluwer Academic Publishers.*

adhesion [4] (the others in this list are: weak boundary layer, electrostatic, and chemical reaction). However, roughening alone is rarely effective; wet-chemical treatments, for example using strong acids or bases, or the sodium/liquid ammonia treatment for fluoropolymers [5], are becoming increasingly unacceptable with time on account of environmental considerations. Furthermore, they tend to have inherent problems of uniformity and reproducibility, criticisms which are also often levelled against flame treatment.

Modification of polymer surfaces by plasma* treatment, both corona and low pressure glow discharges, presents many important advantages and overcomes the drawbacks of other processes mentioned above. Plasma treatment, as we show in later sections of this chapter, can give rise to controlled changes of a polymer's surface chemical composition. This is highly desirable, since it facilitates adhesion via the chemical reaction mechanism cited above, the one which in theory leads to the strongest possible bonds. For example, in polymer metallization, a key application area of adhesion to polymers [6,7], Burkstrand [8], was among the first to show that evaporated metals can react with polymers containing oxygen in their surface structure, via the formation of M-O-C (M=metal) types of chemical linkages. In the case of oxygen-free polymers the required O-containing functionalities at the surface can be created by plasma oxidation [9]. There is now an overwhelming body of evidence, recently reviewed by Liston [10], that plasma treatment for improved bonding is a general and economically attractive process.

In the remainder of this chapter we describe recent trends in corona and low pressure plasma modification of polymer surfaces and interfaces, which we illustrate with results and examples from our own laboratories. Section 2 deals with a brief description of plasma-chemical principles and plasma-surface interactions. In section 3, we describe how surface modification can be characterized in terms of chemical structure, wettability, etc., following which section 4 is devoted to adhesion of specific materials combinations, illustrated by case examples from these laboratories. In the concluding section, we comment on technological aspects and industrial scaleup.

2. Plasma Processes

2.1. GENERAL COMMENTS

The industrial use of plasma processing has been spearheaded by the microelectronics industry since the late 1960s.

a)   for deposition of thin film materials [11],

b)   for plasma etching of semiconductors, metals, and polymers such as organic photoresists [12].

Several dozen plasma equipment manufacturers in North America, Europe and the Far East cater to this market, and their combined annual sales exceed 10^9 US.

Contrary to processes a) and b) above, where material is added to or removed from the surface, respectively, the third type of plasma process (surface modification, of particular interest in the context of this chapter), does neither of these in significant amounts. Instead, the composition and structure of a few molecular layers at the materials' surface are changed by the

---

*A plasma may be succinctly defined as a partially ionized gas, with equal number densities of electrons and positive ions, in which the charged particles are "free" and possess collective behaviour.

plasma. Largely thanks to this third process category, we are currently witnessing a vigorous expansion of industrial plasma processes into areas other than microelectronics, for example into the automotive, aerospace, and packaging sectors.

In spite of this proliferation of applications, there are still many unresolved questions regarding the most efficient use of plasma processing, largely due to the inherent complexity of the plasma state. In order to ensure high quality and reproducibility of a given plasma process, numerous parameters must be controlled with care, such as pressure and flow rate of the reagent gas or gas mixture, discharge power density, surface temperature of the workpiece, etc. Currently, certain important parameters are still only poorly understood, for example frequency of the electrical power source, and plasma-surface interactions, and these are the objects of much ongoing research. In the remainder of this section, we start with a brief outline of low pressure plasma principles, followed by an equally succinct description of corona (atmospheric pressure) plasma.

2.2. PHYSICS AND CHEMISTRY OF LOW PRESSURE PLASMA

In a low pressure ($\leq$ 1 Torr = 133 Pa), high frequency discharge, the heavy particles (gas molecules and ions) are essentially at ambient temperature (~0.025 electron volts), while the electrons have enough kinetic energy (several eV) to break covalent bonds, and even to cause further ionization (that is, to sustain the discharge). Chemically reactive species thus created can partake in homogeneous (gas-phase), or heterogeneous reactions with a solid surface in contact with the plasma. Since this type of plasma chemistry takes place at near-ambient temperature, it is well-suited for processing thermally sensitive materials such as semiconductors and polymers.

As mentioned above, the creation of reactive species (radicals, ions, molecular excited states,...) in a plasma results primarily from inelastic collisions between "hot" electrons with energy $u = (mw^2)/2e$ (w, m and e being the electron velocity, mass and charge, respectively) and ground state atoms or molecules. The rate coefficient C_j for excitation of a particular species or state "j" is given by

$$C_j = [\frac{2e}{m}]^{1/2} \int_0^\infty \sigma_j(u) \ F_0(u) \ u \ \ du \tag{1}$$

where $\sigma_j(u)$ is the particular process cross-section and $F_0(u)$ is the electron energy distribution function (EEDF). It is noteworthy that nearly all processes of importance here possess an energy threshold $u_j = eV_j$ for which

$$\sigma_j(u){\equiv}0 \ , \ u{\leq}u_j \ . \tag{2}$$

The number density $\dot{n}_j$ of species "j" produced per second in the plasma from ground state molecules (of number density N cm^{-3}) clearly also depends on the electron density n, that is,

$$\dot{n}_j = C_j \ Nn \ . \tag{3}$$

The power balance between the applied electromagnetic field (of frequency $f = \omega/2\pi$) and the plasma can be expressed by

$$P_a = \theta nV \tag{4}$$

where P_a is the power absorbed in the volume V of plasma, and θ is the average power absorbed per electron. The parameter θ, which is readily measurable, can also be considered as the power required to sustain an electron-ion pair in the plasma. Ferreira and Loureiro [13] have calculated θ values for the "simple" case of low pressure Argon plasma; their model shows that at constant pressure, θ decreases as f is raised from "low frequency" to microwave (MW) frequencies. In other words, the efficiency of producing electron-ion pairs is greater at MW than at "low" frequency, for a given power density absorbed in the plasma. This can be readily understood from the EEDF shapes, shown in Fig. 1 [14], which have been obtained by solving the Boltzmann equation under steady state and ambipolar diffusion conditions. Figure 1 shows how the EEDF depends on $\dot{\omega}$ (=2πf) for three limiting cases, designated A ($\dot{\omega} \rightarrow 0$, "low frequency" case), H ($\dot{\omega} \rightarrow \infty$, no electron-electron collisions, "microwave" case), and M (dominating electron-electron collisions, Maxwellian distribution). There is strong experimental evidence [14], both for Ar and for more complex molecular gases, that M more closely represents "real" microwave discharges than does H; in any case, the relative numbers of electrons in the body (u < 11.55 eV, the lowest excitation threshold for Ar) and tail (u > 11.55 eV) of the EEDF are clearly quite different for "low frequency" and "microwave" plasmas; again, see ref. 14 for details. Using the three limiting EEDF cases of Fig. 1, ratios of $\dot{n}_j$ values have been evaluated as a function of threshold energy eV_j, under constant absorbed power density conditions. These are shown plotted in Fig. 2. We note that near $eV_j \approx 7.5$ eV, $\dot{n}_j(A) \approx \dot{n}(H) \approx \dot{n}(M)$; however, for all other threshold energy values we observe that

$$\dot{n}_j(M) \gg \dot{n}_j(A)$$
$$\dot{n}_j(M) > \dot{n}_j(H)$$

$$(5)$$

In other words, a microwave (Maxwellian) plasma may be expected to yield substantially higher concentrations of reactive species than a lower frequency plasma at the same power density.

Rather than use either microwave (MW, 2.45 GHz) or radio frequency (RF, 13.56 MHz) power to sustain our plasma, we often combine the two power sources to generate a so-called "mixed" (or dual-) frequency plasma, as shown in a schematic view of the plasma apparatus, Fig. 3 [15]. While MW excitation generates a high concentration of active species in the gas phase, as pointed out above, the role of the RF power is to create a negative DC self-bias voltage V_B on the powered, electrically isolated substrate holder electrode. This causes ions to be accelerated by the potential drop (V_P-V_B) across the RF-induced plasma sheath, to their maximum kinetic energy

$$E_{i,max} = e|V_P-V_B|$$

$$(6)$$

where V_P, the plasma potential, is generally a few tens of volts. In practical situations, at higher working gas pressures ≥ 100 mTorr, the ions lose part of their energy through inelastic collisions. The average energy is then typically

$$\overline{E}_i \approx 0.4E_{i,max} .$$

$$(7)$$

In other words, in dual-frequency (MW-RF) processing, independent control of the RF power allows us to vary the energy of ions bombarding the substrate surface, with values ranging from a few electron volts to several hundreds of eV, and with fluxes up to ~10^{16} ions/cm^2·s [16] (comparable with the operating parameters of low energy ion beam systems). Evidently, the plasma apparatus depicted in Fig. 3 can also be operated in the "pure" MW or RF modes, which in principle allows us to distinguish the effects of excitation frequency and ion bombardment in any given surface treatment experiment. Another important effect, particularly in plasma treatment of polymer surfaces, so far not mentioned here and frequently overlooked by others, is the fact

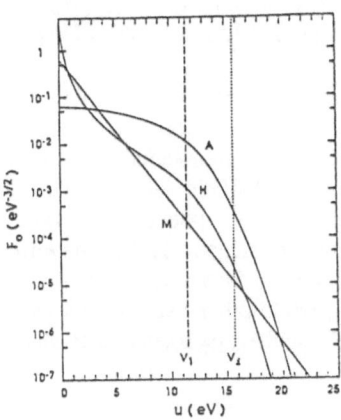

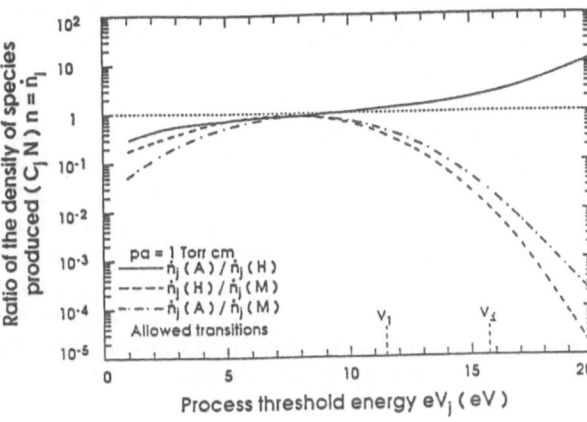

Fig. 1. Normalized EEDFs calculated from a self-consistent model for an argon discharge in a long, cylindrical tube, under ambipolar diffusion conditions, in three limiting cases for pa = 1 mTorr cm. (After ref. 14).

Fig. 2. Ratio of the densities of species produced per second (C_jNn) calculated with the EEDF corresponding to the three limitign situations A (dc), H (microwaves) and M (Maxwellian EEDF), as a function of the process threshold energy eV_j, under constant power density conditions; pa = 1 Torr cm. (After ref. 14).

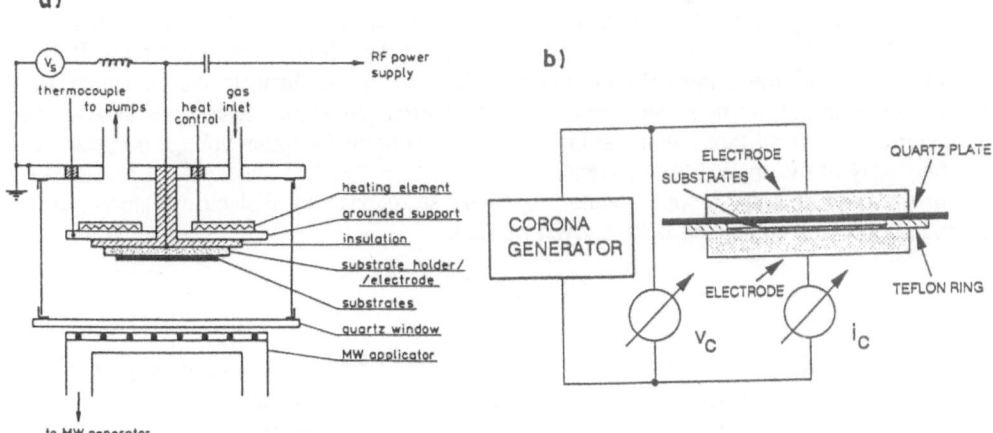

Fig. 3. Experimental apparatus: schematic view of (a) the dual frequency MW-RF plasma system (after ref. 9), (b) the corona treatment system. (After ref. 67).

that plasma can be a very rich source of vacuum ultraviolet (VUV) emission. Liston [10] has recently revived this important subject of VUV photochemistry, after it had suffered a lengthy period of neglect.

Briefly summarizing this subsection, then, low pressure plasma can provide electrons, ions and photons with energies sufficient to break any chemical bond [17]. Much of the plasma chemistry derives from gas phase and surface free radicals, and from plasma-surface interactions which are affected by ion bombardment (sputtering, enhanced surface mobility), radiation (cross linking), and surface temperature (reaction kinetics). On polymer surfaces, all these phenomena tend to occur simultaneously and synergistically; attempts have been made to selectively eliminate the effects of ions, electrons and VUV radiation by exposing the polymer surface only to the plasma effluent in a so-called "downstream" reactor configuration [18], while other experiments were designed to specifically elucidate the role of plasma-emitted VUV photons [9,10,19], or to study the effect of low energy (0-300 eV) ion bombardment [9,20,21].

2.3.   PHYSICS AND CHEMISTRY OF CORONA DISCHARGES

Corona discharges have been known for a long time. Historically, the first recorded form of corona was Saint Elmo's Fire (named after the sailors' patron-saint). It was observed by sailors during stormy weather on the tops of masts and on the end of yardarms, which acted as lightning rods. Because of the masts' rounded head, this light appeared like a crown or halo, whence came the term "corona" — the Latin word for crown. Subsequently, as sources of high voltage were developed, similar phenomena were observed in the laboratory, on high voltage power lines, etc., and they have been traditionally referred to as corona.

Corona discharges occur at atmospheric pressure, usually under highly nonuniform electric fields. A typical corona discharge, as it was first implemented, occurs on electrodes with sharp edges or small radii of curvature. Later, the term was extended to describe discharges between flat electrodes coated with a dielectric material. In this latter situation, an arc cannot develop since the dielectric interrupts the conductive path, and allows only an incomplete breakdown of the gas: Instead of a hot, localized arc, a cooler, diffuse glow develops between the electrodes. Beside a continuous corona current, there also exist avanlanches and plasma channels (arcs, streamers) with a random character. Streamers represent much higher current pulses than avalanches, and they play an important role in surface modification. The so-called corona inception voltage decreases with the frequency of the applied high voltage.

In the case of a flat-topped, cylindrical cavity surrounded by dielectric material, corona inception occurs at an applied field E given by [22]:

$$E = E_b \left(\frac{d_l}{\varepsilon d} + 1 \right) \tag{8}$$

where d_l is the thickness of the dielectric layer, d is the width of the discharge gap, ε is the dielectric constant of insulating material, and E_b is the breakdown field of the gap.

A typical modern corona treatment system consists of an audio-frequency power supply (usually 8 - 32 kHz), a high voltage step-up and matching transformer, and a treater station assembly consisting of a roll, an electrode, and a dielectric buffer. Older systems had a dielectric- or silica-coated roll and a bare electrode, but such systems pose both safety problems and are expensive to maintain when the dielectric coating on the treater roll is damaged. Newer systems, developed since the mid-seventies, are equipped with more reliable ceramic- or quartz-coated electrodes and bare metal rolls. Such electrodes have the advantage of being able to treat both conductive and nonconductive webs without any danger of causing a short circuit. Such systems also permit easy removal of ozone, which can then be catalytically decomposed. The benefits of using adjustable high frequency energy are a quieter discharge, more uniform treatment, a lower power requirement, and the ability to better match the treater system.

Regarding the chemistry of corona discharges, the best-known and largest use for many years has been ozone (O_3) generation [23], either from dry air or flowing oxygen. The corona discharge also emits visible and UV photons; if the feed gas is moist air, nitric acid is also a (undesirable) product. Corona has long been known as a convenient, relatively inexpensive tool for improving the adhesion of coatings and inks to plastics or paper [24,25], but despite many years of industrial use, it still remains more an art than a science.

3.   Surface Modification Related to Adhesion

As already mentioned in earlier sections, energetic particles generated in corona or glow discharge plasma (electrons, ions, photons, free radicals, excited atoms and molecules) interact strongly with the organic polymer surface. These plasma-surface interactions lead to the principal effects of ablation and cleaning, breakage of chemical bonds, localized heating, crosslinking, reaction of surface free radicals and, finally, incorporation of chemical groups originating from the plasma. All these processes affect adhesion, and they can be correlated with all basic adhesion mechanisms. The plasma treatment conditions (for example, the gas composition, treatment power and duration) must usually be optimized for each combination of materials. The reason is that plasma-surface interactions affect the final adhesive bond in a synergistic rather than in a cumulative way, and different agents can affect the polymer at different depths: for example, surface chemical modification occurs well within the first 100 Å [26], while VUV radiation can penetrate significantly deeper into the polymeric substrate [10]. Corona or plasma-induced oxidation is known to enhance polymer-metal adhesion [9], while fluorination [27] and silylation [28] impart improved hydrophobicity. Nitrogen containing plasmas have been used for surface nitridation, to introduce basic groups for applications such as improved dyeability with acidic dyes [29], printability, or cell affinity in biocompatibility [30]. Some of these are discussed in more details in section 4.

In this section we first discuss the effect of plasma modification on surface wettability (section 3.1) then, separately, on the surface chemistry (section 3.2).

3.1.   SURFACE WETTABILITY

Modified wettability is one of the most apparent results of plasma treatment. Polymeric surfaces are usually hydrophobic, displaying contact angles for water drops which typically range from 60° to over 90°. When exposed to plasma, the surface is activated by VUV radiation [10], accompanied by reaction with free radicals arriving from the gas phase (formation of polar groups), and by reactions in the surface leading to branching and crosslinking. The increase in surface energy (or decrease of the contact angle) usually correlates with a better bonding of adhesives, and it has often been used as an estimate of bonding quality. There are, however, exceptions to this rule; for example, good adhesion has been observed for a contact angle as high as 95°, for a polyphenylene sulphide (Ryton® R-4)/epoxy system treated by $CF_4 + 5\%$ O_2 plasma [10]. The reason may be that even a small amount of residual fluorine on the surface can cause the observed high value of advancing contact angle, close to that of the static contact angle, while the concentration of polar (C-O) groups can still be high enough to assure good adhesion: The surface energy is usually determined using measurements of static contact angles of a selected series of liquids [31]. However, real surfaces must be considered heterogeneous: surface wetting is affected by microroughness and, more important, by the relative concentrations of hydrophobic (e.g. C-F, C-H) and hydrophilic molecular groups (e.g. C-O, C-N) and their surface distribution. This leads to contact angle hysteresis, that is, a difference between the advancing θ_a and receding θ_r contact angles. Attempts should always be made to measure the contact angle hysteresis, because the surface energy E_s is derived from θ_a values for hydrophobic surfaces, and from θ_r values for hydrophilic surfaces. For a review of this subject, see reference [32].

The effect of nitrogen concentration (produced by NH_3 plasma treatment) on the contact angle hysteresis of low density polyethylene (PE) and polyimide (PI, DuPont Kapton® H) is illustrated in Fig. 4. Even on the untreated PE surface, the θ_r value is seen to be lower than θ_a, indicating the presence of polar groups, probably surface contaminants. With increasing nitrogen concentration, the θ_r values drop rapidly on the account of the resulting rise in concentration of hydrophilic groups. However, the hysteresis (θ_a - θ_r) remains high, until the concentration of hydrophobic groups has become small, at which point θ_a also decreases. Similar behaviour is seen to apply to polyimide, with the difference that this polymer already contains N and O in its original structure. The measurement of contact angle hysteresis can thus be used to determine relative concentrations of hydrophilic and hydrophobic groups. Measuring only the static contact angle and its relation to adhesion can be misleading, as demonstrated for the above mentioned example of O_2/CF_4 plasma treated Ryton® - R-4 [10]; Figure 5 shows, for the case of plasma-polymerized fluorocarbons, that different F/C surface compositions barely affect θ_a, but have a very major influence on θ_r values [33].

In practical applications, stability of the plasma treated polymeric surface is an important issue. Very often, the plasma-induced increase in surface energy is observed to fade with time. This process, which we refer to as aging, may result from the following three phenomena [34]: (i) inter-reaction of chemical groups on the treated surface, (ii) oxidation and degradation upon exposure to air, and (iii) diffusion of additives and low molecular weight (LMW) material to the surface. The effect of storage time on the wettability of NH_3 plasma treated PE is illustrated in Fig. 6. The initial contact angle values, $\theta_a = 105°$ and $\theta_r = 70°$, drop to $\theta_a = 30°$ and $\theta_r = 5°$ after plasma exposure. These values increase only slightly during the first few days and remain stable, even at the elevated temperature of 90°C. However, some increase in θ_a and θ_r is noted when the samples are exposed to high humidity.

In practice, it is difficult to distinguish between an aging (reversion) process due to the polymeric structure, and an "apparent" aging resulting from the diffusion of additives such as antioxidants, plasticizers, etc. A recent study of high purity polymer films revealed that a molecular relaxation mechanism, driven by entropy, can decrease the surface energy with time after the plasma treatment [35]. The rate of aging appears to depend on the mobility of polar moieties about the polymer chain: The polar groups have a tendency to rotate and "bury" themselves below the surface, so as to reduce the surface free energy. The most pronounced aging process has been observed for polypropylene (PP); polyethylene-terephthalate (PET) and polystyrene (PS) showed a lesser tendency to age [35], while PE [9,35] (see Fig. 6) and PI [9] are quite stable. Attempts have been made to completely suppress aging phenomena, by stabilizing the surface layer via crosslinking using controlled ion bombardment during MW plasma exposure [9,20,21].

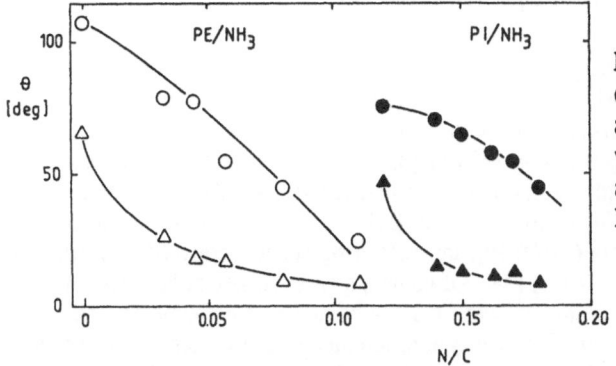

Fig. 4. Effect of XPS atomic concentration ratio N/C on the advancing (o, •) and receding (△,▲) water contact angles for PE (o,△) and PI (•, ▲) treated in NH₃ MW/ RF plasma. (After ref. 20).

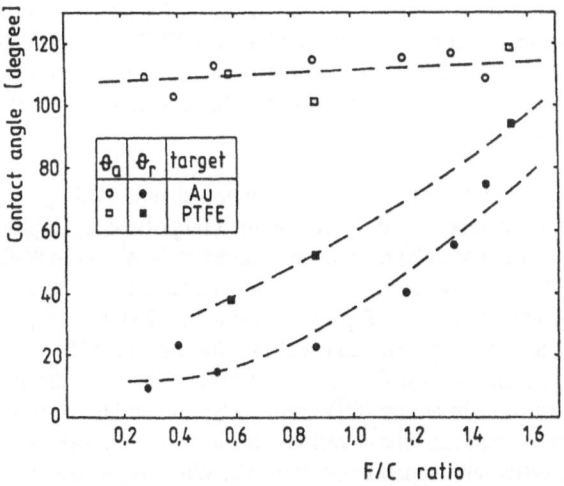

Fig. 5. Advancing (o, ◻) and receding (•, ◼) contact angles of plasma polymerized C_2F_6 with (o,•) and without (◻, ◼) gold clusters, versus F/C ratio. (After ref. 33).

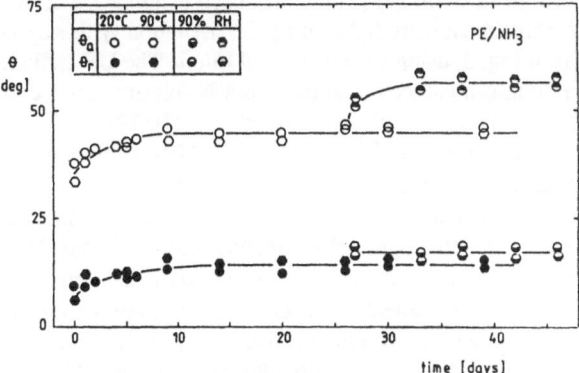

Fig. 6. Advancing (Θ_a) and receding (Θ_r) water contact angles for PE treated at substrate temperatures of 20°C and 90°C in MW plasma of NH₃, plotted versus storage time in ambient atmosphere and at 90% R.H. (After ref. 9).

3.2. SURFACE CHEMICAL STRUCTURE

A variety of surface-specific techniques have been reported in the literature for characterization of polymer surfaces. Among the most powerful and frequently used are X-ray photoelectron spectroscopy (XPS or ESCA) [3,6,8,9,18,21,27], static secondary ion mass spectroscopy (SSIMS) [3,36], Fourier transform infrared spectroscopy (FTIR) [3,7,37], and contact angle goniometry [1,9,20,31,32]. Also very useful are high resolution electron energy loss spectroscopy (HREELS) [6,7], scanning electron microscopy (SEM) [6,7,37], and techniques based on ion beam probes such as elastic recoil detection analysis (ERD) [6], ion scattering spectroscopy (ISS) [38], and Rutherford backscattering spectroscopy (RBS) [38]. Of course, a wide variety of "functional" test methods have been devised to evaluate the effects of a given surface treatment, for example mechanical peel or lap-shear tests, electrical property measurements, and others. These are too numerous to recite here, and the reader is directed to reference [39]. The same applies to plasma diagnostic techniques [40].

Many of the techniques listed above have been used in our past and current work, but in this chapter we report mainly results relating to a subset of these, particularly XPS, contact angle measurements, and various "functional" tests. Regarding XPS, our surface analytical instrument (VG - ESCALAB 3 Mk II) and spectral deconvolution procedures have been described in detail elsewhere [21,27], and will not be repeated here.

3.2.1. *Nitridation.* In the opening remarks of section 3 we already mentioned how basic groups deriving from surface nitridation of polymers can be useful. Two commercial polymers, Kapton® polyimide (PI) and linear, low density polyethylene (PE) have been exposed to MW and MW-RF plasma in pure N_2 gas. Figure 7 shows C(1s) XPS spectra of the untreated, clean polymer surfaces (lower spectra), and of surfaces following MW and MW-RF plasma treatments [21]. For the case of PE, plasma treatment is seen to lead to three new spectral features (designated C2 to C4), beside the original C1 peak at 285 eV associated with C-C and C-H bonded carbon; the new features arise from chemical bonding of nitrogen in amine (C2), imine (C3) and amide (or nitryl, C4) groups [21]. Clearly, the relative concentrations of these various functionalities (proportional to the relative peak areas) show a strong, systematic dependence upon V_B. While N_2 plasma leads to bonded nitrogen primarily in the form of imine groups (C=N, 287.0 eV), amine groups (C-N, 285.8 eV) are in the majority following NH_3 plasma treatment. In the former case, the total bonded nitrogen concentration can exceed 40 at % on PE [21]. It has been estimated that up to 20% of the total nitrogen uptake can be photochemically induced by UV radiation with $\lambda \geq 120$ nm [9]; this was found with the apparatus of Fig. 3, using various optical filters. The UV radiation from NH_3 MW plasma was found to be more pronounced than that from N_2 plasma, on account of intense H_α emission at 121.5 nm, in agreement with other published data [10,19].

Strong effects may also be noted for the case of Kapton® PI {poly (N,N'-P,P'-oxydiphenylene) pyromellitimide}, whose virgin C(1s) spectrum is evidently much more complex than that of PE: The C1 peak is attributed to carbon in oxydiphenylene, while C2 corresponds to carbon singly bonded to oxygen and nitrogen, plus the carbons in the pyromellitic dianhydride ring. The C3 and C4 peaks are associated with carbonyl carbon in imide linkages, and the shake-up satellite owing to aromaticity, respectively. Following N_2 plasma treatment, the spectra show important changes in structure and composition, such as a sharp increase in amide functionality (C3). Particularly noteworthy, however, is surface damage induced by ion bombardment. It is manifested by decreases in C4 and C5 intensities, due to breakage of C = O bonds in the PI structure, and to opening of benzene rings, respectively.

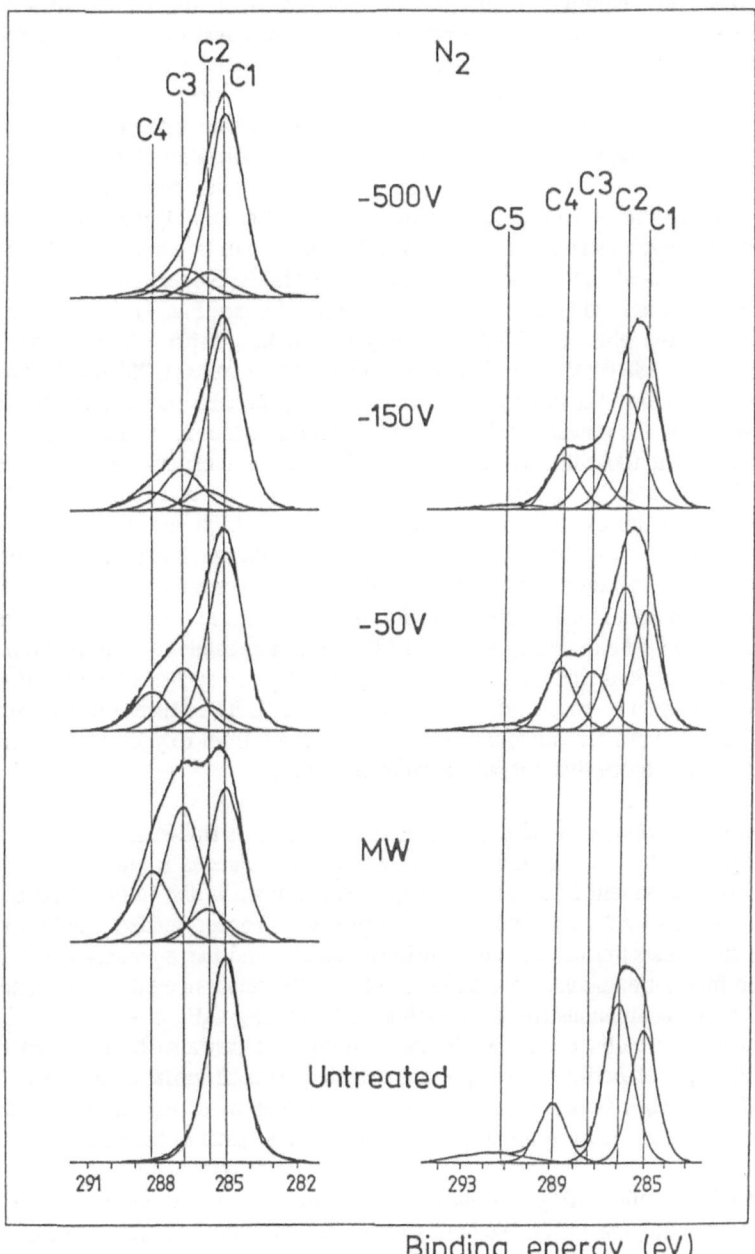

Fig. 7. C(1s) XPS spectra for MW and MW-RF N$_2$ plasma treated polyethylene (left column) and Kapton^R polyimide (right column). (Modified after ref. 21).

3.2.2. *Oxidation.* The vast majority of commercial processes involving corona treatment of polymer surfaces, carried out in ambient atmospheric air, are designed to create polar moieties (ether, carbonyl, hydroxyl ...) by reaction with activated oxygen species [24,25]. As discussed in section 3.1, the presense of such polar groups raises the polymers' surface energy; this permits wetting by printing inks, and substantially higher bond strengths than those obtained with the untreated surfaces, as also pointed out for the case of vacuum metallization. Low pressure air or oxygen plasma may also be used for this purpose, but the need to operate in a partially evacuated chamber tends to add significantly to the treatment cost. Low pressure O_2 plasma is, however, employed extensively in microelectronics for removal ("stripping") of polymeric resists, used for microlithography [41]. Figure 8a represents the curve-fitted C(1s) XPS spectrum of PE treated in air MW plasma; this spectrum (compared with that of "virgin" PE, see Fig. 7) clearly reveals the new features resulting from plasma oxidation, namely C-O (ether, 286.5 eV), C=O or O-C-O (carbonyl or double ether, 288.0 eV), and O=C-OH or O=C-O-C (carboxyl, 289.4 eV). The total surface oxygen concentration C_0 can be controlled by varying process parameters such as the plasma treatment time and the power density. This is discussed further in section 4.3.

Similar modifications to those illustrated by Fig. 8a above are also experienced by chemically cross-linked PE (XLPE) and by epoxy resin surfaces (Fig. 8b), when exposed to "partial discharges" (corona) in air, for example during faulty operation of high voltage equipment [22]. Evidently, this constitutes a highly undesirable example of plasma modification, since the insulating surfaces degrade and erode with time. Indeed, during prolonged corona exposure (tens or hundreds of hours), liquid and solid acidic reaction products (formic, acetic, oxalic acids) are observed to accumulate on the exposed surfaces [42,43]. Another example of undesired oxidation and erosion, mainly by atomic oxygen, occurs on the exposed organic surfaces (thermal blankets, paints, structural composites) of spacecraft in Low Earth Orbit (LEO). Much effort is currently being devoted to finding ways for shielding the organic material from oxygen atom attack, for example by applying inert protective coatings such as SiO_2 [44].

3.2.3. *Fluorination and Silylation.* As already mentioned earlier, it is frequently advantageous to render a surface hydrophobic, that is, non-wetting by water or aqueous solutions. An obvious example is the surface treatment of textiles; perhaps lesser known is the fluorination of inner surfaces of plastic gasoline tanks for automobiles, to provide a barrier against gasoline vapour permeation. For several years this has been done using molecular fluorine, a process which is now becoming outlawed for environmental reasons. Plasma-based fluorination, on the other hand, uses stable, saturated fluorine compounds (perfluorocarbons such as CF_4, C_2F_6, or sulfur hexafluoride SF_6), and these in much smaller quantities. Surface fluorination can also be achieved by the deposition of a thin layer of perfluorinated plasma polymer; here, an unsaturated fluorocarbon feedgas such as C_2F_4 or C_4F_8 is best used [33,45], but excellent water repellency can also be obtained by plasma polymerization of organosilicone "monomers" such as hexamethyldisiloxane [46].

A useful example for illustrating plasma surface fluorination is the case of kraft paper (lignocellulose, a natural polymer) [27]. We have studied plasma etching of kraft paper in O_2/CF_4 gas mixtures [47], but have also examined the accompanying changes in surface composition and water contact angle. Figures 9 a) and b) show the C(1s) XPS spectra of an as-received paper sample, and of the same material following 300 seconds exposure to MW plasma in flowing 0.8 CF_4 + 0.2 O_2 gas mixture. This treatment results in a surface composition containing about 40 atomic % fluorine (nearly the same as from pure CF_4 plasma treatment, and a water contact angle θ_a of 100°.

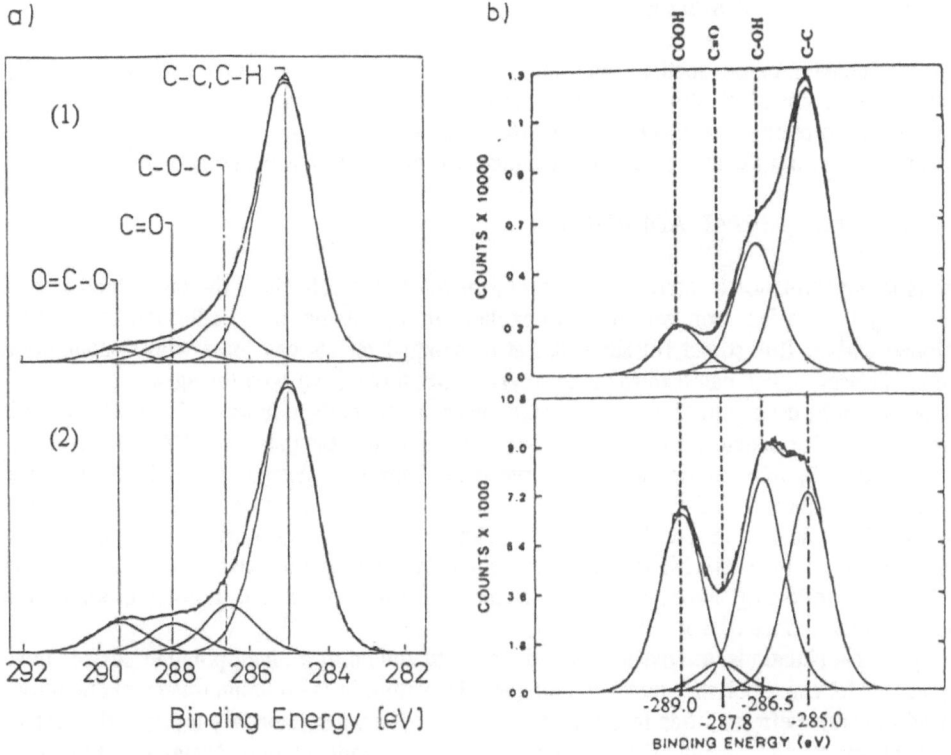

Fig. 8.   C(1s) XPS spectra: PE treated in low pressure air plasma for 30 s ([O] ~ 23 at.%) (1), and in corona discharge in air for 6 s ([O] ~ 25 at.%) (2) (after ref. 57).   Untreated epoxy (3), and epoxy degraded by exposure to corona discharge for 900 hours (4).   (After ref. 43).

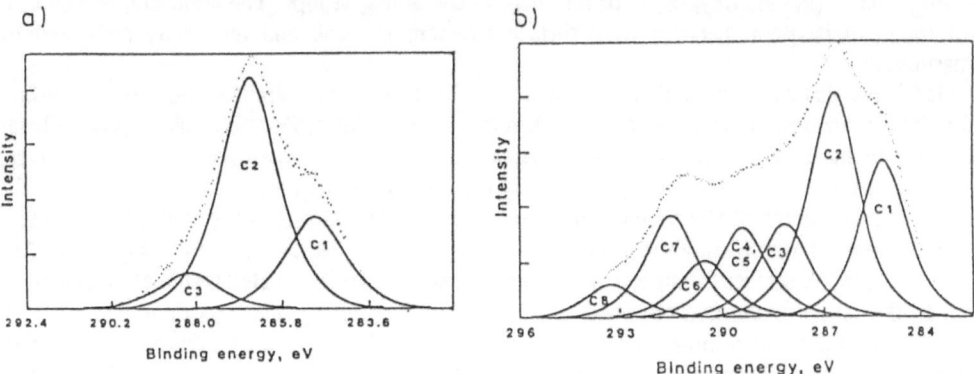

Fig. 9.   C(1s) XPS spectra of kraft paper: a) as received, b) after treatment in 80% CF_4/20% O_2 plasma.   (After ref. 27).

4.   Improvement of Adhesion

As already hinted in earlier sections, adhesion cannot simply be related to the surface wettability and the surface chemical structure, for in most cases it involves the important effects of composition, morphology, and properties of the interphase region in an adhesive joint. In the following, we discuss the role of plasma treatment for different material combinations.

4.1.   POLYMER-METAL ADHESION

There is strong evidence in favour of the chemical reaction mechanism, the one which should theoretically lead to the strongest bonds, to explain strong polymer-metal adhesion. As already mentioned earlier, Burkstrand [8] showed that evaporated metals can react with O-containing polymer surfaces, which can lead to metal-oxygen-carbon (M-O-C) type linkages.

Plasma can be used with success to provide the necessary surface functionalities which result in strong bonds. For example, in situ XPS studies have revealed the presence of Ag-O-C and Ag-N-C linkages [48] after exposing PE to oxygen and nitrogen plasma, respectively. For this particular system, the metal-PE adhesion was found to improve according to the following sequence of plasma gases used: $Ar < O_2 < N_2$. Similar effects were observed in another in situ study of Mg on PP [49]. The highest sticking probability for evaporated Mg atoms was found on a PP surface following exposure to N_2 plasma or, for comparison, following argon ion bombardment at a dose of 5×10^{15} ions/cm^2.

In practical situations, the plasma conditions for treatment of a given polymer surface must be optimized for every polymer-metal combination. Departure from optimum treatment conditions can lead to various effects, often involving the presence of a weak boundary layer: If the surface is insufficiently treated, the ubiquitous contamination layer (lubricant, mould release, antiblocking agents) is incompletely removed. Therefore, even if the surface is found to contain the required chemical groups (-O, -N), adhesion does not improve. On the other hand, the surface can also be "overtreated", which leads to the formation of LMW materials, as has also been clearly shown for the case of corona exposure [50]. In addition, additives from the polymer's bulk can diffuse to the interface and contribute to the formation of a weak boundary layer. A particular example of this are fluorocarbon polymers; they can readily be plasma treated for cleaning, to give good wetting, and to provide oxygen — or nitrogen — containing groups. The adhesion, however, is still found to be poor, because their surface structure is weak and not easily stabilized by crosslinking.

It follows that an "optimum" plasma treatment should yield two simultaneous effects: surface crosslinking, and formation of the requisite chemical functionalities. The crosslinking process leads not only to mechanical strengthening of the interface, but the crosslinked interface sublayer simultaneously provides a barrier against LMW diffusion from the polymer bulk. In practice, the control of crosslinking can be achieved by using plasma gases which strongly emit VUV radiation, the efficiency of which roughly follows the sequence $He > Ne > H_2 > Ar \approx O_2 \approx N_2$ [51]. Therefore, plasma surface treatments using mixtures such as He/O_2 or He/N_2 should, in principle, lead to the most pronounced adhesion improvements. Supporting this important role of VUV radiation, we have found more bound nitrogen on a PE surface following exposure to "filtered" radiation from NH_3 compared to N_2 plasma [9], which we have attributed to the VUV radiation emitted by atomic hydrogen (see also section 3.2.1).

4.2. INTERMEDIATE LAYERS

In some cases improved adhesion cannot be achieved merely by plasma surface modification, and another intermediate layer is needed; it increases the thickness of the interphase, and may also play additional roles. Metal-containing plasma polymers have been suggested as means to improve adhesion of structures such Au-PTFE [33], where chemical bonding is practically excluded. In this case, metal clusters are incorporated by simultaneous sputtering or evaporation during plasma polymerization of a fluorocarbon or hydrocarbon monomer [52]. The metal concentration can be gradually increased from the polymer side towards the metal side, so that adhesion improvement is achieved by complete mechanical interlocking. The composite polymer-metal structure can also be used to tailor the optical properties, for example in decorative applications [52].

Low pressure plasma treatment has been suggested as a solution of adhesion problems of copper to polyimide or to fluorocarbons, as used in multilayer printed circuit boards. In a simple case, polyimide must be pretreated by oxygen plasma, to increase adhesion of a Cr intermediate layer, onto which Cu is deposited by electroless plating [53]. When PTFE is used, on account of its good thermal and dielectric properties, the surface is activated by a H_2 plasma to achieve wettability, following which copper formate is applied by spin coating [54]. The copper formate is then converted to metallic copper by chemical reduction, again in hydrogen plasma.

Metallic intermediate layers of Pd, Pt, Au or Cu, obtained by plasma decomposition of organometallic compounds, have also been investigated for improved adhesion of electroplated Cu to PTFE [55].

Intermediate layers appear to be necessary for good adhesion in corrosive and other chemically active environments. Their role is to suppress diffusion of destructive agents towards the interface. In this capacity, plasma polymerized hydrocarbon (CH_4) films have been found to improve bonding of Pt electrical contacts to parylene or other polymers, used as implantable, biocompatible sensors [56].

4.3. POLYMER-POLYMER ADHESION

Plasma treatment can reinforce polymeric laminated structures incorporating adhesives, but studies are ongoing to achieve good polymer-polymer adhesion only by plasma activation. In the case of adhesives use, the improved strength results mainly from increased wettability, because enhanced surface energy increases the interfacial contact area. As shown in sections 3.1 and 3.2, oxygen or nitrogen plasma treatment is usually satisfactory for this purpose. Conditions which favour microetching, to improve bonding by mechanical interlocking, can lead to further improvement.

Adhesion and interfacial phenomena have recently been studied for the cases of PE/PE and PE/PET laminates without adhesive, following treatment in a low pressure MW air plasma, or in an ambient air corona discharge [57]. Conditions were chosen so as to achieve nominally comparable degrees of treatment, in terms of discharge energy dissipated per unit surface area. Although the concentration of chemically bound oxygen on the polymer surface was found to increase monotonically with the treatment time, the adhesion force exhibits a pronounced maximum between 11 and 14 atomic % oxygen, independent of the treatment type (see Fig. 10). This suggests that the chemical composition of the treated surface plays a major role. To examine this, we have conducted a detailed analysis of the surface chemical structure using high resolution XPS. This revealed systematically varying amounts of oxygen-containing functionalities, namely the ether (C-O-C), double ether (O-C-O) and carbonyl (C=O), and carboxyl (O=C-OH or O=C-O-C) groups, as shown in the spectra of Fig. 8. Comparing the relative concentrations of these

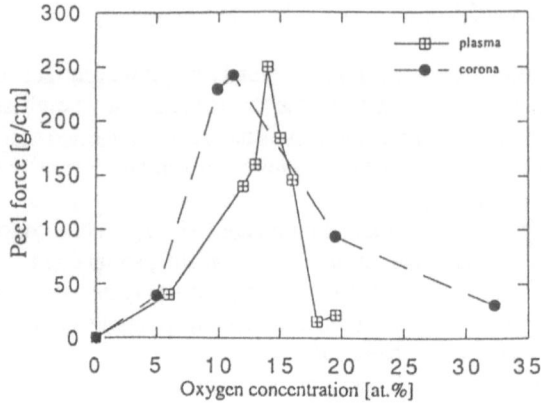

Fig. 10. Peel force of PE/PE laminate as a function of oxygen concentration, after low pressure plasma (⊞) and corona (●) treatment in air. (After ref. 57).

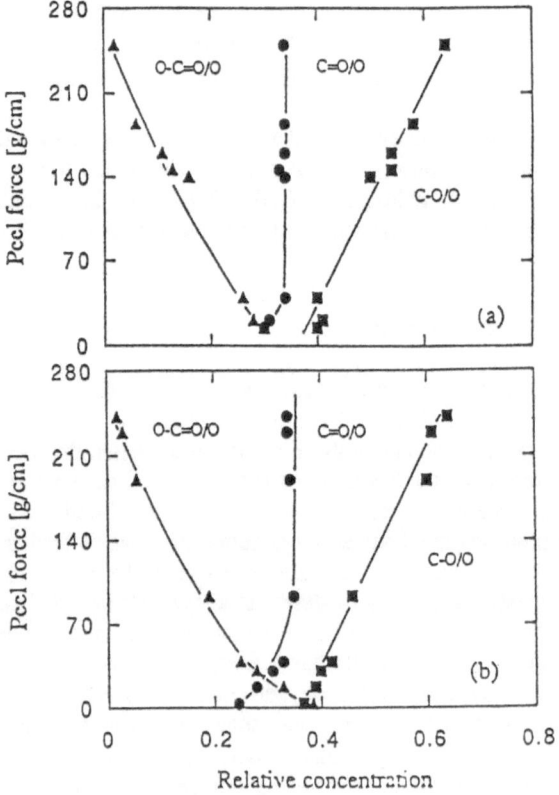

Fig. 11. Peel force of PE/PE laminates plotted versus relative concentrations oxygen-containing groups following (a) low pressure plasma and (b) corona discharge, both in air. (After ref. 57).

various groups with the peel force (see Fig. 11), we conclude that maximum adhesion occurs when the concentration of ether groups is highest and that of carboxyl (acid) groups is lowest. It is noteworthy that the peel force values and the "optimum" concentrations of oxygen and of the chemical groups are the same for both treatment types, suggesting that the same surface interaction mechanism applies to both corona and air plasma in the present case. The maximum adhesion force appears to result when the surface is mechanically stabilized by crosslinking (high ether concentration), and when the effect of a weak boundary layer due to LMW materials is minimal (low carboxyl concentration).

The conclusion here, that the weak boundary layer is composed mainly of acidic LMW material, agrees well with the data presented in section 3.2.2 [42,43]. It has yet to be established whether the maximum adhesive strength, observed above, results in part from direct covalent bonding across the interface between the two laminated surfaces.

The acidity of a strongly oxidized surface suggests that another type of bonding mechanism might be of importance here, namely acid-base interactions [58]. The following example, drawn from our experience, strongly supports this view: adhesion of dye molecules. Non-polar polymers such as PE or polypropylene (PP) cannot be treated with cheap, water-based acid dyes. If, however, basic (Lewis base) moieties are grafted to the surface, the acid-base interactions between these and the dye (Lewis acid) can give rise to strong ("chemical") adhesion at the polymer surface. Surface nitridation (amination) using plasma of N_2, NH_3, or volatile amines has been shown [29] to yield the desired results, which are attractive for both economic and environmental reasons.

4.4. POLYMER MATRIX COMPOSITES

Much research has been done over the years in these laboratories with regard to property enhancement of composite materials by surface modification of the filler material, either by corona or by low pressure plasma pretreatment.

Regarding plasma treatment, particulate filler materials such as $CaCO_3$ powder [59] or mica flakes [60,61] have been exposed to selected plasma gases. The objective was to impart either acidic or basic properties by chemical surface modification, or to deposit strongly-adhering plasma polymer layers, and thereby to enhance interfacial "compatibility" with various polymeric matrix materials. Among those examined were PE, PP, PS, and polyvinylchloride (PVC). In all cases, mechanical properties (for example, tensile strength) could be significantly enhanced for plasma-treated samples, compared with untreated control samples. Depending on the particular materials/treatment combinations, the improved properties could be explained in terms of acid-base interactions [59], or "bridging" via compatible polymer/plasma-polymer homologues [60,61].

Regarding fibre-reinforced composites, aromatic polyamide or "aramid" (DuPont Kevlar®-29) fibres or fabric were also modified or coated with plasma polymer, before being encapsulated in a triazine resin matrix [62]. Here too, very significant increases in the bond strength (peel strength of two-ply laminates) could be attributed to the plasma treatments; this was not entirely surprising, since untreated aramid fibres are known to have superb mechanical characteristics, but are also known to present bonding difficulties.

In a somewhat similar composite system, high strength polyethylene fibres in an epoxy resin matrix (Ciba-Geigy Araldite 6010), corona treatment has also been found to substantially improve the quality of bonding at the fibre-matrix interface [63]. The corona discharge is believed to enhance the interfacial adhesion of the substrate by oxidation (improved wetting and bonding by resin), micro-pitting (mechanical interlocking at the interface), surface crosslinking and chain scission (increased amorphous content which allows better diffusion), and elimination of a weak boundary layer.

The effect of corona treatment on the performance of cellulose/polyolefin composites has been investigated in great detail [64,65]: Corona exposure of PE and/or of the cellulose fibres improved the composites' mechanical properties. Figure 12 shows that yield strengths and strains, and elongations at break, were strongly affected by corona treatment, particularly in the case when the fibre surfaces were modified. Similar results have been obtained for cellulose/PP composites [66].

Sapieha and coworkers [67] found that the degree of surface treatment of cellulose fibres can be determined from the electrical conductance of a distilled water suspension of the treated fibres. The measured conductance is a function of treatment conditions such as discharge current, treatment time and availability of oxygen, and under certain conditions can be related to the mechanical properties of resulting composites, see Fig. 13. Increased conductance obviously results from the increasing concentration of water-soluble ionic species, for example the acids listed in section 3.2.2.

Independent studies have shown that corona treatments increase both the surface acidity and basicity. Typically, acid indexes are increased from 8.7 to 18.9, while surface basicity parameters are increased by corona treatment from 24.5 to 27.4. A major increase in the acid/base parameters would not necessarily affect interactions with a neutral matrix such as PE; on the other hand, acid/base interactions between fibres are practically doubled as a result of corona treatment.

The surface properties of corona-modified components directly affect rheological properties of composites. In the case of cellulose/PE composites, corona treatment of one or both of the constituents result in decreased melt viscosities relative to compounds containing untreated materials. The reason for this may be LMW moieties on the surfaces of both components, which act as lubricants at the cellulose/PE interfaces. Also, the corona-treated fibres have higher maximum packing volumes in PE than do their untreated counterparts. This may result from a reduction of fibre length, when corona-treated fibres are processed under high shear conditions. As a result, these fibres perturb the normal flow pattern in melt processing to a lesser degree than do the longer fibres of untreated cellulose, see Fig. 14.

5.   Scaleup for Industrial Use and Conclusions

The results presented in preceding sections of this review were mostly obtained with small research reactors of the type depicted in Fig. 3, which have typical substrate dimensions of 15 cm. Commercial corona systems exist which can treat webs up to several meters in width. Low pressure plasma apparatus, too, can readily be scaled up for industrial use. For example, regarding our proprietary LMP® (large volume microwave plasma) technology, a larger reactor, also with a single, 30 cm long microwave applicator, has been described elsewhere [68]. In this apparatus uniform plasma treatment of a 30 cm diameter substrate is accomplished by "planetary motion" within the plasma, the purpose being to eliminate possible non-uniformities in surface exposure resulting from the fringing field of the slow-wave applicator [69]. This apparatus is suitable for batch treatment of multiple discrete components. An even larger LMP® machine, incorporating two "counteractivated", 60 cm long slow wave structures, has also been built. This pilot-scale machine has been designed for very uniform plasma treatment of flat, 50 or 60 cm wide substrates, for example "batch-continuous" coating of flexible material (e.g. polymer or metal foil) in roll-to-roll operation. This is accomplished by virtue of the fact that the pair of microwave applicators compensate each others' fringing field inhomogeneities. These various reactor designs

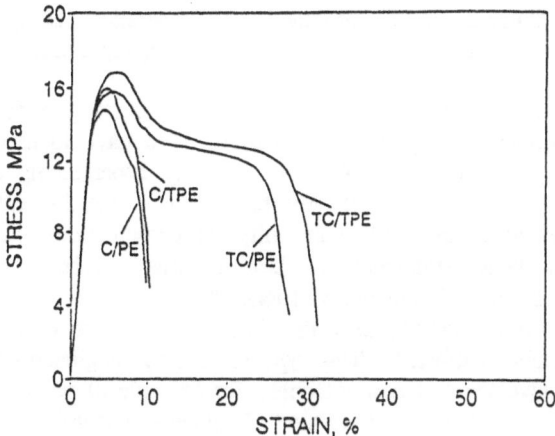

Fig. 12. Stress-strain dependence for corona modified composites. Fiber volume fraction: 28%. C/PE - no treatment, C/TPE - fiber only treated, TC/PE - PE only treated, TC/TPE - fiber and PE both treated. (After ref. 65).

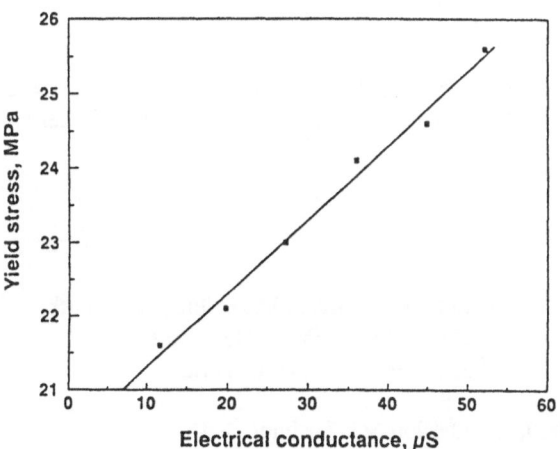

Fig. 13. Yield strength of cellulose/ polypropylene composite as a function of suspension conductance. Both cellulose and polypropylene treated by corona (25 mA). (After ref. 66).

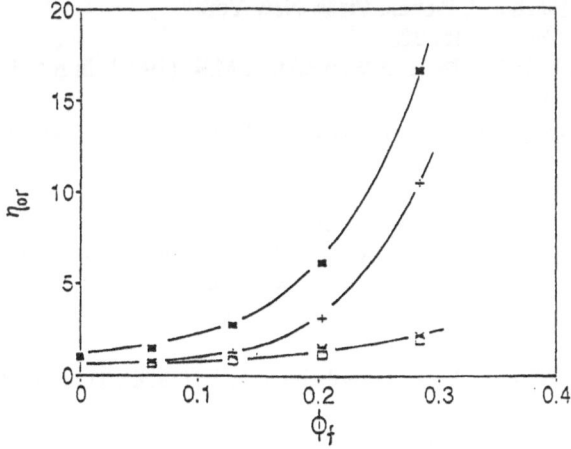

Fig. 14. Relative viscosity at zero shear rate as a function of fiber volume fraction. Composites: ■ - C/PE; + - C/TPE; • - TC/PE; □ - TC/TPE. (After ref. 65).

can, of course, all be made to function in single mode MW or dual MW-RF operation. Design concepts exist for very large industrial LMP® roll coating equipment, incorporating eight or more microwave applicators.

In conclusion, we have shown in this review that corona and plasma treatments can give rise to profound changes in surface and interfacial properties of materials. We have focused specifically on polymeric materials, for which this technology is currently experiencing a remarkable phase of industrial growth. The reasons are not only that plasma processing can achieve controlled surface modification with greater flexibility, uniformity and reproducibility than "conventional" methods (wet chemistry, flame treatments, ...), but that this can be done economically and, increasingly important, in an environmentally "friendly" manner.

Corona and plasma treatments of polymers modify only the near-surface region of the material, leaving bulk properties completely unaffected. This type of "surface engineering" constitutes a very powerful tool for the developers of new materials, for example of polymer matrix composites, as we have tried to illustrate in this text on hand of numerous examples.

Acknowledgements

This work has been supported by grants from the Natural Sciences and Engineering Research Council of Canada (NSERC), and the Fonds "Formation des chercheurs et aide à la recherche" (FCAR) of Quebec.

References

1. Souhang Wu (1982), Polymer Interface and Adhesion, Marcel Dekker Inc., New York.
2. Lee, L.H. (ed.) (1991), Fundamentals of Adhesion, Plenum Press, New York.
3. Clark, D.T. and Feast, W.J. (1978), Polymer Surfaces, Wiley, New York.
4. Mittal, K.L. (1976), J. Vac. Sci. Technol. 13, 19.
5. Siperko, L.M. and Thomas, R.R. (1989), J. Adhesion Sci. Technol. 3, 157.
6. Sacher, E., Pireaux, J.J., and Kowalczyk, S. (eds.) (1990), Metallisation of Polymers, American Chemical Society Vol. 440, Washington, D.C.
7. Mittal, K.L. (ed.) (1991), Metallised Plastics 2, Plenum Press, New York.
8. Burkstrand, J.M. (1978), J. Vac. Sci. Technol. 15, 223.
9. Klemberg-Sapieha, J.E., Martini, L., Küttel, O.M., and Wertheimer, M.R. (1991), in ref. 7.
10. Liston, E.M. (1989), J. Adhesion 30, 199.
11. Mort, J. and Jansen, F. (eds.) (1986), Plasma Deposited Thin Films, CRC Press, Boca Raton.
12. Manos, D.M. and Flamm, D.L. (eds.) (1989), Plasma Etching - An Introduction, Academic Press, Boston.
13. Ferreira, C.M. and Laureiro, J. (1984), J. Phys. D.: Appl. Phys. 17, 1175.
14. Moisan, M., Barbeau, C., Claude, R., Ferreira, C.M., Margot, J., Parasaczak, J., Sá, A.B., Sauvé, G., and Wertheimer, M.R. (1991), J. Vac. Sci. Technol. A3, 2643.
15. Klemberg-Sapieha, J.E., Küttel, O.M., Martini, L., and Wertheimer, M.R. (1990), Thin Solid Films 193/194, 965.
16. Küttel, O.M., Klemberg-Sapieha, J.E., Martini, L., and Wertheimer, M.R. (1990), Thin Solid Films 193/194, 155.
17. Clark, D.T., Dilks, A., and Shuttleworth, D. (1978), in ref. 3.

18. Foerch, R., McIntyre, N.S., Sohdi, R.N.S., and Hunter, D.H. (1990), J. Appl. Polym. Sci. 40, 1903.
19. Hudis, M. and Prescott, L.E. (1972), J. Polym. Sci., Polym. Lett. B 11, 179.
20. Bernier, M.H., Klemberg-Sapieha, J.E., Martinu, L., and Wertheimer, M.R. (1990), in ref. 6.
21. Klemberg-Sapieha, J.E., Küttel, O.M., Martinu, L., and Wertheimer, M.R. (1991), J. Vac. Sci. Technol. A 9, 2975.
22. Bartnikas, R., and McMahon, E.J. (eds.) (1979) Engineering Dielectrics, Vol. I, Corona Measurement and Interpretation, ASTM STP669, Philadelphia PA.
23. Eliasson, D., Hirth, M., and Kogelschatz, U. (1989), J. Phys. D: Appl. Phys. 20, 1421.
24. McLaughlin, Jr., T.F. (1962), Information Bulletin, Du Pont de Nemours & Co.
25. Traver, G.W. (1962), U.S. Patent 3, 018, 189.
26. Clark, D.T., and Dilks, A. (1977), J. Polym. Sci., Polymer Chem. Ed. 15, 2321.
27. Sapieha, S., Verreault, M., Klemberg-Sapieha, J.E., Sacher, E., and Wertheimer, M.R. (1990), Appl. Surf. Sci. 44, 165.
28. Sacher, E., Schreiber, H.P., and Wertheimer, M.R. (1985), U.S. Patent 4, 557, 946.
29. Cop, L., Jordaan, J., Schreiber, H.P., and Wertheimer, M.R. (1987), U.S. Patent 4, 744, 860.
30. Hollahan, J.R., Strafford, B.B., Falb, R.D., and Payne, S.T. (1969), J. Appl. Polym. Sci. 13, 807.
31. Kaelble, D.H., Dynes, P.J., and Cirlin, E.H. (1974) J. Adhesion 6, 23.
32. Morra, M., Occhiello, E., and Garbassi, F. (1991), J. Appl. Polym. Sci. 41.
33. Martinu, L., Pische, V., and d'Agostino, R. (1990), in ref. 6.
34. Spell, H.L., and Christenson, C.P. (1978), Tappi, 283.
35. Morra, M., Occhiello, E., and Garbassi, F. (1991), in ref. 7.
36. Lub, J., Van Vroonhoven, F.C.B.M., Brunix, E., and Benninghoven, A. (1989), Polymer 30, 40.
37. Poncin-Epaillard, F., Chevet, B., and Brosse, J.-C. (1991), Makromol. Chem. 192, 1589.
38. De Puydt, Y., Bertrand, P., and Lutgen, P. (1988), Surface Interface Anal. 12, 486.
39. Mittal, K.L. (ed.) (1983), Adhesion Aspects of Polymeric Coatings, Plenum, New York.
40. Auciello, O., and Flamm, D.L. (eds.) (1989), Plasma Diagnostics, Academic Press, Boston.
41. d'Agostino, R. (ed.) (1990), Plasma Deposition, Treatment and Etching of Polymers, Academic Press, Boston.
42. Gamez-Garcia, M., Bartnikas, R., and Wertheimer, M.R. (1987), IEEE Trans. Electr. Insul. EI22, 199.
43. Hudon, C., Bartnikas, R., and Wertheimer, M.R. (1991), Proc. IEEE CEIDP, IEEE Doc. 91CH 3055-1, 237.
44. Zimcik, D.G., Wertheimer, M.R., Balmain, K.G., and Tennyson, R.C. (1991), AIAA J. of Spacecraft 28, 652.
45. d'Agostino, R., Cramarossa, F., Fracassi, F., and Illuzzi, F. (1990), in ref. 41.
46. Sacher, E., Klemberg-Sapieha, J.E., Schreiber, H.P., and Wertheimer, M.R. (1984), J. Appl. Polym. Sci., Appl. Polym. Symp. 38, 163.
47. Sapieha, S., Wrobel, A.M., and Wertheimer, M.R. (1988), Plasma Chem. Plasma Process. 8, 331.
48. Gerenser, L.J. (1988), J. Vac. Sci. Technol. 15, 2897.
49. Nowak, S., Mauron, R., Dietler, G., and Schlapbach, L. (1991), in ref. 7.
50. Strobel, M., Dunnatov, Ch., Strobel, J.M., Lyons, Ch. S., Perron, S.J., and Morgen, M.C. (1989), J. Adhesion Sci. Technol. 3, 321.
51. Liebel, G., and Bischoff, R. (1987), Kunstoffe - German Plastics 4.

52. Biederman, H., and Martinu, L. (1990), in ref. 41.
53. Blackwell, K.J., Chen, P.C., Knoll, A.R., and Kim, J.Y. (1992), Proc. 35th Ann. Tech. Conf., Society of Vacuum Coaters.
54. Padiyath, R., David, M., and Babu, S.V. (1991), in ref. 7.
55. Meyer, H., Schulz, M., Suhr, H., Haag, C., Horn, K., and Bradshaw, A.M. (1991), in ref. 7.
56. Ratner, B.D., Chilkoti, A., and Lopez, G.P. (1990), in ref. 41.
57. Martinu, L., Klemberg-Sapieha, J.E., Schreiber, H.P., and Wertheimer, M.R. (1991), Le Vide, Supplement 258, 13, and Sapieha, S., Cerny, J., Klemberg-Sapieha, J.E., and Martinu, L. (1992), J. Adhesion, submitted.
58. Fowkes, F.M. (1987), J. Adhesion Sci. Technol. 1, 7.
59. Schreiber, H.P., Wertheimer, M.R., and Lambla, M. (1982), J. Appl. Polym. Sci. 27, 2269.
60. Bialski, A., Manley, R.St.J., Wertheimer, M.R., and Schreiber, H.P., J. Macromol. Sci. Chem. A 10, 609.
61. Schreiber, H.P., Tewari, Y.B., and Wertheimer, M.R. (1976), J. Appl. Polym. Sci. 20, 2663.
62. Wertheimer, M.R., and Schreiber, H.P. (1981), J. Appl. Polym. Sci. 26, 2087.
63. Yuhas, D.E., Dolgin, B.P., Vorres, C.L., Nguyen, H., and Schriver, A. in Ishida, H. (ed.) (1988), Interfaces in Polymer, Ceramic and Metal Matrix Composites, Elsevier, New York.
64. Dong, S., and Sapieha, S. (1991), SPE Technical Papers 37, 1154.
65. Dong, S., Sapieha, S., and Schreiber, H.P. (1992), Polym. Sci. Eng. (in press).
66. Belgacem, N.M., Bataille, P., and Sapieha, S. (1992), J. Appl. Polym. Sci. (in press).
67. Sapieha, S., this conference.
68. Lamontagne, B., Wrobel, A.M., Jalbert, G., and Wertheimer, M.R. (1987), J. Phys. D. 20, 844.
69. Bosisio, R.G., Wertheimer, M.R., and Weissfloch, C.F. (1973), J. Phys. E.:Sci. Instrum. 6, 628.

PLASMAS AND SURFACES - A PRACTICAL APPROACH TO GOOD COMPOSITES

E. M. LISTON
GaSonics/IPC
2730 Junction Avenue
San Jose, California 95134-1909, USA

ABSTRACT. A review is presented of the mechanisms for the initiation of rapid reactions between a plasma and a polymer or graphite surface. It is shown that the initiation mechanism is primarily vacuum ultraviolet photochemistry. Once initiation has occurred normal chain reaction free-radical chemistry will ablate the surface, cleaning it and leaving polar moieties in the surface. The resulting surface will be wettable and bondable. It is shown that there are several general cases in which the correlation between good wetting and good bonding does not hold. It is also shown that, in most cases of commercial polymers, any reversion or loss of wettability is probably caused by the diffusion of additives to the surface of the polymer rather than pure thermodynamic rearrangement of the surface moieties. Data are presented that show significant improvement in the mechanical properties of three different composite systems: graphite/epoxy (55%), polyaramide/epoxy (up to 2400%), and polyethylene/epoxy (380%).

Keywords: plasma/polymer/vacuum ultraviolet photochemistry/wetting/adhesion/ composites/surface treatment/free radical/bonding.

G. Akovali (ed.), The Interfacial Interactions in Polymeric Composites, 223–268.

TABLE OF CONTENTS

1. Introduction

The physics definition of a "Plasma" is an ionized gas with an essentially equal density of positive and negative charges. It can exist over an extremely wide range of temperature and pressure. The solar corona, a lightning bolt, a flame, and a "neon" sign are all examples of electrical plasma.

For the purposes of this paper, the discussion will be limited to low-pressure, 13 to 133 Pa (0.1 to 1 Torr) plasma, or glow discharge, such as is found in the "neon" sign or a fluorescent light bulb. In those two applications the desired result is to produce light. However, for the plasma treatment of materials for cleaning or surface modification, the extremely energetic chemical environment of the plasma is utilized to make chemical changes in the material surface.

Plasma is used extensively in the semiconductor industry to etch patterns in wafers and to remove photoresist. However, in this paper the discussion will be limited to non-semiconductor applications. Also, I will limit the discussion to plasma surface modification. There will be very little on plasma deposition or coating.

It has been known for at least 50 years that plasma could effect desirable changes in the surface properties of materials, particularly polymers. However, the practical application of plasma required the development of commercially available, reliable, and large plasma systems. Such systems are now available and the application of plasma to industrial problems has been increasing rapidly for the past ten years.

In plasma treatment of polymeric materials, all significant reactions are based on free radical chemistry [1]. The glow discharge is efficient at creating a high density of free radicals, both in the gas phase and in the surface of organic materials, even the most stable polymers. These surface free-radicals are created by direct attack by gas-phase free-radicals or by photodecomposition of the surface by vacuum ultraviolet light generated in the primary plasma. The surface free-radicals then are able to react either with each other, or with species in the plasma environment.

Free radicals created by low pressure gas plasma have four major effects on organic substrates: surface cleaning; ablation, a form of dry micro-etching; cross-linking; and surface chemistry modification. These four effects occur concurrently and, depending on processing conditions and reactor design, one or more of these effects may dominate. In all cases, these processes affect only the top few molecular layers (about 10 to 30 nanometers) so they do not change the appearance or bulk properties of the material. The net result of these effects is a major improvement (2 to 10 times) in processes that require adhesion or wettability.

Many different gases and plasma operating parameters are used to surface treat different materials. Studies have been performed on the effect of these different plasmas on various polymers. It has been found that, for best results, different polymers may require different plasma treatment. In some cases it has been found that a plasma which gives excellent results on one polymer may give very poor results on a similar polymer. For example, the best process for perfluoroalkoxy polymer (PFA) gives poor bonding to fluorinated ethylene-propylene polymer (FEP).

This paper will discuss the nature of plasma, the equipment used to generate plasma, plasma chemistry, the interaction between plasma and organic surfaces, the ageing of plasma treated polymers, and the plasma cleaning of inorganic surfaces, the results of plasma treatment on wettability and composite properties, and some practical considerations.

2.   Why Use Plasma?

The basic reason for using plasma processing is to get better parts at a lower cost. There are many features of plasma processing that make this possible, among these are:

- Clean and dry
- Safe for operators
- Material integrity maintained
- Easy to use
- Economical
- No environmental problems
- Multifunctional
- Continuous processing of yarn, fiber and fabric
- Results unavailable by any other means

The utility of plasma and its economic or performance impact on a product, depend on the magnitude of the problems being solved by the plasma processing. Therefore, each of the features of plasma processing will be discussed in relationship to the problem that the feature might solve.

2.1. PLASMA IS CLEAN AND DRY

One of the critical points to be made is that plasma itself is extremely clean. Clean gases are the only material added to the plasma reactor chamber, other than the parts to be cleaned. Even if these gases contain a contaminant, such as oil vapor, the plasma quickly converts it to volatile reaction products that are pumped out by the vacuum system. Therefore, there is no unwanted residual contamination on the cleaned parts from contamination in the plasma itself.

This is in marked contrast to any cleaning process that uses liquids. It will be shown that the residue from the evaporation of liquids containing as little as 10 ppm of non-volatile organics can leave a layer of organic contamination on a surface. If this contamination is strongly adsorbed on the cleaned surface it may not be displaced by an adhesive and it may not be wet by polar liquids such as water or adhesives. Therefore, it is almost always true that parts "cleaned" using liquids, or even a condensing vapor rinse, will have a layer of adsorbed contamination on the surface of the part at the end of the cleaning cycle.

The ability of plasma to clean without leaving a layer of contamination is one of the most important features of plasma processing. Plasma is probably the only cleaning process that potentially leaves no residual contamination.

2.2. PLASMA SYSTEMS ARE SAFE TO OPERATE

These systems, like any vacuum or electronic system, have some potentially dangerous aspects. Accordingly, manufacturers have gone to great lengths to address these potential problems and to make the systems safer to use. In order to insure the operator's safety the following safety rules should always be observed:

1) Never operate a plasma system without having all the sheet metal covers and safety shields in place.
2) Never operate any system with interlocks that have been defeated or bypassed.
3) Observe all the safety rules for the operation of vacuum pumps and for the handling of pressurized gases, especially toxic gases.
4) Follow all of the manufacturer's recommendations regarding safety and routine maintenance.
5) Never attempt to service a plasma system unless you have been specifically trained to do so by the manufacturer. These systems use voltages high enough to kill.

If these rules are observed, and some common sense is used, the systems can be used safely by operators with very little training. With a plasma system it is necessary to do something stupid to get hurt. There are no acids or solvents that can make ordinary accidents dangerous.

2.3. MATERIAL INTEGRITY IS MAINTAINED

Plasma processing affects the top few molecular layers on the immediate surface of a polymer. This layer is only about 10nm (0.4 micro inch) thick. Therefore there are no visual effects on materials that are properly treated. Gross over-treatment can lead to some hazing of some polymers but the parts have to be deliberately over-treated for damage to be visually apparent.

Also, there will be no change in the bulk physical or mechanical properties of the parts. The plasma treated surface layer is just too thin to affect the bulk properties. While it is not possible to affect the bulk material properties, the major use of plasma is to modify the surface properties of materials. This surface modification makes it possible to get greatly improved bonding to surfaces and can lead to greatly improved composite properties. However, in some applications it is not possible to change the physical properties of the surface to a depth sufficient to be useful. For example, it is possible to fully fluorinate some polymers and get a Teflon®-like surface. However, this plasma modified surface cannot be used as a low-friction surface (for most applications) because it is so thin that it is easy to rub off.

It is possible to get much thicker surface layers by plasma deposition of a polymer-like layer or by treating the surface to make it wettable and bondable. That plasma generated surface can then be dip or spray-coated with a coating of the desired properties (such as an insulator, barrier coating, paint, lubricant, etc.).

The extensive use of plasma in the semiconductor and electronics industry has demonstrated that the proper plasma treatment will not damage sensitive electronic parts. The basic reason is that plasma is such a good electrical conductor that it will dissipate or neutralize any static electric charge before it builds up to a dangerous level.

2.4. PLASMA IS EASY TO USE

Commercial plasma systems are designed to be used in production environments. The only thing that a floor operator needs to know is how to load the parts, push the "start" button, and unload the parts. Obviously, someone has to know how to program the system with the proper operating sequence before it can be used in such a simple mode. The programming instructions are in the operating manuals and they should be learned to avoid problems. Remember, "when all else fails, read the directions"! If there are problems, the manufacturer's service department can help.

It is usually not necessary to know any chemistry or physics to use a plasma system. The Applications department of the manufacturer can provide guidelines for developing the necessary processes.

2.5. PLASMA PROCESSING IS ECONOMICAL

Plasma processing is often the least expensive way to get excellent coating, bonding, printing, or wetting. The direct operating cost of even the largest plasma systems is only a few dollars per hour.

In evaluating the economics of changing to a plasma process from an existing process, it is most important to break the cost down further into a per-part cost. This per-part cost can then be compared to that of the TOTAL per-part cost for the existing process.

The total cost should always include the cost of:

- Operator training for the present surface treatment process
- Purchase of acids or solvents
- Special tanks or hoods
- Disposal of used acids or solvents
- Operating fees or licenses
- Safety systems
- Manufacturing steps that can be eliminated by plasma processing

In addition, you should calculate the incremental value added of having:

- Better parts
- Less rejects
- More uniform results, that is, less variability in yield and quality
- The possibility of using better, or cheaper materials
- The possibility of using adhesive bonding systems that cannot be used with the current processing.

Plasma processing is almost always economically advantageous when all of the costs and benefits are included in evaluating the cost of the present process. The cost savings of plasma processing will become even more important as the pressure of environmental problems increases.

2.6. PLASMA PROCESSING HAS NO ENVIRONMENTAL PROBLEMS

At the present time there are no U.S. EPA regulations that cover the emissions from industrial plasma systems. There are controls required on semiconductor systems, but that is because of the extreme toxicity of some of the gases used there.

The gases vented from industrial plasma cleaning systems are primarily oxygen, water vapor, CO, and CO_2 and none of these need to be controlled in the quantities given off by a plasma system. It is not necessary to use controls even when gases like NH_3 or CF_4 are used. These are vented in such small quantities that, with the proper dilution, they are not a problem. There is literally no cost or regulatory problem associated with the disposal of the reagents used in industrial plasma systems. This contrasts with the problems and costs of disposing of used acids and solvents from the normal cleaning, stripping or etching processes.

2.7. PLASMA EQUIPMENT IS MULTI-FUNCTIONAL

A properly designed plasma system will have a very wide operating range in terms of power or pressure range and the variety of gases that can be used. There will be multiple gas inputs available so gas can be changed simply by energizing a different solenoid valve in the controller.

Microprocessor controllers can store many process "recipes" that can be used alone or in sequence. All the operator needs to know is which recipe should be used on any given part. He can then use one machine for treating many different types of parts. Parts of different sizes can be processed by using different spacing between electrodes or even different electrode configurations. Electrodes are easy to change and the rest of the system remains unchanged.

2.8. CONTINUOUS PLASMA PROCESSING OF WIRE, YARN, FILM, AND FABRIC

Historically, most of the applications of plasma have been for the batch treatment of discrete parts. However, continuous air-to-air or cassette-to-cassette processing is now becoming available. The major problems are not in the plasma part of the system, but rather in the vacuum seals around the material entering and leaving the plasma chamber. These problems are being solved and practical, continuous units are now available. Their primary use to date has been the treatment of continuous yarn for the fabrication of composites.

2.9. RESULTS THAT ARE UNAVAILABLE BY ANY OTHER MEANS

All of the features of plasma processing discussed above combine to make plasma a manufacturing tool that gives more yield of better, or less costly, parts than alternative surface preparation processes.

3. Nature Of Plasma

The practical definition of a plasma is "a gas with electricity going through it". This inelegant definition is literally correct for industrial plasma.

The proper physics definition for a plasma is "a partially ionized gas consisting of an equal number of positive and negative charges". Note that this definition says nothing about the temperature or pressure at which a plasma can exist.

Plasma is probably the most common form of matter in the universe. All stars have a mantle of "hot plasma" (tens of millions of degrees) where the gas is so hot that most of the electrons are stripped off the nucleus. Interstellar space is also believed to contain a very low density plasma.

Lightning is considered a "warm plasma" because it is not particularly hot, as plasma goes (10,000°C to 40,000°C), and only a few electrons are stripped off the atoms. However, it is extremely dense. The initial pressure in a lightning bolt is hundreds of atmospheres and the gas is ionized at that high pressure. This hot gas gives off the light from lightning and the expansion of the high pressure core is the cause of the thunder. Multiple strikes can occur on the same path as the original strike because the partially ionized gas from the first strike is a better electrical conductor for later strikes than the surrounding air.

Typical examples of "cold plasma", which will be dealt with in this paper, are "neon" signs and fluorescent light bulbs. The gas in these is usually at room temperature but it may be as hot as 200°C. Also, it is at low pressure, 13 to 133 Pa (0.1 to 1 Torr). At these conditions the charged particles are a few ions (about 10^{-5} of the total number of gas molecules), and an equal number of electrons (which carry almost all the electric current). The presence of the ions and electrons makes the plasma a very good electrical conductor. In the center portion of the plasma, that is, away from the electrodes, it is about as good a conductor as steel. For this reason, once the plasma is lit it only takes a few 10s to 100s of volts to keep it lit.

The ions are essential for the electrical conductivity of a plasma but they only play a limited role in the chemical actions of the plasma. Their role is that of initiating chemical reactions on the surface of parts, especially if there is a static bias on the parts.

In addition to the charged particles, there is about 5% to 20% of dissociated molecules, called "free-radicals", in the plasma. For example, an oxygen molecule (O_2) will dissociate into two oxygen atoms (O). These radicals are not electrically charged but they carry a large amount of chemical energy and can attack chemical bonds on the surface of the parts and cause many different reactions. From a chemical point of view, these radicals are the most important species in a plasma. They are responsible for most of the chemical reactions and for the generation of the ultraviolet light in the reactors.

The other aspect of the chemical action of a plasma is the effect of ultraviolet light. Measurement of the light emitted by plasmas show that almost all of them generate intense, very high energy, ultraviolet light. The parts are immersed in this light source without any intervening windows to absorb the short wavelength, high energy, radiation. Therefore, photochemical reactions are possible in a plasma that are not possible with ordinary ultraviolet light sources. The ultraviolet light reacts with the surface of parts, especially organic surfaces, and the result is that a plasma reactor is largely a photochemical reactor. Experiments have shown that at least half of the action of a plasma on organic surfaces is the direct result of photochemical reactions.

The net result of the presence of ions, free-radicals, and ultraviolet radiation is a gas plasma environment that has an extremely high chemical reactivity. The proper plasma can convert any organic material to volatile products. This is the basis of the use of plasma for cleaning or stripping of organic coatings and the modification of polymer surfaces. Plasma can also be used to etch some metals.

It must be stressed that there is no "universal" plasma process or treatment. Different gases may give different results on the same materials; also, different materials, even ones that are quite similar chemically, may require different gases to get the best results. Some materials cannot be plasma treated to give good results, this may be due to their molecular structure or to the additives used in making them.

4.   Plasma Equipment

Figure 1 is a block diagram of a typical radio frequency plasma system. It consists of 5 modules or functions: vacuum system, power supply, matching network, power monitor, reactor center, and controller.

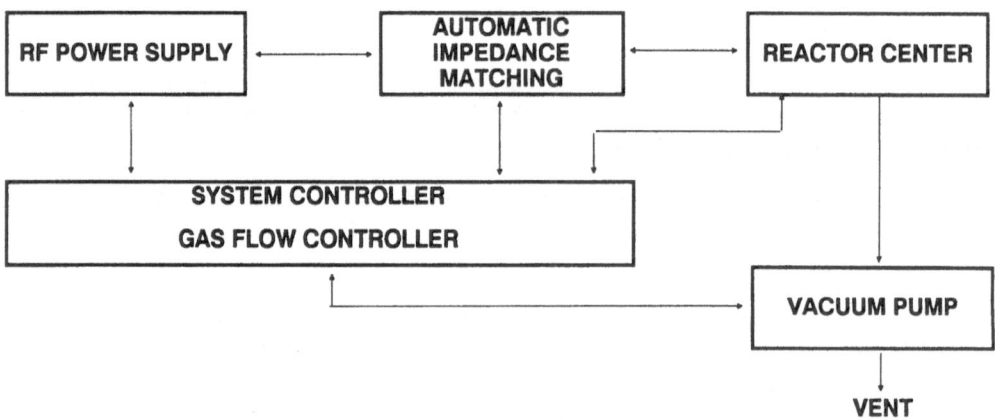

Figure 1. Block diagram of a plasma system.

4.1. VACUUM SYSTEM

Low-pressure plasma systems operate at 13 Pa to 133 Pa (0.1 Torr to 1 Torr) with a continuous gas flow into the reactor. Therefore, the vacuum system must be able to maintain this pressure/flow regime. However, the moderate vacuum level does not require sophisticated pumps. Two stage mechanical pumps are satisfactory. The pump package is usually sized to allow pumpdown in less than 1 minute and to maintain an inlet gas flow of 50 cc/min to 1000 cc/min, depending on the size of the reactor. The most expensive part of the vacuum system is the perfluorinated lubricant required if an oxygen plasma is to be used.

4.2. POWER SUPPLY

The power supply furnishes the electrical power necessary to generate the plasma. The power required ranges from 50 watts to 5000 watts, depending on the size of the reactor. The larger generators are usually cooled, either with air or water.

Plasma reactors have been built utilizing a wide range of frequencies, from DC to microwave. DC plasmas are difficult to use because they require a current limiting resistor to prevent arc formation. This resistor wastes power and must be changed for each different gas and operating condition. Also, ion bombardment becomes a factor in DC plasmas.

Low-frequency plasma (50 Hz TO 500 KHz) is sometimes used because the generators are somewhat less expensive and because they do not require precise impedance matching. However, in our studies of low-frequency plasma we have found that the plasma extinguishes each half cycle and that the reaction rates are significantly slower than at radio frequencies. Also, there is more of a tendency to arcing in a low-frequency plasma.

Radio-frequency plasmas (13.56 MHz — an FCC assigned frequency) are easily generated with equipment that is stable and reliable and has been commercially available for many years. At this frequency it is necessary to use an impedance matching network to match the impedance of the plasma to the output impedance of the generator (usually 50 ohms, resistive). These can be either manually tuned or automatic servo-driven devices. At 13.56 MHz the plasma is very stable and reactive because the quench time of the plasma species is much longer than the time between half-cycles of the excitation.

Microwave plasma (2450 MHz) is often more reactive, that is, faster than RF plasma, especially in small volume reactors. Also, the microwave generator may be less expensive than RF generators. However, if the cost of all the peripheral equipment such as wave guides, power meters, dummy loads, stub tuners, and applicators is included the total system cost appears to be about the same as an RF system. The decision between RF and microwave usually hinges on process considerations and on the availability of an applicator that is tolerant to large impedance variations in the plasma. Also, with a microwave plasma, it is more difficult to get uniform processing in large reactors because of the high dissipation rates of microwave energy in a plasma.

4.3. IMPEDANCE MATCHING NETWORK

This is usually an adjustable transformer or a manual, or servo-driven, pi-network that transforms the impedance of the plasma to the required output impedance of the generator. In a microwave system it is usually a three stub tuner. An impedance match is necessary to prevent excessive reflected power from damaging the generator, and to know how much power is being dissipated in the plasma. The impedance of a plasma can vary from a few ohms to several thousand ohms and can have a very large reactive component, depending on the gas and the operating conditions.

4.4. POWER DISSIPATION IN A PLASMA

Plasma systems usually include two power meters, one to show the power going towards the reactor (the "Forward" power) and the other to show the power being reflected back towards the generator by an impedance mismatch (the "Reflected" power). When there is a proper impedance match, the reflected power is usually less than 1% of the forward power and most of the electrical energy is being absorbed by the plasma.

A recurring problem in specifying plasma processes, or in transferring a plasma process from one type or size of reactor to another, is in defining the "power density". This can be defined in at least three ways: 1) watts per unit area of electrode or shelf, 2) watts per unit volume of primary plasma or, 3) watts per unit volume of the entire reactor. Each of these is deficient in some way. Therefore, it is usually not possible to predict the best process parameters for transferring a process from one reactor to another. It is necessary to start with a reasonable set of parameters and optimize the process experimentally in the new reactor.

4.5. REACTOR CENTER

This is the "heart" of the plasma system. It is a pressure vessel designed to support the pressure/flow conditions of the plasma, couple the electrical energy into the plasma, and contain the material for processing. There are four generic types of reactor chambers: quartz or metal, and batch or continuous.

Quartz chambers are essential in the semiconductor industry because of the requirement for extreme cleanliness and particle-free operation. In that industry plasma is used to strip photoresist from wafers and to etch patterns into the layers on the wafers. Quartz chambers may be used in industrial applications, however, their extreme cleanliness is rarely required, and there is always the danger of breakage if metal parts or fixtures are used. The maximum practical size of a quartz reactor is about 30 cm (12 inches) diameter.

Aluminum is the metal of choice for metal reactors. It has excellent thermal and electrical conductivity and it is not readily attacked by any plasma except the heavy halogens (Cl, Br, I). It has been fabricated into cylindrical reactors (known as "barrel" reactors) and into rectangular reactors with shelf or cage electrodes. It is also possible to make special shaped electrodes for special applications. The only size limitations on metal reactors is the practicality of fabricating large vacuum vessels. The largest reactor that we have made is about one meter diameter by two meters deep. This is used to process automobile parts.

Most commercially available plasma systems are designed for batch operation. That is, they are designed to be loaded with a batch of parts, evacuated, plasma processed, purged to atmospheric pressure, and the parts removed. This has been satisfactory for most applications. However, as polymers are being used in more sophisticated applications, such as composite structures, the requirement for continuous processing of wire, yarn, film, and fabric, is increasing. Therefore, plasma systems have been built for continuous processing. These can be either air-to-air or cassette-to-cassette.

In air-to-air systems there are several sequentially pumped chambers on either side of the reactor chamber, which are connected with some form of "die". These make it possible to continuously bring material from atmospheric pressure to reactor pressure, and back to atmospheric pressure. In this type of system the major engineering problem is the design of these dies.

Figure 2 is a photograph of a continuous air-to-air system. Figures 3 and 4 are drawings that make its construction clearer. This machine has a double sided electrode system that is 0.67 meters (26 inches) square so the film or yarn passes through 1.34 meters of plasma on each pass through the plasma zone. The upper and lower rollers make it possible to run yarn through the plasma several times so sufficient plasma processing can be achieved, even at high line speed.

In cassette-to-cassette systems the source and take-up spools are both under vacuum as is sketched in Figure 5. This configuration is sometimes necessary for the surface treatment of fragile materials like graphite yarn.

In both types of continuous systems the maximum processing speed ("line speed") is determined by the required residence time in the plasma to get the proper treatment. The present data show that, for reasonable power densities and existing equipment, clean polymers (e.g. polyethylene, polyimide, and polyamide) can be treated in less than 10 seconds residence time. This gives a line speed of 150 m/min (500 ft/min) or more.

Graphite yarn takes more plasma time to process because it is usually desired to actually etch the surface of the fiber rather than just change its chemistry. Processing speeds for graphite are usually less than 30 m/min (100 ft/min). However, multiple tows of yarn can be processed in parallel so the net processing speed can be much greater.

There are also discrete-continuous machines. These are systems where discrete parts, or racks of parts, enter the reactor on one side and exit on the other. Typical applications would be the treatment of automobile bumpers or golf balls.

4.6. SYSTEM CONTROLLER

This is the "brain" of a plasma system. It controls all the process variables: type of gas, pressure, gas flow rate, power level, and processing time. It may be as simple as discrete relays, timers, and needle valves or, it might a microprocessor based system with sophisticated displays, fully automated process control, multi-process capabilities, electronic data-output, and alarm systems.

Figure 2. Photograph of a continuous plasma treater.

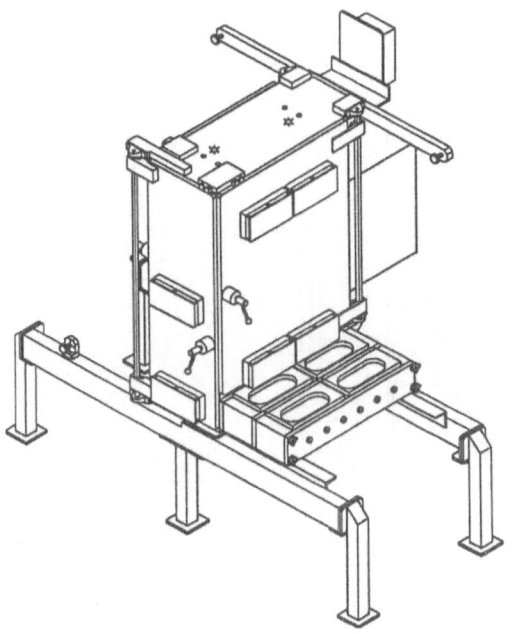

Figure 3. Drawing of a continuous plasma treater.

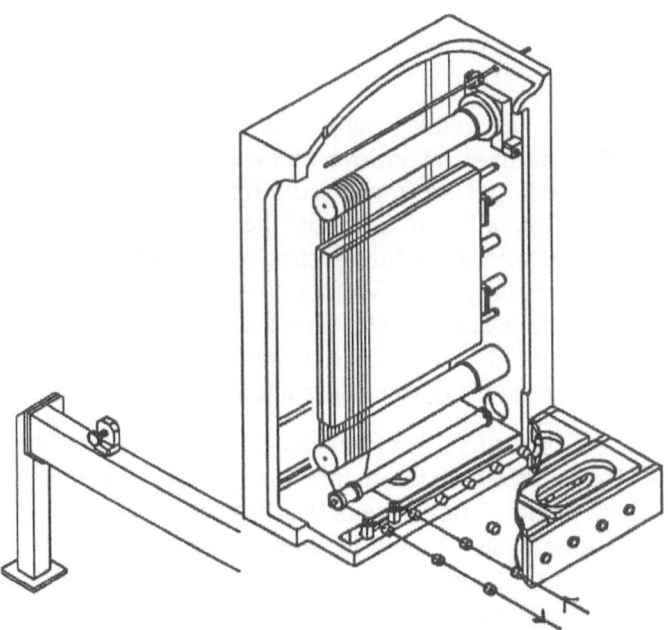

Figure 4. Drawing of a yarn being treated in a continuous plasma treater.

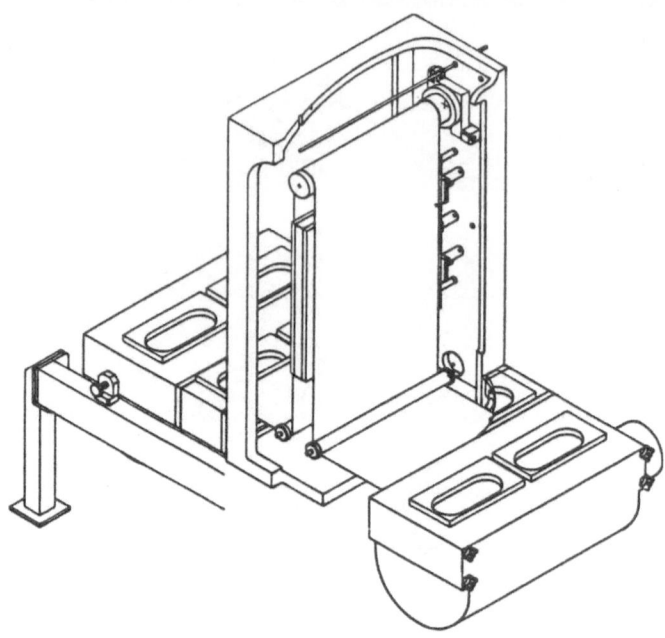

Figure 5. Drawing of a continuous cassette-to-cassette treatment in
a plasma treater.

5. Plasma Chemistry

By definition, plasma is an ionized gas that is spacially neutral. That is, there are an approximately equal number of positive and negative charges in a given volume. It should also be stressed that these are radio-frequency plasmas, not direct current plasmas. The terms "anode" and "cathode" have no meaning, on the time scale of diffusion, because the polarity of the electrodes is reversing every 37 nsec (at 13.56 MHz). Because there is only a few V/cm gradient in the plasma, and because of the enormous mass difference between electrons and ions, the motion of the ions is almost unaffected by the RF field. However, ions may be affected by the build-up of a DC self-bias on the electrodes if the electrodes are of different areas.

For example, an RF field of 10V/cm will cause an ion displacement of only 0.002 mm while the mean free path, due to thermal motion, is about 0.2 mm. The only real effects of radio frequency electrical fields are in the dark spaces over each electrode and they are symmetrical for both electrodes if those electrodes have the same area. Therefore, in the absence of a bias voltage on the parts, the ions have very little kinetic energy and there is almost no ion-etching of the type used in some semiconductor equipment.

Ion attack or sputtering does become an important effect if the electrical energy is capacitively coupled to the electrodes and if the areas of the two electrodes are not the same. In that case the smaller electrode will develop a negative self-bias that can be several hundred volts if there is a large area-ratio. This self-bias will attract ions with sufficient energy to penetrate the boundary layer and to break surface bonds in the polymer surface. Therefore, ions may become an important initiating mechanism if the electrode areas are not the same.

The effects observed in surface treatment of polymers are caused by the chemical energy of free radicals or ions (if there is a bias voltage on the parts) or by the photochemical energy of the ultraviolet light. The plasma gas contains a few parts per million of ions [2], a few percent (2% to 20%) of free radicals [3], and a large amount of extremely energetic vacuum-ultraviolet light (VUV).

There have been many studies of the optical emission of plasmas in the near-ultraviolet and visible region. However, the photon energy in those regions is not sufficient to cause photochemistry to occur. There are almost no data in the literature on the emission from real plasmas in the vacuum-ultraviolet region (VUV is wavelength < 1800Å) or of the effect of that radiation on the chemistry that is occurring in a plasma reactor. One notable exception is the excellent paper by Clark and Dilks [4] in which they show the relative effects of VUV and gas-phase free-radicals or ions on the depth of reaction in the surface of polymers. Their data and other experiments show that the modern plasma reactor is essentially the same light producing device as those used by researchers in vacuum-ultraviolet spectroscopy [5].

We have measured the emission spectra from plasmas using an Acton Research Corporation Model VM-502 Vacuum Monochrometer. This 0.2 meter instrument is equipped with an osmium grating and mirror to maximize short wavelength performance. It is operated in the windowless mode with a turbomolecular pump. To date, measurements have been made on the emission spectra, from 20nm to 450nm, of over 100 plasmas. Many cases have been found of species interactions and of significant spectral changes in mixtures of plasma gases.

It is important to stress the difference between a "pure" plasma and a "real" plasma. Essentially all of the published VUV emission data in the literature, for low pressure discharges, is from spectroscopic studies in which extreme care was taken to ensure pure gases. However, plasma reactors, by definition, are used to process materials and the plasma gas will be contaminated with the by-products of that processing. Therefore, a real plasma gas will never be pure. It will be shown later that there can be 10% to 20% contaminants in the gas. In the case of the oxidation of polymers or photoresist, these will be C, H, O species, all of which can interact, both chemically and energetically.

In one set of experiments, the systems Ar/O_2 and He/O_2 were chosen. Electronically excited Ar (Ar*) can dissociate O_2 to give 2 ground-state O atoms [6]. An argon ion can dissociate O_2 to give a ground state O plus an electronically excited O (O*) which will radiate at 130.5nm. Excited helium (He*) will yield two ground state O's while a helium ion can give two O*.

Figure 6 shows the intensity of the O* radiation (at 130.5nm) from O_2 and He/O_2 plasmas as a function of RF power. The vertical axis is in units of nanoamperes of current from the PMT. These data show a very strong relationship between 130.5nm emission and input power. A three fold increase in power causes almost a 100 fold increase in radiant output.

These data also show that He/5%O_2 plasmas have at least 5 times more radiant output than pure O_2 plasmas at the same input power. This may be due to trapping of the resonant radiation by the large amount of ground state O in the O_2 plasma. It may also be caused by an actual increase in the O* radiation by the reaction between a He + and an O_2 to give two O*.

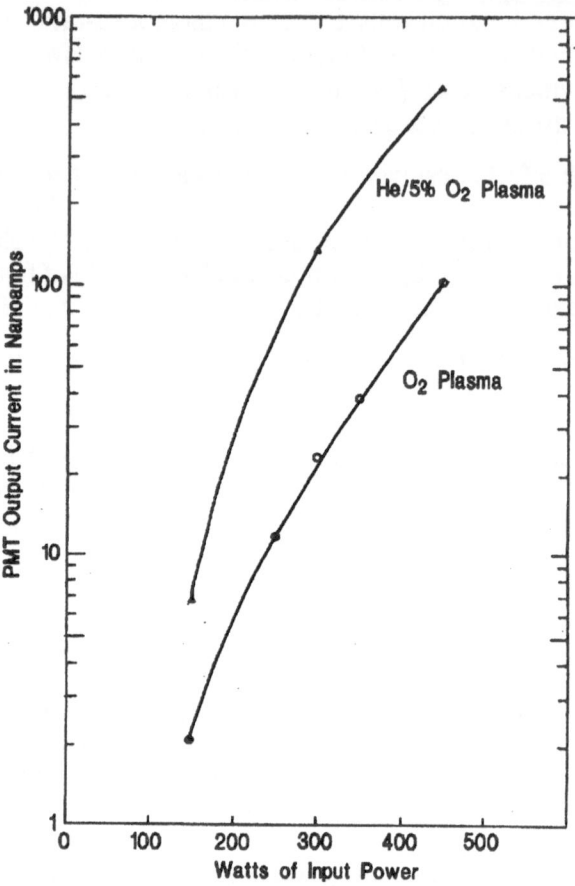

Figure 6. Intensity of 130.5 nm emission from two different plasmas vs. RF power.

Regardless of the mechanisms, these data show that the RF power level has a major effect on the intensity of the VUV reaching the surface being processed.

As was stated, in plasma reactors we are dealing with "real" plasmas. An example of this is shown in Figure 7. This was supposed to be a pure argon plasma. However, there was a small air leak into the plasma reactor during this experiment. It was approximately 2 sccm. This is entirely acceptable for industrial plasma processing because it represents only about 600ppm of O_2 and 6ppm of H_2O as a background. Far more than that comes out of the polymer as dissolved gas or as reaction products during the plasma processing of polymers. However, the effects of this leak on the VUV spectra were dramatic and illustrate the effects of "real" plasma on the emission spectra.

In Figure 7 it can be seen that the intensity of the hydrogen peak at 121.5nm is at least 4 times the intensity of the argon radiation, even though the hydrogen is present only at about 6ppm. These same effects are seen in all real plasmas. The radiation spectrum in the VUV is usually very intense and complicated.

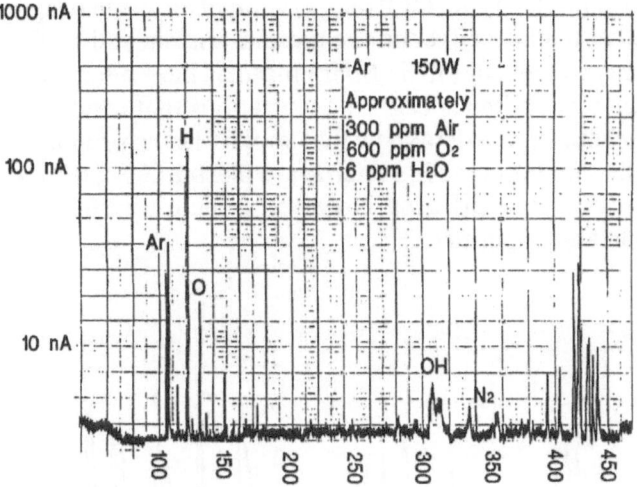

Figure 7. Intensity of emission from a contaminated
argon plasma vs. wavelength.

Figure 8 shows another example of the spectrum of a real plasma. In this case the plasma gas contains a small amount of water vapor (as seen from the OH peak at 310nm). The spectrum shows that almost all of the radiant energy from the plasma is in the energy range beyond the quartz cutoff. Therefore, its presence would not be recognized and its intensity cannot be measured without using a VUV spectrometer. It is stressed that all of this radiation is photochemically active because it exceeds bond energies in organic polymers.

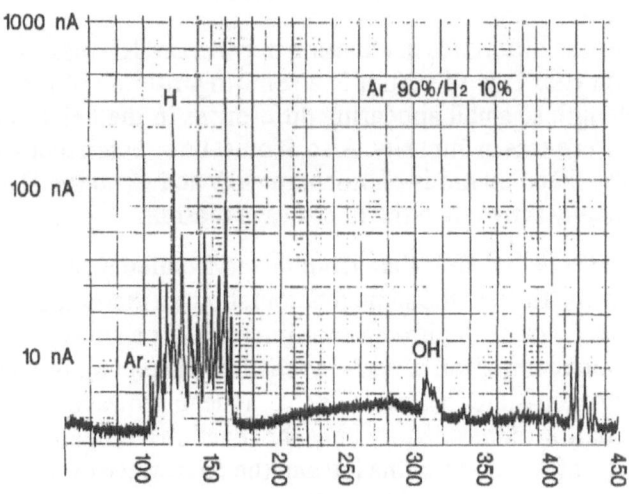

Figure 8. Intensity of emission from an Ar/H$_2$ plasma vs. wavelength.

In addition to the complexity arising from gas composition, the frequency used to excite the plasma also has a major effect on the plasma. Visible and VUV spectral emission data from microwave and RF (13.56 MHz) plasmas have been compared to show the differences in ion and excited species concentrations that result from these two types of excitation. Essentially, the data show that an RF plasma produces a large amount of O^+ ions where few are observed in a microwave plasma. Also, the RF plasma produces more O^* than the microwave plasma. However, a microwave plasma produces a larger amount of O than an RF plasma.

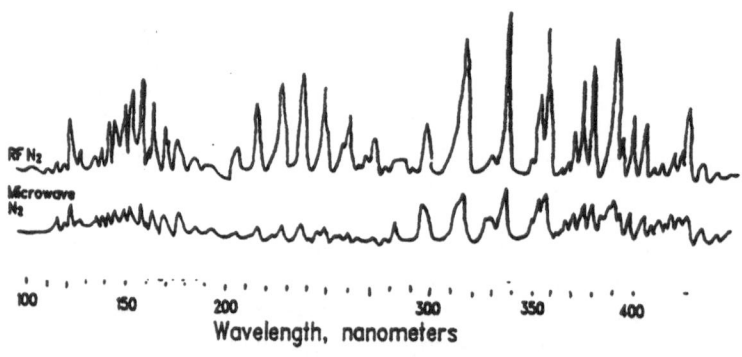

Figure 9. Comparison of RF (top) and microwave (bottom) emission spectra from a N_2 plasma.

Figure 9 illustrates the difference in emission between an RF plasma and a microwave plasma in nitrogen gas. This spectrum is a semi-logarithmic plot of photomultiplier current vs. wavelength so small appearing differences in the height of lines are really very significant differences in intensity. Also, it must be stressed that all of the radiation at wavelengths less than 200 nm is photochemically active, so the RF plasma is much more photochemically active than the microwave plasma.

Vacuum ultraviolet spectroscopic data from mixed frequency plasma (microwave and RF plasma at the same time, in the same reactor) show an increased vacuum ultraviolet radiation over what is found in either RF or microwave plasmas alone. This radiation will increase the rate of initiation of surface reactions on polymers, while the increased concentration of ground-state free-radicals (from the microwave) will increase the rate of free-radical oxidation of the polymer once the reaction has been initiated. Therefore, there is a synergistic effect between the RF and the microwave excitation of the plasma. This effect, along with the effect of RF self-bias, helps explain the increased rates of reaction that have been reported for mixed frequency plasma [7].

6. Interaction Of Plasma With Organic Surfaces

It is being proposed that, because of the large number of collisions that occur in the boundary layer, the concentration of excited free radicals from the plasma is greatly reduced during the process of diffusion to the surface being treated. As a result, these excited free radicals cannot initiate reactions on the surface. Also, in the absence of a negative surface charge on the polymer, ions cannot diffuse to the surface, either through thermal motion or by motion induced by the radio frequency field.

Therefore, the initiation reactions can only be caused by VUV photons or ions (if there is a negative surface charge). Once the initiation has occurred, the surface oxidation can progress through free-radical chain reactions with other species in the boundary layer, such as ground state O or O_2.

The interaction between a plasma and the surface of a polymer takes place in four interrelated steps:

1) The initial breaking of surface and sub-surface bonds.
2) The volatilization and reactions of organic fragments.
3) The reaction of the polymer surface with the boundary layer gas (not the plasma).
4) The reaction of the residual polymer surface after the plasma is off.

6.1. INITIAL BREAKING OF SURFACE BONDS

It can be shown thermodynamically that ground state O or O_2 do not react easily with "ground state" polymers [8]. It has also been shown experimentally by Golub [9] that the oxidation of a polymer in an oxygen plasma, where VUV and ions are present, is about 100 times faster than it is downstream of the plasma where there are no ions and very little VUV. The polymer must either be heated, or contain accessible free-radicals or a surface carbonyl for direct reaction with low-energy oxygen species.

This is well known in the semiconductor industry. It is necessary to heat silicon wafers to about 230°C to get rapid oxidative removal of photoresist in "down-stream" strippers. In this equipment there is a high concentration (10% or more) of ground state O at the wafer but there are no ions or VUV. The stripping rate follows a typical first order Arrhenius reaction.

Therefore, in the absence of high temperature, there must be some other initiation process for the oxidation of the surface. There are at least three possibilities: electronically excited species, ions, or VUV photons.

The electronically excited species (e.g., O^*) usually have too short a lifetime (10^{-9} sec) to penetrate the boundary layer directly from the plasma. However, a ground state O, deep in the boundary layer, can absorb a 130.5nm photon and be excited. If the excited atom is close enough to the surface to reach it before radiational decay, the O^* can break organic surface bonds and initiate the oxidation of the polymer by O or O_2. Note however that this is fundamentally a VUV photochemical process.

The second possibility is the reaction of ions with the polymer surface. This is thermodynamically favorable, if the ions can reach the surface.

The term "dark space" is used to designate the volume of plasma between the plasma that is glowing (and generating ions) and the surface of the polymer. It is important to stress that ions are formed outside the dark space, not in it [10]. Any ions in the dark space have diffused into it with diffusional velocity plus any velocity imparted by the self-bias on the surface of the polymer. The dark space is essentially governed by electrical processes. In practice it is about 2mm thick.

The term "boundary layer" is used to designate that volume of gas between the uniform gas concentration in the plasma and the surface of the polymer. It is in this layer that the first steps of oxidation occur after the organic fragments break away from the surface of the polymer. It is essentially governed by chemical/diffusional processes. In reality, the boundary layer must be thicker than the dark space because it must overlap that volume where ions and free radicals are being formed. It is probably several millimeters thick but, for the purposes of this discussion, it will be assumed to be the same thickness as the dark space, 2mm. The thickness and processes in the boundary layer and dark space are strongly interrelated because of the collisions the occur in the boundary layer.

If it assumed that there are 1 ppm of ions in the plasma [2], the flux of ions to the interface between the plasma and the dark space will be approximately $(1.6) \times 10^{14}/cm^2$-sec. However, in the absence of a static surface charge, these ions will have to diffuse through the dark space where no new ions are being formed. If it is assumed that the net diffusional drift is 50 cm/sec [11], and that the collision frequency is 10^6/sec [12], an ion will undergo 4000 collisions during the diffusion across the dark space.

The ultimate by-products of the plasma oxidation of a polymer will be H_2O, CO, and CO_2. If it is assumed that the ablation rate is 100 nm/min and that the O_2 flow into the reactor is 200 sccm, the average atom percent concentration of by-products in the plasma will be about 15% (depending on the total surface area exposed). The concentration of these by-products in the boundary layer will be much higher.

The by-product contamination, in the boundary layer, will be in the form of C, O, H compounds such as H_2O, CO, CO_2, O, OH, HO_2, COH, CH, CH_2, CH_3, etc. If it is assumed that 15% of the total species in the boundary layer are by-product species and that 10% of these contaminating species will be free radicals or other easily reactive species, then, an ion diffusing through the dark space will undergo 60 collisions, each of which has at least a 0.5 probability of reacting with the ion.

Therefore, the net probability of an ion diffusing through the dark space (unchanged) is $(0.5)^{60}$ or $(10)^{-18}$. The products of these collisions could be other ions. But, these would probably not be other O_2 ions because most of the species in the boundary layer have ionization potentials lower than O_2. Collisions with free radicals or electrons could destroy the ions, probably forming excited free radicals, which would then decay through other collisions. In short, in the absence of a static surface charge, an ion cannot diffuse through a contaminated boundary layer, it undergoes too many destructive collisions.

No data have been found on the surface charge of "floating" polymer surfaces so the importance of surface-charge accelerated ions has not been established. However, three points should be made: 1) The ions will carry positive charge to the negative surface and would tend to neutralize the surface charge; 2) The VUV photons carry sufficient energy to cause photoemission of electrons from the polymer surface and create positive ions in the surface, therefore, there are at least two mechanisms for neutralizing a negative surface charge on a polymer surface; and 3) Experimental data [13] show that when only VUV is present the reaction rates is still 60% to 80% as fast as when both VUV and ions are present.

It should also be stressed that, even if ions can reach the surface of the polymer they will not have a great deal more energy than VUV photons and they will be present at about the same flux as the VUV. The VUV photons have sufficient energy to initiate the surface reactions, ions are not necessary from a chemical standpoint.

The calculated VUV photon flux, at the entry slits of our VUV spectrometer, to give 100 nA of PMT current, is $(10)^{13}$ photons/cm^2-sec. If it is assumed that the absorption coefficient for O (at 130.5nm) is $(10)^4$/cm-atm [14], [15], the boundary layer will absorb less than 10% of the radiation from the plasma. Even if the absorption coefficient is assumed to be $(10)^5$/cm-atm, only about 50% of the radiation will be absorbed, and some of that will be reradiated toward the polymer surface. Therefore, VUV radiation can propagate through the boundary layer with little attenuation. These photons have more than enough energy to break any organic bond (130.5nm = 219 Kcal/mole, 121.5nm = 235 Kcal/mole). When the VUV photon reaches the surface it will break surface bonds and form free radicals in the surface. It can also cause photoemission of electrons and form ions in the surface.

Figure 10 shows the VUV absorption spectrum for polyethylene [16]. This is typical for polymers. The absorption coefficients are extremely high so the VUV radiation is absorbed in a very shallow depth in the surface of the polymer. The absorbed photons will break surface bonds and generate surface free-radicals which can then react with the boundary layer gas, which will have a different composition than the plasma gas.

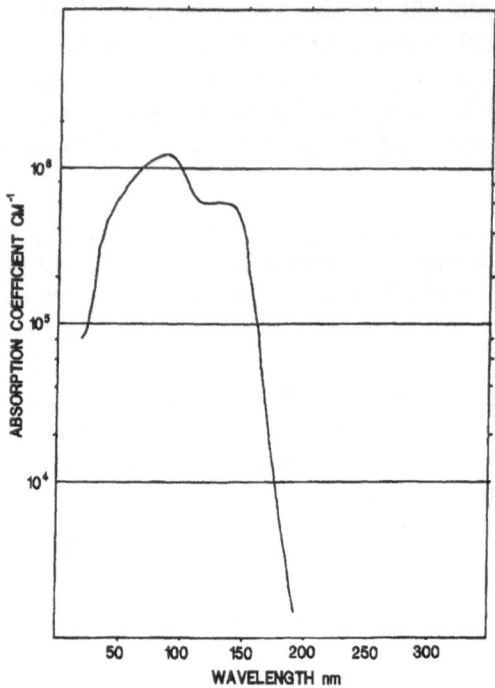

Figure 10. Absorption coefficient of polyethylene vs. wavelength.

Table I gives some examples of the 1/e depth (the depth for 63% absorption of radiation) for various wavelengths and materials. These calculations assume a pressure of 40 Pa (0.3 Torr), and pure gases, which is not the case in real plasmas, but they do indicate that the VUV radiation can propagate through the plasma with very little attenuation.

TABLE 1
ABSORPTION DEPTHS IN MATERIALS AT VARIOUS WAVELENGTHS

INCIDENT RADIATION

Radiating Species	He	Ar	H	O
Wavelength - nm	59.0	104.8	121.5	130.5
Photon Energy - eV	21.0	11.8	10.2	9.5
Photon Energy - Kcal/mole	484	273	235	220

ABSORBING MATERIAL

	He	Ar	H	O
O_2 [17]	—	—	11,260cm	280cm
Ar [5]	3.2cm	(NA)	(NA)	(NA)
NH_3 [18]	6cm	9cm	15cm	10cm
Polyethylene [16]	15.4nm	10.5nm	16.9nm	16.7nm
Polyimide [19]	11.8nm	18.2nm	47.6nm	66.7nm

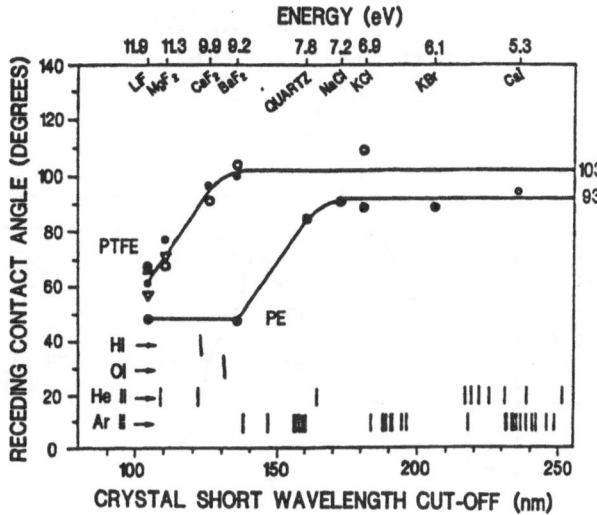

Figure 11. Effect of short wavelength cutoff on receding water
contact angle (Egitto 1990).

The data in Table 1 also show that there are differences in the absorption spectrum of polymers. This may make it possible to tailor the plasma emission spectra to maximize the photochemical effect on different materials.

For example, data from Egitto [20] are shown in Figure 11. These data illustrate why different plasma gases are required to treat different polymers. An O_2 plasma emits very strongly at 130.5nm. This radiation is strongly absorbed by polyethylene (PE) and O_2 plasma is very efficient for treating PE. In Figure 12 the data for the absorption cross-section for PE has been overlaid on the Egitto data. It is obvious that there is a very strong relationship between the two sets of data. The VUV energy must be strongly absorbed for there to be good treatment of a polymer.

However, an O_2 plasma is not efficient for treating PTFE. One reason for this is also shown in Figure 11, the 130.5nm emission from an O_2 plasma is not strongly absorbed by PTFE and does not cause decomposition of that polymer. It is necessary to use a plasma gas that radiates at a shorter wavelength than O_2. For example, an H_2 containing plasma radiates at 121.5nm. This radiation will be absorbed by the surface of the PTFE and will break the surface bonds. Also, the H can abstract the surface fluorine to form HF and expose carbon free-radicals to further attack by other species in the boundary layer.

Thus, one of the primary mechanisms operating in a plasma reactor is the VUV photochemistry of the surface. This is supported by literature references to the cleaning [21], cross-linking [13], fluorination [22], functionalization [23], and free-radical

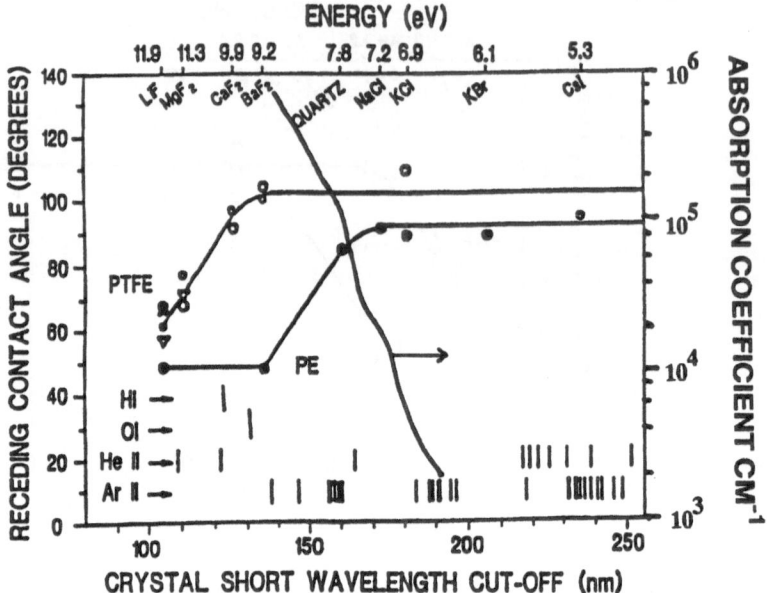

Figure 12. Effect of short wavelength cutoff on receding water
contact angle with overlay of polyethylene absorption
spectrum (after Egitto 1990 and Painter 1980).

generation [24] in polymer surfaces separated from the plasma by a VUV transparent
window. The window allows the VUV radiation to illuminate the surface but does not
allow the energetic radicals or ions to reach the surface. The reported results showed
that VUV radiation alone can cause as much as 80% of the rate of surface reactions
that are caused by complete immersion in the plasma.

The essential point is that the oxidation of a polymer surface cannot be initiated by
ground state atoms or molecules, because they do not have sufficient energy. Nor can
electronically excited species initiate the oxidation because either their lifetime is too
short or there will be too many collisions with reactive species in the boundary layer.
Ions cannot initiate the oxidation because (in the absence of a static bias voltage) they
also undergo too many collisions in the boundary layer. However, the oxidation can be
initiated through VUV photochemical processes, or through the diffusion of ions, if
there is a negative surface charge on the polymer.

The net conclusion is that the reaction between a plasma and an organic surface is not
driven by the energetic species diffusing from the plasma to the surface but rather, the
reaction is driven by the free radical species in the surface of the polymer reacting with
the ground state species that can diffuse through the boundary layer or with species in
the boundary layer that have originally come from the surface of the polymer. This

makes the theoretical analysis of plasma reactions extremely difficult because we know so little about the composition and energy state of the polymer surface. Also, we know very little about the real composition of the boundary layer.

6.2. VOLATILIZATION AND REACTION OF ORGANIC FRAGMENTS

During plasma treatment of polymers, bonds may be broken at both ends of small segments of polymer chains and organic fragments will be released into the boundary layer over the surface. These fragments will react, either with other free radicals in the boundary layer, or with the VUV flux. In either case, in an O_2 plasma, they decompose towards the ultimate by-products of CO, CO_2, and H_2O. However, in the process of this decomposition many transient species are formed that greatly complicate the chemistry of the boundary layer. It is these species that react with the polymer surface and that will interfere with the diffusion of active species from the plasma to the polymer surface.

The counter-current flow of the polymer decomposition products and the plasma species results in a very large number of reactions occurring within a few millimeters of the surface. This is the zone where the plasma/polymer interaction is really occurring. In reality, the polymer is not reacting with the "plasma" but rather, it is reacting with the VUV and other species that can diffuse through the boundary layer and with the many species that are generated in the boundary layer. It is also a zone in which it is very difficult to do analytical chemistry, but, analyses done outside this zone are almost meaningless. All important chemistry is occurring within a few mean-free-paths of the surface..

The transients will also contain terminal or branch free-radicals that will be very reactive. If they deposit on a surface before reacting with some gas-phase radical terminator they will form a cross-linked surface layer. This is the basis for plasma polymerized deposition. To maximize deposition a very low pressure and low power are used. These process variables maximize wall collisions and minimize organic fragmentation.

If deposition is not desired, as with cleaning or surface modification, higher power and pressure are used. Also, for any removal process it is essential to use a plasma gas that will form permanently volatile reaction products. For example: O_2 containing plasma will ultimately yield CO, CO_2, and H_2O; N_2 plasmas yield CN or HCN; and H_2 plasmas yield low molecular-weight hydrogenated organic fragments. If a non-reactive gas is used, for example, argon, there will be very slow cleaning or surface modification. There will also be redeposition of the molecular fragments on all exposed surfaces. This will form a very tenacious cross-linked varnish. Therefore, pure argon plasmas should not be used for cleaning processes.

6.3. REACTION OF SURFACES WITH THE BOUNDARY LAYER GAS

Four major effects of plasma on surfaces are normally observed. Each is always present to some degree, however, one effect may be favored over another, depending on substrate chemistry, reactor design, gas chemistry, and processing conditions. The effects are: cleaning of organic contamination from the surfaces; material removal by ablation (micro-etching) to increase surface area or to remove a weak boundary layer; cross-linking or branching to cohesively strengthen the surface; and surface chemistry modification to improve chemical and physical interactions at the bonding interphase.

Surface modification alone, or in combination with any or all the competing reactions, provides a means to dramatically improve the strength of adhesive bonds, as will be discussed later.

Cleaning — Cleaning of surfaces is one of the major reasons for improved bonding to plasma treated surfaces. Most other cleaning procedures leave a layer of organic contamination that interferes with adhesion processes.

For example, it is known that as little as $0.1 \, ug/cm^2$ (a single molecular layer) of organic contamination on a surface can interfere with bonding [25]. This amount of contamination is the residue from $0.2 \, drops/cm^2$ of a liquid containing 10 ppm non-volatiles. It is extremely difficult to get solvents or water with less than 10 ppm non-volatiles so, almost by definition, a surface will remain contaminated after any cleaning process that finishes with a liquid rinse.

Plasma is capable of removing molecular layers from polymers and all organic contamination from inorganic surfaces. Figure 13 shows the Auger analysis of the surface of vapor degreased solder, before and after plasma cleaning [26]. These results show that, by using plasma, it is possible to get hyperclean inorganic surfaces that are much cleaner than the best that can be obtained using wet cleaning processes. These plasma cleaned surfaces will also be much more wettable by adhesives, and will have much stronger adhesive bonding, than liquid cleaned surfaces.

Plasma cleaning will also expose polymer surfaces that are really the polymer and not the surface of some contamination on the polymer. Therefore, these surfaces give very reproducible bonds and, in most cases, make stronger bonds than normally "cleaned" surfaces.

However, it is critically important to plasma clean a polymer for a sufficiently long time to remove all of the contamination from the surface. Almost all polymer films, and most molded parts, contain additives or contaminants such as oligomers, anti-oxidants, mold release agents, solvents, anti-block agents, etc. which are oils or waxes. Most of these are deliberately incorporated into the polymer formulation to improve its properties or manufacturability and are designed to "bloom" to the surface of the polymer and coat that surface.

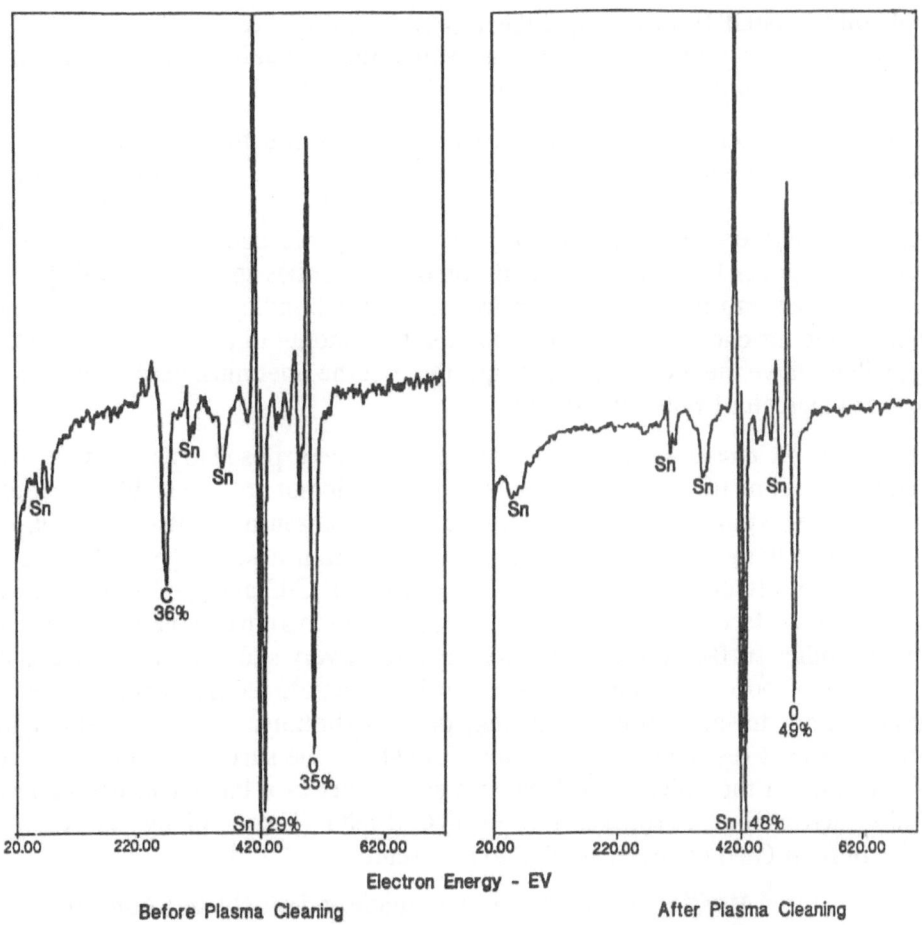

Figure 13. Auger analysis of solder surface before and after plasma cleaning.

These materials often have the same, or close to the same, chemistry as the base polymer. Therefore, they are often difficult to detect with ESCA or other analytical techniques. Typically the contaminants can be present in layers 1nm to 10nm thick, even after solvent cleaning. They just continue to diffuse to the surface after solvent cleaning. Polyethylene is a particularly bad polymer for this problem.

The surface contamination will react with the plasma in the same way that the polymer will. That is, if the plasma treatment is not of sufficient duration to remove the contamination, the contaminant will become wettable and will have a modified ESCA pattern similar to that of the polymer. However, it will still be a plasma-treated contaminant layer, not a plasma-modified polymer surface. At normal power levels it is necessary to clean most polymers for several minutes. A treatment of a few seconds is not long enough (unless very high power densities are used) to remove the contaminants, but it is long enough to plasma treat the contaminants and give a surface that is wettable and that appears to be properly treated, but it is not.

Ablation — Ablation is important for the cleaning of badly contaminated surfaces, for removal of weak boundary layers formed during the fabrication of a part, and for the treatment of filled or semicrystalline materials. Amorphous polymer is removed many times faster than either crystalline polymer or inorganic material, consequently, a surface topology can be generated with the amorphous zones appearing as valleys. This change in surface morphology can improve mechanical bonding as well as increase the area available for chemical interactions. A limited amount of ablation of reinforcing fibers will improve the properties of composites, but the fiber must not be significantly reduced in diameter by over treatment.

Cross-linking — Cross-linking occurs in polymer surfaces exposed to plasmas which are effective at creating free radicals in the polymer, but do not provide stable moieties at the radical sites. Noble gas plasmas, such as helium and argon, are cross-linking plasmas if they are used in the absence of oxygen or other free radical scavengers. The ions and the VUV light attack the polymer surface and break C-C and C-H bonds, leaving radicals in the surface. Once free radicals are created in this environment they can only react with other surface radicals and are, therefore, very stable [24]. If there is any flexibility in the polymer chain, or if the radical can migrate on the chain, there can be recombination, unsaturation, branching, or cross-linking. The latter effect may improve the heat resistance and adhesive strength of the surface by forming a very cohesive skin on the polymer surface. It may also act as a barrier layer, hindering diffusion across the interphase. The term CASING (Cross-linking via Activated Species of Inert Gas) has been applied to this treatment [27].

Surface Chemistry Modification — The most dramatic and widely used effect of plasma is the surface modification of polymers, where the surface layer of a polymer is altered to create chemical groups capable of interacting with adhesives or other materials deposited on the polymer. The inherently low surface-energy of untreated polymers hinders the wetting and interaction with adhesive systems [28-29] or deposited metals. Typically, plasma is used to add polar functional groups which dramatically increase the surface energy of polymers.

For example, Figure 14 shows the low angle ESCA analysis of a polystyrene surface before, and after, treatment with a water vapor plasma [30]. This demonstrates the spectacular change in chemistry of the treated surface that is typical of most polymers exposed to an oxygen containing plasma. This surface will be very polar, completely water wettable, and receptive to reactive adhesives. It is believed that, during curing, the adhesive can react with the surface oxygen species and covalently bond to the plasma treated interphase.

Reference 31 presents what is believed to be the first direct experimental evidence for the formation of covalent bonding between plasma-generated surface functionalities

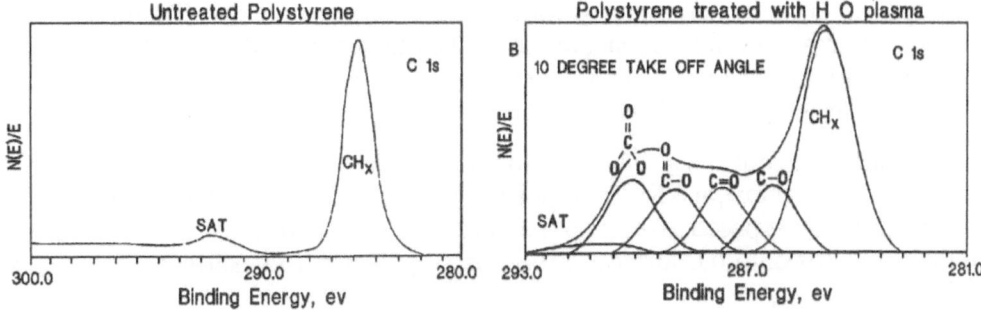

Figure 14. Effect of plasma treatment on the surface composition
of polystyrene.

and an epoxy monomer. The proof of this covalent bonding is extremely important because it would predict greatly improved hot-wet stability of adhesive bonds to plasma treated polymers.

It has been shown [32] that the plasma treatment of a polymer increases the adhesion of a deposited metal layer. This may be due to the direct reaction between the depositing metal or possibly just because the polymer surface was hyperclean.

However, it has also been shown [33] that the very acid nature of an oxygen plasma treated polymer surface inhibits electroless plating unless the surface is neutralized with hydroxide before plating.

Depth profiling has shown the plasma treatment affects only the top 10nm to 30nm of the polymer. Therefore, it is important that the surface be cleaned before plasma treatment and that the treatment last long enough to remove any weak boundary layer. This shallow depth of treatment is the reason why proper plasma treatment does not affect the optical, physical, or mechanical properties of the part.

Normally, plasma treatment will leave 5% to 20% of the surface carbons with some form of organic oxygen species [34]. However, studies of the fluorination of polymers have shown almost complete replacement of the surface H with F or CFX [35-36]. More work needs to be done in this area.

6.4. REACTION OF THE RESIDUAL SURFACE AFTER PLASMA IS TURNED OFF

As has been discussed, a polymer exposed to a reactive plasma will be undergoing simultaneous surface modification reactions and surface removal reactions. If the removal rates dominate or if a non-reactive plasma is used, the polymer surface will contain a large number of free radicals when the plasma is turned off. These radicals may react with themselves to form cross-linking or unsaturation, they may react with the plasma gas as it decays after the power is turned off or, they will react with atmospheric O_2 or water vapor to form surface oxygen moieties. Therefore, these post-plasma reactions may dominate the final surface composition of plasma treated parts. Post-plasma reactions are a source of the oxygen species found in the ESCA spectra of He, Ar, or N_2 plasma treated polymers.

Post-plasma reactions can also be used for grafting desired species onto the surface of polymers. For example, acrylic acid vapor will graft to surface free-radicals if the plasma treated surface is exposed to that vapor before it is exposed to air. Also, a large amount of OH will graft onto the surface of the treated polymer if the surface is exposed to water vapor.

7. Wettability And Bonding Of Polymers

The most obvious result of the plasma treatment of a polymer is the improved wettability of the surface. For most untreated polymers the surface energy is 25 to 50 dynes/cm, with a water contact angle of 95° to 60°. After plasma treatment, with an oxidizing plasma, the contact angle decreases to less than 40°, with some surfaces being so wettable that the contact angle is difficult to measure.

Table 2 shows typical advancing contact angle measurements for a number of polymers, both before and after plasma treatment. Note that in all cases there is a significant improvement in wetting. Table 3 shows typical values for lap-shear bonding of polymers before and after plasma surface treatment. These data show the great improvement that can be obtained, and the reasons why plasma treatment is used on polymers.

It must be stressed that the data in Tables 2 and 3 are TYPICAL data and are only valid for the batch of polymers, and processing, used by those investigators. The data obtained from any other test or batch of polymer, even polymers that are nominally the same, may be different because of differences in processing variables or differences in polymer additives.

The surfaces also become wettable by many adhesives. If there is an improvement in the wetting of the surface by the adhesive there is usually a reduction of voids in the bond line and much better bonding. However, there is not always a good correlation between wetting and bonding.

Table 2
TYPICAL POLYMERS COMMONLY TREATED IN LOW-PRESSURE GAS PLASMA FOR SURFACE MODIFICATION (partial list)

		Initial Surface Energy (dynes/cm)	Initial Water Contact Angle (degrees)	Final Water Contact Angle (degrees)
HYDROCARBONS:				
PP	Polypropylene	29	87	22
PE	Polyethylene	31	87	22
PS	Polystyrene	38	72	15
ABS	Acrylonitrile/butadiene/ styrene copolymer	35	82	26
–	Polyamide (Nylon)	36	63	17
PMMA	Polymethyl metharylate	36	–	–
PVA/PE	Polyvinyl acetate/polyethylene copolymer	38	–	–
–	Epoxy	36	59	12
–	Polyester	41	71	18
PVC	Rigid polyvinylchloride	39	90	35
PF	Phenolic	–	59	36
FLUOROCARBONS:				
ETFE	PTFE/PE copolymer	37	92	53
FEP	Fluorinated ethylene propylene	22	96	68
PVDF	Polyvinylidene fluoride	25	78	36
ENGINEERING THERMOPLASTICS:				
PET	Polyethylene terephthalate	41	76	17
PC	Polycarbonate	46	75	33
PI	Polyimide	40	79	30
–	Polyaramid	–	–	–
–	Polyaryl etherketone	36	92	3
–	Polyacetal	36	–	–
PPO	Polyphenylene oxide	47	75	38
PBT	Polybuytlene terephthalate	32	–	–
–	Polysulfone	41	76	16
PES	Polyethersulfone	50	92	9
–	Polyarylsulfone	41	70	21
PPS	Polyphenylene sulfide	38	84	28
ELASTOMERS:				
SR	Silicone	24	96	53
FLUOROELASTOMERS:				
FPM	Fluorocarbon copolymer elastomer	36	87	51

Table 3
EXAMPLES OF LAP-SHEAR BONDING IMPROVEMENT

	Control PSI	After Plasma PSI
Polyimide (PMR®-15)/graphite	420	2600
Polyphenylene sulfine (Ryton® R-4)	290	1360
Polyether sulfone (Victrex® 4100G)	130	3140
Polyethylene/PTFE (Tefzel®)	very low	3200
HDPE	315	3125
LDPE	370	1450
Polypropylene	370	3080
Polycarbonate (Lexan®)	410	928
Nylon	850	4000
Polystyrene	570	4000
Mylar A®	530	1660
PVDF (Tedlar®)	280	1300
PTFE	75	750

It is common practice to measure the wettability of a treated surface to evaluate its suitability for bonding. In many cases this is an excellent quality control tool. However, there are dangers in relying solely on wettability to predict bondability because it does not always give the correct answer.

7.1. PROBLEMS WITH WETTABILITY

There is usually a good correlation between bonding and wetting, but not always! Two different cases where this correlation can break down are: 1) where the surface is wettable but the structure beneath the surface is too weak to have a good bond strength and 2) where the surface is not wettable by water but there is still excellent bonding. Two examples of the first are the bonding of PTFE and the bonding of a waxy or oily surface.

Polytetrafluoroethylene (PTFE) can be plasma treated to give good wetting by water or adhesives. However, when this surface is bonded the measured bond strength is about half of that obtained using a sodium etch. Andrews and Kinloch [37] have shown that the surface structure of PTFE is very weak because there is almost no cross-linking. The top layer of the polymer will shear off with the adhesive, even if the surface is treated by plasma to give good wetting. To get good bonding to PTFE it is necessary to use a surface treatment that cross-links a significant depth of the polymer (1μ or more), such as a sodium etch. Plasma treatment only affects the top 0.01μ and the resulting surface is just not thick enough to give a strong bond even though it is wettable and bondable by the adhesive.

As a second example of a weak surface layer; it is easy to plasma treat a waxy or oily surface and make it completely wettable and bondable by adhesives. However, these bonds will show almost no strength because the adhesive is not bonded to the substrate, only with the surface contamination layer. This is the ultimate example of a weak boundary layer. It is also the primary danger of using a "wetting test" as a quality control test for plasma treatment. The apparent surface of the part may be completely wettable but still give very poor bonding because that surface is really a layer of cross-linked contaminant.

Figure 15 shows data for the second case where the correlation between wetting and bonding breaks down. These data are for bond strength vs. water contact-angle for a polymer that has been plasma treated using seven different gases. The treated pieces of polyphenylene sulfide (Ryton® R-4) were bonded using an epoxy adhesive (Dexter-Hysol EA 9330) with a ½ in. by ½ in. overlap. In all cases the polymer samples were plasma treated for 5 minutes, in the same plasma system, with all variables held constant except the composition of the plasma gas.

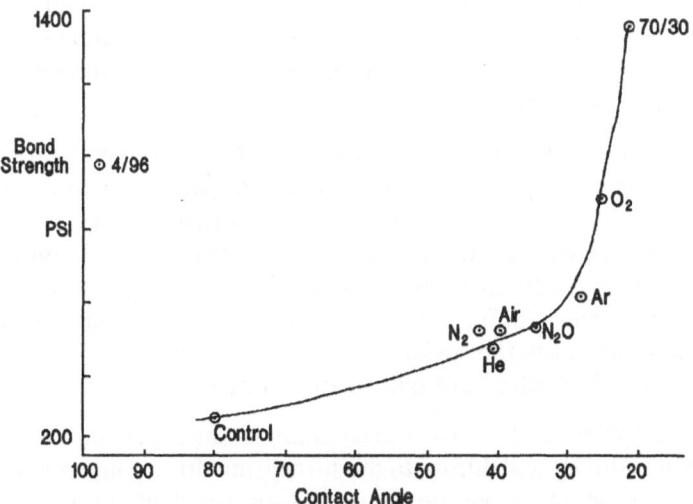

Figure 15. Relationship between adhesive bond strength and water contact angle as a function of plasma gas composition.

Most of the data in Figure 15 show the relationship that is intuitively anticipated; the bond strength increases as the wetting increases (note: 70/30 is 70%O_2/29%CF_4/1%Ar). However, the data for treatment with 4/96 (4%O_2/96%CF_4) illustrate the problem with depending on wetting to predict the best treatment for bonding. This gas mixture gives the second strongest bonding, of those gases tested, even though the surface has a water contact-angle of 104°. These data are real and reproducible. We have recently begun analyzing this surface to try to find out why it bonds so well. The preliminary results indicate that there may be a small amount of C-O or C-O-F that can bond with epoxy but that there is so much CF_2 on the surface that water cannot wet the surface. These studies are continuing [38].

Therefore, it is dangerous to depend solely on wetting tests to show that the surface of a part has been adequately treated for bonding. The surface may still be contaminated even though it has been made wettable by plasma treatment or, the surface of a part may be very bondable, even though it has been made hydrophobic by plasma treatment.

Wetting tests should be used only for routine quality control after it has been established, through bonding tests, that there is a good correlation between wetting and bonding for the specific process involved and then, only if the pre-clean procedures are strictly applied and controlled.

8. Ageing Of Plasma Treatment

One of the major problems in the practical use of plasma treatment of polymers is the large variability in the longevity of the plasma treatment, either in terms of wettability or of bonding. It is often observed that the surface treatment seems to fade with time after treatment [39]. Unfortunately the rate of fading can be much different for different investigators looking at what is nominally "the same polymer", and these differences are not trivial. One investigator may find a plasma treated polypropylene sample loses its wettability in a few weeks while another investigator has polypropylene samples that remain wettable for years.

The problem is that almost all polymers have some contamination on the surface. This comes accidently from handling or storage contamination, or, most importantly, from the diffusion of oily or waxy additives from the bulk polymer to the surface of the polymer. These additives can be antioxidents, anti-blocking agents, mold release, plasticizers, lubricants, etc. The diffusion of the additives usually cannot be stopped because they are designed to diffuse to the surface to perform their desired role. Unfortunately, when they reach the surface they cover-up the surface moieties left by the plasma treatment and the effects of that treatment may disappear.

The thickness and composition of this surface contaminant layer is almost never known, but it is there. These additives will diffuse to the surface and form a layer of oily or waxy contamination. This surface layer must be removed, using precleaning and a sufficiently long plasma treatment, before there is treatment of the "real" polymer surface. If the contamination layer is not completely removed you are treating and analyzing an unknown contaminated surface.

It is often not possible to use ESCA to analyze for this type of contamination because, in the most insidious case, it is coming out of the polymer as an oligomer or as a material that is chemically quite similar to the polymer. Therefore, it will be completely distributed through the polymer and it will have a very similar ESCA spectrum to the polymer. But, it will not be a polymer, it will be an oily or waxy material that can interfere with the use of plasma treatment if it is not completely removed from the surface.

You cannot use a wettability test to determine if a surface has been sufficiently processed to remove all the contamination because the contaminant will become as wettable as the plasma treated polymer because the lack of cross-linking will make it easier to treat.

One way to determine if there is a problem with contamination is to rub the polymer on the surface of plasma-cleaned glass (O2 plasma for at least 10 minutes) and see if there is a non-wettable smear left on the glass. Another test is to wrap plasma-cleaned glass microscope slides in the polymer film or yarn for a week and see if the glass surface becomes non-wettable. Either of these tests will show the presence of surface contamination on the polymer, if it is there.

These problems with surface contamination make the investigation of the longevity of plasma treatment very difficult. However, they are "real world" problems; commercial polymers are not pure materials, they often contain additives. The variability in the amount and type of additives also makes it impossible to compare the results from different investigators, or even the same investigators using different batches of materials. The presence of these contaminants also makes it impossible to use published test results to predict the lifetime of another commercial polymer sample. It is necessary to perform your own longevity tests on your own polymer and then ensure that you always have the same amount and type of additives. This is sometimes difficult to do because additives present at fractions of a percent can adversely affect wetting or bonding but formulators do not consider materials present at these low levels to be significant. Your supplier may make small changes in the additives and not tell you until you start having bad results.

In the studies of very high purity polymer films (which do not contain mobile additives), there appears to be a molecular relaxation mechanism, that is probably driven by entropy, that can cause a decrease in surface energy with time after plasma processing [40]. The rate of this decrease depends on the mobility of the polar moieties around the polymer chain. The tendency is for the polar group to rotate and "bury" itself to reduce the surface energy.

I have seen no data on polymers containing additives where it has been possible to differentiate between this "true" reversion and the "apparent" reversion caused by the diffusion of additives to cover up the surface polar groups.

If it is determined that the polymer that you are working with does contain additives that are diffusing to the surface and interfering with the plasma treatment results, there are three approaches that can be used to eliminate the problem: 1) do the subsequent processing step (bonding, printing, painting, etc.) shortly after plasma treatment so it is completed before the diffusion can interfere with the results. The definition of "shortly" will depend on the diffusion rate of the interfering material. 2) A variation of the first approach is to coat the surface immediately after plasma treatment with a primer that will bond well with the plasma surface and will not be affected by the subsequent diffusion of the contamination. This approach works well with silicone rubber because there are good primers available for that surface. 3) If both of the first two approaches fail, or cannot be used for some reason, the only choice left is to change the polymer formulation. This may not be as bad as it sounds at first because the use of plasma treatment makes it possible to use polymers that normally could not be used at all.

High temperature "engineering" polymers usually do not show any loss of wettability or bondability with time because they do not contain any mobile additives and are so heavily cross-linked that there is no mobility of polar species in the surface.

Another major source of "apparent" reversion is the transfer of hydrophobic material to the polymer surface from the storage material used after plasma treatment. Polyethylene bags are particularly bad for this form of recontamination. Storage materials must be very carefully tested, using something like the microscope slide test, to be sure that these materials are not recontaminating the surface of the plasma treated polymer.

9.   Examples Of The Improvement Of Composites By Using Plasma

9.1. INTRODUCTION

In the previous discussion it has been shown that the plasma environment will have three effects on the surface of fibers that can affect the performance of a composite made from those fibers:

1) Ablation of the surface.
2) Cleaning of the surface.
3) Chemical modification of the surface.

Each of these effects will contribute an improved surface that will change the properties of the composite.

Ablation of the surface will remove the surface and reduce the diameter of the fibers. This will weaken the individual fiber but it may increase the net strength of the composite because the plasma will preferentially remove small defects in the surface of the fibers. This will increase the critical flaw size of the fibers.

Most fibers or cloths will have some sizing or other contamination of the surface of the fibers. Most of this can be removed by washing or cleaning but there is always some residual contamination from those processes. The proper plasma cleaning can remove all of the contamination and leave the real surface of the fiber exposed for bonding to the matrix.

The residual chemical modification of the surface makes the surface wettable by the matrix (reducing skips, voids, and bubbles) and it provides a surface that the matrix can covalently bond to. This improves both the uniformity of the bonding and mechanical properties of the composite, and it improves the hydrothermal stability of the part.

A general comment should be made regarding the toughness of composites. In most composites, part of the toughness is derived from the energy dissipation that occurs when the bond between the fibers and the matrix fails. After plasma treatment of the fibers the bond between the fiber surface and the matrix is usually improved greatly. Therefore, there is an improvement in the peel and flex properties of the composite but there is also a decrease in the toughness because the energy loss mechanism, from bond failure, is no longer available. Therefore, the toughness must be built in through some other mechanism. The easiest way is to use a toughened (lower modulus) matrix. This will not only add an energy dissipation mechanism, it will also increase the critical flaw size and thus increase the strength of the composite.

The following data provide examples of each of these processes. Again, it must be stressed that these data are only *typical* results and they should not be used for anything except indications of what can be achieved. There are large variations among reinforcing materials and matrix materials so detailed tests *must* be performed on each proposed composite system to obtain reliable engineering data.

9.2. EFFECT OF PLASMA TREATMENT ON GRAPHITE COMPOSITES

Ismail [41] has reported a detailed analysis of the surface area and morphology of graphite fibers after plasma treatment and high temperature oxidation (HTO). He found that the Active Surface Area (ASA) of the fibers, as measured by O_2 absorption, increased from 0.029 m^2/g to 0.154 m^2/g (a 430% increase) with only a 2% burn-off in an O_2 plasma. This is essentially the same increase as was found for a HTO of much longer duration. The BET area, as measured using the cryogenic adsorption of Kr, increased only slightly for either treatment. He attributed the apparent differences in area to chemical adsorption of the O_2 by active sites left on the surface of the fibers by the plasma treatment.

He found that the results for pitch fiber could not be generalized to other graphite fibers. For example, he found that plasma treatment of graphitized rayon increased the BET area by almost an order of magnitude while plasma treatment had almost no effect on the area of carbonized PAN. However, the HTO of PAN increased the BET area by three orders of magnitude with no pit formation.

Ismail also examined the surface of the pitch fibers using SEM. He found that the plasma ablated fibers were smooth while the HTO fibers had pits in the surface whose size correlated with the length of the oxidation. He proposed that "...with plasma, oxygen atoms are simultaneously attacking the edge and basal plane carbon atoms, peeling off one complete outer basal layer after another. On the other hand, the formation of pits with HTO samples indicates that molecular oxygen was mainly attacking the edge carbon atoms of the crystallites at the twist and tilt boundaries".

He recommended that, if bonding is primarily chemical, plasma treatment should be used on pitch carbon while, if the bonding is primarily physical, HTO should be used.

Jang [42] has reported the results of tests on a high modulus graphite/epoxy composite using both an O_2 and a N_2 plasma treatment. He found that, after a one minute N_2 plasma treatment of the graphite, the composite shear strength increased by 55%, the flex strength decreased by 7%, and the toughness decreased by 24%. However, he also found that the toughness could be restored by the addition of less than 5% CTBN rubber to the matrix.

Smith [43] also reports the results of the plasma treatment of high modulus graphite in high modulus epoxy (Thornel T-300 in DER 332/T403). He found the following T-peel results for different plasma gases: dry air, +41%, Ar, +32%, NH_3, +22% and, O_2, -18%.

These reported results show that it is possible to get significant improvements in composite properties from simple plasma treatment of graphite fibers.

9.3. EFFECT OF PLASMA TREATMENT ON POLYARAMIDE COMPOSITES

There are two excellent papers on polyaramide/epoxy composites that are both somewhat difficult to obtain so I will abstract the author's findings.

Allred [44] used commercial Kevlar® cloth but he first performed several cleaning steps to ensure that he had very clean cloth to work with.

SEM examination of the surface showed that the fibers were very smooth before and after NH_3 plasma treatment. From his analysis of the Weibull statistics of the fiber strengths before and after plasma cleaning, he concluded that the plasma treatment was probably healing small surface flaws in the fibers. He also found that, after plasma treatment, there was a change in the failure mode of the composite to fiber splitting, fibrillation, separation of skin and core of the fiber, and crack propagation through the epoxy matrix. From these observations he concluded that there was covalent bonding between the epoxy and the fibers.

He performed a chemical derivitization analysis of the plasma treated surface using a dye and concluded that there was about one NH_2 group per polymer-repeating unit after one minute of N_2 plasma. Oxygen contamination of the NH_3 plasma interferes with the incorporation of NH_2 into the surface, probably because of competition for the active sites by oxygen species. He found no ageing of the NH_2 concentration on the surface, even after 18 months storage in air.

After one minute of NH$_3$ plasma treatment he found a 114% increase in T-peel, a 31% increase in interlaminar strength and, a 5% decrease in toughness. However, he also found the major changes in failure mechanisms already mentioned and attributed the low loss in toughness to those changes.

The plasma treatment also reduced the water absorption of the composite by a factor of 3 and converted it from capillary (wicking along the fiber-matrix interface) to Fick's law absorption through the body of the matrix. The rate of water absorption in the composite was less than it was in a block of neat epoxy the same size as the composite. The difference in absorption rate was the same as the volume fraction of epoxy in the composite. This change in absorption mechanism would predict that there should be a great improvement in the hydrothermal stability of plasma treated composites.

Smith [43] has also reported the results of a series of T-peel tests on three different types of polyaramides each treated for one minute in four different plasma gases. Table 4 shows his results.

Table 4

Kevlar®		Percent Improvement In Plasma			
Type	Denier	O$_2$	Air	Ar	NH$_3$
49	7100	318	218	204	169
49	380	129	154	169	83
29	1500	2383	–	–	2486

These data show several things. The most important is that the best plasma gas depends on the type of material being processed. Also, the amount of improvement depends on the material. The Type 29 Kevlar® has a lower modulus and greater elongation than the Type 49. Therefore, once the Type 29 is plasma treated so that it can be bonded to the matrix, it has much greater peel strength than the best results for the Type 49, 5.25 N/mm vs. 2.70 N/mm (30 lb/in vs. 15.4 lb/in). This is another indication for the need for a lower modulus material in composites that have good bonds between the matrix and the fiber.

The first material, 49/7100, showed no ageing or loss in performance after 300 days of storage in air.

Again, these data show the spectacular improvements that can be made in composite performance, with no increase in weight, after plasma treatment.

9.4. EFFECT OF PLASMA TREATMENT ON POLYETHYLENE COMPOSITES

Smith [43] has also given data on the improvement in the T-peel performance for two different polyethylene yarns processed in four different plasma gases. Table 5 shows his results:

Table 5

Spectra® Type	Denier	Percent Improvement In Plasma			
		O_2	Air	Ar	NH_3
900	1200	360	380	220	150
1000	660	153	100	87	45

Again, these data show that the best process for a material will depend on exactly what that material is.

The data in Table 5 are for medical grade material, the "standard" grade material had so much contamination on the surface that plasma treatment made no difference in its composite performance.

The data in these references illustrate the great improvements that are seen in most composites after plasma treatment of the reinforcing material.

10. Cost Of Plasma Treatment

Almost the entire cost of plasma treating a yarn of fabric, on a continuous basis, is in the amortization of the capital cost of the plasma equipment. That will be about $20 to $50 per hour, depending on the size of the system and the vacuum pumps, and on the utilization rate. Assuming a line speed of 160 m/min (which has been used for the continuous treatment of Kevlar® yarn), this is between 200 and 400 meters per dollar to at least double the strength of a composite.

The direct operational cost of a large plasma system is about $4/hour.

The maintenance cost is also very low because there are almost no moving parts in a plasma system. The major maintenance cost is the cost of pump repair.

11. Summary

There are several major points reviewed and presented in this paper:

1. The reactions between a plasma and the surface of a polymer are free-radical reactions.

2. The initiation of the free-radical chain reaction is primarily through vacuum ultraviolet photochemistry on the surface of the polymer. Ions may play a part, if

there is a static space charge on the surface, but electronically excited atoms or molecules are probably not an important initiating species.

3. Plasma processing of a polymer usually oxidizes the top few molecular layers of the polymer, making it very polar and both wettable and bondable.

4. There are several general cases in which the correlation between good wetting and good bonding does not hold.

5. Almost all polymers contain additives that can cause an "apparent" reversion of the surface and a loss of wettability. There may also be a "true" reversion of the surface in very pure polymers. In "real" polymers these two effects probably cannot be isolated.

6. Plasma treatment of composite reinforcing yarn of fiber will almost always increase the strength of the final composite. The amount of improvement will depend on the type of fiber, the type of matrix, and the type of plasma treatment.

12. References

1. Hollahan, J.R. and Bell, A.T., "Techniques and Applications of Plasma Chemistry", New York, N.Y. (1974)

2. Gousset G., Panafieu P., Touzeau M. and Vaille M., 'Experimental Study of a DC Oxygen Glow Discharge By Vacuum Ultraviolet Absorption Spectroscopy', Plasma Chem. and Plasma Processing, 7, 409-427 (1987)

3. Egitto F.D., Emmi F. and Horwath R.S., 'Plasma Etching of Organic Materials: I. Polyimide in O2-CF4', J. Vac. Sci. Tech., B3(3), 893-904 (1985)

4. Clark D.T. and Dilks A., 'ESCA Applied to Polymers, XV. RF Glow Discharge Modification of Polymers Study by Means of ESCA in Terms of A Direct and Radiative Energy Transfer Model', J. Polym. Sci., Polym. Chem. Ed., 15, 2321-2345 (1977)

5. Samson, J.A.R., "Techniques of Vacuum Ultraviolet Spectroscopy", John Wiley & Sons, New York, NY (1967)

6. McNesby, J.R. and Okabe H. "Vacuum Ultraviolet Photochemistry" in Advances in Photochemistry, Vol. 3, 157-239 (1964)

7. Martinu, L., Klemberg-Sapieha, J.E., and Wertheimer, M.R., 'Duel-Mode Microwave/Radio Frequency Plasma Deposition of Dielectric Thin Films', Appl. Phys. Lett., 54(26), 2645-2647 (1989)

8. Ranby, B. and Rabek, J.F., "Photodegradation, Photo-oxidation and Photostabilization of Polymers", John Wiley & Sons, New York, NY (1975)

9. Golub, M.A. and Wydeven, T., 'Reaction of Atomic Oxygen (O3p) With Various Polymer Films', Polym. Degrad. Stab., 22, 325-338 (1988)

10. Cobine, J.D., "Gaseous Conductors", Dover Pub., New York, NY (1958)

11. Dushman, S. and Lafferty, J.M., "Scientific Foundations of Vacuum Technique", John Wiley & Sons, New York, NY (1962)

12. Benson, S.W., "The Foundations of Chemical Kinetics", McGraw-Hill, New York, NY (1960)

13. Hudis, M. and Prescott, L.E., 'Surface Crosslinking of Polyethylene Produced by the Ultraviolet Radiation From a Hydrogen Glow Discharge', Poly. Letters, 10, 179-183 (1972)

14. Okabe, H., "Photochemistry of Small Molecules", Wiley-Interscience, New York, NY (1978)

15. Mitchell, A.C.G. and Zemansky, M.W., "Resonant Radiation and Excited Atoms", Cambridge Univ. Press, New York, NY (1971)

16. Painter, L.R., Arakawa, E.T., Williams, M.W. and Ashley, J.C., 'Optical Properties of Polyethylene: Measurement and Applications', Rad. Res., 83, 1-18 (1980)

17. Watanabe, K., Zelikoff, M. and Inn, E.C.Y., "Absorption Coefficients of Several Atmospheric Gases", AFCRL Technical Report No. 52-23, Geographical Research Papers No. 21, June 1953 (NTIS AD19700)

18. Sun, H. and Weissler, G.L., J. Chem. Physics, 23(6), 1160 (1955)

19. Arakawa, E.T., Williams, M.W., Ashley, J.C. and Painter, L.R., 'Optical Properties of Kapton®: Measurement and Applications', J. Appl. Physics, 52(5), 3579-3582 (1981)

20. Egitto, F.D. and Matienzo, L.J., 'Modification of Polytetrafluoroethylene and Polyethylene Surfaces Downstream from Helium Microwave Plasma', Polym. Degrad. Stab., 30, 293-297 (1991)

21. DeLollis, N.J., "The Use of R.F. Activated Gas Treatment to Improve Bondability," Sandia Labs, Albuquerque, NM, Report No. SC-RR-71 0920 (1972)

22. Cohen, R.E., Baddour, R.F. and Corbin, G.A., 'Surface Fluorination of Polymers in a Glow Discharge Plasma: Photochemistry', 6th International Symposium On Plasma Chemistry, IUPAC, Montreal, Canada, Paper Number B-8-2, 537-541 (1983)

23. Shard, A.G. and Badyal, J.P.S., 'Plasma versus Ultraviolet Enhanced Oxidation of Polyethylene', Polym. Commun., 32(7), 217-219 (1991)

24. Yasuda, H., 'Plasma Modification of Polymers', J. Macromol. Sci.-Chem., A10(3), 383-420 (1976)

25. Jackson, L.C., 'Improving Adhesion by Gas-Plasma Contaminant Removal', Adhesives Age, Sept. 1978

26. Smith, M.D., 'RF Plasma Cleaning of Ceramic Insulators', Insulations/Circuits, May 1980

27. Hansen, R.H. and Schonhorn, H., 'A New Technique for Preparing Low Surface-Energy Polymers for Adhesive Bonding', J. Polymer. Sci., Polym. Lett. Ed., B(4), 203-209 (1966)

28. Hook, T.J., Gardella, J.A. and Salvati, L., 'Multitechnique Surface Spectroscopic Studies of Plasma-Modified Polymers II: H2O/Ar Plasma Modified Polymethylmethacrylate/Polymethacrylicacid Copolymers', J. Mater. Res., 2(1), 132-142 (1987)

29. Everhart, D.S. and Reilly, C.N., 'Chemical Derivitization in Electron Spectroscopy for Chemical Analysis of Surface Functional Groups Introduced on Low Density Polyethylene Film', Anal. Chem., 53, 665-671 (1981)

30. Anonymous, "Plasma Modified Polymers: Angle Resolved ESCA Analysis," Applications Note No. 8401, Perkin-Elmer Corp., Physical Electronics Division, Eden Prairie, MN.

31. Webster, H.F. and Wightman, J.P., 'Effect of Oxygen and Ammonia Plasma Treatment on Polyphenylene Sulfide Thin Films, and Their Interactions With Epoxy Adhesives', J. Adhesion Sci. Technol., 5(1) 93-106 (1991)

32. Tanigawa, S., Ishikawa, M. and Nakamae, K., 'Adhesion of Vacuum Deposited Metal Thin Film on Poly(ethyleneteraphthalate) Film Pretreated With RF Plasma', J. Adhesion Sci. Technol., 5, 543 (1991)

33. Mance, A.M., Waldo, R.A. and Dow, A.A., 'Interaction of Electroless Catalysts With Plasma-Oxidized Surfaces of Polystyrene Based Resins', J. Electrochem. Soc., 136(6), 1667-1671 (1989)

34. Penn, L.S., Byerley, T.J. and Liao, T.K., 'The Study of Reactive Functional Groups in Adhesive Bonding at Aramide-Epoxy Interfaces', J. of Adhesion, 23, 163-185 (1987)

35. Corbin, G.A., Cohen, R.E. and Baddour, R.F., 'Kinetics of Polymer Surface Fluorination: Elemental and Plasma Enhanced Reactions', Polymer, 23, 1546-1548 (1982)

36. Anand, M., Cohen, R.E. and Baddour, R.F., 'Surface Modification of Low-Density Polyethylene in a Fluorine Gas Plasma', Polymer, 22, 361-371 (1981)

37. Andrews, E.H. and Kinloch, A.J., 'Mechanics of Adhesive Failure. I and II', Proc. Roy. Soc., A332, 385-399 and 401-414 (1973)

38. Personal communication from Jolanta Klemberg-Sapieha, École Polytechnique, Montreal, Que., Canada

39. Sowell, R.R., DeLollis, N.J., Gregory, H.J. and Montoya, O., "Effects of RF Activated Inert Gas Plasma on Critical Surface Tension and Bondability of Polydimethyl Silicones and Polyethylene," Sandia Laboratories, Albuquerque, NM, Report No. SC-RR-0483, 1971

40. Morra, M., Occhiello, E. and Garbassi, F., "Dynamics of Plasma Treated Polymer Surfaces: Mechanisims and Effects," Paper presented at The International Conference on Polymer-Solid Interfaces, Namur, Belgium, Sept. 1991

41. Ismail, I.K. and Vangsness, M.D., 'On The Improvement of Carbon Fiber/Matrix Adhesion', Carbon, 26, 749-751 (1988)

42. Jang, B.Z. and Das, H., 'Plasma Treatments of Fiber Surfaces for Improved Composite Performance', in "Interfaces in Polymer, Ceramic, and Metal Composites", Elsevier Sci. Pub., New York, NY, 1988

43. Smith, M.D., "Surface Modification of High-Strength Reinforcing Fibers by Plasma Treatment", Allied-Signal Aerospace Company, Technical Communications No. KCP-613-4307, March 1990 (NTIS DE91016481)

44. Allred, R.E., "Surface Chemistry and Bonding of Plasma-Aminated Polyaramide Filaments", Proc. of ACS Symp. on Composites, Mar. 21-25, 1983, Polymer Preprints, 24, 223-263 (1984)

PLASMA POLYMERIZATION OF ACETYLENE: A COATING TECHNIQUE FOR FIBRE REINFORCEMENT OF COMPOSITES

W. WEISWEILER
Institute of Technical Chemistry, University of Karlsruhe
Kaiserstr. 12
D-7500 Karlsruhe/Germany

ABSTRACT. In this work, the plasma polymerization is applied as a versatile method for controlled deposition of polymeric films to create a reactive fibre surface. Films of different composition were deposited continuously on carbon and glass fibres using the r.f. sputtering technique and acetylene/air mixtures as reactive gases. The plasma polymeric films were characterized by IR-spectroscopy, gas phase analysis while depositing, X-ray photoelectron spectroscopy, elemental analysis, thermogravimetric analysis and contact angle measurement. The influence of polymer layer thickness on mechanical data such as tensile strength, Young's modulus and strain to failure of single fibres was investigated. To examine the fibre/matrix adhesion, composites of different carbon fibres and epoxy resin were manufactured. The interlaminar shear strength (ILSS) is found to be about 25 % increased for the coated fibres in the composites. So it should be concluded that the reactive chemical groups are responsible for the improvement of the fibre/matrix adhesion according to the acid/base concept.

1. Introduction

One of the main features for fibre reinforcement of a polymer matrix has been found to be the improvement of interfacial adhesion. Therefore, a variety of surface treatments and coating techniques have been developed. For example oxidation of carbon fibre surfaces with gaseous or liquid oxidizing agents /1-2/ allows an activation of the surface by introduction of chemical groups containing oxygen. Such functional groups can improve interfacial adhesion between the fibres and particular matrices. But as a disadvantage of gaseous oxidation a substantial decrease in tensile strength due to damage of the fibre surface is known to occur whereas wet oxidation results in treatment times of several hours.

On the other hand, coating techniques do have less influence on the fibre surface. Nevertheless, brittle coatings such as SiC deteriorate mechanical properties of the fibres. As an example, reactive sputtered SiC layers improve the interfacial shear strength of carbon fibre reinforced epoxy resin /3/. Recently, Dagli and Sung /4/ reported about the coating of carbon fibres by plasma polymerization of acrylonitrile and styrene using an inductively coupled plasma.

G. Akovali (ed.), The Interfacial Interactions in Polymeric Composites, 269–285.
© 1993 *Kluwer Academic Publishers.*

2. Experimental details

2.1. DEPOSITION OF PLASMA POLYMER FILMS

Plasma polymerization represents a coating technique whereby a monomer gets fragmented inside a plasma and polymerize on a substrate surface.

Such a plasma consists of ions, atoms, molecules, neutral particles, electrons and metastable as well as exited states of these particles. It can be distinguished between "hot plasma" and "cold plasma". In a "hot plasma" all particles are in thermal equilibrium whereas in a "cold plasma" the electrons have got temperatures around 10^5 K while that of ions, atoms and neutral particles are lying between 600 and 1000 K. A "cold plasma" is generated in a vacuum of about $1 - 1*10^{-3}$ mbar. Fig. 1 shows some plasma techniques. Glow discharge is a technique that is commonly used to produce plasmas. Almost all types of electrical discharges are build up of three elements in common: they are sustained by a source of electrical power (d.c., a.c., r.f. or microwave); this power can be coupled by means of a resistive, capacitive or inductive mechanism into the plasma. The coating by plasma polymerization is affected by operating parameters like pressure, power, nature of monomer, monomer flow rate, additional gas, carrier gas and reactor arrangement.

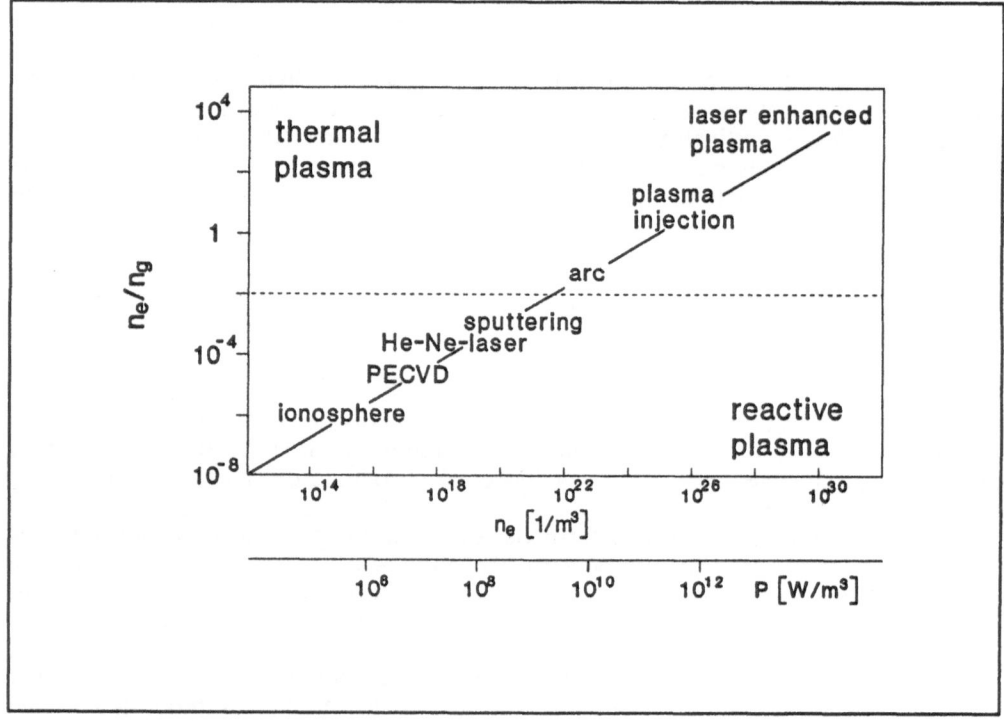

FIG. 1. Some plasma techniques

In this work, a r.f. sputtering unit (13.56 MHz) with planar electrodes in a diode arrangement was used for continuous deposition of plasma polymer layers on multifilament glass and carbon fibres (FIG. 2). Graphite was applied as target material bonded on a watercooled steel cathode with a diameter of 8 in.. As supports on the anode different materials like glass slides, KBr for IR measurements or metal platelets such as aluminium foil could be chosen for analytical purposes. The fibres were electrically insulated from the surrounding electrodes and electromechanically transported through the plasma. The distance between cathode and anode amounted to 8 cm. Before deposition was started the reactor chamber was pumped down to a total pressure of $4 \cdot 10^{-3}$ Pa. The analysis of the residual gas phase as well as of the gas mixture was performed by a quadrupole mass spectrometer during the polymerization process.

In this work only two mixtures of monomer gases, acetylene C_2H_2 (purity 99.5 %) and air (purity of N_2: 99.999 %, O_2: 99.995 %) were fed into the reactor through gas flowmeters. The r.f. input power varied between 0.1 and 1.5 kW resulting in a power density up to 4.6 W/cm^2 at the cathode. The peak voltage ranged from 1.0 to 3.2 kV.

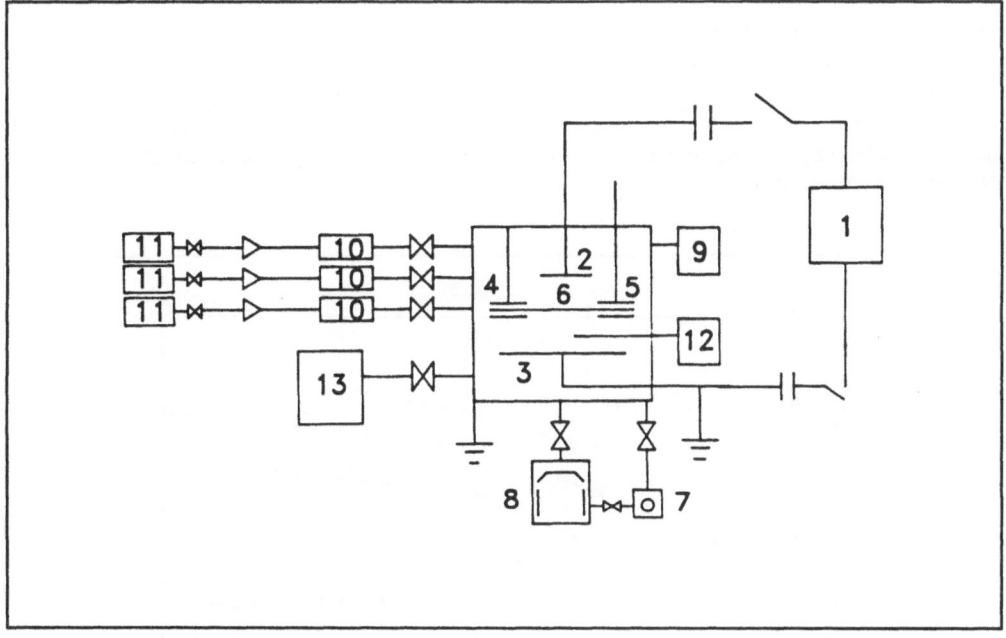

FIG. 2 Schematic of the r.f. sputtering set-up. Legend: 1, r.f power supply; 2, carbon target; 3, stainless steel anode; 4,5, take up and supply spools; 6, carbon fibre bundle; 7,8, rotary and diffusion pumps; 9, pressure gauge; 10, gas flow controllers for monomer gas and additional gases; 11, gas supplies; 12, thickness monitor; 13, mass spectrometer and computer.

2.2 CHARACTERIZATION OF COATINGS

The deposition rate of the polymer films was monitored by quartz crystal oscillating technique and controlled by multiple beam interferometry according to Tolansky's

272

method /5/. The thermal stability of the polymers was examined thermogravimetrically ranging from 25 to 900 °C while the split-off gases were analysed by gas chromatography. For IR analysis KBr powder was coated for 2 or 3 h and pressed to compacts.

The measurements of tensile strength σ_z, Young's modulus E and strain to failure ϵ of coated carbon and glass fibres were performed with monofilaments fixed in a clamping frame of 30 mm inner length. To get average values of the mechanical data at least 50 measurements were used.

Composites containing 60 vol.% carbon fibres in an epoxy resin of type LY556 cured with diamin HT972 (both from Ciba Geigy Comp.) were manufactured to investigate adhesion between fibres and resin.

3. Results and discussion

3.1. DEPOSITION RATE OF PLASMAPOLYMERIZED ACETYLENE

The deposition rate of plasma polymer layers was estimated as a function of the flow rate of the monomer and power. As a result of the apparatus used, sputtering of the graphite target was superimposed on the plasma polymerization process. As depicted in FIG. 3 the sputtering rate of the graphite target was in the range between 3 and 10 nm/min. The deposition rate of carbon is known to be very low because of its high energy of sublimation /6/.

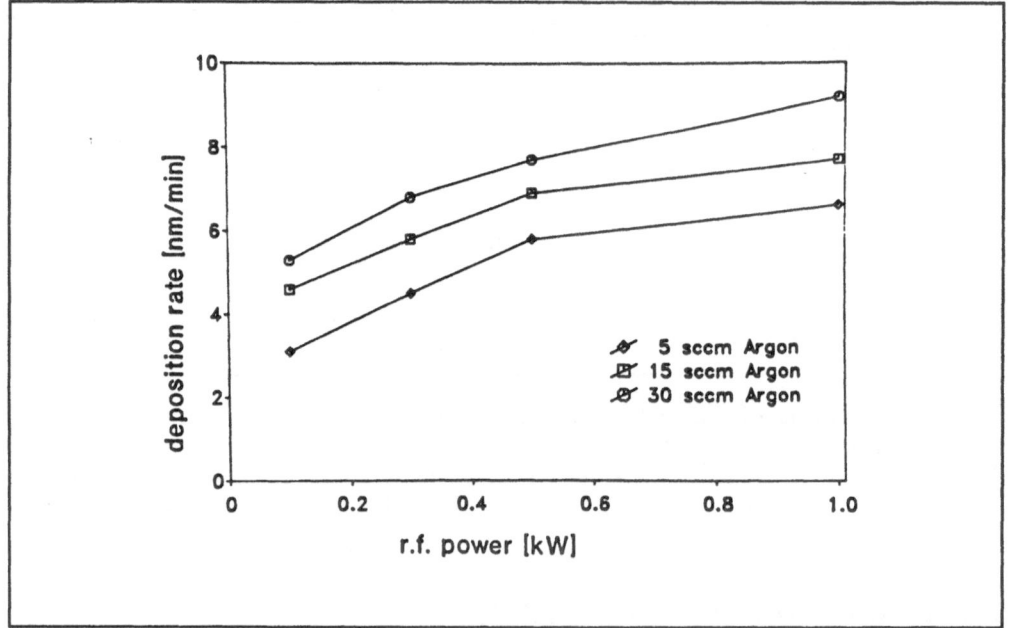

FIG. 3 Deposition rate of carbon originating from sputtering of the graphite target as function of r.f. power and argon flow rate as parameter.

In contrast, with acetylene higher deposition rates are achieved. Adding 5 standard cm³ min⁻¹ (in the following abreviated: sccm) acetylene to 15 sccm air improved the deposition rate by about four times even at low power levels (FIG. 4). An increase in power level does not affect strongly the deposition rate. The small improvement may be due to superimposed sputtering of carbon. Higher amounts of acetylene led to higher deposition rates of 25 to 160 nm/min.

To estimate the maximum of deposition rate a model was used developed by Denaro et al. /7/. The authors assumed the following reactions to describe plasma polymerization process:

$$R_n + M \longrightarrow R_{n+1} \tag{I}$$
$$R_n + R_m \longrightarrow P_{m+n} \text{ (or } P_m + P_n) \tag{II}$$
$$R_n \longrightarrow R_n \text{ (trapped)} \tag{III}$$

R - radical; M - monomer; P - polymer.

Only reaction (I) contributes to the chain propagation. All reactions lead to a loss of radicals so that the deposition rate can be described as

$$r_P = r_R \frac{v_1}{v_1 + v_2 + v_3} \tag{IV}$$

r_P — deposition rate
r_R — rate of radical formation
v_1, v_2, v_3 — reaction rates of reactions I, II and III

Mathematical transformation leads to the expression

$$\frac{1}{r_P} = A \frac{1}{p} + \frac{1}{r_R} \tag{V}$$

p - total pressure.
A is a constant which includes v_1 - v_3.

A plot of $1/r_P$ versus $1/p$ should result in a linear relation. The ordinate value ($1/p = 0$; infinite pressure) is a measure for the maximum deposition rate (FIG. 4).

It is found that the measured depositon rates are about two-thirds of the calculated rates (TABLE 1). This is obviously due to the limitation of pressure in our apparatus.

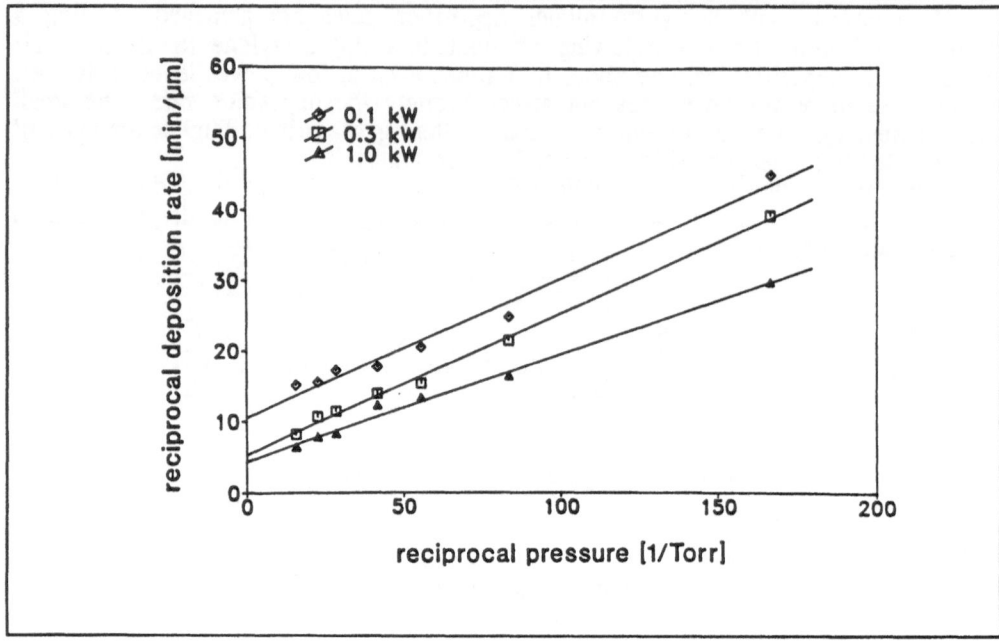

FIG. 4 Reciprocal deposition rate of acetylene in dependence of reciprocal pressure with variation of the r.f power

TABLE 1: Deposition rate of plasmapolymerized acetylene

power [kW]	calculated [nm/min]	measured [nm/min]
0.1	95.1	66.6
0.3	188.7	122.0
1.0	230.4	160.5

To estimate the influence of power on the deposition rate it is assumed that the deposition rate and power are connected via equation (VI)

$$r_P = b \cdot P^x \qquad\qquad (VI)$$

P - power; b - preexponential factor

A plot of log (r_P) versus log (P) should give a straight line with a gradient of x (FIG. 5).

It is found that with increasing monomer flow rate the influence of power on the deposition rate increases. However, the influence is not very strong (x is in the range between 0.2 and 0.4) because of the very low pressure which is between 10^{-2} to 10^{-3} Torr.

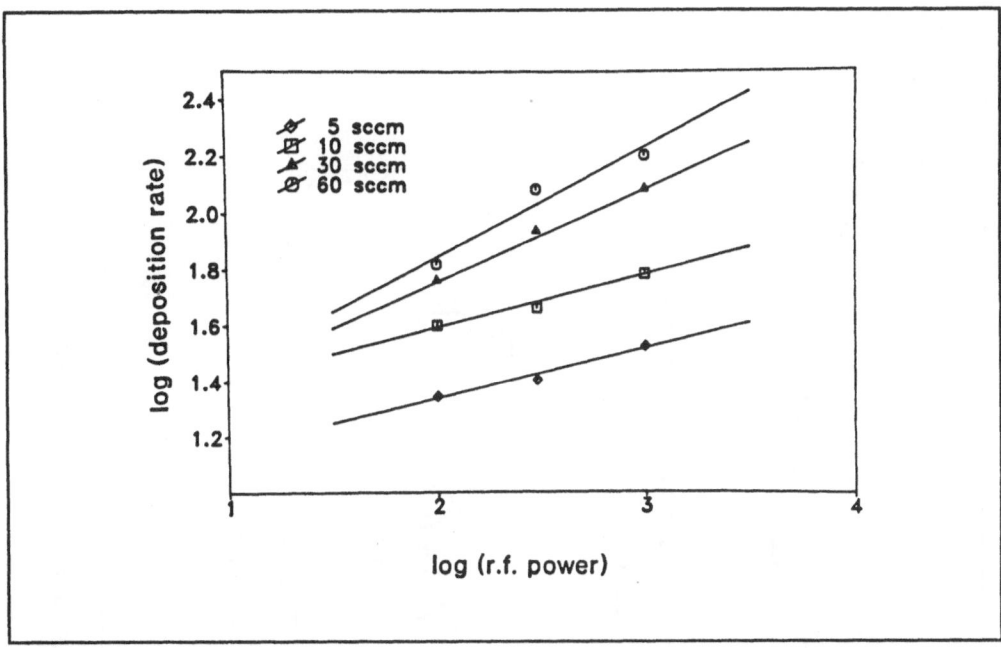

FIG. 5 Dependence of the deposition rate of acetylene as a monomer on the r.f. power

3.2. IR SPECTRA

In FIGS. 6 a and 6 b (and TABLE 2) IR spectra of plasma polymers are depicted when applying acetylene in combination with argon and air.

Plasmapolymerized acetylene using inert argon causes almost no functional groups. The weak absorbance between 1750 and 1630 cm^{-1} (carbonyl or hydroxyl groups) may be due to oxygen which is incorporated during the exposure to the laboratory atmosphere. Replacement of argon by air results in a functionalization of the polymer layer (FIG. 6 b). The most intense peak is found at 2172 cm^{-1} which belongs to the C≡N group. Shoulders between 1755 and 1630 cm^{-1} can be related to carbonyl groups, secondary or tertiary amides. Between 900 and 600 cm^{-1} peaks resulting from vibrations of aromatic groups are detected in accordance to literature /8/.

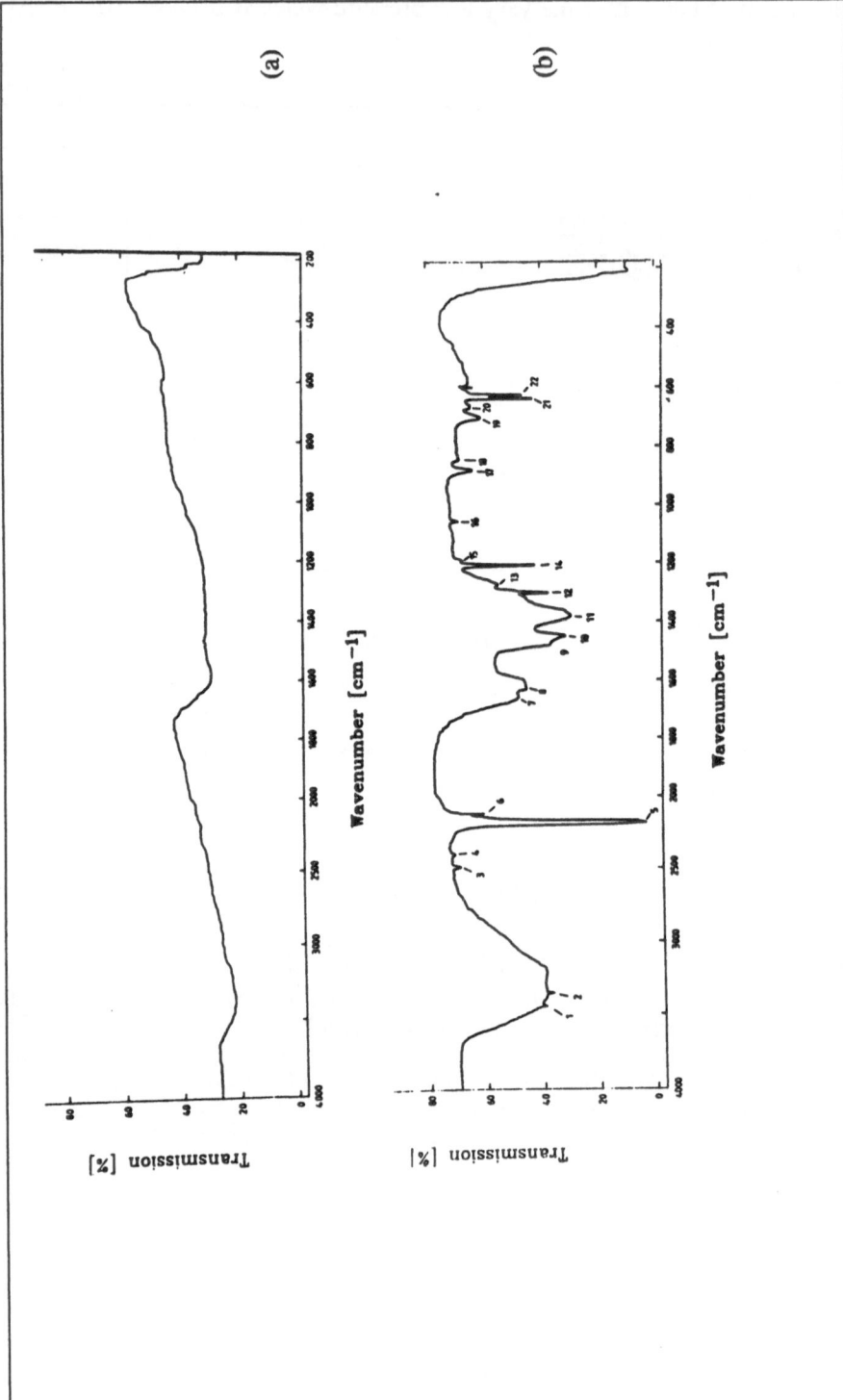

FIG. 6a, b IR spectra of plasmapolymerized acetylene using (a) acetylene/argon with a mixing ratio of 15 sccm/15 sccm and (b) acetylen/air with a mixing ratio of 5 sccm/15 sccm

TABLE 2: IR spectra of plasmapolymerized acetylene

no.	relative intensity	wavenumber [cm^{-1}]	tentative assignment
1	m	3450	-O-H intermolecular bonded
2	m	3358	-O-H stretching
3	m	2492	$=NH_2^+$, $\equiv NH^+$, N-H stretching
4	m	2400	$=NH_2^+$, $\equiv NH^+$, N-H stretching
5	vs	2172	-C$\equiv$N stretching
6	m	2115	-C$\equiv$C-H, C$\equiv$C stretching
7	w	1658	-C=O stretching
8	w	1630	-N-H deformation
9	w	1478	-C-H stretching
10	s	1448	-CH$_2$ deformation
11	m	1371	-C-N stretching
12	s	1300	-C-OH, C-O stretching
13	m	1270	-C-OH, O-H deformation
14	s	1205	-C-O-C, C-O stretching
15	w	1188	-C-OH, C-O stretching
16	m	1058	-C-O-C deformation
17	s	863	R$_2$C=CHR, out of plane
18	w	830	metadisubstituted benzene
19	s	701	RHC=CHR, out of plane
20	m	668	C-H deformation in C$_6$H$_5$-R
21	s	638	$\equiv$C-H deformation
22	s	625	$\equiv$C-H deformation

vs - very strong;   s - strong;   m - medium;   w - weak

3.3. GAS PHASE ANALYSIS

In FIGS. 7 and 8 the composition of the gas phase is shown in dependence of the power level for the mixture containing 5 sccm acetylene/15 sccm air. The nitrogen peak is defined as 1000 since this peak does not vary with increasing power level. Most of the acetylene is cracked even at low power levels (FIG. 6). This is also proved by a pressure drop in the range of the acetylene partial pressure after glow discharge has been struck. Inversely to the decrease of acetylene concentration, the concentration of radicals increases (FIG. 8). The radicals are such as CHO·, COO·, OH·, O· and N· and molecules like H_2, H_2O and $(CN)_2$ which originate from radicals by recombination. The concentration of the radicals is influenced by the power level to a smaller extent. Therefore, incorporation of functional groups into the plasmapolymerized layer should not be strongly affected by the power level applied.

3.4. XPS ANALYSIS

The deposited polymer films were studied by XPS analysis to determine the chemical state of the surface. The spectra corresponding to the bond energies of C 1s, O 1s and N 1s in plasmapolymerized acetylene were used. The bond energy of C 1s level exhibits a chemical shift of approximately 5 eV from 284 to 289 eV which can be

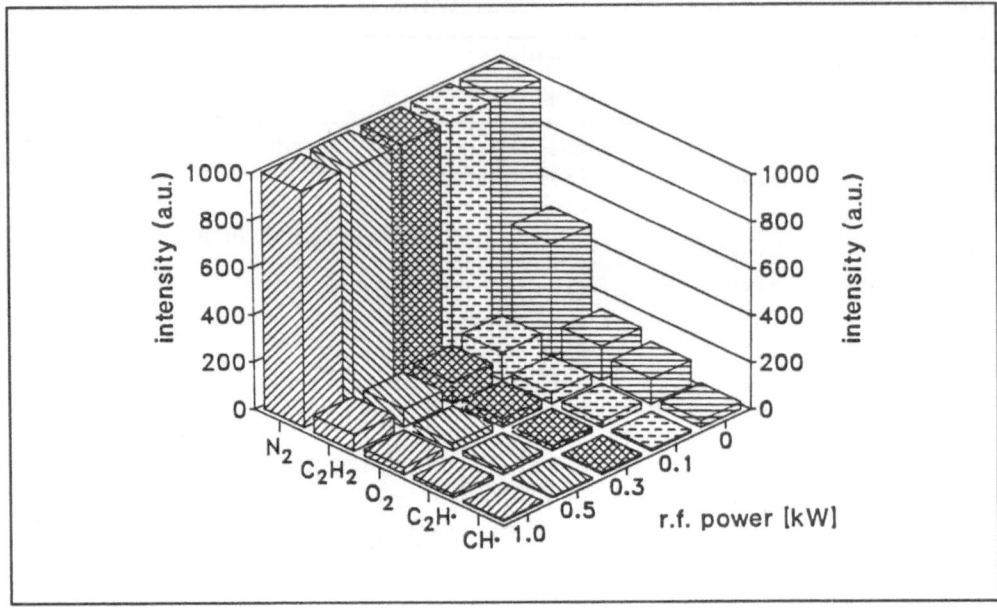

FIG. 7 Influence of r.f. power on the mass peaks of nitrogen, acetylene, oxygen and fragments using a gas mixture of 5 sccm acetylene/15 sccm air

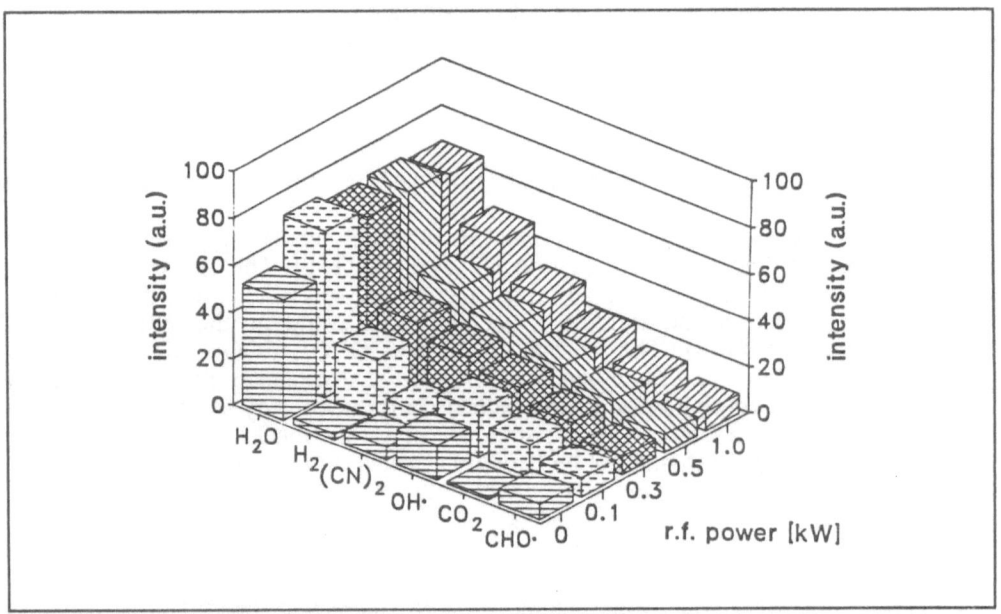

FIG. 8 Influence of r.f. power on the mass peaks of some molecules and radicals occuring during plasma polymerization

correlated to -C-OH, -C=O and -COOH groups. For the O 1s state a binding energy of about 537.5 eV was measured in accordance with carbonyl and hydroxyl groups. Also the N 1s peak was shifted to about 399 eV, typical for nitrile groups.

3.5. ELEMENTAL ANALYSIS

Nitrogen and oxygen were incorporated into the plasma polymers (TABLE 3). It was found that the polymers contain less hydrogen relative to the hydrogen content of the monomers.

This result is contrary to that of Yasuda /9/ who deposited acetylene in a nitrogen atmosphere by using an inductively coupled plasma. These differences may be explained by the different method by which the plasma was generated.

Deposited polymer films originating from 1:3 acetylene/air ratio yielded in a higher amount of heteroatoms than those from the 1:1 gas mixture. Increasing the power level did not affect significantly the elemental composition of the polymer layers in agreement with conclusions already drawn from mass spectra.

TABLE 3: Elemental analysis of plasmapolymerized acetylene using different gas mixtures

flow rate [sccm]	power [kW]	elements found [%]				empirical formula
		C	H	N	O	
5 C_2H_2/ 15 air	0.3	55.4	3.7	13.9	27.0	$C_2H_{1.6}N_{0.4}O_{0.7}$
5 C_2H_2 15 air	1.5	57.1	3.3	16.7	22.9	$C_2H_{1.4}N_{0.5}O_{0.6}$
15 C_2H_2 15 air	0.3	61.7	3.6	14.8	19.9	$C_2H_{1.4}N_{0.4}O_{0.5}$
15 C_2H_2 15 air	1.5	61.9	3.1	14.4	20.6	$C_2H_{1.2}N_{0.4}O_{0.5}$
15 C_2H_2 15 nitrogen	0.3	68.0	4.3	17.8	9.9	$C_2H_{1.5}N_{0.5}O_{0.2}$

According to the literature /10,11/ plasma polymer layers contain free radicals which will react with oxygen and water in the atmosphere. In order to examine this fact plasmapolymerized acetylene was deposited using nitrogen instead of air. The elemental analysis of these films revealed less oxygen than those deposited in air. This proves that most of the oxygen is incorporated into the layer during plasma polymerization.

3.6. THERMOGRAVIMETRIC ANALYSIS

About 60 wt.% of the polymer layer resisted temperatures up to 900 °C under an argon atmosphere. Similar results are reported by Thompson and Mayhan /12/ who investigated plasmapolymerized styrene. Endothermic reactions due to degradation are occuring.

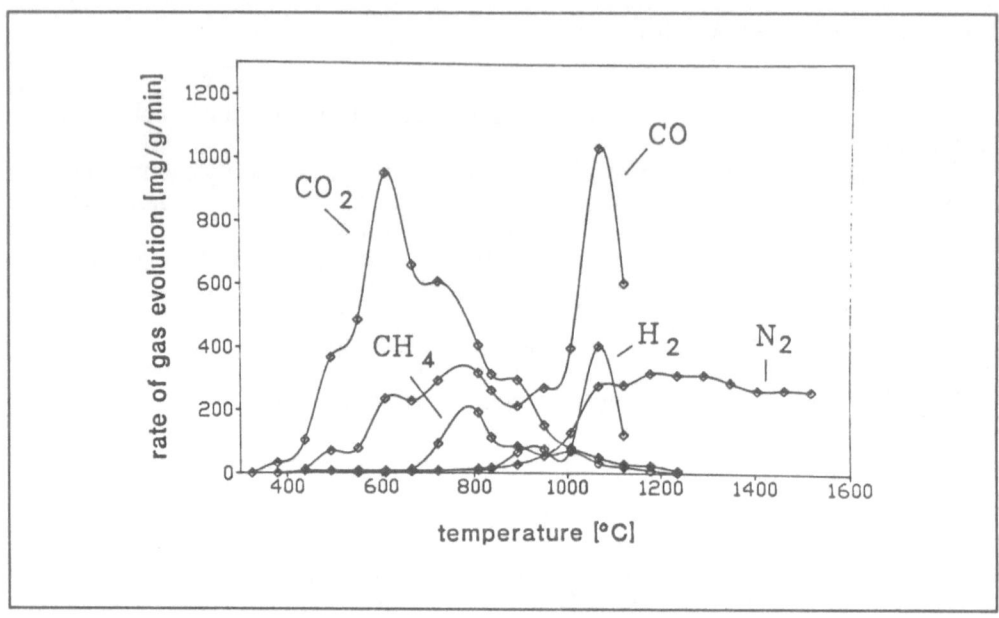

FIG. 9 Gas evolution of plasmapolymerized acetylene

Examination of the degradation products in the temperature range between 400 and 1500 °C was performed by heat treating plasma polymers in a graphite tube under a helium atmosphere. Gaseous products split off were analyzed by gas chromatography. Up to 800 °C mainly CO_2 is generated while at higher temperatures CO and N_2 are produced (FIG. 9). The increased CO content and a decrease in CO_2 may be due to the degeneration of the layer with regard to oxygen but also to the Boudouard equilibrium. Nitrogen does not occur at temperatures below 800 °C, indicating that nitrogen is strongly bonded into the polymer layer.

TABLE 4: Contact angle measurements of plasmapolymerized
layers. Water as liquid; flow rate of air 15 sccm

power [kW]	5 sccm C_2H_2	15 sccm C_2H_2	60 sccm C_2H_2
0.1	32.5° ± 3.2°	56.8° ± 5.0°	63.5° ± 4.7°
0.3	30.5° ± 2.5°	50.5° ± 3.9°	60.6° ± 5.0°
1.0	29.8° ± 3.1°	46.3° ± 4.1°	58.8° ± 4.3°

3.7. CONTACT ANGLE MEASUREMENTS

To estimate the surface energy of the deposited films, glass slides were coated. Then the wettability against water was measured. According to Young's equation the contact angle was used to characterize the surface energy of the films.

It was found that the wettability of the surfaces slightly increases with increasing power, using mixtures of acetylene/air. Increasing the flow rate of acetylene leads to an decreasing wettability of the surfaces. These two effects are due to a higher amount of heteroatoms in the layers (TABLE 4).

Using argon only, high contact angles were measured. This shows that the surface energy of the deposited films depend strongly on the gas used to sustain the glow discharge.

4. Coating of fibres

To estimate the influence of layer thickness on the mechanical data of fibres such as tensile strength σ_z, Young's modulus E and strain to failure ϵ, carbon fibre bundles of about 10 cm length were coated discontinuously.

4.1. MECHANICAL DATA OF COATED FIBRES

To investigate the influence of the deposited layer on the tensile strength of fibres, commercially available carbon fibres (Grafil XAS, Hercules AS4 and Toho HT7) and one type of glass fibre (Gevetex EC) are coated in the range between 50 and 500 nm. The deposition was performed at 0.3 kW resulting in rates of 21 and 27 nm/min for 5 and 15 sccm acetylene/15 sccm air, respectively. Both acetylene/air mixtures led to the same results. This fact implies that the mechanical data of the plasmapolymerized films are determined rather by the carbon skeleton than by the functional groups on the surface.

In FIG. 10 it is shown for the glass fibre that the tensile strength decreases with increasing layer thickness whereas in the case of the carbon fibres the tensile strength increases till a film thickness of about 50 nm is reached (FIG. 11).

These results may be explained by the strain to failure of coated glass and carbon fibres (FIG. 12). The strain to failure of the glass fibre decreases with increasing layer thickness while the carbon fibres exhibit a small improvement.

An explanation for the damage behaviour of the glass fibre is found in the crack initiation within the polymer layer which is more brittle than the fibre itself. The expanding cracks cause surface damages of the glass fibre and, according to Griffith's theory /13/, lead to a decrease of tensile strength.

On the other hand, carbon fibres are more brittle than the deposited polymer layers. Since the layers are bearing a part of the load applied, the tensile strength of the fibre/layer compound will increase. With increasing volume of the layer, the number of surface flaws is growing due to Griffith's theory, resulting in a decrease of tensile strength above about 50 nm layer thickness. Therefore Young's modulus of coated carbon fibres will decrease with increasing layer thickness.

To study the adhesion of the polymer layer on the fibres scanning electron microscopy was applied. As an example, FIG. 13 (a) shows the fracture of a coated HT7 carbon fibre. For sake of demonstration a coating of 1.2 μm thickness was applied. It can be concluded that there is a good adhesion between fibre and the polymer layer. The surface morphology of the layer was found to be

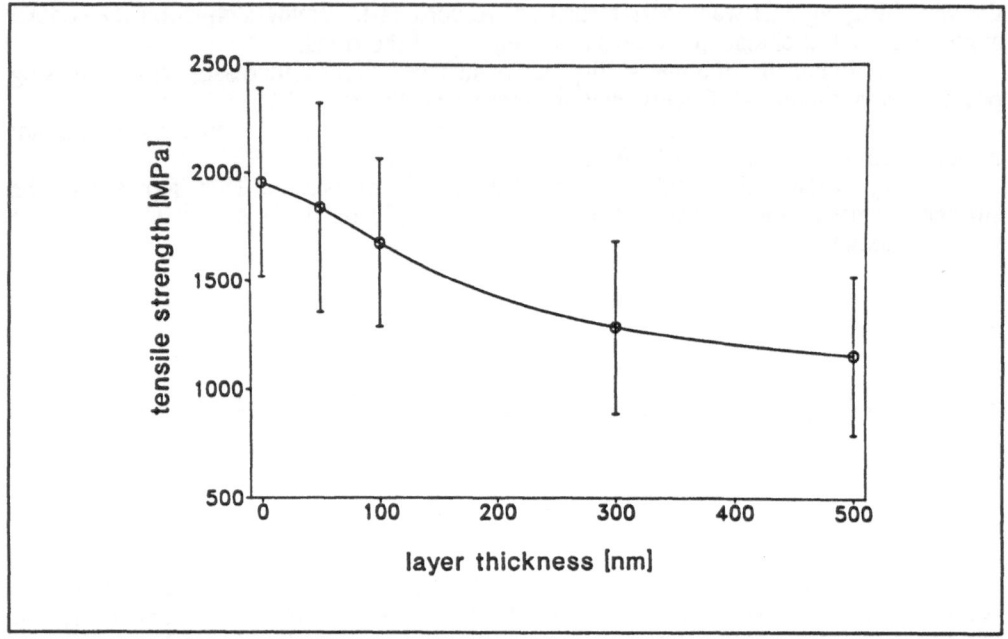

FIG. 10   Tensile strength of EC glass fibre as a function of thickness of the plasmapolymer layer using 0.3 kW r.f. power and mixtures of 5 sccm or 15 sccm acetylene/15 sccm air

a copy of the fibre morphology. Additionally, FIG. 13 (b) shows the topography of a plasmatreated fibre. The layer seems to be dense and covers the fibre completely.

Similar results were found for glass fibres.

4.2 ADHESION BETWEEN CARBON FIBRES AND EPOXY RESIN

To investigate the adhesion between fibres and matrix, composites containing 60 vol.% of coated Hercules AS4 and Grafil XAS carbon fibres in a matrix of epoxy resin were manufactured.

In FIG. 14 the interlaminar shear strength (ILSS) of the two types of composites is illustrated in dependence of the deposition time of the plasma polymer. Applying the mixture of 5 sccm acetylene/15 sccm air it is found that the ILSS is increased by about 25 % relative to that of the virgin fibre. This result is explained by the improvement of adhesion between fibre and matrix resulting from chemical bonds between epoxy resin and the functional groups of the plasmapolymerized acetylene.

The rising ILSS with increasing deposition time is attributable to the increase of tensile strength of the fibres. The gas mixture with the higher air/acetylene ratio leads to higher ILSS values (FIG. 15), although this gas mixture results in lower deposition rates than the mixture with 15 sccm acetylene/15 sccm air (FIG. 8).

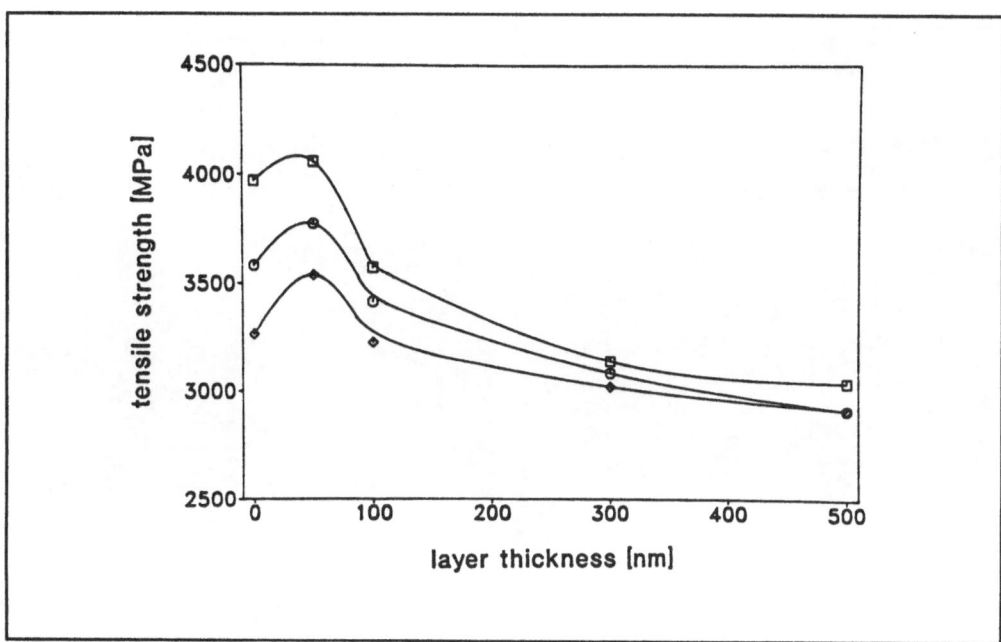

FIG. 11 Tensile strength of carbon fibres as a function of thickness of the plasmapolymer layer. Parameters see in FIG. 10

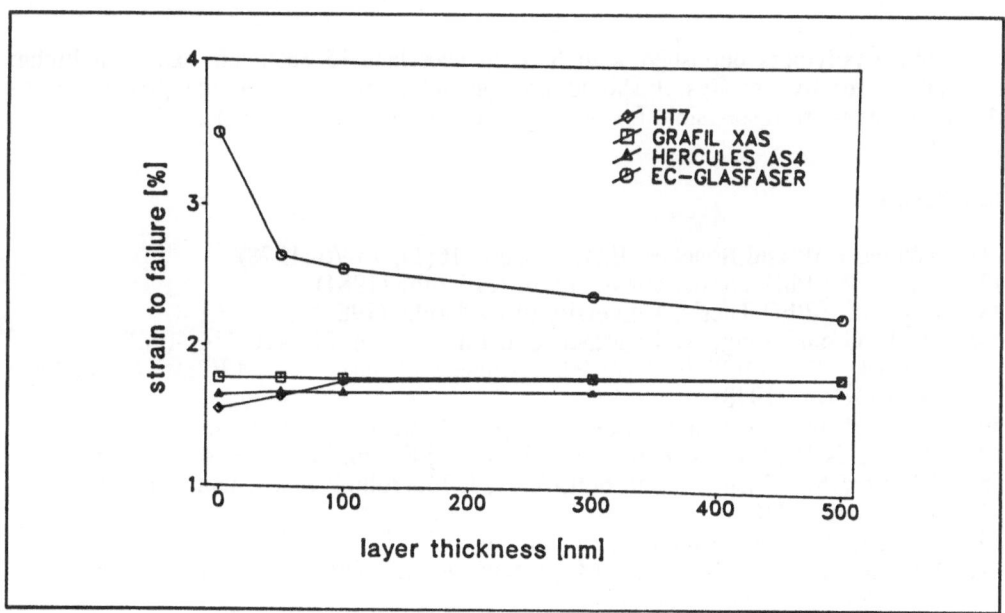

FIG. 12 Strain to failure of various fibres as a function of the thickness of the plasmapolymer layer. Parameters see in FIG. 10

284

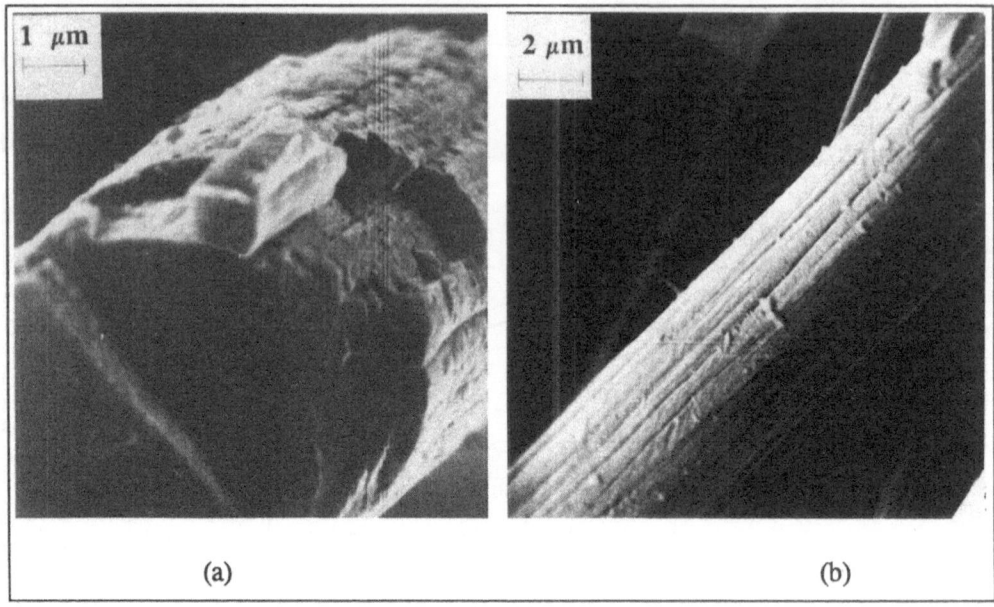

(a)   (b)

FIG. 13 a,b Scanning electron micrographs of plasmapolymer-coated carbon fibres:
(a) fracture and (b) topography of an HT 7-type fibre

Since polymers deposited with 5 sccm acetylene/15 sccm air exhibit a higher
amount of reactive groups it should be concluded that the reactive groups in the
deposited layer are responsible for the improvement of fibre/resin adhesion.

5. Literature

/1/   Adams, L.B. and Boucher, E.A.; Carbon 16(1), 75-76 (1978).
/2/   Weiss, R.; PhD Thesis, University of Karlsruhe (1984).
/3/   Nagel, G.; PhD Thesis, University of Karlsruhe (1987).
/4/   Dagli, G. and Sung, N.H.; Mater. Sci. Eng. Polym. 56, 410-414 (1987).
/5/   Tolansky, S.; "Multiple Beam Interferometry of Surfaces and Films" Clarendon
       Press Oxford (1948)
/6/   Laegreid, N. and Wehner, G.K.; J. Appl. Phys. 32, 365 (1961).
/7/   Denaro, A.R., Qwens, P.A., Crawshaw, A.; Europ. Polym. J. 4, 93 (1968).
/8/   Tibbitt, J.M., Shen, M. and Bell, A.T.; J. Macromol. Sci. Chem. A10(8),
       1623-1648 (1976).
/9/   Yasuda, H. J.; Macromol. Sci. Chem. A10(3), 383-420 (1976).
/10/   Simionescu, C.I., Denes, F., Macoveanu, M.M., Totolin, M. and Cazacu, G.;
       Acta Polym. 33(1), 26-30 (1982).
/11/   Yasuda, H. and Marsh, H.C.J.; Appl. Polym. Sci. 19(11), 2981-2990 (1975).
/12/   Thompson, L.F. and Mayhan, K.G.; J. Appl. Polym. Sci. 16, 2317-2341
       (1972).
/13/   Griffith, A.A.; Phil. Trans. Roy. Soc. London A221, 163-198 (1920).

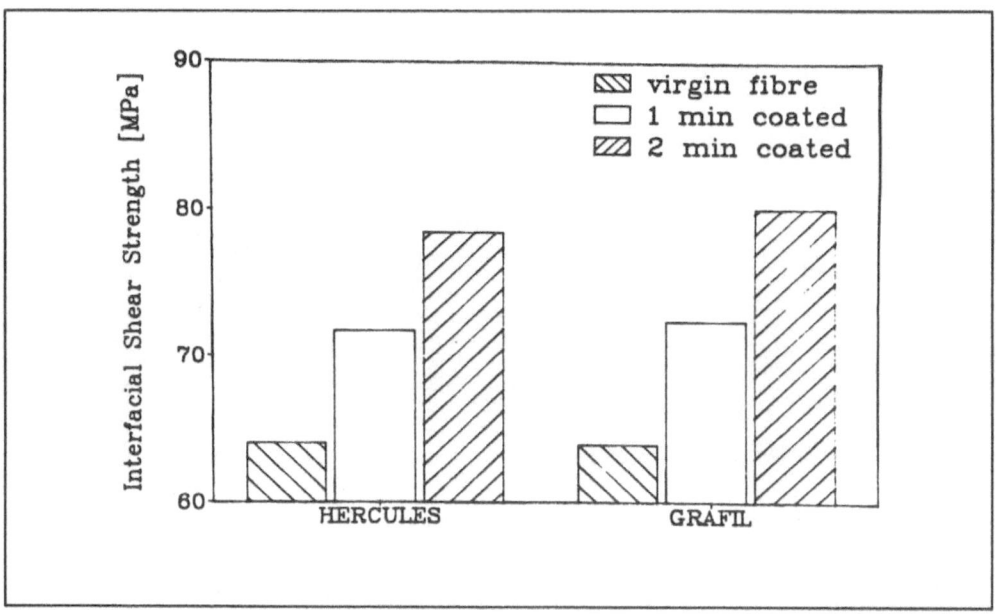

FIG. 14 Interlaminar shear strength of epoxy resin composites in dependence of the two types of fibre. Parameters as in FIG. 10

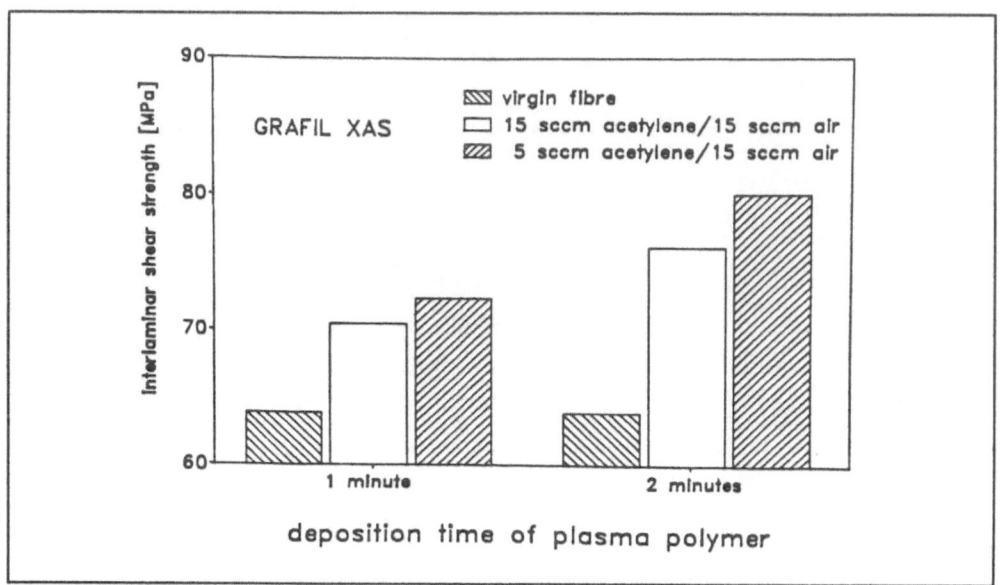

FIG. 15 Interlaminar shear strength of epoxy resin composites in dependence of deposition time of the plasmapolymer

PLASMA ENHANCED CVD OF AROMATICS: SURFACE TREATMENT OF CARBON FIBRES TO OPTIMIZE FIBRE/MATRIX ADHESION

E. EBERT and W. WEISWEILER
Institute of Technical Chemistry, University of Karlsruhe
Kaiserstrasse 12, D-7500 Karlsruhe/Germany

ABSTRACT. In this work, we shall describe plasma polymerization of typical aromatics like benzene, aniline and pyridine that form polymeric films on different carbon fibres. For a given type of aromatic monomer, plasma polymerization was performed with a radio frequency sputtering apparatus using argon as carrier gas and additionally for special experiments with gas like ammonia, water vapour, sulphur dioxide or air as reactive component. The plasma polymeric films were characterized by IR-spectroscopy, X-ray photoelectron spectroscopy, elemental analysis, mass spectrometry, electron spin resonance, grazing incidence diffraction as well as contact angle determination. The influence of polymeric films on the mechanical properties of fibres were investigated using tensile strength measurements. For testing the application of the plasma polymer coated fibres composites containing 60 % fibres in a matrix of epoxy resin and poly-carbonate as well, were manufactured. Interlaminar shear strength tests and three point bend tests were conducted on these composites. In the case of epoxy resin matrix in particular, an improved type of composite was obtained.

1. Introduction

An objective in the field of materials science is to get composite materials with more specific and qualitatively better characteristics that are not obtained in pure materials alone. A natural composite like wood or straw, an artificial composite like concrete are examples of a combination of two materials into one, with characteristics and critical values quite different from that of the individual constituents.

The polymeric composite consists of fibres for reinforcement on the one hand and of the matrix on the other hand. One of the main problems encountered here is the contact or adhesion between the fibres and the polymeric matrix, the so-called interphase problem. The adhesion between fibre surface and polymeric matrix depends

G. Akovali (ed.), The Interfacial Interactions in Polymeric Composites, 287–307.

on the following types of interactions:

a) chemical,

b) physical and

c) mechanical.

There are several methods to improve fibre/matrix adhesion.

1) Oxidative methods

Functional surface groups are formed by chemical or electrochemical reactions.

a) Thermal oxidation in oxidizing atmospheres containing gases like O_2, O_3, CO /1, 2, 3, 4/.

b) Chemical oxidation using liquids such as H_2O_2, HNO_3, H_2SO_4 /5, 6/.

c) Anodic oxidation in acids, alkalies or salt solutions /7/.

d) Electrochemical polymerization /8/.

2) Surface coating processes

Surface coating in vacuum is an important industrial process. The techniques can be divided into PVD (physical vapour deposition) and CVD (chemical vapour deposition):

a) PVD processes /9, 10/

- sputtering of metallic, covalent, ionic as well as intercalation compounds. Examples are Au, Ag, SiC, Al_2O_3, TiN, respectively.

- vacuum evaporation using volatile substances like Ag, C, salts.

The advantage of PVD processes lies in the low substrate temperature.

b) CVD processes /11, 12/

- thermal decomposition and chemical transport reaction (SiC, TiN, W and others). High substrate temperatures above 800 °C are disadvantageous and limit the choice of substrates.

Plasma enhanced CVD, i.e., the so-called PECVD, as combination of CVD and PVD, allows chemical vapour deposition at low substrate temperature.

The PECVD or plasma polymerization represents a new technology that enables the production of thin films with manifold properties. Plasma polymerized layers are insoluble in organic solvents, indicative of the highly three dimensional crosslinked structure. The properties of such films can be influenced by parameters like pressure, flow rate, nature of monomer, carrier or reactive gases, power input, reactor configuration, substrate location, frequency (r.f. or microwave). In a "cold plasma" the particles are not in thermical equilibrium. The temperature of the electrons goes up to 10^5 °C, that of neutral particles and ions reaches about 300 °C. The monomers get fragmented in the plasma and polymerize on the fibre surface.

2. Experimental

2.1. DEPOSITION OF PLASMA POLYMERIC FILMS

In this work a radio frequency (13.56 MHz) power generator was used as plasma generator. A schematic diagram of the apparatus is shown in FIG. 2 in the ealier paper of Weisweiler in this book. The electrical power with values up to 1.5 kW is coupled into the plasma by means of a capacitive coupling mechanism. Two watercooled platelike electrodes, one as graphite cathode and the other as stainless steel anode, are mounted in a stainless steel chamber with electrode spacing of about 8 cm. The pumping system for evacuating the chamber consists of a rotary pump with a capacity of 11 l/s and an oil-diffusion pump with a capacity of 1200 l/s. The pressure in the chamber is controlled by an ionization gauge. The working pressure ranges from $4*10^{-5}$ mbar for the evacuated chamber state up to $5*10^{-2}$ mbar during the plasma polymerization process.

The monomer feed tank is placed in a silicon oil bath whose temperature is controlled to provide a suitable vapour pressure of the monomer. Likewise, the tubes between the monomer tank and the vacuum chamber had to be heated to maintain the vapour status of the monomer. The flow rate of the monomer vapour is regulated by a throttle valve and controlled with a flow meter. The partial pressure of a gas component in the vacuum chamber is proportional to its flow rate. The gas flow of the carrier gas argon as well as that of the reaction gases are regulated using flow meters.

In the experiments multifilament carbon fibres were continuously coated with plasma polymers. Further substrates were chosen for characterization of the deposited films comprising glass slides for GID and for thickness measurements, KBr-compacts for IR-measurements, graphite for XPS and single carbon fibres for tensile strength and adhesion measurements. The monomers used are benzene on the one hand, pyridine and aniline on the other hand, to yield nitrogen-free and nitrogen containing plasma polymers respectively. Additionally, for the benzene monomer, reactive gases like air, ammonia and sulphur dioxide was fed into the plasmareactor to functionalize the plasma polymer film through the incorporation of polar chemical groups. A quadrupole mass spectrometer was also installed to analyze the residual gas as well as the gas mixture during the plasma polymerization process.

2.2. CHARACTERIZATION OF PLASMA POLYMERIC FILMS

The deposition rate was determined by multiple beam interferometry according to Tolansky´s method /13/. To simplify the procedure, a small glass slide was fixed on a bigger one. After coating the glass slide arrangement with plasma polymers for a certain time, the smaller glass slide was weighed to get the mass deposition rate. The bigger partially coated slides with their layer margins help to determine the layer thickness in according with Tolansky´s method. Another method could be used for evaluating the deposition rate of the plasma polymer film. This method is based on an oscillating quartz crystal with a frequency of 6 MHz. The frequency change due to the deposition of plasma polymers on the quartz crystal; this is a measure of the deposition rate, and can be monitored if the density of the plasma polymer is known.

Elemental analysis was conducted with the plasma polymers to determine the elemental composition. For IR measurements, KBr powder was coated with plasma polymers for several hours. Then the KBr powder was pressed in a mould to form pellets.

The films were characterized by XPS. This photoelectron spectroscopy gives information about the specimen surface to a depth of several nanometers. For this purpose the carbon platelets coated with the films were excited with Al-K_α X-rays to the energy of 1486.6 eV. To get more information about the structure of plasma polymers, glass slides were coated and the layers were investigated in a grazing incidence diffractometer (GID). The angle of incidence between the X-ray and the polymer surface was varied from 0.5 to 1.5°.

The Wilhelmy technique was employed to study adhesion on the coated carbon fibres.

The powder of plasma polymer material was investigated by electron spin resonance (ESR) to get information about the spin concentration corresponding to the amount of unpaired electrons or radicals in the sample.

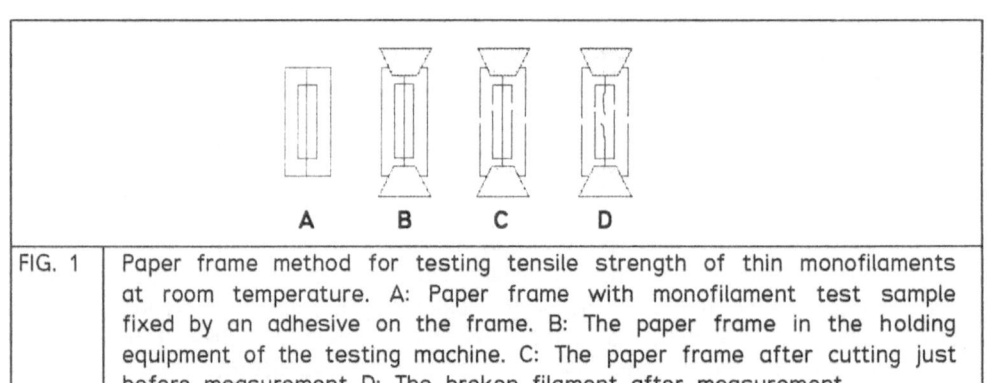

| FIG. 1 | Paper frame method for testing tensile strength of thin monofilaments at room temperature. A: Paper frame with monofilament test sample fixed by an adhesive on the frame. B: The paper frame in the holding equipment of the testing machine. C: The paper frame after cutting just before measurement D: The broken filament after measurement. |

The coated carbon fibres were tested as monofilaments mounted in a paper frame as demonstrated in FIG. 1. The gauge length is amounted to 30 mm.

Tensile strength σ_z, strain to failure ε and tensile modulus E are results of these tests. The meanings and dimensions in the equations (1) to (3) are as follows: — F: Tensile force at failure (N); A: Fibre cross-section (m^2); Δl: Elongation of fibre (m); l_o: Gauge length of specimen (m).

$$\sigma_z = \frac{F}{A} \qquad [MPa] \qquad (1)$$

$$\varepsilon = \frac{\Delta l}{l_o} * 100 \qquad [\%] \qquad (2)$$

$$E = \frac{\sigma_z}{\varepsilon} \qquad [GPa] \qquad (3)$$

Some specifications of the carbon fibres used in this work are listed below in TABLE 1.

TABLE 1. Specifications of carbon the fibres used

fibre type	density	dtex	number of filaments	fibre cross-section	mean diameter
Hercules AD 584	1.77 g/cm^3	0.881 g/m	12000	41.3 μm^2	7.25 μm
Grafil XAS	1.82 g/cm^3	0.797 g/m	12000	37.4 μm^2	6.82 μm

The influence of coating on mechanical properties of uncoated and coated carbon fibres, embedded in epoxy resin and polycarbonate matrix, respectively, were conducted by a three point bend test applied to the composites.

3. Results and Discussion

3.1. GAS ANALYSIS

The nature of coatings generated by plasma polymerization is affected by operating parameters such as pressure, r.f. power, type of monomer and carrier gas flow rate. Since the conditions vary from system to system, basic investigations were performed to determine the influence of these parameters.

The residual gas, before and during the plasma polymerization process, was controlled by mass spectrometry. The mass peak 2, corresponding to hydrogen, is found to be the most intensive peak of all mass spectra.

FIG. 2 shows the intensity of fragmentation of the three monomers benzene, aniline and pyridine before and after the ignition of the plasma. The main parameters are: 300 W r.f. power and an argon flow rate of 15 standard cm^3 min^{-1} (sccm). The degree of fragmentation of the original monomer molecules exeeds 95 %. As example, the fragmentation during the plasma polymerization of benzene is depicted in FIG. 3. The dependence on the r.f. power demonstrates that a sufficient fragmentation of monomers could be obtained at low power levels of 50 to 300 W. Higher power levels are disadvantageous because the plasma polymerization process will be affected by increased sputtering of elemental carbon, the material of which the cathode is made. The fragmentation of benzene was found to increase with carrier gas flow rate and to decrease with monomer gas flow rate, at constant r.f. power in each case.

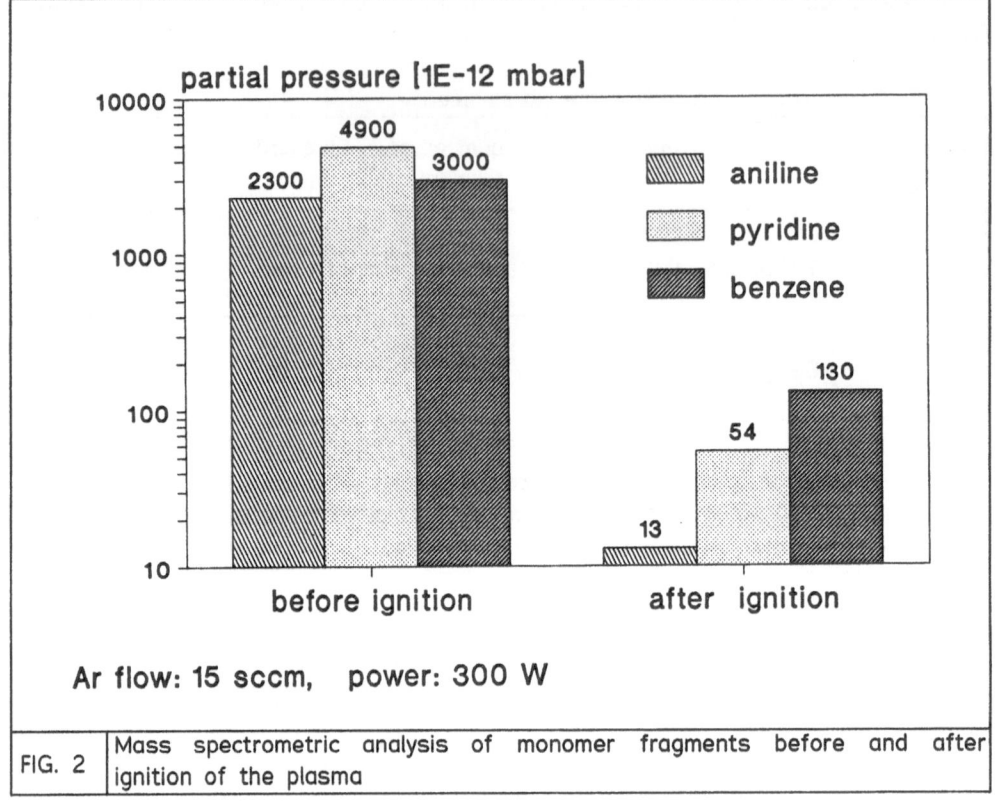

| FIG. 2 | Mass spectrometric analysis of monomer fragments before and after ignition of the plasma |

3.2. DEPOSITION RATE

In TABLE 2 the experimental results concerning the deposition rate of plasma polymers are summarized. The deposition rates of plasma polymerized benzene (B), aniline (A) and pyridine (P) can be studied. High deposition rates were achieved at r.f. power levels of 50 and 100 W. The deposition rate increases from low power levels of 50 W to a maximum of roughly 100 W and then decreases at higher power levels. The flow rate of the monomers pyridine and aniline are calculated by the partial pressure of these monomers, whereas the flow rate of benzene (5 standard cm^3 min^{-1} corresponding to a partial pressure of $3*10^{-3}$ mbar) is regulated by a mass flow meter. The application of the monomers aniline and pyridine led to lower deposition rates. This can be explained by the fact that the nitrogen of the monomer molecule is bound only in part to the polymer film, the rest getting lost as stable N_2 and HCN to be found in the residual gas. It is clear from TABLE 2 that the addition of reactive gases like ammonia and water vapour do not affect the deposition rate.

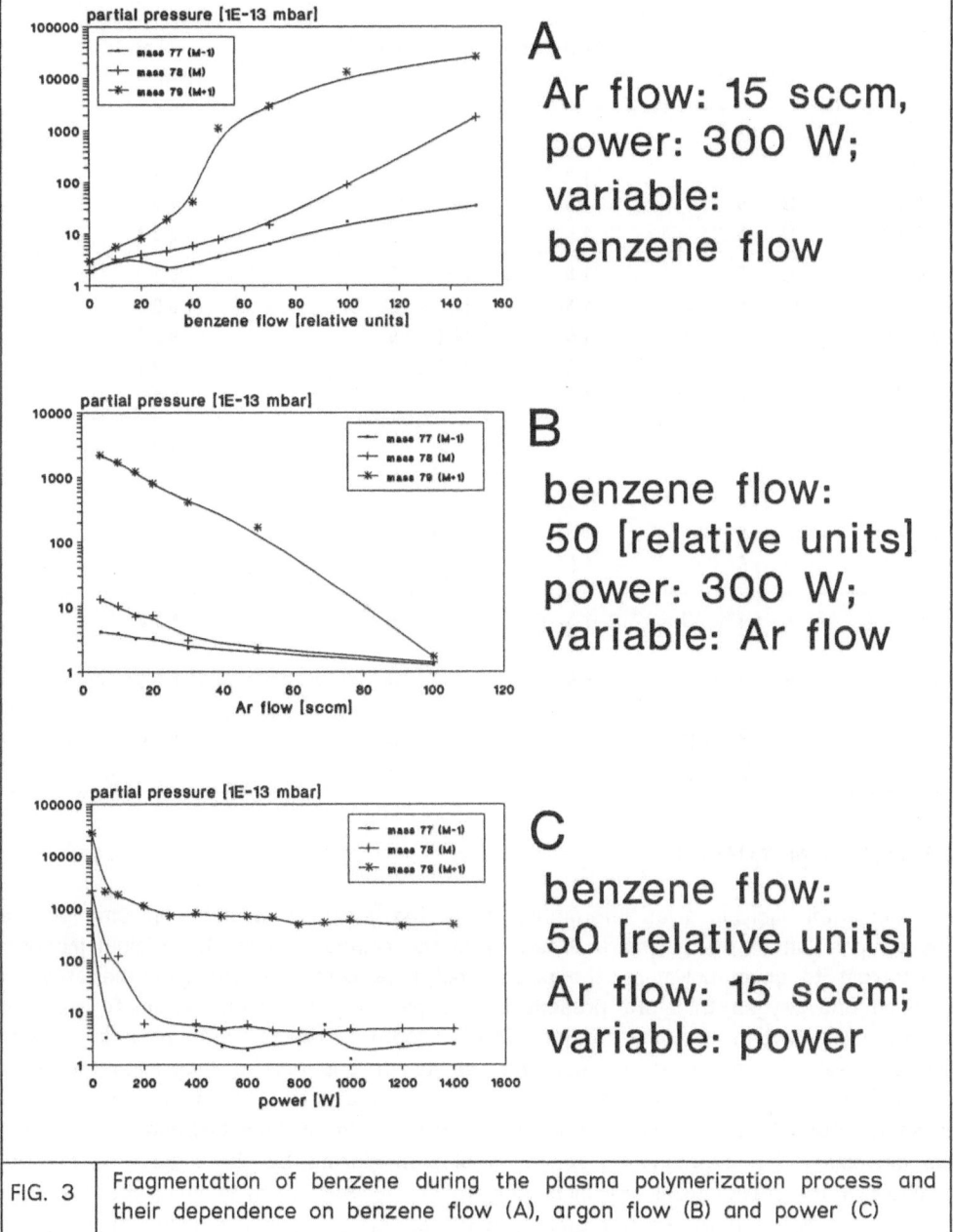

FIG. 3 | Fragmentation of benzene during the plasma polymerization process and their dependence on benzene flow (A), argon flow (B) and power (C)

TABLE 2. Deposition rate of individual plasma polymers

power (W)	flow rate (standard cm^3 min^{-1}) monomer	argon	reactive gas	deposition rate (nm h^{-1})
30	B 5	5	—	436
100	B 5	15	—	1200
300	B 5	15	air 5	420
600	B 5	15	air 5	354
50	B 5	15	NH$_3$ 5	510
100	B 5	15	NH$_3$ 5	1008
200	B 5	15	NH$_3$ 5	654
600	B 5	15	NH$_3$ 5	522
900	B 5	15	NH$_3$ 5	120
50	B 5	15	H$_2$O 5	972
100	B 5	15	H$_2$O 5	677
200	B 5	15	H$_2$O 5	655
300	B 5	15	H$_2$O 5	531
500	B 5	15	H$_2$O 5	390
100	A $1.5*10^{-3}$ mbar	15	—	435
100	P $1.7*10^{-3}$ mbar	15	—	293
300	P $5.6*10^{-3}$ mbar	15	—	316

3.3. ELEMENTAL ANALYSIS

The elemental analysis gives information about the composition of the plasma polymer. The composition of the polymers differs from the composition of the original monomer. In contrast to usual polymers, the plasma polymers contain additionally the atoms of nitrogen and oxygen that are present in the plasma volume and originate from the atmosphere. Oxygen, because of its biradical nature, reacts with plasma polymers during formation, and later on too. The results of quantitative determination of the elements C, H, N and O in the plasma polymerized benzene, aniline and pyridine are listed in TABLE 3. These plasma polymerized layers contain less hydrogen compared to the monomer. This experimental result is also supported by the mass spectrometric analysis of the residual gas. By increasing the r.f. power which is in accordance to an increasing proportion of excited states, (e.g. radicals) a higher amount of oxygen is incorporated into the layers. In contrast, nitrogen containing monomers react to plasma polymers with less nitrogen. This remarkable loss of nitrogen leads to the formation of molecular nitrogen and to prussic acid in the residual gas. Both reaction products are thermodynamically stable.

TABLE 3. Elemental analysis of plasma polymers

partial pressure monomer (mbar)		power (kW)	elements found (%)				empirical formula
			C	H	N	O	
benzene	$5.6 * 10^{-3}$	0.3	84.7	4.7	3.0	7.4	$C_6H_{4.0}N_{0.2}O_{0.4}$
benzene	$5.5 * 10^{-3}$	1.0	75.9	3.9	5.0	15.2	$C_6H_{3.7}N_{0.3}O_{0.9}$
pyridine	$1.1 * 10^{-3}$	0.3	76.9	4.7	6.4	12.0	$C_5H_{3.7}N_{0.4}O_{0.6}$
aniline	$1.1 * 10^{-3}$	0.3	78.0	5.5	4.4	12.1	$C_6H_{5.5}N_{0.3}O_{0.8}$

3.4. X-RAY PHOTOELECTRON SPECTROSCOPY (XPS)

In particular, bonding energy shifts of core electrons in plasma polymer surfaces were studied by the XPS technique. Conclusions pertaining to the chemical state of the elements C, N and O can be made according to the measured bond energies of C_{1s}, N_{1s} and O_{1s} electrons. For example, the energy of the C_{1s} electron in graphitic carbon is found to be 284.3 eV in literature /14/. If C is bonded to an electronegative element such as O there is an energy shift of C_{1s} to higher values. Because of the positve partial charge of the C atom, the core electrons are bonded more strongly to the C nuclei.

Likewise, bond energies for the sum of the 1s-electrons of the elements O, C and N in the deposited plasma polymers are measured and summarized in TABLE 4. It should be noted that the polymer films were deposited with 15 sccm Ar as carrier gas.

TABLE 4. Bond energies of core electrons in plasma polymeric films measured by XPS

Monomer	add. gas	power (W)	O_{1s} (eV)	C_{1s} (eV)	N_{1s} (eV)
benzene	–	300	532.5	285.2	–
benzene	–	600	532.8	284.85	–
benzene	air	300	533.85	286.2	400.35
benzene	NH_3	300	532.95	285.6	400.05
aniline	–	300	532.8	285.3	399.9

All the plasma polymers contain oxygen although the monomers used do not contain oxygen in their molecule. The reason for this fact is that free radicals of the polymer surface react after exposing on to the atmosphere with atmospheric oxygen to form peroxo, carbonyl and hydroxyl groups. The bond energies of 532 to 534 eV are typical for carbonyl and hydroxyl groups. But nitrogen of atmospheric origin is not incorporated on to the polymer surfaces, the originally molecular nitrogen is physically adsorbed. The N_{1s} peak in the range around 400 eV can be correlated with amine and nitrile groups.

3.5. IR SPECTROSCOPY

In FIG. 4 the spectra of three plasma polymers obtained under different parameters are shown. The absorption bands are labeled by number 1 to 18 and explained in TABLE 5. The IR spectrum of the non-aged plasmapolymer (a) exhibits no functionalization with air. This state can be influenced even in a short time, as demonstrated in FIG. 5. By aging of the plasma polymerized benzene for one month or more the OH-stretching vibration peak and the carbonyl peak increase. The two other spectra (b) and (c), shown in FIG. 4, are produced with addition of the gases air and ammonia that functionalize the plasmapolymers. Both spectra are similar in the type and degree of funtionalization, although in the case of ammonia no oxygen was present during plasma polymerization.

TABLE 5. Explanation of IR bands as found in plasma polymers (see FIG. 4 and 5)

line	wavenumber (cm^{-1})	relative intensity	tentative assignment
1	3450	m	-O-H intermolecular bonded
2	3250	m	-O-H stretching
3	2920	w	CH_2 asymmetric vibration
4	2492	w	amine vibration
5	2175	vs	-C≡N stretching
6	2115	w	-C≡C stretching
7	1600	s	=N-H stretching, NH_2/N-H bending
8	1450	s	-CH_2 deformation, -C-OH in plane deformation
9	1380	s	-O-H deformation
10	1300	w	-C-OH stretching
11	1270	s	-O-H deformation
12	1205	w	-C-O-C/CO stretching
13	863	s	R_2C=CHR out of plane
14	830	m	metadisubstituted benzene
15	700	w	RHC=CHR out of plane
16	668	m	-C-H deformation in C_6H_5-R
17	638	m	≡C-H deformation
18	625	m	≡C-H deformation

vs: very strong;   s: strong;   m: medium;   w: weak

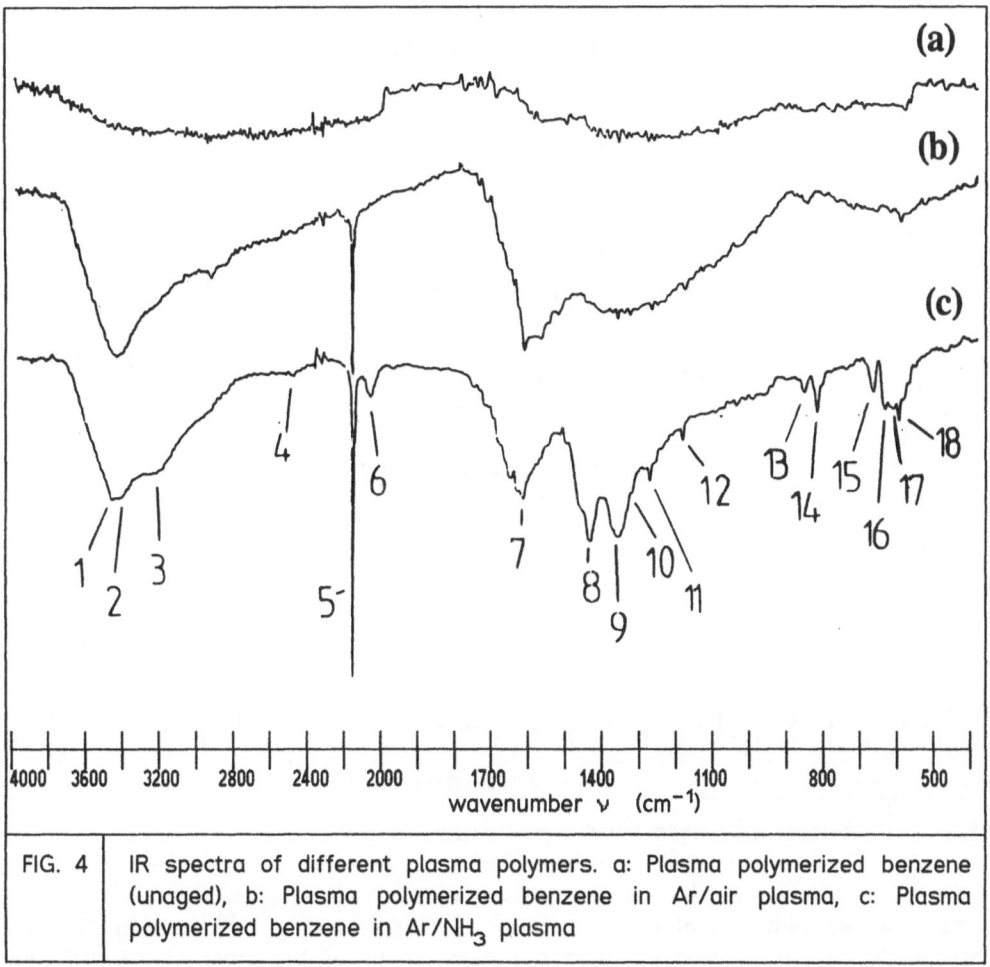

FIG. 4 | IR spectra of different plasma polymers. a: Plasma polymerized benzene (unaged), b: Plasma polymerized benzene in Ar/air plasma, c: Plasma polymerized benzene in Ar/NH$_3$ plasma

3.6. GID AND ESR MEASUREMENTS

Generally, plasma deposited polymeric films are described in literature as amorphous. The investigation using grazing incidence diffraction gives information about the layer. Pyridine, for example, led to a non-uniform structure of the polymer film. When angles of incidence of 0.5° and 1.5° were applied, no reflexes resulted, whereas at 1° corresponding to the bulk of the polymer film a sharp peak occurred. This signifies that the polymer film is of amorphous structure around its two surfaces; but in its bulk of approximately 5 nm interlayer distance that is independent of the total layer thickness, it exhibits a texturated growth.

To explain the chemical reaction of plasma polymer films with oxygen the electron spin resonance method was used to determine the spin concentration due to the free radicals. To get quantitative results the ESR peak area of plasma polymer samples

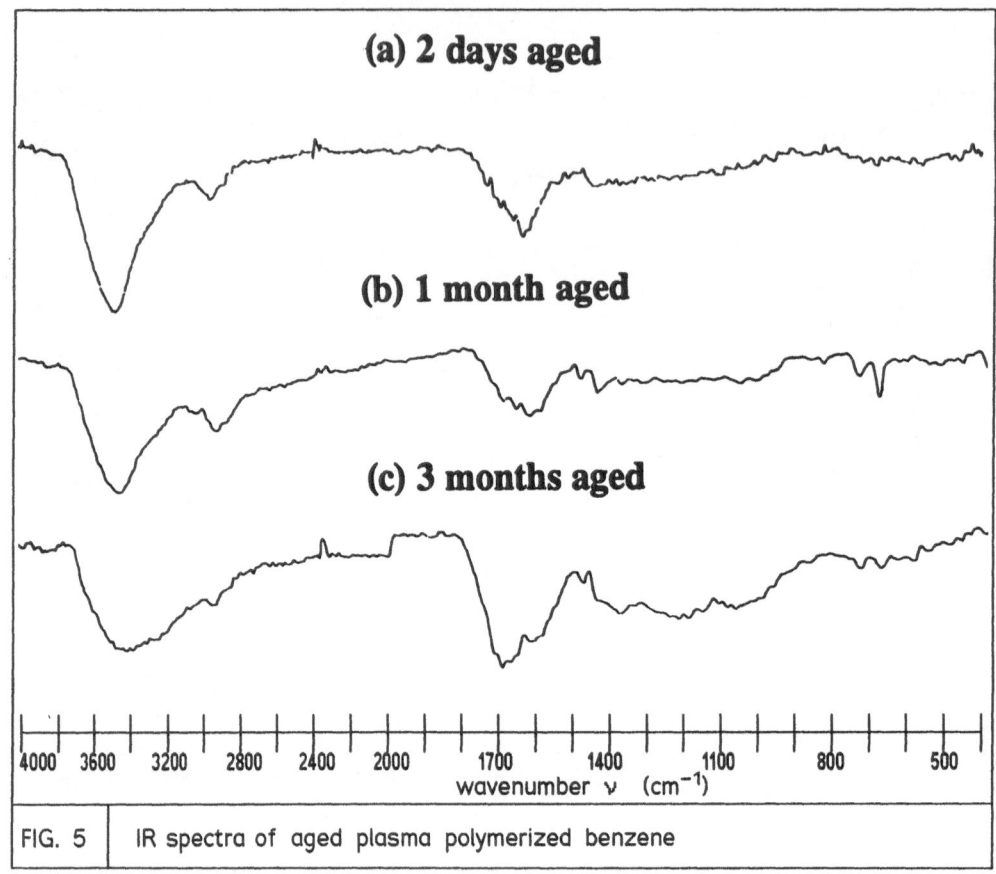

FIG. 5 | IR spectra of aged plasma polymerized benzene

were compared with that of an anthracite/KCl standard having 10^{15} spins/g. The spin concentration of the plasma polymers are found to be of the order of $2.6*10^{18}$ and $3.4*10^{19}$ spins/g for plasma polymerized benzene and pyridine respectively. One month aged plasma polymerized pyridine showed a decreasing spin concentration ($7.5*10^{18}$ spins/g). These results confirm the high affinity of plasma polymers to oxygen.

3.7. WORK OF ADHESION AT PLASMA POLYMER/LIQUID INTERPHASE

For measurement of the work of adhesion to the surface of the fibres the Wilhelmy method is known to be appropiate /15/. Although here the contact angle cannot be measured directly. But the method allows us to measure the wetting force F_W and the tear-off force F_{TO} with an electrical balance. The contact angle can be obtained according to one of the following two equations.

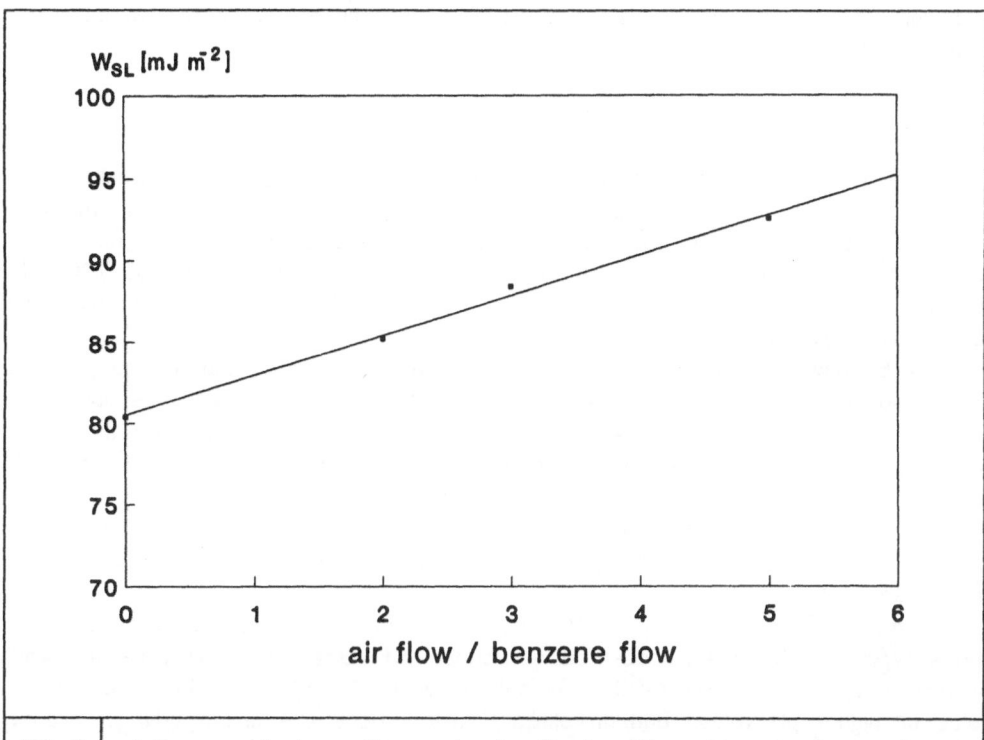

FIG. 6	Influence of air on the work of adhesion W_{SL}. As example specimens plasma polymerized with a constant benzene flow rate of 5 sccm and a variable air flow rate are investigated

$$\cos \Theta = \frac{F_W}{U \gamma_L} \qquad (4)$$

or

$$\cos \Theta = \frac{F_W}{F_{TO}} \qquad (5)$$

Θ = contact angle, U = circumference, γ_L = surface tension of test liquid, F_W = wetting force, F_{TO} = tear-off force.

By measuring the tear-off force the contact angle is directly obtained without a separate determination of the radii of the fibres. With the value of the contact angle, the work of adhesion W_{SL} is calculated:

$$W_{SL} = \gamma_L (1 + \cos \Theta). \qquad (6)$$

The work of adhesion at a solid–liquid interphase can be attributed to two interaction terms /16-19/:

$$W_{SL} = W_{SL}^D + W_{SL}^{AB}. \tag{7}$$

The term W_{SL}^D results from dispersion interactions corresponding to London forces, whereas W_{SL}^{AB} includes non-dispersion interactions which, according to Fowkes and Hüttinger /20/, are due to acid base interactions. The pH dependence on the work of adhesion can be used to draw conclusions regarding the surface groups /20/. For determination of the non-dispersive interactions W_{SL}^{AB} alone, the so-called two liquid method /21/, was applied.

FIG. 6 shows the influence of the air/benzene ratio during the plasma polymerization process on the work of adhesion W_{SL}. Plasma polymerized benzene and double destilled water of pH 7 are used as materials for the solid/liquid interphase. For fibres coated with plasma polymerized benzene, the W_{SL} value reached 80.4 mJ m^{-2}. A linear relation between W_{SL} and the concentration of air during the plasma polymerization process was found to be due to surface functionalization (–NH$_2$, –C≡N, C=O, C–OH, C–O–C etc.). Obviously, the non-dispersion interactions are responsible for the increase of W_{SL}.

To arrive at conclusions about the functional groups on coated carbon fibres, the terms W_{SL}^{AB} or W_{SL} were measured at various pH values from pH 1 to 14. Acidic surface groups like carboxyl (-COOH) or sulfonic acid (-SO$_3$H) should increase the W_{SL} value at higher pH values and, in similar, basic groups like amine (-NH$_2$), hydroxyl (-OH) or carbonyl (-C=O) should increase W_{SL} at lower pH values.

FIG. 7 shows the pH dependence of the non-dispersion interactions W_{SL}^{AB} (determined by the two-liqid-method with liquid 1, cyclohexane and liquid 2, water) for coated Hercules carbon fibres. One of the carbon fibres was coated in a benzene/air plasma, the other in a benzene/ammonia plasma. Both reactive gas components added during the plasma polymerization process influence the term W_{SL}^{AB} in a similar way. This may be caused by the consecutive reactions between the plasma polymers and atmospheric oxygen due to the high concentration of free radicals on the film surface. This explanation is in accordance with the results of IR analysis.

TABLE 6. Determinations of W_{SL} and contact angle on coated Hercules (H) and Grafil (G) fibres. Coating conditions: benzene flow rate: 5 sccm, argon flow rate: 10 sccm, SO$_2$ flow rate: 5 sccm, r.f. power: 300 W. Contact liquid: double destilled water.

value	fibre type	pH 1	pH 4	pH 7	pH 10	pH 14
contact angle (°)	H	77.8	69.3	69.6	67.3	60.0
W_{SL} (mJ m^{-2})	H	88.1	98.5	98.0	100.9	109.2
contact angle (°)	G	68.0	65.3	68.9	55.6	48.9
W_{SL} (mJ m^{-2})	G	100.1	102.7	98.9	113.5	120.5

It should be mentioned that the decrease of W_{SL} at values of pH > 10 are due to basic groups on the surface.

While reactive gases like ammonia and air cause predominantly basic surface groups, sulphur dioxide was applied to obtain sulfonic acid groups. For comparison, two types of carbon fibres were coated under the same conditions but with the addition of sulphur dioxide. As a result, an increasing work of adhesion at higher pH values was obtained for both types of fibres due to acidic groups on the surface (TABLE 6). But with coated Grafil (G) fibres generally higher W_{SL} values are realized than with coated Hercules (H) fibres. The morphology of the original fibre surface is copied during the plasma polymerization (FIG. 8). Grafil fibres show a rather rugged surface (FIG. 8a) while Hercules fibres exhibit a smoother surface (FIG. 8b). Thus, one may conclude that a rugged surface yields higher values of W_{SL}.

One of the main conclusions of the W_{SL} measurements is that only sulphur dioxide as reactive component causes an acidic surface. In all the other plasma polymer surfaces, basic groups are more dominant.

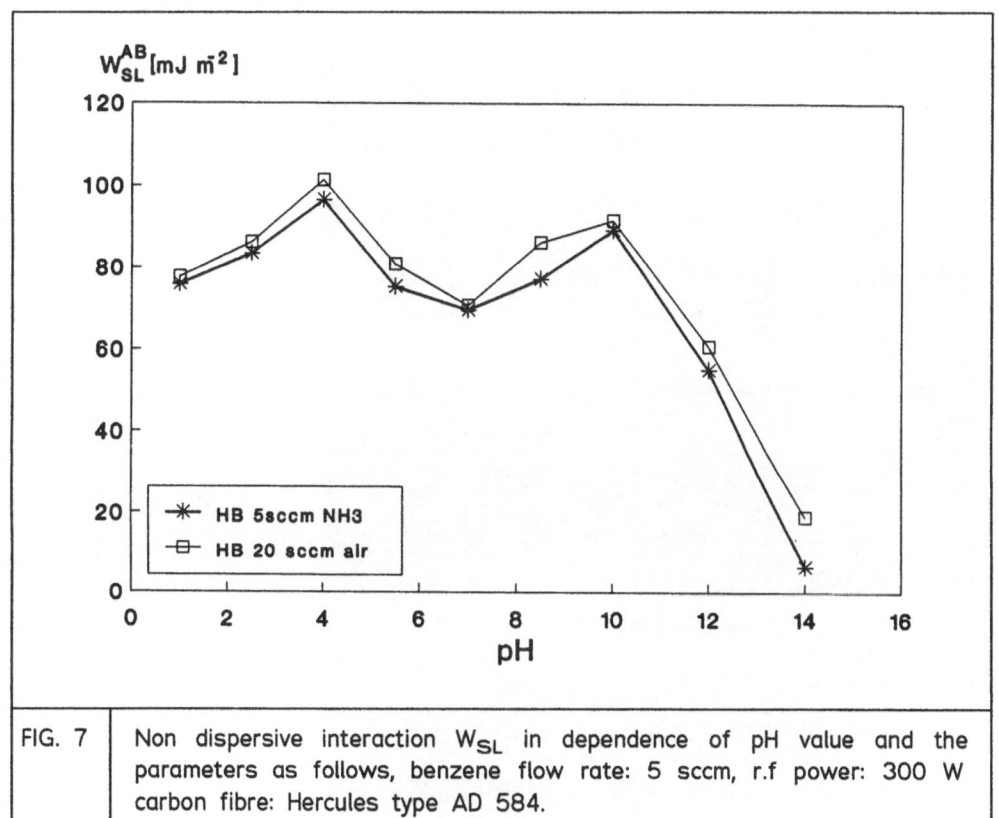

| FIG. 7 | Non dispersive interaction W_{SL} in dependence of pH value and the parameters as follows, benzene flow rate: 5 sccm, r.f power: 300 W carbon fibre: Hercules type AD 584. |

302

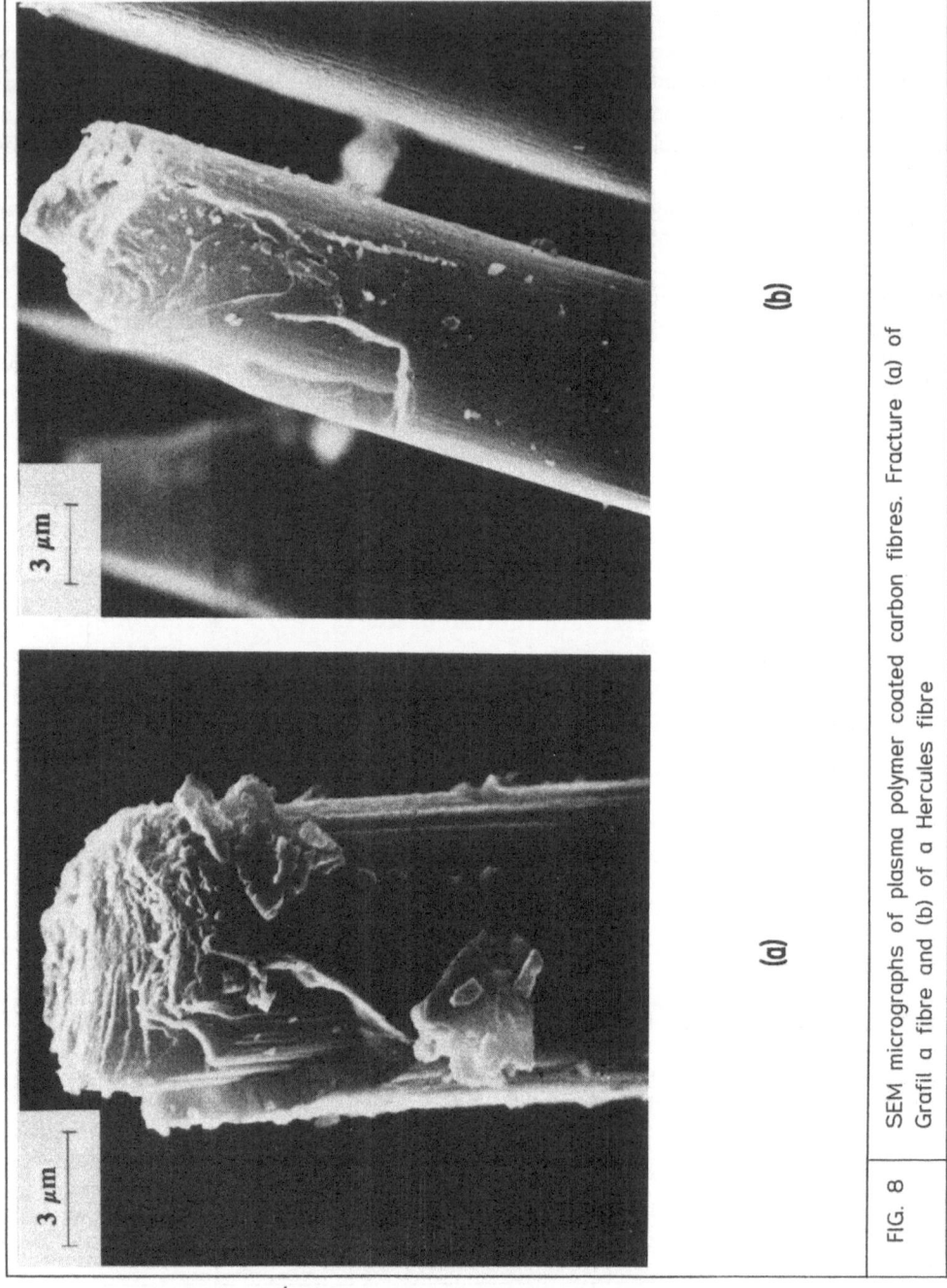

FIG. 8 | SEM micrographs of plasma polymer coated carbon fibres. Fracture (a) of Grafil a fibre and (b) of a Hercules fibre

3.8. MECHANICAL CHARACTERISTICS OF COATED FIBRES

To understand the influence of plasma polymer films on mechanical data, the two coated types of fibres, Hercules AD 584 (H) and Grafil XAS (G), were tested. In particular, tensile strength tests on monofilaments were conducted as described in FIG. 1. To get average values of the mechanical data, at least 40 measurements were performed. The results of such measurements are listed in TABLE 7 and TABLE 8. The tensile modulus E is found to be independent of the geometry of the sample, but depends on the integral structure and composition of the materials. Therefore, the relative mean standard deviation of E is limited to about 8 %, while that of σ_z and ε extend to about 21 and 19 %, respectively. The higher relative standard deviation of σ_z and ε are caused by the statistical distribution of the fibre diameters ($\pm$ 10 %).

It is well known since Griffith that fracture in brittle materials is initiated by tensile stresses far below the intrinsic strength of the material, due to notches. These notches can be reduced by decreasing the volume of the material or by healing effects. The three monomers used for the plasma polymerization process lead to higher values of σ_z and ε for coated Hercules and Grafil fibres, because of healing mechanism during the plasma polymerization. σ_z and ε also increase with deposition time ranging from 1 to 12 minutes.

The adhesion of plasma polymers on carbon fibres is found to be rather good (see FIG. 8). The layers are dense, and cover the fibres in most cases completely. The values of ε are more increased than that of σ_z.

TABLE 7. Results of tensile strength tests of coated Hercules fibres.

monomer	deposition time (min)	power (W)	σ_z (MPa)	ε (%)	E (GPa)
—	0	300	3299 ±712	1.488 ±0.301	221 ±10
pyridine	1	300	3688 ±833	1.681 ±0.333	219 ±15
pyridine	2	300	3570 ±714	1.615 ±0.260	220 ±16
pyridine	6	300	3482 ±686	1.587 ±0.282	219 ±11
pyridine	12	300	3335 ±600	1.557 ±0.260	214 ±13
aniline	2	300	3690 ±664	1.707 ±0.270	216 ±11
aniline	6	300	3857 ±609	1.760 ±0.287	219 ±15
aniline	12	300	3569 ±475	1.650 ±0.211	217 ±13
benzene	1	600	3596 ±503	1.765 ±0.242	204 ±12
benzene	2	600	3545 ±585	1.761 ±0.269	201 ±10
benzene	12	600	3782 ±734	1.811 ±0.335	209 ±11
benzene	1	300	3847 ±607	1.787 ±0.263	215 ±11
benzene	2	300	4173 ±751	1.983 ±0.352	211 ±12
benzene	12	300	3933 ±770	1.879 ±0.303	208 ±13

TABLE 8. Results of tensile strength tests of coated Grafil fibres.

monomer	deposition time (min)	power (W)	σ_z (MPa)	ε (%)	E (GPa)
—	0		3590 ±625	1.624 ±0.278	221 ±13
pyridine	1	300	3838 ±829	1.748 ±0.302	218 ±16
pyridine	2	300	4025 ±765	1.771 ±0.290	227 ±17
pyridine	12	300	3987 ±869	1.829 ±0.313	217 ±17
aniline	2	300	3791 ±804	1.714 ±0.319	221 ±16
aniline	12	300	4043 ±898	1.837 ±0.408	219 ±18
benzene	1	600	3766 ±918	1.752 ±0.368	214 ±14
benzene	2	600	3561 ±762	1.713 ±0.319	208 ±14
benzene	12	600	4134 ±946	1.926 ±0.399	213 ±12
benzene	1	300	4204 ±828	1.960 ±0.333	213 ±12
benzene	12	300	3733 ±675	1.728 ±0.283	216 ±10

Thus, the E moduli slightly decrease, and are lower for coated fibres than for uncoated fibres. The mechanical properties of fibres coated at a r.f. power level of 600 W compared to that coated at 300 W change to lower values.

4. Composites

Unidirectional carbon fibre reinforced epoxy resin as well as polycarbonate composites were manufactured with a fibre volume fraction of around 60 %. The main steps in manufacturing a composite are the impregnation, the moulding and the curing by thermoset matrices. Composites should be non-porous, crack-free and stress-free, with homogeneously distributed fibres embedded in the matrix.

4.1. MATERIALS TREATMENT AND COMPOSITE FABRICATION

Hercules AD 584 carbon fibres as a bundle of 12000 monofilaments were continuously coated with different plasma polymers. Then the fibres were wound on a special rack to get unidirectional alignment. The next step was the impregnation with the liquid matrix. The wound carbon fibres were impregnated in a mould by pouring the epoxy resin LY 556 together with the curing agent HT 972 (both chemicals are produced by CIBA-GEIGY Comp.) into the mould. The mould was compressed in a hot press system at 80 °C to a final thickness controlled by a spacer, then cured at 80 °C for 4 hours and cured afterwards at 120 °C for 2 hours. After cooling, the composites pieces, having a dimension of 2 * 6 * 200 mm were removed from the mould and cut into smaller pieces.

Polycarbonate composites were manufactured in a similar way by means of the same rack and mould. The polycarbonate (Makrolon made by BAYER AG) was dissolved in methylenechloride. The fibres were impregnated in the mould and pressed at 270 °C.

Three point bend tests were conducted on the composites in order to obtain information about the influence of the plasma polymeric films on mechanical characteristics of the composites.

4.2 COMPOSITE TESTING

Under three point bend loading of a composite (beam), cracks may be developed due to tensile stresses at the lower stratus of the specimen as well as compression stresses at the upper one, or due to interlaminar shear. The type of failure depends on the ratio of span to depth (L/D). Short beam specimens usually fail in shear and long ones by tensile or compression stresses. For interlaminar shear strength (ILSS) tests, a L/D = 5 was chosen (ASTM-D-2344-76). In case of flexural strength tests, this ratio was fixed to 40 (DIN 29971).

4.3. RESULTS AND DISCUSSION

The results of three point bend tests can be studied in TABLE 9 for epoxy composites (EC) and in TABLE 10 for polycarbonate composites (PC). For determination of ILSS at least 20 measurements and for that of flexural strength at least 5 measurements were conducted.

TABLE 9. Summary of the results of three point bend tests for epoxy composites

Coating of fibres	IL SS (MPa)	flexural strength (MPa)	flexural modulus (GPa)	edge fibre elongation (%)
uncoated	64.0 ± 2.1	1493 ± 34	104.0 ± 1.2	1.43 ± 0.04
B-Ar	67.2 ± 2.9	1940 ± 125	114.0 ± 8.0	1.84 ± 0.14
B-Ar-air	67.6 ± 2.2	2053 ± 36	114.0 ± 8.0	2.03 ± 0.09
B-Ar-NH$_3$	71.8 ± 1.3	2043 ± 169	119.1 ± 3.0	1.81 ± 0.18
A-Ar	70.1 ± 3.7	1678 ± 82	101.7 ± 0.7	1.79 ± 0.23
P-Ar	68.7 ± 1.8	2077 ± 99	126.3 ± 2.1	1.64 ± 0.06

One aim of this research was to obtain a surface treatment that would increase the interlaminar shear strength of the composites without decreasing the tensile strength of the fibres.

Plasma treatment with benzene monomer, with pyridine monomer as well as with aniline monomer led to an improvement of interlaminar shear strength of 5 to 12 % in epoxy resin composites. The highest improvement in ILSS tests on epoxy composites was achieved by treatment of the fibres in a benzene/argon/air plasma. This treatment led also to better values of 38 % in flexural strength and the edge fibre elongation increased to 41 %. The reason for the improvement of the fibre matrix adhesion lies

TABLE 10. Summary of the results of three point bend tests for polycarbonate composites

Coating of fibres	ILSS (MPa)	flexural strength (MPa)	flexural modulus (GPa)	edge fibre elongation (%)
uncoated	51.7±1.5	1215±120	103.3±6.6	1.18±0.12
B-Ar	51.3±1.7	1308±180	115.4±5.7	1.13±0.10
B-Ar-air	62.0±3.0	1448±130	103.4±9.9	1.41±0.09
B-Ar-NH$_3$	61.4±2.4	1245±158	103.2±3.4	1.25±0.18
A-Ar	51.2±2.3	1260±123	109.5±2.1	1.15±0.12
P-Ar	51.1±1.6	1315±159	102.4±9.2	1.29±0.18

in the introduction of polar groups and in the high amount of free radicals through the plasma polymerization. The epoxy resin can react with free radicals and polar groups to form chemical bonds.

In contrast to epoxy resin composites, the fibre matrix adhesion in polycarbonate composites is formed by physical and mechanical interactions. Therefore, only with the plasma treatment by using additional gases like ammonia and air an improvement of ILSS resulted. The plasma treatment of the fibres with benzene/argon/air led to an improvement of ILSS of 20 %. Also the mechanical properties like flexural strength and edge fibre elongation are improved. The enhancement in interlaminar shear strength and in mechanical characteristics might be due to the introduction of polar compounds through the plasma treatment.

In almost all flexural tests the specimens failed in the upper stratus, due to compression stresses. This can be explained by means of the mechanical properties of carbon fibres. The compression strength of carbon fibres are about 90 % of the value of the tensile strength. Therefore, cracks are developed by compression stresses.

Further investigations should be realized in plasma treatment of carbon fibres in relation to acidic surface groups. An acidic surface could be produced in a benzene/argon/sulphur dioxide plasma. Because of the rather basic matrix of the polycarbonate, an acidic fibre surface should improve the adhesion to a basic matrix due to acid–base interactions.

5. References

1.   *R. Mollyre, M. Bastik;* Proc. 4th London Int. Carbon and Graphite Conf., Soc. Chem. Ind. London (1974) 190
2.   *J.W. Herrick, P.E. Gruber Jr., F.T. Mansur;* Surface Treatment for Fibrious Reinforcement AFML-TR-66-178 Part I, Air Force Materials Laboratory (1966)
3.   *L.B. Adams, E.A. Boucher;* Carbon 16 (1978) 75
4.   *I.L. Kalnin;* U.S. Patent 3623607 (1966)
5.   *D.W. McKee;* Carbon 8 (1970) 131 and 623
6.   United Aircraft; U.S. Patent 3642153 (1972)
7.   *H.P. Rensch;* Ph. D. Thesis, University of Karlsruhe (1990)
8.   *R.V. Subramanian;* Advances in Polymer Science 33 (1979) 34
9.   *A.R. Nyalesh, L. Holland;* Vacuum 31,7 (1981) 315
10.   *H. Kausche;* "Neue Kathodenzerstäubungsverfahren", Lehrgang Techn. Akademie Esslingen (1978)
11.   *M. Sahebkar;* Ph. D. Thesis, University of Karlsruhe (1973)
12.   *K. Brennfleck, E. Fitzer, G. Schoch, M. Dietrich;* Proc. 9th Int. Conf. on CVD, Extended Abstract, Pittsburg/USA (1984) 649
13.   *S. Tolansky;* Multiple Beam Interferometry of Surfaces and Films, Clarendon Press, Oxford (1948)
14.   *T. Carlson;* Photoelectron and Auger Spectroscopy, Plenum Press, New York (1975) 351
15.   *S.K. Li, R.P. Smith, A.W. Neumann;* J. of Adhesion, 17 (1984) 105-122
16.   *F.M. Fowkes, D.C. McCharty, M.A. Mostafa;* J. Colloid Interface Sci. 78 (1980) 200-206
17.   *F.M. Fowkes;* in: Physicochemical Aspects of Polymer Surfaces, K.L. Mittal (Ed.), Vol. 2, Plenum Press, New York (1983) pp. 583-603
18.   *F.M. Fowkes;* J. Adhesion Sci. Technol. 1 (1987) 7-27
19.   *F.M. Fowkes;* in: Acid-Base Interactions: Relevance to Adhesion Science and Technology, K.L. Mittal and H.R. Anderson; Jr. (Eds), VSP, Zeist, The Netherlands (1991) 93-115
20.   *K.J. Hüttinger, S. Höhmann-Wien and G. Krekel;* J. Adhesion Sci. Technol. 6, 3 (1992) 317-331
21   *J. Schultz, K. Tsutsumi, J.B. Donnet;* J. Coll. Interf. Sci. 59 (1977) 272

SOME NOTES ON SURFACE MODIFICATION
BY PLASMA

Guneri AKOVALI
Dept. of Chemistry and Polymer
Science and Technology Program
Middle East Technical University
06531 Ankara TURKIYE

ABSTRACT

For modification of polymer/filler surfaces to improve properties like adhesion, fluid absorbancy and wetting properties; "exposure of the surfaces to proper electrical glow discharge method" is usually and succesfully employed. This technique can also be employed in polymer composite systems to modify interfaces and interphases to improve properties.

In this short communication, some results of several studies done with SBR-Carbon black and PET-PVC composites will be summarized.

G. Akovali (ed.), The Interfacial Interactions in Polymeric Composites, 309–320.
© 1993 *Kluwer Academic Publishers.*

1. INTRODUCTION

There is a growing demand for new resins and resin systems with specific performances as well as with high (performance/cost) ratios. One practical way to reach at these systems is possible by use of multicomponent polymeric systems obtained by proper matching of already existing polymers (1). Hence, polymer blends and polymer composites are formed. Homogeneous as well as heterogeneous blends comprising mixtures of two or more polymers, as well as mixtures of polymers with reinforcing agents and fillers, provide a route to combinations of properties not otherwise available.

Different polymers are blended mainly to obtain an advantageous combination of properties. Some properties of a multicomponent material are roughly additive, but synergestic interaction can also occur to yield properties and performances superior of the individual constituents, which obviously makes polymer blending so attractive. This concept, beginning from the last decade, is continuously drawing considerable interest which already resulted at more than 4500 patents (2).

If a polymeric blend system is homogeneous (3), it can be linked to alloys of metals where components mix completely on a molecular scale. If it is heterogeneous, and form a two- or multi phase structure, depending on degree of phase separation; there may be different phases either at macro or micro level (4). Heterogeneous blend systems are of growing interest, where, to utilize physical properties of the components efficiently; a good stress-transfer between the components is essential; hence, proper attraction and adherence of different components with each other is needed.

The interactions and adhesion involved is basically a surface phenomena, and the nature and characteristics of interfaces and interphases are of prime importance. They can be promoted by use of proper compatibilizing agents (5) or directly, by modifying surface properties (6). Permanent chemical modification of polymer surfaces is a rather difficult task, and for polyolefins, which are poorly reactive; this difficulty is even greater. On the other hand, if activated, polyolefins can react with oxygen of the atmosphere easily which is a chain reaction with autocatalytic nature which can get out of control damaging the substrate properties. Other radioactive and/or chemical processes to modify surfaces have turned out to be either incontrollable, or industrially deficient; or even too destructive. However, plasma process offers a highly controllable and efficient treatment for polymer surfaces (7,8,9). In this communication, some examples will be given for several multicomponent-multiphase polymer blend systems where plasma surface treatment was applied prior to blending to one of the components to create proper interfaces/interphases.

2. EXPERIMENTAL

Following two multicomponent-multiphase polymer systems (with and without plasma surface modification) will be presented:

(a) PET (Polyethylene Teraphthalate)-PVC

(b) SBR (Poly[Styrene-Butadiene] Rubber)-Carbon Black

In all of these systems, surface of one of the component (PET and Carbon Black; respectively) were modified by plasma.

In plasma, an electric discharge is produced under appropriate conditions which ionizes the gap of gas producing a plasma in which electrons, positive ions, excited species and radiation are present. Since there is no equilibrium (glow-discharge) type of plasma employed here, the advantage of having "high energy electrons at normal gas temperatures" was used and benefited. The plasma energy distribution is quite sufficient to penetrate up to 50 microns into the film surface, and to break some chemical bonds, giving rise to free radicals etc., which can then be used in modification.

(a) Studies with (PET-PVC) Composite Systems:
Aim of Study:

PET and PVC packaging materials are frequently discarded after their use to the environment, which causes pollution. Their recycling in the form of blends is of interest, but; if one tries to prepare blends from the melt, the results obtained are negative; because:

    i) PET and PVC are mutually insoluble,

    ii) At the melting point of PET, which is rather high (250-260°C), PVC (even if stabilized), decompozes.

On the other hand, the use of PET powder in PVC matrix as filler is very attractive- but since they are highly incompatible, surface modification of PET powder would be very promising which will help to produce new composite systems meanwhile decrease pollution.

Method:

Plasma surface modification of PET powder was carried out by using vinyl chloride (VC) plasma, operating at 13.56 MHz for 15 minutes, at 10 Watts, in the tubular plasma reactor system, (Figure 1).

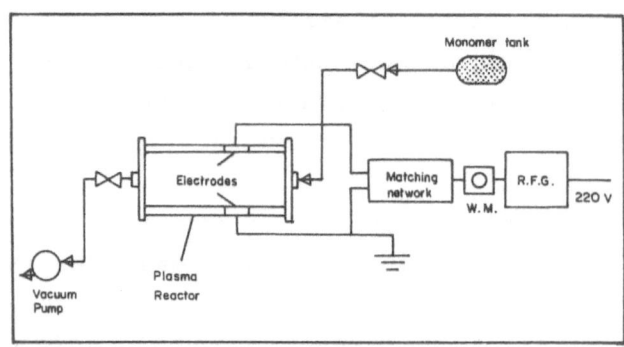

Figure 1. Plasma System Used

Mixtures of PET/PVC (with virgin PET and surface modified PET, separately) were prepared by solution blending technique at different compositions and the values of (Δb) factor were obtained from comparision of experimental interaction coefficient for the mixture of components (b_{12}) with the theoretical (b_{12}^*) as computed from Williamson-Wright equation (10).

$$\Delta b = b_{12} - b_{12}^*$$

The positive values of Δb indicate attractive interactions, hence compatibility; whereas negative values refer to incompatibility. The results obtained for liquid solutions can then be directly extrapolated to the solid state (11,12).

Results (13):

The plot of intrinsic viscosities versus compositions for the mixtures tested (Figure 2) show that although for both PET/PVC and modified PET (M.PET)/PVC mixtures, there is considerable deviation from ideality (the latter is represented by straight lines); the results obtained with M.PET are much closer to ideality.

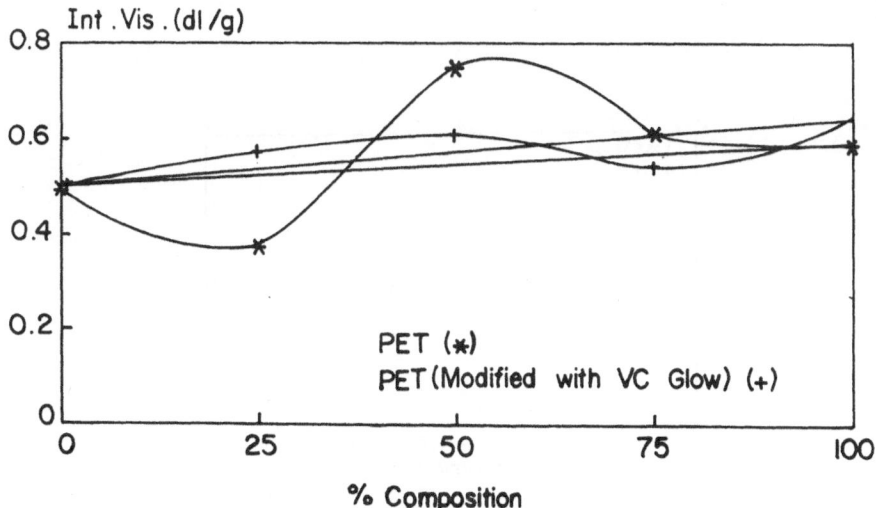

Figure 2. Intrinsic Viscosities for PET/PVC and M.PET/PVC Systems with Different Compositions

314

     Correspondingly, the (Δb) values obtained (Figures 3 a and b) result in the same trend: although all (Δb) values are negative showing incompatibility, its degree is considerably different and much smaller for the M.PET case. In addition, 25 and 50 % M.PET mixtures show similar degree of incompatibilities.

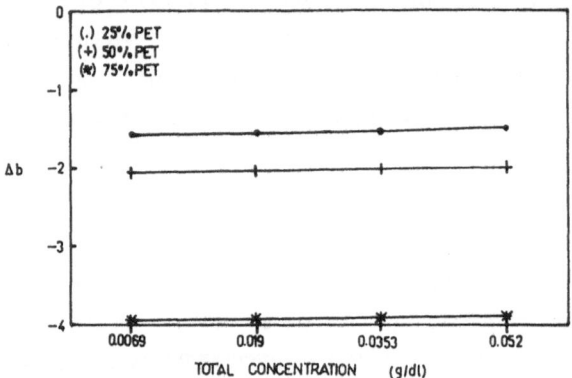

Figure 3(a). Δb versus Total Concentration for PET/PVC Mixtures

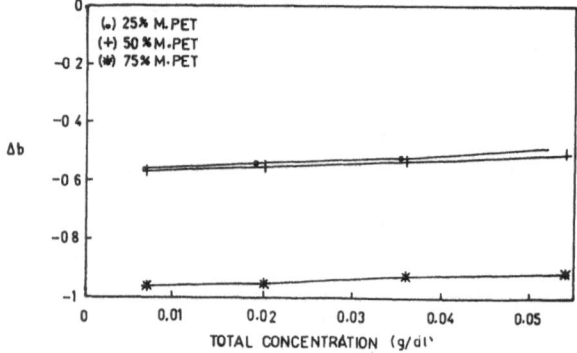

Figure 3(b). Δb versus Total Concentration for M.PET/PVC mixtures

The results obtained from this rather introductory study which was tried with one type of monomer (VC) at one plasma condition only clearly showed the importance and effect of interfaces and interphases upon the final performances. These results are very promising, and yet there are a number of parameters to be checked. Our compatibility studies with this system is continuing, and the results of this first phase is already submitted for publication (13).

(b) Studies with (SBR/Carbon Black) Composite Systems:
Aim of Study:

Carbon black is often used in rubbers for reinforcement purposes. With it's unique pore structure which results in high (surface area/size) ratio, it effectively enters the structure of the rubber to reinforce the system further. The ultimate mechanical properties of rubbers are strongly influenced by the size-surface area-structure and surface characteristics of the fillers used. Hence it is of prime interest to see the effects of surface modification of carbon black on the composite's performances without changing it's surface area appreciably. Glow discharge (plasma) is the best suited method for this purpose, because it gives very thin polymer films with controllable thicknesses on yhe substrate surfaces selected (14). In addition, our previous studies have shown that, surfaces of carbon black coat easily and rather homogenously by thin plasma polymer films without taking any special precautions in the reactor, being independent of whether they are located at the surface layer of the bulk or at the bottom (14).

The purpose of this study is to explore the effects of plasma polymerized styrene (PPS) butadiene coating on carbon black in SBR matrices.

Experiment:

The same plasma system (Figure 1) was employed. To have optimum match between the matrix and the filler, the surface of the latter (carbon black) was treated with "butadiene" and "styrene" plasma, separately. Some characteristics of the matrix and carbon black are given below:

SBR used was a product of Petkim, Izmit (coded as SBR 1502) emulsion type with 23.5 7 % bound styrene in it. It had a $\bar{\bar{M}}_w$ of 320-400 K and $\bar{M}_n$ of 80-110 K.

Carbon Black used was also a product of Petkim (coded as Petkara N 220) with surface area 115 m^2/gr and average particle size of 20-25 nm. It was used in SBR matrix with the following characteristic recipe and it was vulcanized by the conventional technique.

SBR	100		
Filler	35		
ZnO	3	MASTER	FINAL
		BATCH	BATCH
Stearic acid	2		
Antioxidant	2		
			BATCH SIZE = 45
Sulfur	2		(based on phr)
Accelerator	1.3		

Results:

Results showed that, the tensile strength of filled vulcanizates increase up to 30 minute plasma duration, then stays constant for PPS; while the maximum is obtained at 15 minute plasma duration for PPB (15). At 30 minute plasma, it was shown that surface area of carbon black stays the same for PPS case, while it is some 6 % lower for the PPB - which is somewhat expected because of differences involved in their polymerization kinetics in plasma. The T_g of system with PPS coated sample is found to be about the same with that of uncoated (Figure 4). While PPB coating on carbon black is seen to decrease the T_g of the system appreciably (16). These results, without further speculation, obviously point out the effect of interfaces/interphases in characterizing the final properties.

It is interesting to see the differences between SEM pictures of uncoated (Figure 5), PPS modified (Figure 6) and PPB modified (Figure 7) Carbon Black containing SBR vulcanizates . The pictures were taken from fracture surfaces.

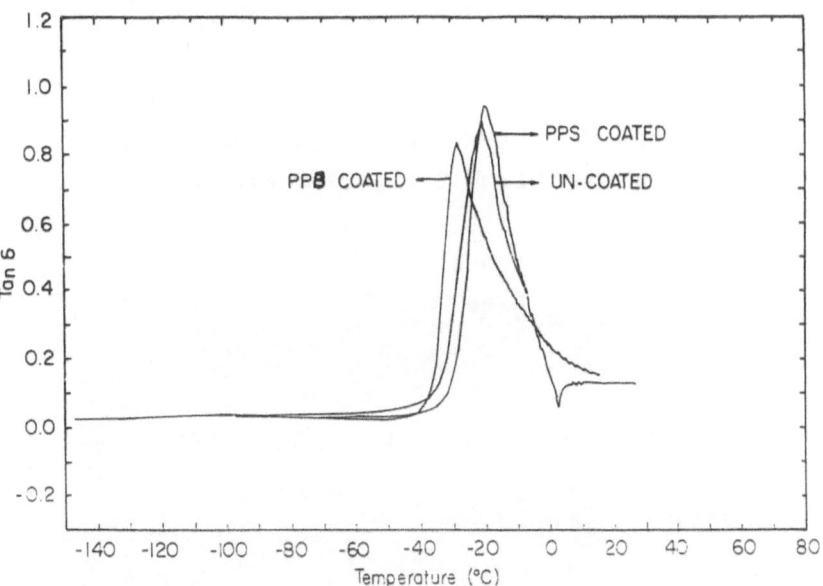

Figure 4. DMA results of SBR vulcanizates with uncoated PPS and PPB coated charcoal

318

Figure 5　SEM photograph　of uncoated Carbon
Black filled SBR vulcanizate .

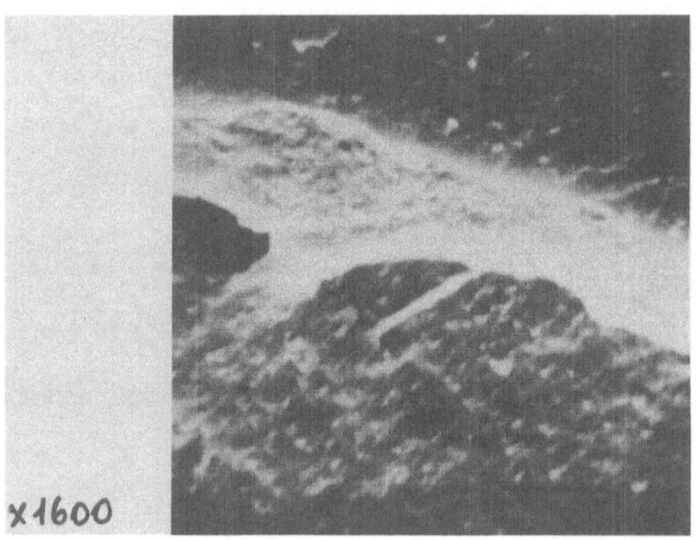

Figure 6　SEM photograph　of PPB modified Carbon
Black filled SBR vulcanizate

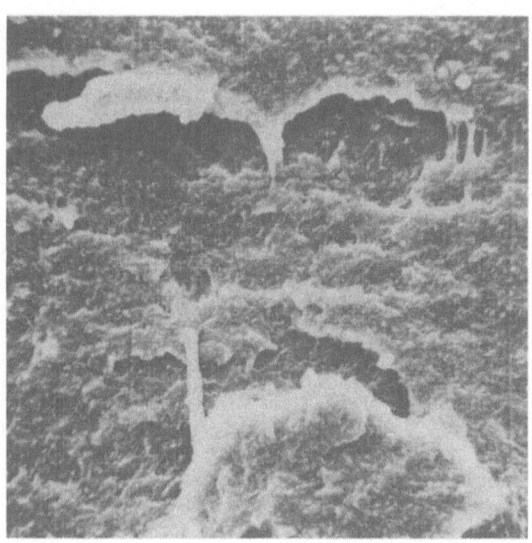

x1600

Figure 7   SEM photograph of PPS modified Carbon
Black filled SBR vulcanizate

Refences:

(1) D. H. Paul, S. Newman, "Polymer Blends" Academic Press.
NY, 1978.

(2) B. D. Favis, The Canadian J. of Chem. Eng. 69. (June
1991)

(3) W. Siol Macplas. 33. 1987

(4) D. J. Walsh, I. S. Higgins, A. Maconnachie, Eds.
"Polymer Blends and Mixtures" Nato-ASI Series 89 (E),
(Apll. Sci); 1985. Martinus-Nijhoff Publ.

(5) Anon, Plastics Techn. 67. (Feb. 1989)

(6) B. E. Koel, R. G. Windhaum "Chemical Modification of
Surface Properties"; Proceedings, Ind. Univ. Advanced
Materials Conference, Denver (Ed: J. G. Morse), 1987

320

(7) J. R. Hollahan, A. T. Bell Techniques and Applications of Plasma Chemistry, Wiley, N.Y. 1974

(8) G. Akovali, F. Takrouri J. Appl. Poly. Sci. 42. 2717 (1991)

(9) "Surface Engineering - A Consultancy Report" UN Ind. Dev. Org. Adv. in Materials Techn.: MONITOR. Issue 24/25 (Feb. 1992)

(10) G. R. Williamson - B. Wright - J. Poly. Sci. A. 3. 3885 (1965)

(11) S. Aslan, MSc Thesis (The Compatibility of waste PET with PVC), Middle East Technical University, Dept. of Chemistry. Ankara (Feb. 1991)

(12) K. S. Shih and C. L. Beatty British Polymer J. 22. 11 (1990)

(13) A Compatibility Study of Waste PET with PVC - G. Akovali, S. Aslan Submitted to "J. Appl. Poly. Sci." For Publication

(14) G. Akovali, N. Hasirci J. Appl. Poly. Sci. 29. 2617 (1984)

(15) N. Kajouei "Mechanical Properties of Plasma Surface Modified Carbon Black/SBR Composites" MSc. Thesis, Middle East Technical University, Ankara (Sept. 1992); Paper is in Preparation.

(16) I. Ulkem "Static and Dynamic Mechanical Properties of (Bauxide) and (Carbon Black) Filled SBR Vulcanizates Effects of Filler Surface Modification on Properties" Ph.D. Thesis, Middle East Technical University, Dept. of Chem., Ankara, (July 1990), Paper is in Preparation.

SCIENCE AND TECHNOLOGY OF POLYMER COMPOSITES

L. NICOLAIS, J.M. KENNY, A. MAFFEZZOLI, L. TORRE, A. TRIVISANO
Department of Materials and Production Engineering, University of Naples, Piazzale Tecchio, 80125 Naples, ITALY

ABSTRACT. The polymeric matrix and its interaction with a reinforcing phase, in the form of continuous or discontinuous high strength and stiffness fibers, is one of the major controlling factors in the processing and property characteristics of composites. Although traditionally the matrix has been thought to play a passive role in composite performance and development with major emphasis being placed on the fiber properties alone, the demanding recent uses of composites require that polymeric matrix plays an increasingly important role in composite performance.
The first part of this work is dedicated to the description of the mechanical properties of polymer composites, starting with a brief description of the behavior of the components: matrix and fibers. Then the properties of particulate, long fiber and laminate composites are described through the different models generated in literature. The second part is focused on the processing of polymer composite materials, highlighting the fundamental operations associated with the matrix behavior. This section, concentrated on high performance thermoset matrix composites, includes aspects related with reaction kinetics, chemorheology, fluid flow, and heat transfer, proposing a general approach able to describe and predict their processing behavior under different processing conditions.

1. Introduction

Composite materials with polymeric matrices have emerged as strong candidates for load bearing structural applications in the commercial airplane and automotive industries. Furthermore, in view of the energy shortage, the composite's low weight coupled with energy efficient operations during their processing makes them one of the most desiderable materials for use in the future. However, unlike the metals which the composites are replacing in structural applications, their processing characteristics and their final properties can be greatly affected by chemical composition, the load conditions and the various environments under which they perform. Consequently, the problems of understanding and predicting the influence of these effects on the composite material performance has attracted attention from both academic and industrial researchers due to its complexity and relatively prominent role in providing short-and long-term solutions to many needs of the transportation industry.
The state of development of composite materials is quite unique in the scientific world

G. Akovali (ed.), The Interfacial Interactions in Polymeric Composites, 321–357.
© 1993 *Kluwer Academic Publishers.*

with simultaneous advances being made both in their usage and basic understanding. The complexity and high technology required in manufacturing structural parts with these materials as well as the need for fundamental description of their processing and property characteristics necessitates a close collaboration between different scientific areas. The fact that the transportation industry with its current international character has a vital interest in composite materials for weight savings applications has provided a strong incentive for extending the fundamental research in the field of mechanical properties and processing operations.

The polymeric matrix and its interaction with a reinforcing phase, in the form of continuous or discontinuous high strength and stiffness fibers, is one of the major controlling factors in the processing and property characteristics of composites. Although traditionally the matrix has been thought to play a passive role in composite performance and development with major emphasis being placed on the fiber properties alone, the demanding recent uses of composites require that polymeric matrix plays an increasingly important role in composite performance. The wide variety of polymeric materials available both of thermoplastics and thermosetting nature that can be tailor-made to meet specific performance characteristics, necessitate a clear identification of property and processing requirements of the material to be used as a matrix in composites. Accordingly, the content of this chapter address explicitly the effects of processing and properties of the polymeric matrix on the behavior of high performance composite materials.

The first part is dedicated to the description of the mechanical properties of polymer composites, starting with a brief description of the behavior of the components: matrix and fibers. Then the properties of particulate, long fiber and laminate composites are described through the different models generated in literature.

The second part is focused on the processing of polymer composite materials highlighting the fundamental operations associated with the matrix behavior. This section, concentrated on high performance thermoset matrix composites, includes aspects related with reaction kinetics, chemorheology, fluid flow, and heat transfer, proposing a general approach able to describe and predict their processing behavior under different processing conditions.

2. Mechanical Properties of Composite Materials

2.1 MECHANICAL PROPERTIES OF POLYMERIC MATRICES

A relevant contribution to the composite properties depends on the matrix characteristics. The major role of the matrix in a fiber reinforced composite is to transfer stresses between the fibers to provide, a barrier to adverse environment and protect the surface of the fibers from the mechanical and chemical aggression. While the contribution of the fibers mainly affects the tensile properties of the composite, the matrix is responsible of off-axis properties and interlaminar shear as well as in plane shear properties. Furthermore the processability and the eventual defects of a composite depend strongly on the physical characteristics of the matrix such as viscosity, glass transition temperature and processing conditions (1-10). Composite matrices traditionally are crosslinked polymers but recently a growing interest is devoted towards thermoplastic polymers taking advantage of their easier processability and higher fracture toughness characteristics.

Polymeric materials are structurally much more complex than ceramic or metals. Depending on their chemical structure, polymeric matrices can be divided in two broad categories: thermoplastics and thermosets. Thermoplastic polymers are constituted of linear molecules and are held together by weak secondary bonds. On heating they can be softened, melted (if crystalline) and reshaped as many time as desired.

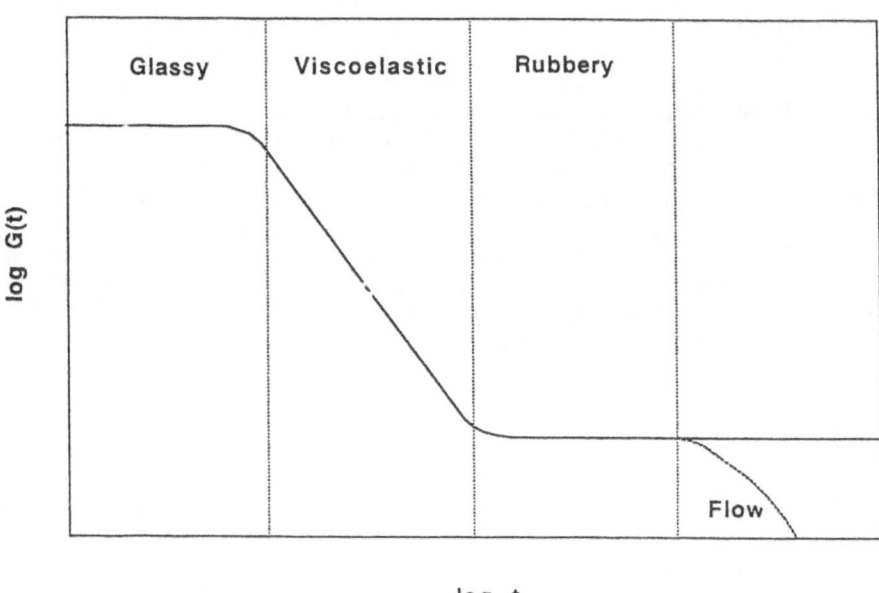

Figure 1: Creep modulus of a polymer as a function of time.

In a thermoset polymer the molecules are chemically joined together by crosslinking forming a rigid structure. Once these crosslinks are formed during polymerization the polymer can not be melted or reshaped anymore.

The peculiar molecular structure of polymers, thermosets or thermoplastics, is responsible of different types of behavior under different level of stresses obtained with constant or variable loads. In particular the viscoelastic behavior differentiates polymers from other traditional materials.

The elastic solid is defined as a body with a definite shape that can be deformed into a new equilibrium shape reversibly. On the other hand a viscous liquid has no definite shape and is able to flow irreversibly under the action of external forces. One of the most interesting features of polymers is that they can exhibit an intermediate behavior between the two above mentioned, depending on the experimental time scale and on the test temperature (5,6).

The effect of the time scale of the experiment is one of the main consequences of viscoelasticity and must be taken into account when mechanical properties of a viscoelastic material are measured. Figure 1 shows the creep modulus as a function time, here at very short time the polymer behaves as glassy solid characterized by a time independent modulus. At intermediate times the polymer is in the viscoelastic region and the modulus becomes time dependent. At long times again, a time independent behavior is shown in the rubbery region. Finally at very long times thermoplastics can flow while thermosets show a rubbery behavior.

324

As shown up to now the mechanical properties of polymers are greatly affected by the time scale of experiment and by the temperature. A great variation in stress-strain behavior of polymers as measured at constant rate of strain is shown in Fig. 2. Curve A is for a hard and brittle material, curves B is typical of hard ductile polymer showing uniform extension, curve C is also for a hard ductile polymer but is typical of a material which cold-draws with necking and curve D is typical of elastomeric materials. All these types of curves could be obtained from a single polymer by just changing the test temperature over wide enough range. At low temperature the elongation to break is low and there is no yield point (curve A). At higher temperatures first appears the yield point (curve A) then necking and cold drawing are possible (curve C) and finally above T_g the typical rubber behavior is shown (curve D). Therefore increasing temperature a polymer show a transition from a brittle to a ductile behavior.

Finally Table 1 summarize some properties of the most widely used matrices for composites including thermosets (epoxy and phenolics), amorphous thermoplastics (Poly-ether-imide and poly-amide-imide) and semicrystalline thermoplastics (Poly-ether-ether-ketone, poly-phenylene-sulfide and poly-imide).

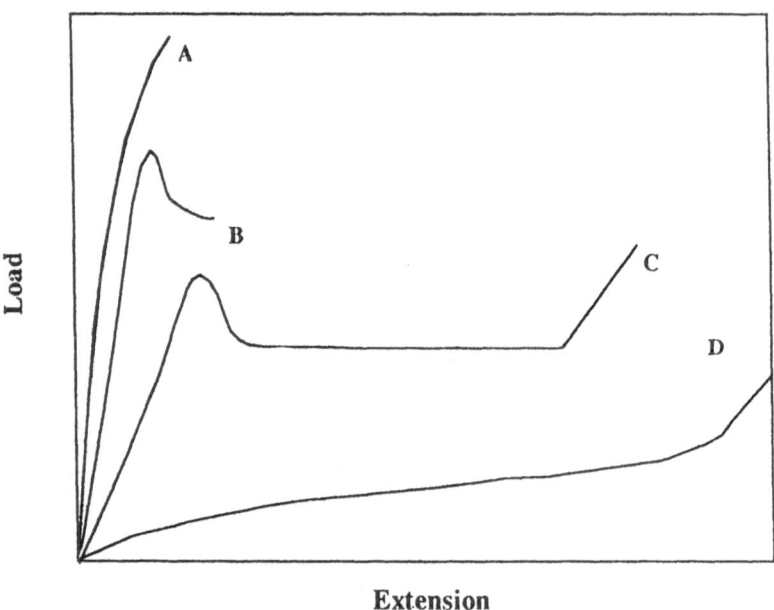

Figure 2: Different stress-strain behaviors of polymers.

2.2 PROPERTIES OF REINFORCEMENTS

Reinforcements are usually in the form of long fibers, flakes, whiskers, particles, discontinuous fibers or fabric. However in high performance composites the choice is oriented toward long fibers and fabrics.

The use of fibers for high-performance engineering materials is due to their unique characteristics (11, 12). First a small diameter respect to the grain size allows the higher fraction of the strength to be attained than that possible in a bulk form. This is the so called "size effect", the smaller is the size the lower is the probability to have an imperfection in the material. In Fig. 3, is shown the decrease of strength when the diameter of the fiber increase. Therefore the exceptionally mechanical properties of fibers are a consequence of this size effect.

Second the high aspect ratio (the ratio l/d length/diameter), enhances the capability to transfer the load to the other fibers through the matrix. The interfacial bond strength is one of the main characteristic that have to be taken into account when designing a composite with the desired mechanical properties. In fact non-catastrophic failure requires a balance between the values of the tensile strength of the fibers and the interfacial bond strength. Too high interfacial bond strength, for a given fiber tensile strength, will produce a more brittle composite that of course will be flaw sensitive and with lower tensile strength. Too low interfacial bond strength will produce lower shear and tensile strength in the composite but lower flaw sensitivity. In order to improve the adhesion between fiber and matrix, the surface of the fibers is treated by different processes, among these the most important for carbon fibers is the oxidative process. Furthermore fiber for high performance application would require to have lowest density and a very high degree of flexibility that allows the composite to be processed with the widest number of techniques so that many different shapes can be obtained.

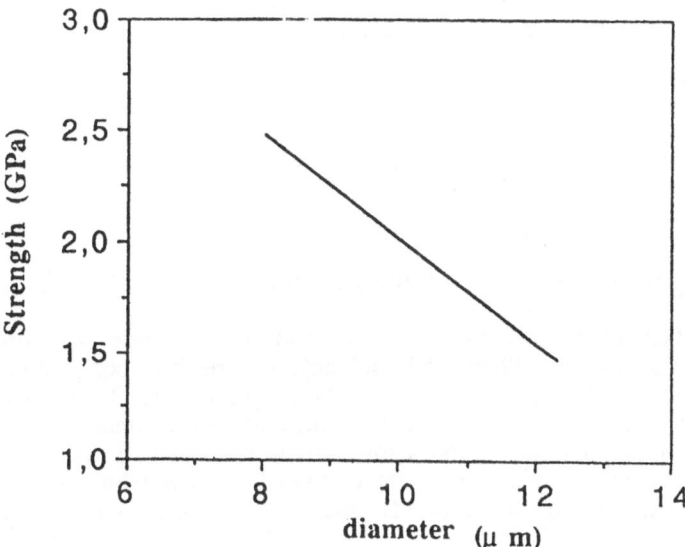

Figure 3:   Strength of a fiber as a function of the diameter.

Thermal expansion coefficient is also an important property for reinforcements. The thermal expansion coefficient of fibers is usually lower than that one of the matrix causing residual stress after processing. However this difference determines lower thermal expansion of the composites and consequently a better dimensional stability after processing operations. Finally the wide range of fibers available gives the possibility to design composites with different electrical conductivities.

Reinforcing fibers can be divided in three main groups:

- Graphite fibers. Carbon fibers owe their success in high performance composites to their extremely high tensile modulus-weight and tensile strength-weight ratios, high fatigue strength and low coefficient of thermal expansion coupled with a low ratio cost-performance. Carbon fibers are commercially available with a variety of modulus ranging from 270 GPa to 600 GPa. They are produced following two different processes depending on the type of precursors namely textile precursor and pitch precursor. A flow diagram of the carbon fibers processing is given in Fig 4.

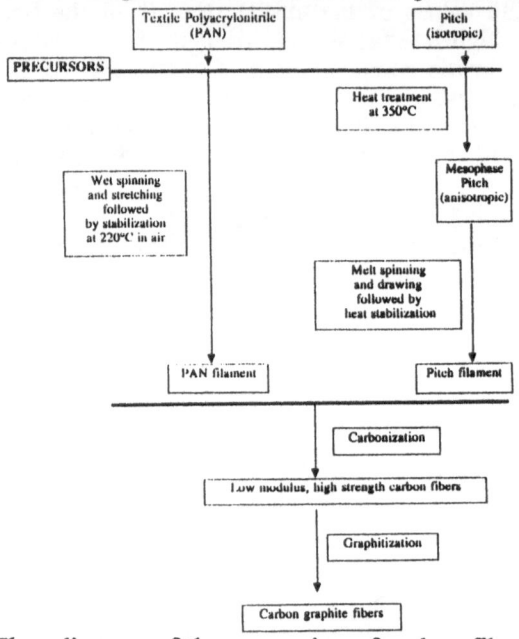

Figure 4: Flow diagram of the processing of carbon fibers.

- Glass fibers. Glass fibers are the most common of all reinforcing fibers for polymeric matrix composites. Their main advantages are low cost, high tensile strength, high chemical resistance and good insulating properties. On the other hand they display a low tensile modulus, a relatively high density compared to the other fibers, a high sensitivity to wearing and a low fatigue resistance. Depending on the chemical composition of the glass they are commercially available in 3 different grades: E, S and C. Among these the E fiber are cheapest and present lower mechanical properties. The C fiber are mainly used when corrosion is the main concern.

- Boron fibers. These fibers are characterized by an extremely high tensile modulus coupled with a large diameter offering an excellent resistance to buckling that contributes to a high compressive strength of the composites. Their high cost is due to the processing operations. For this reason boron fibers find application only in aerospace and military application.

- Aramid fibers. The most common aramid fibers available are the Kevlar® 49. These fibers are composed by a highly oriented crystalline polymer and present the highest tensile strength-weight ratio. On the other hand the disadvantages they present are the low compressive strength, difficulty of manufacturing and a sensitivity to ultraviolet light and to water. However Kevlar® fibers find applications in sporting goods and aircraft industry.

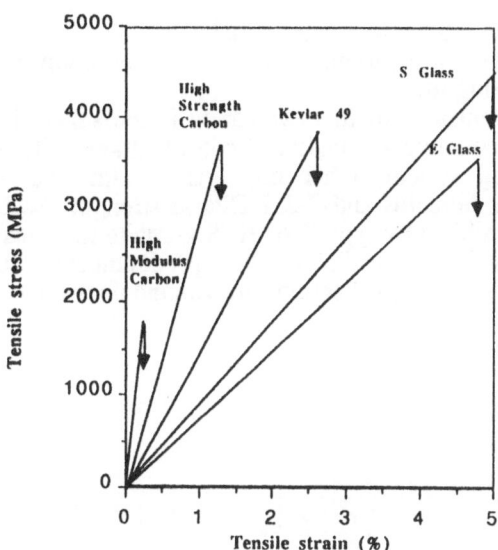

Figure 5: Stress strain behavior of the most common fibers.

2.3 MECHANICS OF COMPOSITE LAMINATES

2.3.1 General Aspects

Technical composites are multiphased, anisotropic bodies. Efficient and reliable use of such materials necessitates optimization of a laminated composite with respect to all of its directional properties (particularly, transverse tensile, shear, and longitudinal compression, as well as the earlier emphasis upon longitudinal properties).

Employing current design engineering characterization and design procedures, we are in a position today to evaluate the performance capability of composites for a variety of filament ranging from E-glass through the very high-modulus, high-strength graphite systems. These observation may be summarized as follows (9):

- The longitudinal strength of unidirectional composites are relatively independent of the specific reinforcement
- Engineering design levels are the product of laminate stiffness and the smallest ply strain (usually the strain transverse to the fiber direction) required to damage the composite
- Matrix properties and filament anisotropy exert a strong influence upon design capability.
- Filament which do not exhibit a longitudinal composite strain capability at failure of at least + 0.005 to 0.007 or more cannot be used as a general engineering

material.

\-   Anisotropic graphite and organic polymeric filament possess small but negative expansion coefficient transverse to the fibers. As a consequence, special attention must be paid to residual thermal stress in fabrication, but this penalty is offset by the potential of developing absolute dimensional stability in the plane of a composite plate.

\-   All filaments except glass are fatigue insensitive; and

\-   High modulus matrix systems must produce a transverse ply strain capability of +0.003 to be of serious engineering use.

One of the more interesting conclusions of this evaluation is the suggestion that for general structural applications, economic and technical considerations will dictate only one continuous filament prepreg system, other than glass, with fiber possessing properties of roughly 19-26 GPa modulus and 2-2.5 GPa in strength. Such a system will be close to the properties exhibited by type II or HTS graphite fiber materials. Of course, other fiber systems will be available, but predominantly for special applications and at a relatively high price, reflecting low-volume demand.

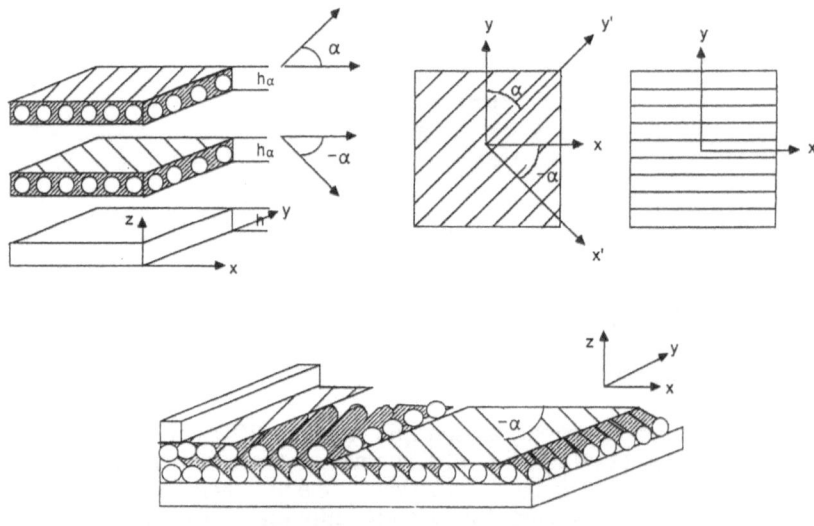

Figure 6: Illustration of a continuous fiber composite laminate.

2.3.2. Laminate Properties: Strength and Stiffness

Current attitudes regarding composite materials (14-16) emphasize the relationship of structural performance to the properties of a ply. A "ply" is a thin sheet of material consisting of an oriented array of fibers embedded in a continuous matrix material (Fig. 6). These plies are stacked one upon other, in a defined sequence and orientation, and bonded together yielding a laminate with tailored properties. The properties of the laminate are related to the properties of the ply by the specification

of the ply thickness, stacking sequence, and the orientation of each ply. The properties of the ply are, in turn, specified by the properties of the fibers and matrix, their volumetric concentration, and geometric packing in the ply. Generally, the ply material is preformed and can be purchased in a continuous compliant tape or sheet form which is in a chemically semicured condition. The basis for engineering design of such a material is then the properties of a cured ply or lamina as it exists in a laminate. This ply is treated as a thin two-dimensional item and is mechanically characterized by its stress-strain response to loading in:
- The direction of the filaments which exhibit a nearly linear response up to a rather large fracture stress;
- In the direction transverse to the filament orientation which exhibits a significantly decreased moduli and strength; and
- The response of the material to an in-plane shear load.
By the contrast with isotropic metallic materials, an oriented ply, in the form of a thin sheet, is anisotropic and requires four elastic (plane stress) constants (14-16) to specify its stiffness properties in its natural orientation:

$$\sigma_1 = Q_{11} \epsilon_1 + Q_{12} \epsilon_2$$
$$\sigma_2 = Q_{12} \epsilon_1 + Q_{22} \epsilon_2 \tag{1}$$
$$\sigma_6 = Q_{66} \epsilon_6$$

where $\sigma_6 = \tau_{12}$ and $\epsilon_6 = \tau_{12}$, or in matrix form:

$$
\begin{array}{ccc}
\sigma_1 \\
\sigma_2 \\
\sigma_6
\end{array}
=
\begin{array}{ccc}
Q_{11} & Q_{12} & 0 \\
Q_{12} & Q_{22} & 0 \\
0 & 0 & Q_{66}
\end{array}
\begin{array}{c}
\epsilon_1 \\
\epsilon_2 \\
\epsilon_6
\end{array}
\tag{2}
$$

where the plane-stress stiffness moduli are:

$$Q_{11} = E_{11}/(1 - \nu_{12} \nu_{21})$$
$$Q_{22} = E_{22}/(1 - \nu_{12} \nu_{21}) \tag{3}$$
$$Q_{12} = \nu_{21} E_{11}/(1 - \nu_{12} \nu_{21})$$
$$Q_6 = G_{12}$$

Where ν_{ij} is the Poisson ratio defined as: $-\epsilon_i / \epsilon_j$.
If, however, the ply is rotated with respect to the applied stress or strain direction two additional moduli appear, which results in the direction indicated shear coupling rotation in simple extension:

$$
\begin{array}{c}
\sigma_1 \\
\sigma_2 \\
\sigma_6
\end{array}
=
\begin{array}{ccc}
Q_{11}^* & Q_{12}^* & Q_{16}^* \\
Q_{12}^* & Q_{22}^* & Q_{26}^* \\
Q_{16} & Q_{26} & Q_{66}
\end{array}
\begin{array}{c}
\epsilon_1 \\
\epsilon_2 \\
\epsilon_6
\end{array}
\tag{4}
$$

where:

$$Q_{11}^* = U_{11} + U_2 \cos(2\Theta) + U_3 \cos(4\Theta)$$
$$Q_{22}^* = U_1 - U_2 \cos(2\Theta) + U_3 \cos(4\Theta)$$
$$Q_{12}^* = U_4 - U_3 \cos(4\Theta)$$
$$Q_{66}^* = U_5 - U_3 \cos(4\Theta) \tag{5}$$
$$Q_{16}^* = -1/2 \, U_2 \sin(2\Theta) - U_3 \sin(4\Theta)$$
$$Q_{26}^* = -1/2 \, U_2 \sin(2\Theta) + U_3 \sin(4\Theta)$$

the invariants U_i to the rotation are:

$$U_1 = 1/8 \ (3 \ Q_{11} + 3 \ Q_{22} + 2 \ Q_{12} + 4 \ Q_{66})$$
$$U_2 = 1/2 \ (Q_{11} - Q_{22})$$
$$U_3 = 1/8 \ (Q_{11} + Q_{22} + 6 \ Q_{12} - 4 \ Q_{66}) \qquad (6)$$
$$U_4 = 1/8 \ (Q_{11} + Q_{22} + 6 \ Q_{12} - 4 \ Q_{66})$$
$$U_5 = 1/8 \ (Q_{11} + Q_{22} - 2 \ Q_{12} + 4 \ Q_{66})$$

In addition, lamination can result in up to 18 elastic coefficients and increased deformational complexities. But the additional coefficients can all be derived from the four primary coefficients using the concept of rotation and ply-stacking sequence (14, 15). These complications are the result of geometric variables. If the laminate is properly constructed, the in-plane stretching or stiffness properties can still be specified by four elastic coefficients. We shall consider laminates of this nature.

The flow diagram of a typical calculation is shown in Fig. 7. Note that both short and continuous fiber are handled in the same manner. These calculations, while tedious, are analytically simple. The "plane stress", the Q_{ij} terms, are employed because lamination neglects the mechanical properties through the ply thickness. These stiffnesses are sometimes regrouped into new constants called "invariants", the U_i terms, for analytical semplicity.

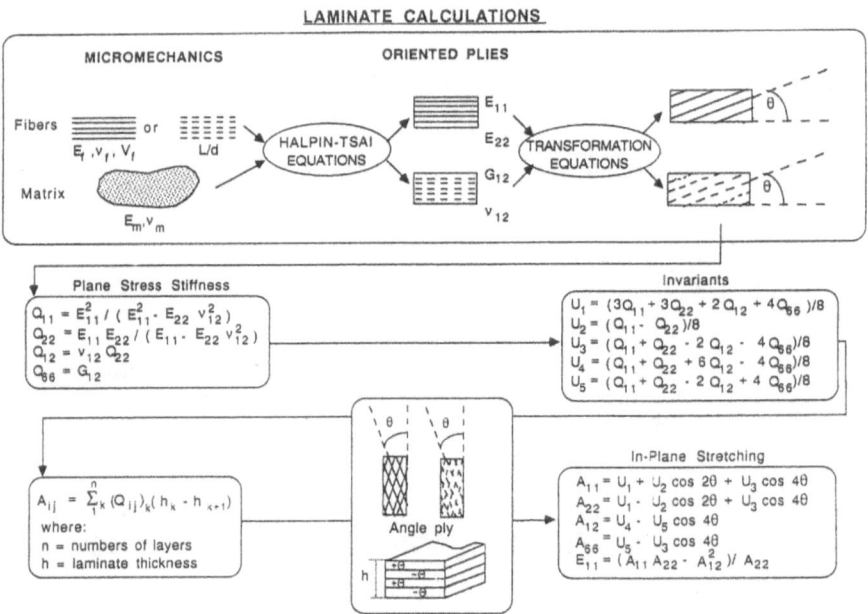

Figure 7:   Laminate calculations.

To compute the properties of the laminate one then sums the ply properties through the thickness of the laminate, weighted by the thickness (h_k) of each oriented ply:

$$A_{ij} = \sum_{k=1}^{N} (Q_{ij})_k h_k \tag{7}$$

For a balanced (same number of $\pm\Theta$) and symmetrical ($+\Theta$ or $-\Theta$ at same distance above and below the mid plane) the laminate solution is:

$$\begin{aligned}
A_{11} &= U_1 + U_2 \cos(2\Theta) + U_3 \cos(4\Theta) \\
A_{22} &= U_1 - U_2 \cos(2\Theta) + U_3 \cos(4\Theta) \\
A_{12} &= U_4 - U_3 \cos(4\Theta) \\
A_{66} &= U_5 - U_3 \cos(4\Theta)
\end{aligned} \tag{8}$$

Note the inverted terms A_{ij} yield the required elastic properties of the laminate in terms of the individual ply properties E_{11}, E_{12} and G_{12}.

$$\begin{aligned}
E_{11} &= (A_{11} A_{22} - A_{12}^2)/A_{22} \\
E_{22} &= (A_{11} A_{22} - A_{12}^2)/A_{11} \\
\nu_{12}/E_{11} &= A_{12}/(A_{11} A_{22} - A_{12}^2); \qquad G_{12} = A_{66}
\end{aligned} \tag{9}$$

These calculations have been thoroughly tested and agree closely with experiment (Fig. 8). In this figure the dots are the experimental points and the lines are the theoretical prediction for a nylon fibers reinforced rubber. The angle-ply laminate is predicted from the ply properties.

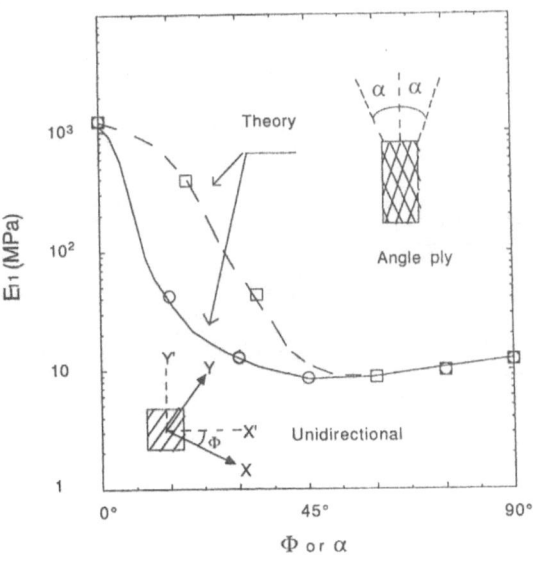

Figure 8: Young's modulus vs. fiber orientation for nylon-fiber reinforced rubber. $E_f = 2.0$ GPa, $E_m = 2.1$ MPa $\nu_f = 0.2$, $\nu_m = 0,4999$.

The ply properties are in turn correlated with the transformation equations and the micromechanics. The micromechanics employed in this demonstration are based upon the "self-consistent-method" developed by Hill (17). Hill rigourously modeled the composite as a single fiber, encased in a cylinder of matrix; with both embedded in an unbounded homogeneous medium which is macroscopically indistinguishable from the composite. Hermann (18) employed this model to obtain a solution in terms of Hill's "reduced moduli". Halpin and Tsai (19) reduced Hermann's solution to simpler analytical form and extended its use for a variety of filament geometries:

$$
\begin{aligned}
E_{11} &= E_f V_f + E_m V_m \\
\nu_{12} &= \nu_f V_f + \nu_m V_m \\
p/p_m &= (1 + \eta \xi V_f) / (1 - \eta V_f)
\end{aligned}
\tag{10}
$$

where

$$
\begin{array}{lll}
\eta = (t_f/t_m - 1) / (p_f/p_m + \xi) & & \\
\xi_E = 2 \, (e/d); & \xi_{G12} = 1; & \xi_{G23} = 1/(3 - 4 \nu_m) \\
p = E_{22}, G_{12}, G_{23}; & p_f = E_f, G_f; & p_m = E_m, G_m
\end{array}
$$

These equations are suitable for single calculation and were employed previously for the single-ply and angle-ply properties. The short fiber composite properties are also given by the Halpin-Tsai equations where the moduli in the fiber orientation direction is a sensitive function of aspect ratio (l/d) at small aspect ratios, and has the same properties of a continuous fiber composite at large but finite aspect ratios.

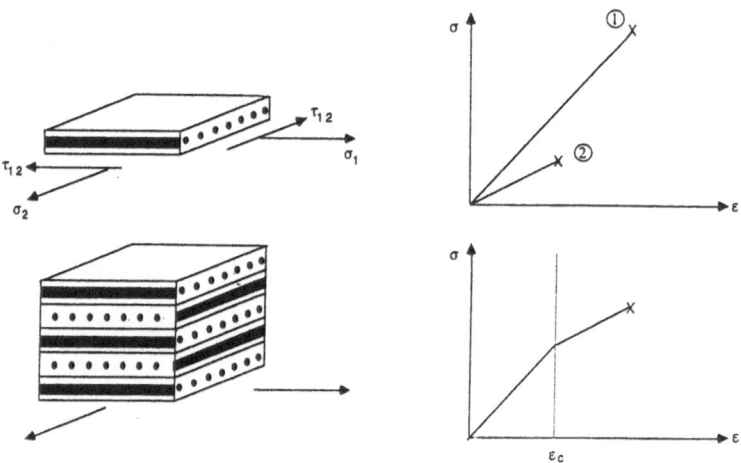

Figure 9: Determination of laminate properties.

If the ply illustrated in Fig. 9 is used in the construction of a balanced and symmetrical 0/90 laminate and is mechanically tested, a bilinear stress-strain curve is obtained and the stiffness is the sum, through the thickness, of the plane-stress stiffness of each layer. As the laminate is deformed, each ply possesses the same in-plane strain, ϵ, and when the strain on the 90-deg layers in the laminate prevents the 90-deg layer from carrying their share of the load, Q_{ij} (90 deg) = 0. This load is transferred to the unbroken layers, the 0-deg layers for our illustration and results in a loss of laminate stiffness or modulus. Continual loading will ultimately produce a catastrophic failure of the laminate when the strain capability of the unbroken, 0-deg, layers is exceeded. For a 0/90 construction, employing the glass/epoxy material, the ratio of the ultimate failure stress to the crazing stress, the ratio of the ultimate failure stress to the crazing stress, (the knee in Fig. 9) is 6.1. Experimental data and a theoretical stress-strain curve are shown in Fig. 10 for a $\pi/4$ (0-deg / $\pm$45-deg / 90-deg) glass/epoxy laminate. Note a change in stiffness as the 90-deg and then the 45-deg layers and the correspondence of the theoretical ultimate strength of 356 MPa with the experimental results of 346 MPa. While the strain for transverse ply failure is constant from laminate to laminate, the stress required to craze the system as well as final failure load is a function of laminate geometry because the construction of the laminate specifies the stiffness properties (crazing stress= stiffness x allowable transverse ply strain). It must be noticed that the area under the stress-strain curve is proportional to the impact energy. Therefore, lamination permits the engineer to tailor a fixed prepreg system to meet the conflicting stress/strain demands at different points in a structure. A further point, the crazing stress of threshold is generally at or below the creep fracture or fatigue limit for all classes of composites (for glass/epoxy the fatigue limit lies between 0.25-0.30 of static ultimate strength). Boron and graphite are fatigue insensitive filaments, thus no fatigue damage is realized below first ply failure.

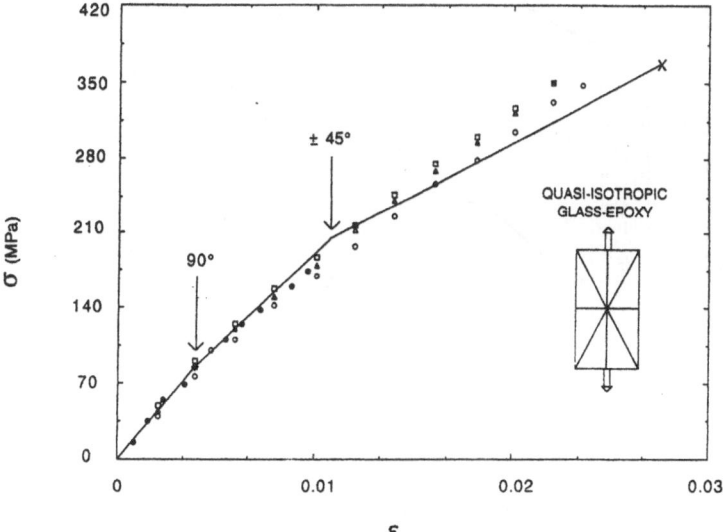

Figure 10: Stress-strain curve for a $\pi/40$ (0 deg/$\pm$45 deg/90 deg) glass/epoxy laminate. The solid line is a theoretical prediction.

Thus, the material properties of a laminate are specified in terms of the ply engineering module; E_{11}, E_{22}, ν_{12} and G_{12}; the engineering strains to failure; ϵ_1, ϵ_2 and ϵ_6; and the thermal expansion coefficients; e_1 and e_2.

2.4 SHORT-FIBER COMPOSITES

Random or nearly random distributions of fibers, finite in length and arranged in matrix, constitute many naturally occurring and synthetic materials. In the majority of cases, the spatial orientation of the discontinuous fibers is intermediate between a truly random array in three dimensions and two-dimensional random array of fibers. Generally these materials posses internal orientations which are independent of the thickness. In some systems of technological interest, the distribution of fibers may vary through the thickness. Halpin and Pagano (14,19) proposed that such materials can be modelled mathematically as laminated systems. The laminate model consists of layers of unidirectional composites, in our case short-fiber composites, with the fiber volume fractions in a layer oriented at an angle being governed by the percentage of fibers at angle Θ in the actual material. If the material to model exists in sheet form wherein the planar thickness is considerably less than the average lengths of fibers, then the reinforcement may be considered as two-dimensional random array of fibers. This situation is described in laminated plate theory as "quasi-isotropic".

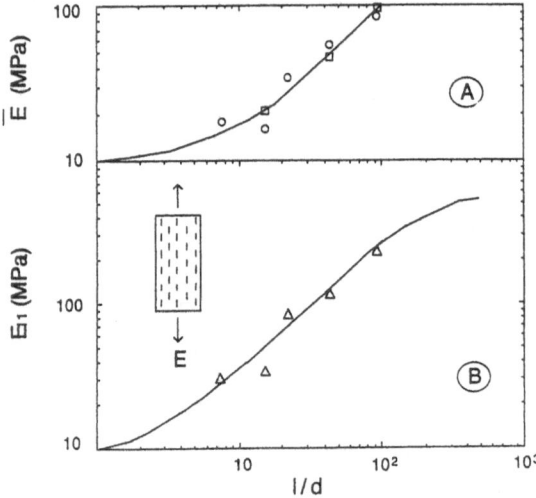

Figure 11: Dependence of longitudinal modulus on aspect ratio for random, o ; quasi-isotropic, □ ; and oriented, ◇ , short nylon fiber/rubber composites. $E_f/E_m = 973$, $V_f = 0.35$. The lines are theoretical predictions.

The success of the lamination approximation is strongly dependent upon the assumption of physics volume averaging in real material systems combined with an ability to estimate the stiffness for an oriented short-fiber sheet (Fig. 11B). Employing the Halpin-Tsai equation, indicated earlier, we are able to predict the strong dependence of E_{11}, on aspect ratio (20). Note that as the aspect ratio becomes large uniaxial stiffness E_{11} becomes identical with oriented continuous filaments. The other moduli, E_{22}, G_{12} and possibly ν_{12} are not sensitive functions of aspect ratio and may be approximated by the continuous fiber result. This approximation is necessary as no three dimensional elastic micromechanic solution exists for this problem. The strength properties do not approach the continuous filament composite properties as was shown by Riley (21).

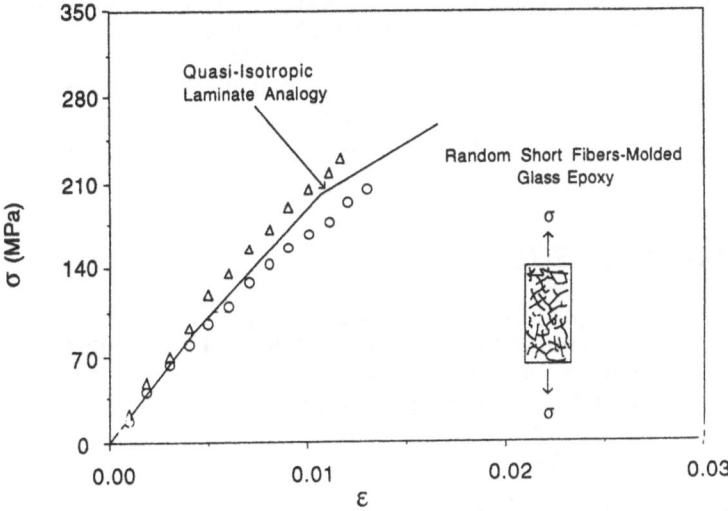

Figure 12: Stress-strain curve of randomly oriented short glass-fiber/epoxy composites with the maximum strain theory prediction.

The properties of random short-fiber composites are shown in Fig. 11A as circles. To test the laminate model, "quasi-isotropic" laminates of two different aspect ratios were made and tested. The stiffness of these two laminates are comparable to the random composites. Confirmation of the model is also achieved by noting the agreement between the data and the theoretical calculation.

Maximum strain theory may be modified to predict the strength of randomly oriented short-fiber composites (22). The Halpin-Tsai equations (14) have established relations for the stiffness of an oriented short-fiber ply from the matrix and fiber properties. These equations show that the longitudinal stiffness of an oriented short-fiber composite is a sensitive function of the aspect ratio.

The short-fiber stiffness asymptotically approaches the continuous filament stiffness at large aspect ratios. Strength, like stiffness, is a function of aspect ratio and approaches an asymptotic limit as the aspect ratio becomes large. However, the strength limit for short fibers does not approach the continuous filament strength.

Thus, the oriented short-fiber material fails at an ultimate longitudinal strain which is less than the ultimate longitudinal strain of the continuous-fiber material. Previous workers have shown that the asymptotic short-fiber strength limit is less than the continuous-filament composite strength (23). A finite element analysis of the discontinuous fiber by Barker and McLaughlin (24) shows a stress concentration factor due to the fiber ends. This stress concentration factor becomes constant at sufficiently large aspect ratios. Chen (23) and Riley (21) also report that a plateau strength is reached once the fibers become sufficiently long.

The strength of high aspect ratio short-fiber randomly oriented composites can be predicted with maximum strain by reducing the longitudinal strain allowable to reflect the reduced strength in the fiber direction due to discontinuous reinforcements. Chen's data and analysis (23) show that at sufficiently large aspect ratios, the strength of short-fiber glass/epoxy composites will be 60 percent of the strength of the continuous-filament composite. This fact is incorporated into maximum strain theory by reducing the continuous filament longitudinal strain allowable by 60 percent. Figure 12 shows the experimentally measured stress-strain curve for a randomly oriented short-fiber composite and the prediction from maximum strain theory and the laminate analogy. The effect of fiber volume fraction of the quasi-isotropic strength can be included by using the Halpin-Tsai equations to calculate the orthotropic short-fiber ply moduli.

2.5 PARTICULATE COMPOSITES

The introduction of another phase in polymer such as a particle, or a simple filler, generally modifies the morphological order of the polymer. This phenomenon can be exploited in order to obtain a materials with the desired properties. Generally the main reasons of using particles in a composite are (13):

- to rise the rigidity of a system;
- to reduce the viscous flow;
- possibility to manipulate the thermal expansion coefficients;
- possibility to vary the permeability of the composites to gas or liquids;
- to rise the resistance to the abrasion;
- to modify the electric properties of the materials;
- to modify the rheological characteristics of the materials;
- to lower the cost of the material;

The use of fillers modifies the material properties as a consequence of changes of the microstructure of the material.

Depending on the adhesion between the filler and the polymer the particles can act as reinforcing agent or simply as stress concentrators. Usually the bond between the filler and the matrix is created by the pressure increase subsequent to the shrinkage during polymerization for thermosets matrices, or to the molding process for thermoplastics.

When spherical particles are included in the polymeric matrix the behavior of such material is isotropic and the elastic properties can be easily predicted by using the Kerner equation (15) shown below or the Halpin-Tsai equations for G_{12} and E_{22} discussed earlier:

$$E_c = E_m \frac{1 + ACV_f}{1 - CV_f} \qquad (11)$$

where:

$$A = \frac{7 - 5\nu_m}{8 - 10\nu_m} \qquad C = \frac{E_f/E_m - 1}{E_f/E_m - A} \qquad (12)$$

ν_m is the Poisson's ratio of the matrix, V_f is the volume fraction of filler and E_f and E_m are the Young's moduli of the filler and the matrix respectively.

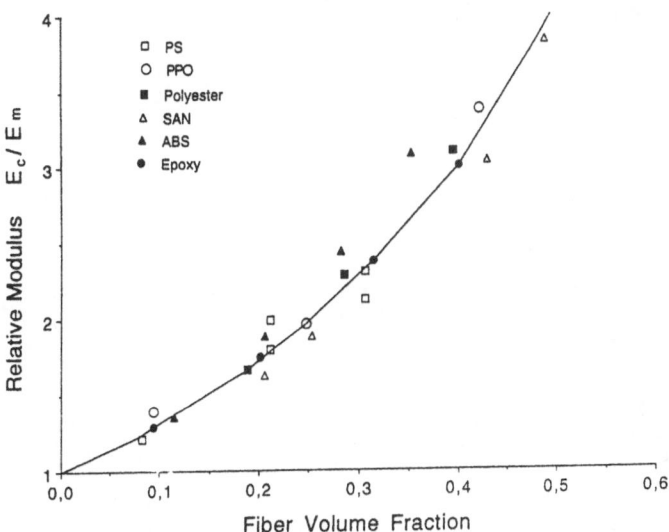

Figure 13: Relative moduli Ec/Em vs volume fraction of glass beads for various composites (24). The solid line is the plot of the Kerner equation.

In Fig. 13 experimental data relative to styrene acrylonitrile (SAN), acrylonitrile butadiene-styrene (ABS), polystyrene (PS), epoxy, and polyester/glass bead composites are reported together with the plot of the Kerner equation. In addition to the elastic moduli, the other tensile properties of this class of materials are sensitive to the properties of the matrix and the adhesion between filler and matrix. In fact when a matrix is able to craze, it contributes an inhomogeneous deformational mechanism which leads to an increase of elongation and work to break for the resulting composites due to formation and propagation of crazes through the polymer. Accordingly to Kambour (25), the crazes normally observed in thermoplastic materials are not cracks, but rather localized regions of highly oriented polymer. The high elongation observed (26) in the glass-bead filled SAN can be explained assuming that the growth of crazes can be terminated by the glass beads or vacuoles around them. In fact, if the propagating craze encounters a glass sphere to which the matrix is not strongly adherent, interfacial debonding can effectively blunt the tip of the craze and prevent, or at least slow down, further craze propagation. The filler particles act

338

as stress risers allowing multiple volume elements to reach the critical stress for craze formation. This situation is reflected in the stress- strain response by the appearance of a knee in the deformation curves at a well defined value of stress which is practically independent of filler content. In Fig. 14 a typical stress-strain curve for such composites is reported showing a sudden change in the slope at a fixed value of stress which is practically independent of filler content. The strength of these composites decreases as the volume content of beads increases. In general particulate composites, even if they do not have structural applications, can be used in many applications due to the fact that the addition of rigid particulate fillers to a polymeric matrix leads to a material with higher modulus, lower creep, more resistance to abrasion, different rheological and electrical properties, different deformational mechanism, and lower cost.

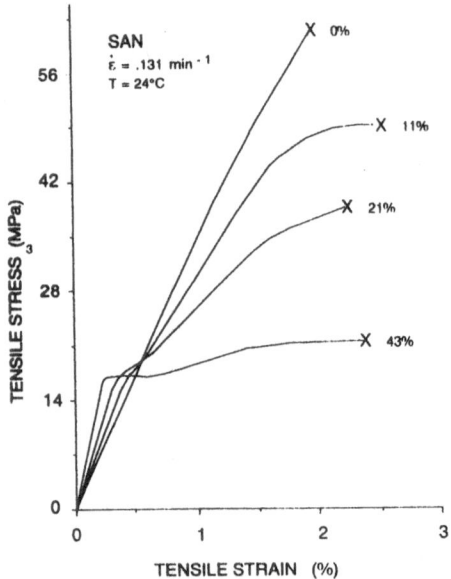

Figure 14: Stress-strain curve for SAN/glass bead composites at various filler contents.

2.6 FRACTURE MECHANICS IN COMPOSITE MATERIALS

2.6.1 General Aspects

Good fibrous reinforcement are generally brittle in character: they deform linearly to failure without yielding. This attribute creates a situation in which, in the presence of a notch or hole under static tension or compression test conditions, the fiber reinforced composite behaves more like a brittle material than metal. This issue has been a source of concern in the materials development and selection activities, as well as in engineering design (27).

Theoretical investigations into phenomenon of the notch sensitivity of composites have tended to rely on classical fracture concepts. These efforts have taken two forms: micro-and macro-mechanical representations. In the micro-mechanics format, local fracture processes (debonding, matrix cracking, fiber breaking, etc.) are studied in the hope that a mechanical treatment can be developed for a non-homogeneous multiangular laminate or moulded part. This approach faces serious difficulties and has not yet matured. The macro-mechanical approaches use a simplified model of the composite and classical fracture mechanics for homogeneous isotropic materials. The simplified composite model is the plane stress laminate theory (14,28,29) which converts the non-homogeneous laminated anisotropic solid into an anisotropic homogeneous solid. Within the macroscopic approaches, two lines of activity exist: a fracture mechanics argument (30-32); and a blending of classical fracture concepts and notch theory (33,34).

2.6.2 Theoretical Considerations

Strength concepts generally imply that rupture results because the spatial average stress or strain exceeded some critical value (generally an empirical criterion) which characterizes the mechanical stability of a solid. Such an attitude is useful if the microstructural perturbations to the local stress/strain fields inside a solid are of a small mean dimension with a low dispersion around the mean size. On the other hand, there are conditions in which discrete flaws, substantially larger than the uniform size distribution normally present, can exist in a material. Because they are discrete, usually relatively sharp, and larger than the surrounding disturbances, they induce additional stress concentrations and provide the site of cohesive fracture initiation. If these inherent flaws are cracklike, ordinary elastic stress-concentration factors can not be used because theoretical results predict an infinite concentration factor multiplying the average stress in the crack vicinity. Thus, the local stress value will exceed the finite allowable stress (or strain), experimentally measured for the base material containing only a reasonable uniform distribution of inherent flaws.

3. Processing of High Performance Composite Materials

3.1. INTRODUCTION

The widespread use of polymer matrix composite materials has been facilitated by the development of several specific technologies for the processing of parts with different geometries using different raw materials. However, the understanding of the fundamental engineering principles associated with each technology still represents a challenge for research activities in the field of the processing of high performance composite materials characterized by improved material quality and reliability requirements. Today the choice of processing parameters in the industrial environment is often based on extensive and costly testing. In addition, specific testing results may not be applicable if raw materials or part geometries are changed. In the last decade several studies focused on the analysis of the fundamental aspects of the processing of high performance composites have been reported (2,3,35-39), allowing a more rational choice of processing conditions and reducing the experimental work needed to determine the proper fabrication cycle. From a materials point of view, resin chemistry, storage and processing conditions should be selected to give the most uniform and reliable fabrication process, with low defects and a composite part obtained in the shortest possible time. In practice, factory operations, material quality and cost must be considered in a compromise to design an optimum processing cycle.

340

3.2 GENERAL ASPECTS OF PROCESSING TECHNOLOGIES

Thermoset based composites are commonly fabricated using epoxy, polyester or other thermosetting resins reinforced with carbon, glass or aramidic fibers. Main advanced processing technologies include lamination, resin transfer molding, pultrusion and filament winding. A brief description of this technologies, based on the available literature (40) may be useful to understand the role of the polymeric matrix on the processing behavior of high performance composites.

3.2.1   Autoclave Lamination

Thermoset based composite laminates are generally produced by Autoclave/Vacuum Degassing Lamination Process (38, 39). The characteristics of this process are shown in Fig. 15. In this process, prepreg plies of desired shape are laid up in a prescribed orientation to form a laminate. The laminate is covered with successive layers of an absorbent material (glass bleeder fabric), a fluorinated film to prevent sticking, and, finally, with a vacuum bag. The entire system is placed upon a smooth metal tool surface into an autoclave, vacuum is applied to the bag and the temperature is increased at a constant rate in order to promote the resin flow and polymerization. The autoclave process will be used along this section as a case study to describe the influence of the matrix characteristics on the processing behavior of high performance composites.

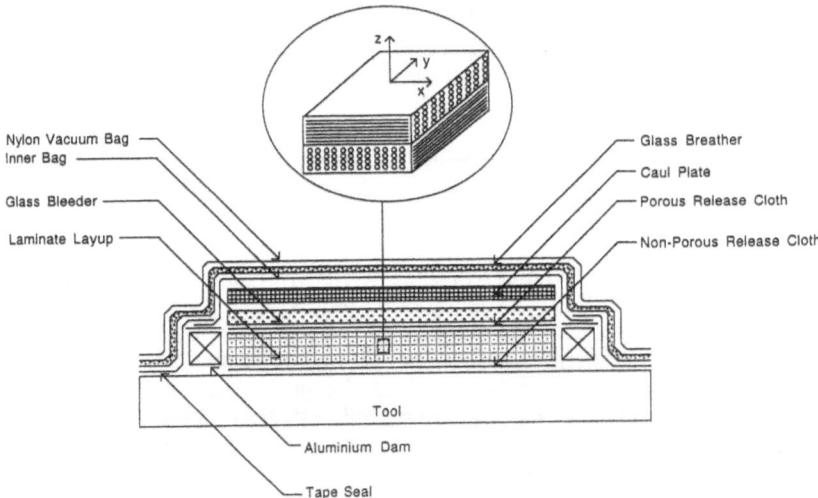

Figure 15:   Schematic diagram of a laminate lay-up. Insert shows a microscopic view of the fiber orientation within the laminate. (After Dave' et al., ref. 38).

3.2.2   Pultrusion

In the last years a growing interest has been devoted to the fabrication of high performance composites applying the pultrusion technology, offering, among others, the advantages of a continuous production and the integration of fiber impregnation and composite consolidation in the same process. As shown in Fig. 16 during pultrusion, fibers in form of tape, woven and/or mat are driven through a resin bath where a good impregnation can be achieved using a resin of proper viscosity. After the resin excess is removed in preforming guides, the fiber/resin system acquires the desired shape and go through the cure process in a heated die, acting as a continuous reactor. Usually, different heating zones are provided along the die depending on several factors such as, among others, the type of resin, the pulling speed and the length of the die. The application of the dragging force to the pultruded parts is performed using a pulling device able to impose the desired processing speed. Finally, the pultruder is equipped with a sawing system to cut off the continuous composite produced. The viscosity changes of the matrix determines the pulling force and the fiber wettability while the polymer reactivity determine the pulling speed in order to obtain a complete cured composite at the end of the process.

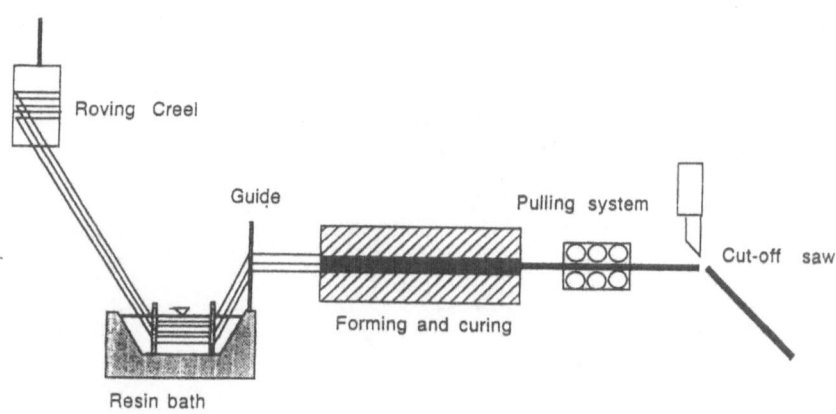

Figure 16:   Schematic diagram of the pultrusion process.

3.2.3   Resin Transfer Molding (RTM)

This technology is characterized by preplacement of dry reinforcement in the mold before the mold is closed and resin is injected (Fig. 17). Normally a low viscosity resin is used and either low pressure or low vacuum employed to assist resin flow and wetout of reinforcement. Complex preform shapes are generated by shaping of mats or fabrics, pretreated with thermoplastic binder with heat or pressure. RTM, presenting the ability to combine precise control in the placement of a high concentration of fibers with rapid processability, provide many of the fundamental requirements to meet the economic needs of a mass production industry. The main characteristic of RTM as an alternative of autoclave processing of high performance composites are given by the capability to produce large integrated parts with complex geometries, including box sections using foam cores and other sandwich structures. In addition, lower investment costs and no storage problems of unstable B-staged prepregs are involved.

Current research in RTM technology is addressed to further development of preform technology to optimize cycle times, fiber wetout, mechanical properties and surface appearance. A wide range of reinforcements and resins has been developed recently fulfilling the specific requirements of RTM: high reactivity and low viscosity systems for fast fill, complete penetration of the preform and good wetout of the fibers, low void formation, lack of movement of the reinforcement fibers, and avoidance of resin-rich areas.

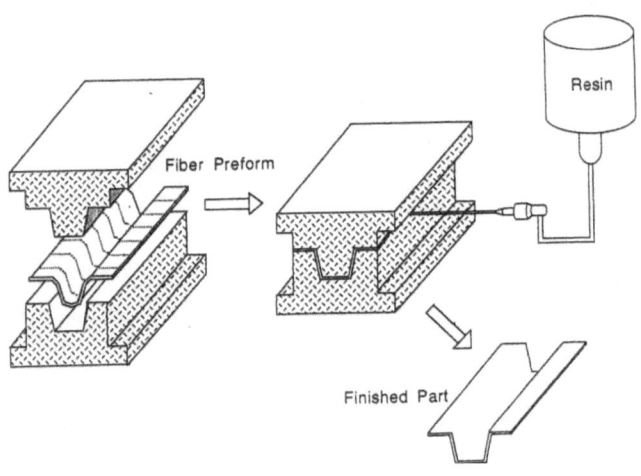

Figure 17:   Schematic diagram of the RTM process.

3.2.4   Filament Winding

In filament winding (Fig. 18) a filamentous yarn or tow is first wetted by a resin and then uniformly and regularly wound around a rotating mandrel along a predescribed path. Also preimpregnated tapes can be used in alternative. After the wound stage, the composite is cured by heating at a given temperature in an oven or autoclave or by exposure to IR radiation, and the mandrel is removed. Typical products range from a simple pipe, to an aircraft fuselage, while typical materials include glass, carbon or aramid fibers coupled to polyester, vinyl ester, or epoxy resin. The principal advantage of filament winding over other methods for composite fabrication are the possibility to adopt automation and robotic procedures. The greatest disadvantage is the geometric limitation of available tools, including the inability to wind on negatively curved (concave) surfaces.

Also in filament winding the processing behavior is strongly dependent on resin characteristics. The final fiber content is a function of the radial motion of the fiber with respect to the resin during winding. This motion is a consequence of the forces acting on the fibers: the imposed tension and the friction between the fiber and the resin that is a function of matrix viscosity.

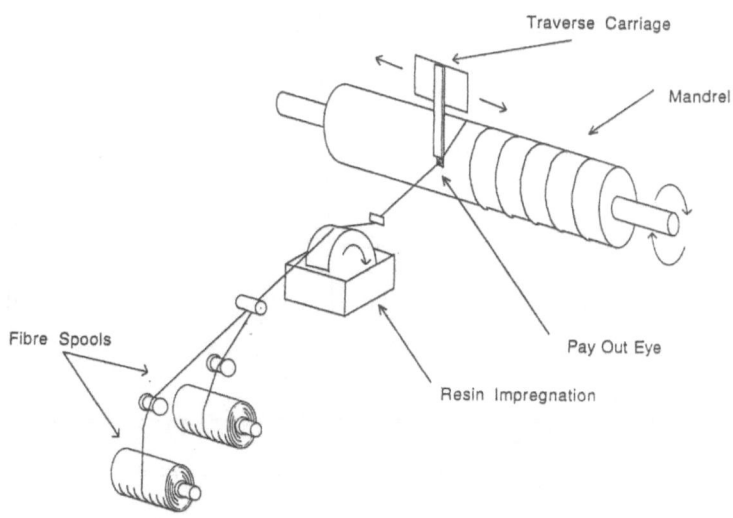

Figure 18: Schematic diagram of the filament winding

3.2.5   Physico-Chemical Behavior

Although the described technologies have very different characteristics, the behavior of the thermoset matrix can be described by applying the same fundamental principles that are discussed in this section. During the processing of thermoset based composites, shaping operations are accompanied by polymerization reactions (*curing*) and rheological changes of the matrix that strongly influence the final properties and the quality of the composite part. Moreover, the cure process is not only associated with significative variations of the material viscosity but is also coupled with a strong heat generation due to the exothermic nature of the thermosetting reactions. The relative rates of heat generation and transfer determine the values of the temperature, and therefore, the values of the advancement of the reaction and the viscosity through the thickness of the composite.

An uncontrolled polymerization may cause undesired and excessive thermal and

344

rheological variations that could induce microscopic defects in the network structure of the matrix phase such as different crosslinking densities and macroscopic defects such as voids, bubbles, and debonded and broken fibers (35, 41). Processing of polymeric composites based on thermoset matrices needs, therefore, optimization of the cure cycle parameters as well as adequate formulation of the reacting system as a function of the geometry of the part. The analysis presented in this review is referred to the production of carbon fiber/epoxy matrix laminates for aeronautic applications. However, the approach used is completely general and can be easily applied to other processing technologies for thermoset composites like resin transfer like molding, pultrusion, filament winding.

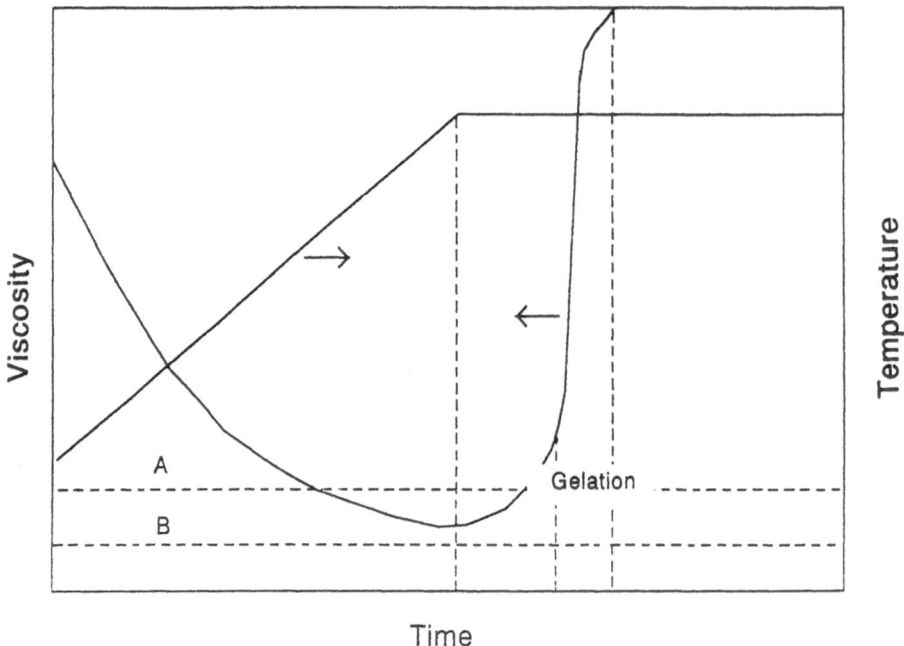

Time

Figure 19: Temperature and pressure as a function of time for a typical TGDDM-DDS prepreg cure cycle and the corresponding change in viscosity. Lines A and B corresponds to the maximum and minimum of viscosity that are allowed at the low viscosity part of the cure cycle. (After Kenny et al., ref. 2).

As discussed before the role of the matrix on the processing behavior of high performance composites fabricated in the autoclave process will be analyzed in detail, as a case study, in this section.
The principal physical events occurring during a typical cure cycle are illustrated in Fig. 19 (37,39,40). The viscosity initially decreases with time as the temperature is increased, but the reactions are still not activated. During this first stage, sorbed volatiles start to diffuse out of the resin. When the viscosity reaches its minimum trapped bubbles must be allowed to leave the composite. Pressure, which drives the resin flow, should be correctly applied in a limited range of viscosity in order to properly remove the excess resin and trapped bubbles, and to consolidate the plies. Figure 19 qualitatively shows the viscosity limits (A and B) that must be matched by the resin. The lower limit (B) is imposed by the flow characteristics of the system to

avoid an excessive loss of resin from the composite and to ensure the flow forces necessary to mobilize the bubbles. A too viscous liquid (upper limit A), on the other hand, does not allow a sufficient flow of resin and consolidation of the system. Other important physical events, such as void nucleation and volatiles diffusion and desorption, must be considered.

3.3 Modelling of Composites Processing

As mentioned before, in the early 1980's, research activities on process modelling started in the field of processing of thermosets and thermoset matrix composites [59-66]. The final objective of these activities has been the construction of a general processing model that could be adapted to different specific processes. In order to develop such general model several submodels are needed as shown in Fig. 20. The first submodel should describe the kinetics of the matrix chemical transformations, responsible of the final structure of the composite. The thermokinetic model predicts the exothermal heat of reaction and the degree of cure as a function of process time and temperature.

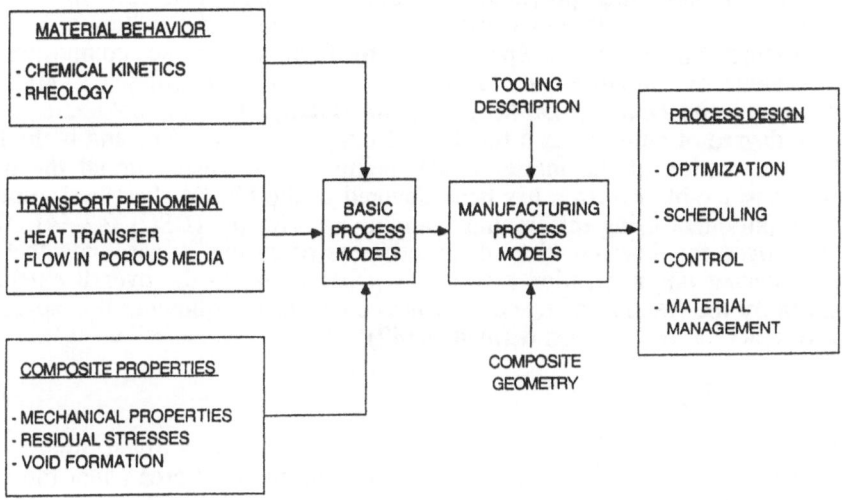

Figure 20: Schematic of the constituent submodels in a general composites processing model.

Also a rheological model to describe the viscosity evolution as function of time and temperature, is needed. Viscosity will also depend on the degree of cure. Therefore, the rheological model has to be combined with the thermokinetic model. Together, they form the chemorheological model allowing the prediction of the viscosity of the reactive matrix as a function of the degree of cure and of the temperature during the polymerization process. The chemorheological model includes also the prediction of the time required for the resin to reach the gelation point once the processing conditions have been fixed.

A flow model is the third submodel needed. This model should be able to predict resin content distribution and final composite thickness.

Since the final objective of this approach is concerned with the cured composite characteristics, void formation should also be included in the modelling effort. The void model should ideally be able to predict the conditions needed to avoid the formation of voids. In addition, it is also of interest the prediction of the volume fraction and of the size distribution of voids in the cured composite, for specified processing conditions.

For non-isothermal conditions also a heat transfer model is needed. If the heat transfer model is combined with the chemorheological and cure kinetics models, the degree of cure, temperature and viscosity, as a function of time and position in the composite, can be predicted.

A more detailed examination of the modeling of composite processing will be given in the following sections where each of the submodels mentioned in this section, integrated into a general master model, will be discussed.

3.4 THERMOKINETIC MODEL

3.4.2 Reaction Kinetics

In the cure of thermosetting polymers, a series of independent reactions occur, involving monomers and molecules composed of a sequence of two (dimers), three (trimers) and more structural units. Epoxy resins for high performance composites are commonly based on mixtures of *Tetra-Glicidyl Diamino Diphenyl Methane* (**TGDDM**) epoxy and *Diamino Diphenyl Sulfone* (**DDS**). The thermokinetic model describes the degree of cure, α, as a function of temperature and time and is the first step in the construction of the master model being a prerequisite for all the other submodels. Considerable research has been devoted to the kinetic characterization of thermosetting polymers using differential scanning calorimetry (DSC) (42-44). DSC has been employed for determination of the progress of curing by assuming that the heat evolved during polymerization reaction is proportional to the overall extent of reaction given by the fraction of reactive groups consumed. Following this approach the degree of reaction, α, has been defined as (42):

$$\alpha = H(t)/H_T \tag{13}$$

where H(t) is the heat developed between the starting point and a given time, t; and H_T is the total heat developed, calculated by integrating the total area under the DSC curve. The reaction rate is thus given by the following expression:

$$d\alpha/dt = 1/H_T \, dH/dt \tag{14}$$

where dH/dT is the rate of generation of heat as measured directly from the DSC thermogram. Although processing of thermosetting matrices involves very complex reactions, several simple equations have been proposed to describe their general behavior as an overall kinetic process in the form:

$$d\alpha/dt = K \, f(\alpha) \tag{15}$$

where K is the temperature-dependent rate constant and f(α) is a function to be determined by best fitting of the experimental results. The temperature dependence is normally considered through the rate constant K given by an Arrhenius type equation:

$$K = K_0 \exp(E_a/RT) \tag{16}$$

K_0 is the preexponencial factor (frequency factor), R is the gas constant and T the absolute temperature. Physically, K is a measure of the velocity of the reaction and E_a is the activation energy which magnitude depends on the operating chemical reaction mechanism and on catalyst chemistry.

On the other hand the form of f(α) depends on the particular reaction mechanism. The research work on reaction kinetics of thermosets has been extensively dedicated to the study of high performance epoxy systems. Kenny and Trivisano have recently proposed (44) a new model for the kinetic behavior of TGDDM-DDS systems that takes into account the later diffusion controlled effects, and have analyzed the correlation between isothermal and dynamic DSC results testing the kinetic model under the complex thermal conditions characteristic of the processing of epoxy-based composites. Incomplete reaction during isothermal processes is attributed to diffusion control owing to the loss of mobility of the reacting molecules within the developed network. Structural changes produced by polymerization are associated with an increase of the glass transition temperature, T_g, of the reactive polymer. When the T_g approaches the isothermal cure temperature, mobility is strongly reduced. When the system reaches vitrification, the reaction becomes diffusion controlled and eventually ceases, or perhaps decreases to a minimum value (44,45,46) and stops. The model can be modified mathematically to provoke zero prediction of the reaction rate at the vitrification point:

$$d\alpha/dt = K (\alpha_m - \alpha)^n \tag{17}$$

This clearly predicts the expected behavior: the reaction rate during an isothermal process will be zero when the degree of reaction equals α_m. The average value of the total heat H_T developed in dynamic tests was used as a reference to determine the final degree of reaction during the isothermal tests ($\alpha_m = H_i/H_T$). For modelling purposes, it is convenient to determine the behavior of α_m as a function of the isothermal test temperature. The linear dependence of α_m on T illustrated in Fig. 21 recalls the dependence between T_g and α for a reactive polymer. As discussed above, the T_g value reached by the polymeric matrix can be assumed to be of the same order as the isothermal test temperature. It has been found a simple linear dependence to express the empirical dependence of α_m on T:

$$\alpha_m = p T + q \tag{18}$$

The values of the kinetic parameters of the general model given by eqs. (17)-(18) are listed in Table 1. The ability of the model to represent the kinetic behavior has been tested by comparison with experimental results (44). Reaction rate data plotted as a function of time, obtained from isothermal DSC experiments and from eq. (18) predictions, are shown in Fig. 21. Theoretical curves (full lines) were computed using the average parameter values listed in Table 8. The good fit corroborates the soundness of the procedure used to formulate the model. Close fits can be seen also in Fig. 22, where the reaction rate is shown as a function of time for different heating

348

rates. Dotted lines correspond to DSC experimental results, full lines corresponds to model predictions from the values in Table 1.

TABLE 1: Kinetic parameters of the model represented by eqs. (17)-(18), obtained from the thermal characterization of a typical TGDDM-DDS matrix prepreg (44).

Hr (kJ/g)	Ea (kJ/mol)	ln Ko (1/s)	n	q	p (K^{-1})
456	62.4	10.4	1.07	-1.96	.00635

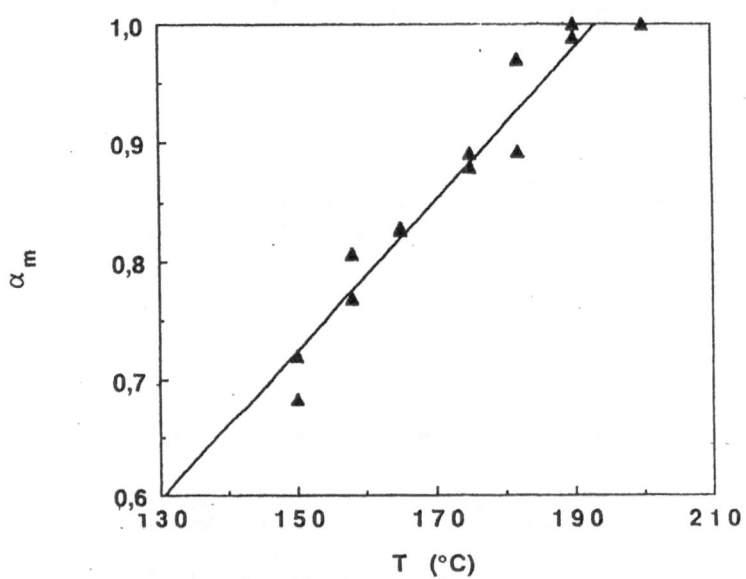

Figure 21: Maximum degree of cure of cure obtained in isothermal experiments vs. isothermal test temperature. (After Kenny et al., ref. 44).

However, the application of the model to industrial processes should also consider the complex thermal processes associated with the polymerization reaction as a consequence of the geometry of the laminate, and of the heat transfer from the air and tool to the prepregs. In this case, a simple scaling of DSC results is not sufficient and the kinetic model must be included in a general model considering the characteristics of the lamination process (2).

3.5. CHEMORHEOLOGY

The processing and final properties of thermosets depend on their composition as well as on the network structure generated before the gel point. The viscosity reflects the molecular distribution and can be considered one of the most important properties in polymer processing. An accurate predictability of the material properties, such as the polymerization kinetics and related changes in viscosities, implies the knowledge of the basic phenomena occurring during the overall process.

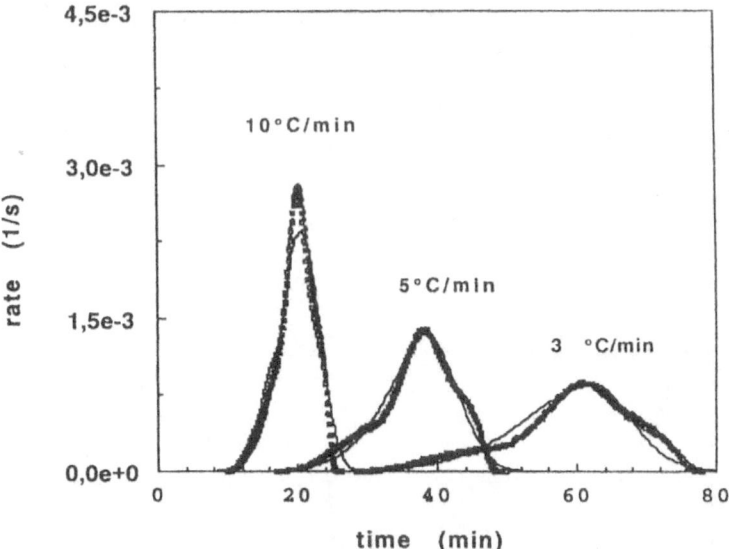

Figure 22: Reaction rate vs. time for different heating rate dynamic tests: comparison between experimental DSC data (points) and model predictions (full lines). (After Kenny et al., ref. 44).

From the physical point of view two major events must be analyzed in relation with their technological influence: gelation and vitrification. The gelation event is of great technological importance since after that no flow is possible and void diffusion and further consolidation of the composite can no longer occur. Gelation is a constant conversion event, meaning that irrespective of temperature, gelation always occurs at the same degree of cure that can be determined theoretically from Flory's branching theory (47), provided stepwise polymerization is the only mechanism and the functionalities of the reactive molecules are known.

The second physical event of great importance is vitrification. As of the curing reaction progresses, the mobility of the participating molecules is steadily decreased raising the Tg of the system. If the Tg approaches the curing temperature the reaction stops before completion.

The viscosity model should be able to predict the viscosity at any combination of temperature and time. Since the viscosity depends on the degree of cure, the viscosity model has to be combined with the thermokinetic model to really become useful. The combination of the thermokinetic and the viscosity model describes the chemorheological behavior of the thermoset.

The viscosity changes in a reacting system continuously heated up to the final cure temperature can be derived by considering both the temperature dependence of the viscosity, according to the WLF equation where the glass transition temperature of the system is a function of the degree of reaction and the molecular weight dependence of the viscosity. Then the final chemorheological model can be written in the following way:

$$\mu(T,\alpha)/\mu(To) = \left(\frac{gMw(\alpha)}{Mwo}\right)^{3.4} \frac{\exp\{C1(Tr\text{-}Tgo)/(C2+Tr\text{-}Tgo)\}}{\exp\{C1(Tr\text{-}Tg(\alpha))/(C2+Tr\text{-}Tg(\alpha))\}} \tag{19}$$

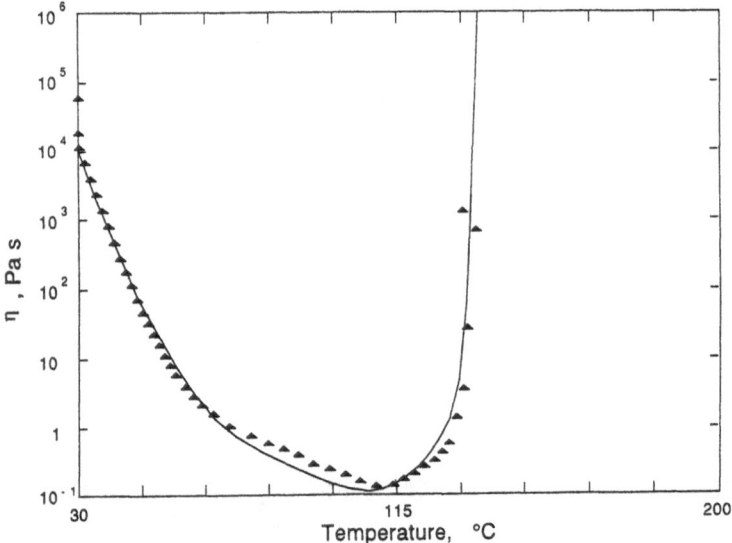

Figure 23:   Relative viscosity vs. time for a typical TGDDM-DDS system during a dynamic test: comparison between experimental data (points) and model predictions (full lines). (After Kenny et al., ref 2).

The rate of variation of the glass transition temperature with the degree of cure can be experimentally determined by differential scanning calorimetry. Theoretical models accounting for the influence of the degree of cure on the polymer glass transition can be also used (48):

$$(Tg(\alpha) - Tgo)/Tgo = K\,\alpha/(1 - \alpha) \tag{20}$$

where K is a constant reported to be between 1.0 and 1.2 (48).

Viscosity measurements are preferably performed in a dynamic mechanic spectrometer for fluids provided with disposable parallel plates. The location of the gelpoint can be determined from infinite viscosities obtained in constant shear or in dynamic tests (49). Figure 23 show model and experimental values of the viscosity as a function of time for a typical commercial epoxy system. Excellent agreement was found in the first part of the curve (unreacted resin) where the viscosity depends only on the temperature and is described by the unmodified WLF equation. Also, the minimum of the viscosity and the gelation limit are well described by the proposed model, which gives an important tool to predict and/or control the rheological behavior during processing of epoxy laminates. The difference between the numerical and experimental results can be mainly attributed to the approximations on the simple kinetic equation utilized.

For thermosets with complicated reaction mechanisms or where the composition and functionalities of the molecules in the resin mixture are unknown, an empirical approach is necessary. In order to describe the viscosity of polyester matrices Kenny et al. (3) adopted a model similar to the one originally used by Castro and Macosko for polyurethanes viscosity (50):

$$\mu(\alpha, T) = A_\mu \exp[E_\mu/RT] \left[\alpha_g/(\alpha_g-\alpha)\right]^{(A+B\alpha)} \tag{21}$$

where α_g is the extent of reaction at the gel point and A_μ, E_μ, A and B are constants that have to be determined by regression analysis of experimental data.

3.6. HEAT TRANSFER

For isothermal processing conditions, the models present so far will give all the information needed to predict the behavior of the material at a given processing temperature. However, as discussed earlier the cure of a thermoset is always associated with a significant development of heat. The temperature distribution inside the composite will depend on the competition between heat generation and heat diffusion through the thickness. The reacting system must be viewed as a non-isothermal bulk reactor with volumetric heat generation and transfer for the initial heating and for the dissipation of the heat of reaction. The temperature and degree of cure profile inside the composite can be computed taking into account the system geometry, the thermal diffusivity of the composite, and the resin reaction rate. This can be done by solving the energy balance together with an appropriate expression for the cure kinetics. The following assumptions are normally introduced:

1.- The simple case of absence of resin flow is considered and then only heat transfer by conduction is assumed.
2.- The laminate thickness is small compared to the other two dimensions. Then, only conduction of heat in the transverse direction is considered.
3.- The density ρ, the specific heat Cp and the thermal cnductivity k are computed as proper averages of the single resin and fiber property values.

Kenny et al. (2) introduced dimensionless numbers to facilitate the numerical solution and to generalize the model to other processing technologies. A characteristic time of the material behavior is defined: the isothermal gel time, t_g, represents the time interval in which the material is changing from the liquid to the rubbery state. Taking the imposed processing temperature Te as the reference temperature, tg is obtained by integrating the kinetic model. If eq. (51) is considered the following equation for t_g is obtained:

$$t_g = [(1-\alpha_g)^{(1-n)}-1]/[(n-1) \, A \, \exp(-E_a/RT_e)] \tag{22}$$

The dimensionless variables of the model are defined as:

$$\Theta = (T-T_0)/(T_e-T_0) \quad t^* = t/t_g \quad Z = z/(h/2) \tag{23}$$

The final equations then become:

$$\delta\Theta/\delta t = De \, \delta^2\Theta/\delta Z^2 + St \, d\alpha/dt^* \tag{24}$$

$$d\alpha/dt^* = A \exp[(E(\Theta-1)(T_e-T_0)/(\Theta+T_0)] (1-\alpha^n) \tag{25}$$

In eq. 24 De is a modified dimensionless diffusion Deborah number, defined in this case as the ratio of a time scale of the morphological changes occurring during reaction, to the characteristic time for diffusion. In the case of thermosets processing the time scale is characterized by the isothermal gel time t_g. The heat diffusion time scale is the square of the characteristic dimension of the laminate, h/2, divided by the heat diffusivity $k/\rho C_p$. Hence the Deborah number is expressed as:

$$De = K \, t_p/\rho C_p(h/2)^2 \tag{26}$$

In eq. 24 also the Stefan number St, is introduced:

$$St = H_T/[(T_e-T_0) \, C_p] \tag{27}$$

and may be considered as the relationship between the latent heat associated with the chemical reaction and the accumulation of heat in the material. Typical Stefan numbers for this process are on the order of 1 (2).
Finally, dimensionless kinetic constant (A) and activation energy (E) were used in eq. (25):

$$A = [(1-\alpha_g)^{(1-n)}-1)]/(n-1) \tag{28}$$

$$E = E_a/RT_e \tag{29}$$

3.6.1. Application of Processing Models

Numerical results reported (2) on a typical TGDDM-DDS matrix laminate, assuming that the prepregs are suddenly exposed to the cure temperature, are shown in Fig. 24 (a,b,c) as the variation of the temperature, degree of reaction and viscosity as a function of the processing time, both on the skin and on the core of the laminate. Input data of the full model are given in Table 9 (2). Due to the contribution of the thermal conductivity of the fibers the temperature at the center of the laminate rapidly reaches the external imposed temperature and increases as a consequence of the imbalance between the rate of heat generation and the thermal diffusivity of the composite (Fig. 24a). When these two quantities are comparable, the temperature profile reaches a maximum.
The values of temperature (Fig. 24a) and degree of cure (Fig. 24b) have been used to compute the viscosity as a function of cure time by means of eq. 68 (Fig. 24c). The high temperatures developed and the coexistence of gelled and ungelled regions could induce, on cooling, undesired stresses and, therefore, the adiabatic like condition

should be avoided.

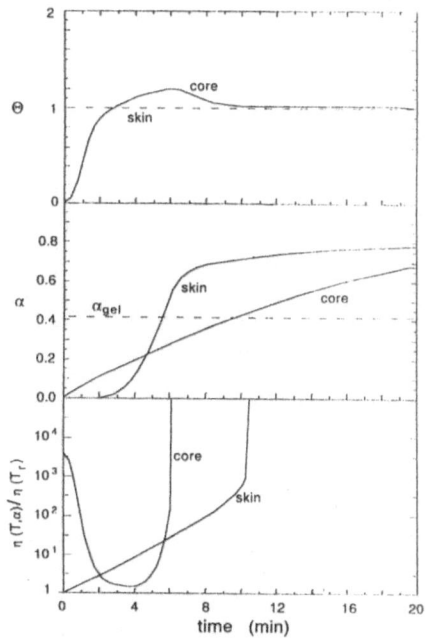

Figure 24: Dimensionless temperature (a), degree of reaction (b), and viscosity (c), vs, processing time, computed at the skin and at the center of a typical epoxy matrix/carbon fiber laminate of 7mm half-thickness, cured isotermally at 177 C. (After Kenny et al., ref. 2).

The processing of polyester matrix composites has also been modelled (3). In reference (3) the resin transfer moulding process (RTM) was modelled and the predictions were experimentally verified. Different from epoxy based materials, in this case the polyester resin gelled from the skin to the core as shown in Figure 25.

3.3. CONCLUDING REMARKS

This review has described the general features of models for composites processing. A more detailed treatment has been given presenting the individual submodels that are integrated into a general processing model. The strong influence of processing conditions on composite laminate structure and properties as well as the economy of the composites product is not apparent. Recent work has demonstrated how the use of a scientific approach can be a valuable tool for the solution of practical processing problems such as the choice of matrix characteristics and processing conditions.

Recently developed processing models are based on an understanding of the physics and chemistry ruling the behavior of the polymeric matrix. In order to solve practical processing problems, an approach is suggested which combines resin characterization and modelling with experiments in the actual processing environment.

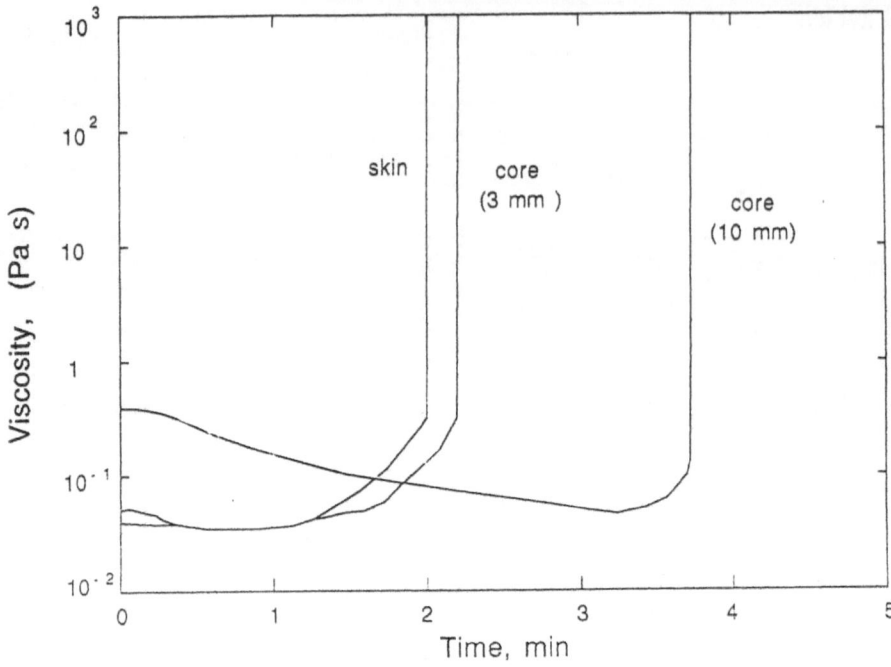

Figure 25: Viscosity vs. time at the skin and at the core for two glass fiber/polyester composites with different thickness cured in a mold at 60°C. (After Kenny at al., ref 3).

TABLE 2: Parameters of the thermo-chemo-rheological model
of the autoclave lamination process (2)

Heat of reaction (J/g)	H_T	473
Activation energy (KJ/mol)	E_a	138
Kinetic constant (s^{-1})	$\ln K_0$	30.0
Reaction order	n	1.7
1st constant of eq. (68)	C_1	40.5
2nd constant of Eq. (68) (K)	C_2	52
Epoxy molecular weight	M_{we}	422
Amine molecular weight	M_{wa}	218
Epoxy functionality	f_e	2
Amine functionality	f_a	4
Epoxy content	M_e	1
Amine content	M_a	1
T_g for the unreacted polymer(K)	T_{go}	273
Composite density (Kg/m^3)		1500
Thermal conductivity (W/mK)	k_x	0.4
Specific heat (J/g^{o}C)	C_p	1.67

4. REFERENCES

1. J.C. Seferis and L. Nicolais, "The role of the polymeric matrix in the processing and structural properties of composite materials", Plenum Press, NY, (1981).
2. J.M. Kenny, A. Apicella and L. Nicolais, *Pol. Eng. Sci.*, 1989, **29**, 973
3. J.M. Kenny A. Maffezzoli, L.Nicolais, *Compos. Sci. Tech.*, 1990, **38**, 339.
4. J.M. Kenny, A. Maffezzoli, *Pol. Eng. Sci.*, 1991, **32**, 607.
5. I.M. Ward, "Mechanical Properties of Solid Polymers", Wiley Interscience, London, 1971.
6. L.E. Nielsen, "Mechanical Properties of Polymers and Composites", M. Dekker, New York, 1974.
7. T. Murayama, "Dynamic mechanical analysis of polymeric material", Elsevier, NY, 1978.
8. A. Maffezzoli, J.M. Kenny, L. Torre, L.Nicolais, in "Proceedings of the 11th SAMPE European Chapter Conference, Basilea, Switzerland, 1990", p. 307.

9.   A. Maffezzoli, J.M. Kenny, L. Nicolais, in "Proceedings of the 49th SPE-ANTEC, Montreal, 1991", p. 2079.
10. J.M. Kenny and L.Torre, in "Proceedings of the 49th SPE-ANTEC, Montreal, 1991", p. 2108.
11. R.J. Diefendorf, in "Carbon fibers and their composites" ed. E. Fitzer, Springer-Verlag, New York, 1985.
12. K.K. Chawla, "Composite Science and Engineering", Springer-Verlag, New York, 1987.
13 L. Nicolais, *Polym. Eng. Sci.*, 1975, 15, 137
14 J.E. Ashton, J.C. Halpin and P.H. Petit, "Primer on Composite Materials: Analysis", Technomic Publishing Co., Stamford, Conn., 1984.
15 S.W. Tsai, J.C. Halpin and N.J. Pagano, "Composite Materials Workshop", Technomic Publishing Co., Stamford, Conn., 1968.
16 J.C. Halpin and L. Nicolais, *Ing. Chim. Ital.*, 1971, 7, 173.
17 R. Hill, *J. Mech. Phys. Solids*, 1964, 12, 199.
18 J.J. Hermans, "Proc Konigl. Nederl, Akad, Weteschappen Amsterdam", 1970, B70, 1.
19 J.C. Halpin and N.J. Pagano, *J. Comp. Mat.*, 1969, 3, 720.
20 J.C. Halpin, *J. Comp. Mat.*, 1969, 3, 732.
21 V.R. Riley, *J. Comp. Mat.*, 1968, 2, 4366.
22 K.L. Jerina, J.C. Halpin and L. Nicolais, *Ing. Chim. Ital.*, 1973, 9, 94.
23 P.E. Chem, *Polym. Eng. Sci.*, 1971, 11, 51.
24 R.M. Barker and T.F. MacLaughlin, *J. Comp. Mat.*, ,1971, 5, 942
25 R.P. Kambour, *J. Polym. Sci.*, 1965, A-2, 3, 1713.
26 R.E. Lavengood, L. Nicolais, and M. Narkis, *J. Appl. Polym. Sci.*, 1973, 17, 1173.
27 G. Caprino, J.C. Halpin and L. Nicolais, *Composites*, 1979, 4, 223
28 P.H. Petit and M.E. Waddoups, *J. Comp. Mat.* 1969, 2, 3.
29 J.C. Halpin and J.L. Kardos, *Pol. Eng. Sci.*, 1978, 18.
30 M.E. Waddoups, J.R. Eisenmann and B.E. Kaminski, *J. Comp. Mat*, 1971, 5, 446.
31 J.C. Halpin, K.L. Jerina and T.A. Johnson, "Analysis of the Test Methods for High Modulus Fiber and Composites", ASTM STP, 1973, 521, 5.
32 M.E Waddoups and J.C. Halpin, *Computers and Structures*, 1974, 1, 4.
33 J.M. Whitney and R.J. Nuismer, *J. Comp. Mat.*, 1974, 8, 253.
34 R.J. Nuismer and J.M. Whitney, "Fracture Mechanics of Composite", ASTM STP, 1975, 593, 117.
35 J.C. Halpin, G.I.. Kardos and M.P. Dudukovic, *Pure & Appl. Chem.*, 1983, 55, 893
36 A.C. Loos and G.S. Springer, *J. Compos. Mater.*, 1983, 17, 135.
37 J.C. Halpin, A. Apicella and L. Nicolais, in "Polymer Processing and Properties", G. Astarita and L. Nicolais eds., Plenum Press, New York, 1984, p. 143.
38 R. Dave', J.C. Kardos and M.P. Dudukovic, *Polym. Compos.*, 1987, 8, 29.
39 J.M. Kenny, A. Trivisano and L.A. Berglund, *SAMPE J.*, 1991, 27(2), 39.
40 P.K. Mallick and S. Newman Eds., "Composite Materials Technology, Processing and Properties", Hanser Publishers, Munich, 1990.
41 J.L. Kardos, M.P. Dudukovic and R. Dave', in "Advances in Polymer Science, 80: Epoxy Resins and Composites IV", K. Dusek, Springer Verlag, Berlin, 1986, p. 101.
42 R.B. Prime R.B., in "Thermal characterization of polymeric materials", E.A.

357

Turi ed., Academic Press, New York, 1981, chap. 5.

43   J.M. Barton, *Makromol. Chem.*, 1973, **171**, 247
44   J.M. Kenny and A. Trivisano, *Polym. Eng. Sci.*, in press
45   J.B. Enns and J.K. Gillham, *J. Appl. Polym. Sci.*, 1983, **28**, 2567
46   K.P. Pang and J.K. Gillham, *J. Appl. Polym. Sci.*, 1990, **39**, 909.
47   P.J. Flory, "Principles of polymer chemistry", Cornell Univ. Press, Ithaca, New York, 1953.
48   A.T. Di Benedetto, *J. Appl. Polym. Sci.*, 1987, **25**, 1949
49   A. Apicella in "Developments in Reinforced Plastics -5", G. Pritchard ed., Elsevier, New York, 1986, p. 151.
50   J.M. Castro and C. Macosko, *AIChE J.*, 1982, **28**, 251

REINFORCING FIBERS FOR COMPOSITES

J. P. WIGHTMAN
Chemistry Department
Center for Adhesive and Sealant Science
Center for Composite Materials and Structures
NSF Science and Technology Center
Virginia Institute for Materials Systems
Virginia Polytechnic Institute and State University
Blacksburg, VA 24061 U.S.A.

ABSTRACT. The paper focuses on carbon fibers including carbon fiber preparation, the physical properties of carbon fibers, functional groups present on carbon fiber surfaces and the relationship of surface chemistry to composite properties. Specific topics include thermal treatment of PAN-based carbon fibers, carbon fiber structure, tensile breaking strength and modulus, surface area and surface energy, XPS analysis, and chemical derivatization.

1. Introduction

Carbon fibers produced from polyacrylonitrile (PAN) precursor are finding increased usage in fiber-reinforced plastics [1]. Despite their present relatively high cost, carbon fibers are used where weight savings are more important than cost considerations. As new applications for carbon fiber composites are being found, new demands of the composite mechanical performance are occurring. These demands are resulting in a wider choice of carbon fiber and polymeric matrix mechanical properties.

The first generation of composites using carbon fibers was made with thermosetting resins such as epoxies [2]. It is now becoming apparent that these composites are too brittle for many current design applications [3-5]. Recent trends in composite development are toward composites that can withstand impact loads and still function properly. These demands in composite performance are being met by improving the toughness of the matrix resin.

Methods used to increase matrix toughness have included modifying existing epoxy formulations by adding a second phase (such as rubber or a thermoplastic resin) that can absorb energy [3-7], using thermoplastic resins [8-11], and depositing a ductile material on the fiber surface [12-14].

359

G. Akovali (ed.), *The Interfacial Interactions in Polymeric Composites*, 359–385.
© 1993 *Kluwer Academic Publishers.*

Although the newer materials being used as matrix materials have increased toughness, it is difficult to predict the mechanical properties of a composite based on the properties of the individual components along [11,15]. The main reason for this is that the interaction of the fiber with the matrix also has an important effect on the mechanical properties of composites [15-17]. The interaction between fiber and matrix includes adhesion and wetting as well as the effect of the fiber on the morphological characteristics of the polymer.

Adhesion between fiber and matrix can be altered by surface treating the fiber [16,17]. Surface treatments of carbon fibers have been developed for epoxy systems, but the optimum surface treatment for epoxy systems may be inadequate for newer resin systems. It may also be possible that by tailoring the interface between fiber and matrix, the mechanical properties of the composite can be controlled. To tailor the interface, it is necessary to understand the nature of carbon fiber surfaces and their reactions when surface treated.

This paper reviews carbon fiber preparation and discusses physical properties of carbon fibers. Emphasis is placed on functional groups present on carbon fiber surfaces and the relationship of these functional groups to adhesion in polymeric matrices.

2. Carbon Fiber Preparation

Most of the presently available carbon fibers are synthesized from polyacrylonitrile starting materials. Although several other precursors do exist such as rayon and pitch [18], PAN precursor fibers have the best mechanical properties for structural applications. The technology of carbon fiber synthesis is protected very strongly by carbon fiber producers. However, the basic chemistry of carbon fiber synthesis is known. A brief review is included here.

2.1. THERMAL TREATMENTS FOR PAN-BASED CARBON FIBER SYNTHESIS

The processes involved in the synthesis of carbon fibers from PAN have been outlined by Deifendorf and coworkers [19,20]. These processes include spinning of PAN into fiber form, oxidizing the fiber at 200-300°C and carbonizing the fiber at 1000-2500°C in an inert atmosphere, surface treating, and sizing. The strength, modulus, and structure of the fiber can be controlled by stretching the fiber during the process as well as by changing the heating rates, the extent of oxidation, and the final carbonization temperature.

The chemical changes occurring during carbon fiber formation from PAN have been reviewed by Watt [21] and by Goodhew et al. [22]. Coleman and coworkers [23,24] have proposed a scheme of the chemical changes that occur during oxidation of PAN at 200°C (Fig. 1). The first step is cyclization within the polymer backbone to form a ladder structure. This ladder structure stabilizes the polymer for heating to higher temperatures. The polymer is stretched during cyclization to maintain alignment of the polymer molecules in the fiber direction. The ladder structure is then oxidized.

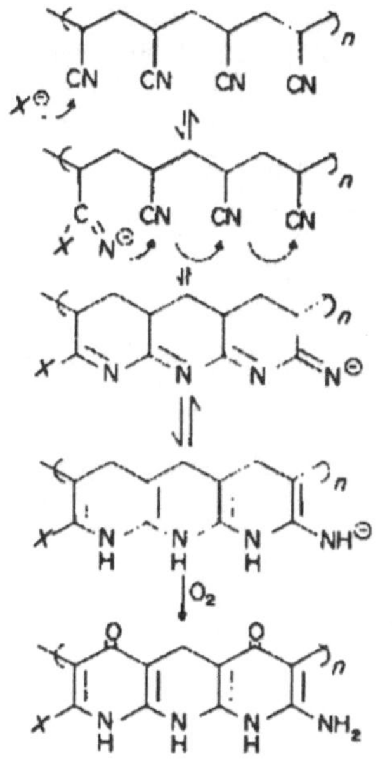

Figure 1. Presently accepted mechanism for cyclization and oxidation of polyacrylonitrile. (Reprinted with permission from Ref. 23. Copyright 1981 Pergamon Press.)

The chemical changes occurring during carbonization of the fiber are shown in Figure 2. Although the fiber is carbonized between 1000 and 2500°C, reactions begin to occur at much lower temperatures as the fibers are being heated to the carbonization temperature. At 400-600°C, the cyclic molecules begin to link together, resulting in loss of hydrogen and probably oxygen. This is followed by nitrogen loss and further linking at 600-1300°C to form graphitic sheets.

2.2. CARBON FIBER STRUCTURE

After carbonization, the carbon is in a sheet form as shown in Figure 3. The carbon is in an sp^2 hybrid state. There is an unbonded electron in an orbital perpendicular to the graphite plane. The unbonded electron coupled with unbonded electrons from adjacent carbon atoms will cause the formation of the conduction band of electrons between the carbon layers. The structure shown in Figure 3a is an idealized model for the molecular structure of graphite carbon [25]. However, the carbon in a carbon fiber will contain some discontinuities [25]. Figure 3b represents an imperfect graphite sheet which is probably more typical of the structure of carbon fibers. The carbon fiber is made up of many sheets that will coalesce to form aggregates similar in structure to the graphite unit cell. Since the carbon sheet is imperfect, the ideal graphite structure shown in

362

Figure 4a cannot be formed. Instead, we find a disordered crystal structure, the turbostratic structure, as shown in Figure 4b [26]. The turbostratic crystals can form aggregates of 0.03 μm (several hundred angstroms) in size. The layers of carbon twist and undulate along the length of the fiber as shown in Figure 5 [27]. Between the ordered areas there are areas of amorphous carbon [28].

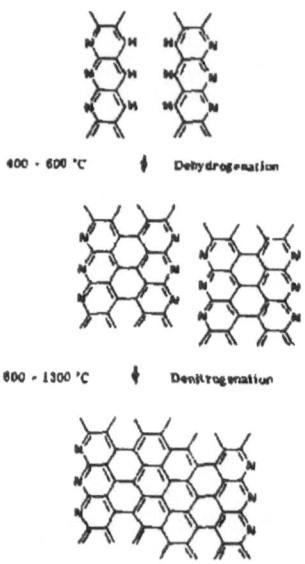

Figure 2. Proposed mechanism for carbonization of cyclicized PAN fiber into aromatic carbon sheets. (Reprinted with permission from Ref. 22. Copyright 1975 Elsevier).

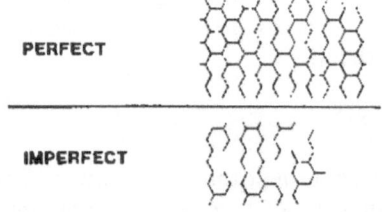

Figure 3. Schematic diagrams of (a) perfect graphite sheets and (b) imperfect sheets, which are more indicative of the structure in carbon fibers [25].

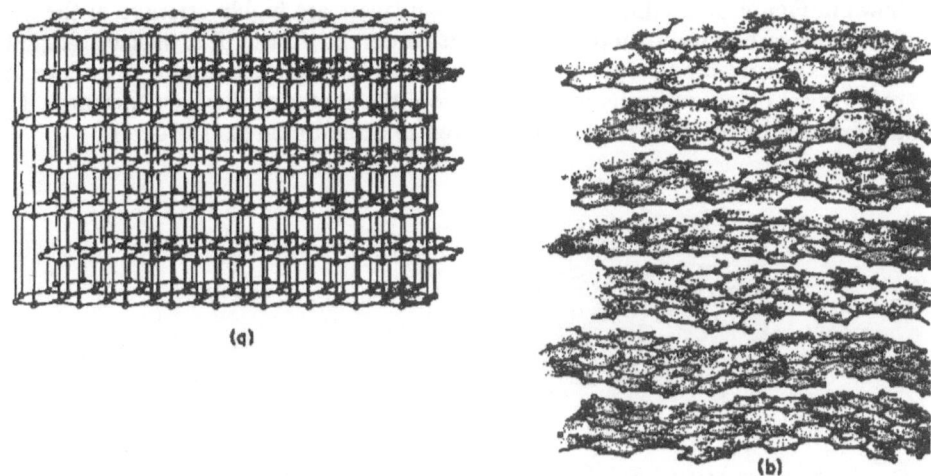

(a)

(b)

Figure 4. Structural models of graphite and carbon structure: (a) perfect graphite crystal and (b) turbostratic model. (Reprinted with permission from Ref. 26. Copyright 1969 Marcel Dekker).

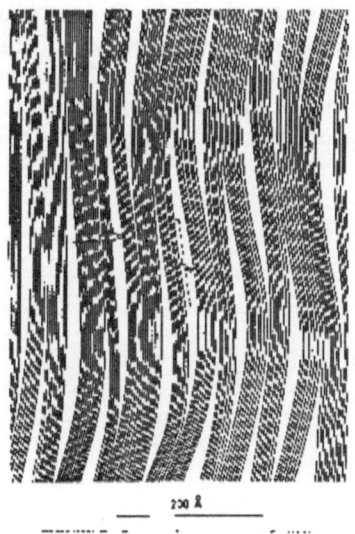

200 Å

Figure 5. Proposed structure of PAN-based and carbon fibers showing layers undulating in and out of crystalline regions: L_c is the width and the turbostratic crystals and L_{ap} is the length of the crystal. (Reprinted with permission from Ref. 27. Copyright 1970 International Union of Crystallography).

The structure from the core to the surface of the fiber varies also. A model for the PAN-based carbon fiber structure proposed by Diefendorf and reported by Drzal [29] is shown in Figure 6. In this model, the carbon layers are highly oriented at the fiber surface. The carbon layers in the core are less ordered. At the fiber surface, graphitic basal planes are oriented perpendicular to the outer fiber surface. In the fiber core, the graphitic basal planes are oriented radially from the center of the fiber outward. This model for carbon fiber morphology is referred to as an onion skin structure.

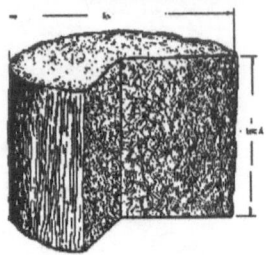

Figure 6. Proposed structure of carbon fiber morphology showing higher order at the fiber surface. Outer layers have basal planes oriented normal to the fiber surface. (Reprinted with permission from Ref. 29. Copyright 1977 Pergamon Press.)

The degree of order of the fiber surface was shown by Bennet and others [30,31] to depend on the carbonization temperature. The sketches of Figure 7 represent longitudinal sections of carbon fibers examined by Bennet with the transmission electron microscope (TEM). The structural order of the fiber surface increases with increasing carbonization temperature. Thus fibers formed at higher temperatures adhere to polymeric resins with difficulty.

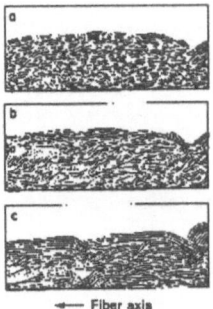

Figure 7. Sketches of TEM images of carbon fiber surface after carbonization at (a) 1000°C, (b) 1500°C, and (c) 2500°C. The crystalline order increases with carbonization temperature [30].

2.3. CARBON FIBER SURFACE TREATMENT

After the carbon fibers come out of the carbonization furnace, they are surface treated. This surface treatment serves several purposes: (a) to remove the outer layer of the carbon fiber surface, which is believed to be disordered carbon and of low shear strength, and (b) to oxidize the fiber surface, thus fixing on the fiber surface functional groups that will promote adhesion to the polymer matrix used for making composites.

Possible surface treatment mechanisms include anodization [32-34], plasma and flame treatment [35], solution oxidation [36,37], gas phase oxidation, and high temperature oxidation. Some of these treatments have been reviewed by Donnet and coworkers [18,38]. The most practical surface treatment for commercial production of carbon fibers is anodization. This is because anodization (electrolytic oxidation) can be performed continuously on carbon fibers. Typical anodizations have been performed in aqueous acidic or basic solutions. Electrolytes include sodium hydroxide, potassium hydroxide, sulfuric acid, nitric acid, and solutions of amine salts. Amine salts have an added advantage in that, after treatment, excess electrolyte can be removed simply by heating the fiber to high temperatures (250°C).

2.4. CARBON FIBER SIZING

After surface treatment, the fibers are heated to remove from the fiber surface volatile materials that would otherwise create voids in the composite during high temperature processing. The fibers are sized to protect the fiber surface from surface damage during handling and to protect the surface chemistry created by the surface treatment [39].

3. Physical Properties of Carbon Fibers

3.1. TENSILE BREAKING STRENGTH AND MODULUS

The tensile modulus E and strength σ_B of carbon fibers are shown as a function of carbonization temperature in Figure 8 [21]. The carbon fiber modulus increases with increasing carbonization temperature. This increase in modulus is caused by increased graphitization of the carbon at higher temperatures, since the more perfect graphite has a higher modulus than the less ordered carbon sheets.

Carbon fibers formed at higher temperatures (>2000°C) are referred to in the literature as high modulus or type I fibers. Fibers formed at lower temperatures (1000-1600°C) are referred to as low modulus or type II fibers. Recent developments in carbon fiber synthesis have resulted in carbon fibers with a tensile modulus intermediate between type I and II but with a tensile strength similar to type II [25]. These newer fibers have been called "intermediate modulus." Many improvements in the processing of PAN fibers are being made. A wide range of mechanical properties is available for specific design applications.

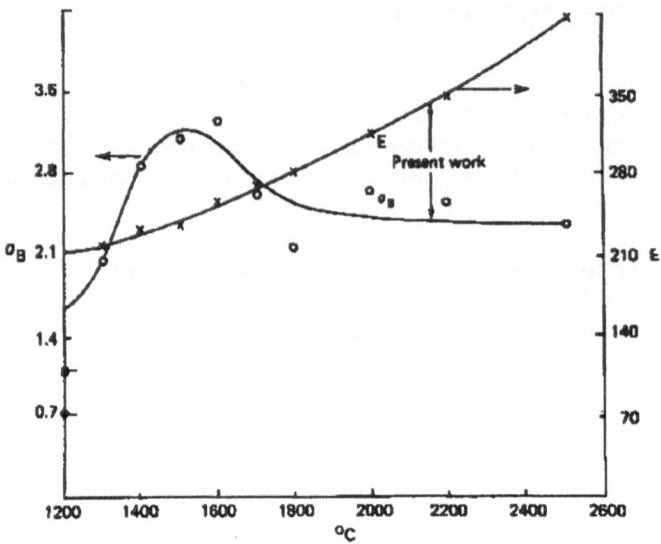

Figure 8. Dependence of PAN-based carbon fiber tensile strength σ_B and modulus E on carbonization temperature. (Reprinted with permission from Ref. 21. Copyright 1985 Elsevier.)

The tensile breaking strength of carbon fibers is dominated by flaws within the fiber and on its surface. Some of the flaws that can affect the fiber strength include:

Discontinuities in the crystal or fibrillar structure of the fiber.
Variations in the thickness of the onion skin layer.
Variations in the overall thickness of the fiber.
Microscopic impurities in the precursor.
Surface defects due to handling and processing.

Since the breaking strength of the fiber is controlled by the presence of flaws, it is expected that as the length of the fiber decreases, the breaking strength will increase. The dependence of strength on length is expected, since there is a lower probability of encountering a defect in the shorter fiber.

Surface treatment has been observed to change the breaking strength of carbon fibers. Bahl et al. [40] and Fitzer and Weiss [41] have observed that treatment of carbon fibers in nitric acid initially increases the fiber tensile strength. Continued anodization results in a loss in strength caused by fiber damage. This initial increase in strength can be explained by removal from the fiber surface of defects, which can initiate fracture.

3.2. SURFACE AREA

A key material property of powders and fibers in particular is the specific surface area. The extent of adsorption from the vapor and liquid states on a solid surface is determined in part by the specific surface area of the solid. Typically, to determine the specific surface area, a gas adsorption isotherm is measured; for example, the adsorption of nitrogen is measured on the substrate of interest at 77°K, the boiling point of nitrogen. The experimental isotherm is then analyzed by the BET (Brunauer, Emmett and Teller) model [42,43] to determine the monolayer capacity of the substrate. The specific monolayer capacity multiplied by the cross-sectional area of the adsorbed gas molecule gives the specific surface area. Amorphous silica gel may have a specific area of 200-300 m^2/g while carbon fibers may have a value of around 0.1 m^2/g.

3.3. SURFACE ENERGY MEASUREMENT

There are many techniques for probing the chemical and physical properties of a solid surface to predict the bonding of organic polymers to solid surfaces. The electronic structure of solid surfaces has been studied by measuring the thermodynamic interaction of the solid surface with simple liquids of known molecular structure. Experimental techniques for measuring the thermodynamic interaction between solid and liquid include contact angle measurement, calorimetry, and gas chromatography. Some of these techniques are discussed below. Specific techniques related to characterization of carbon fiber surfaces are also discussed.

3.3.1. *Contact Angle Measurement*. When a liquid drop is placed on a solid surface, the liquid will either spread on the surface or form a drop. Between itself and the solid, this drop will have an angle that is indicative of the interaction between the two materials [44]. In addition, the liquid will have a vapor pressure with which the solid surface will be in equilibrium. The forces in the drop are balanced as shown in Figure 9 [45]. These forces include the tendency of the drop to minimize its surface area by forming a sphere, and the tendency to spread on the solid surface and thus increase the extent of interfacial contact. This balance of forces has been described by the Young equation:

$$\gamma_{sl} - \gamma_{sv} + \gamma_{lv} \cos\theta = 0 \tag{1}$$

where γ_{sl} is the surface energy between solid and liquid, γ_{sv} is the surface energy between solid and vapor, γ_{lv} is the surface energy between liquid and vapor, and Θ is the angle of the drop between solid and liquid. By measuring the angle between the liquid drop and the solid surface, the interaction between solid and liquid (γ_{sl}) can be estimated.

Adhesion is defined thermodynamically by the change in free energy when two materials come into contact. The work of adhesion in the contact angle experiment has been defined [44] by Eq. (2).

$$W_a{}^T = \gamma_{lv}[1 + \cos(\theta)] \tag{2}$$

where W_a^T is the total work of adhesion.

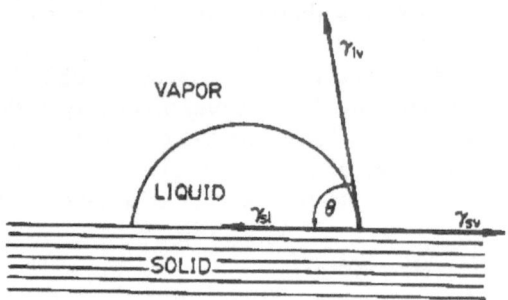

Figure 9. Diagram of contact angle experiment [45]: γ_{lv} = surface energy of the liquid-vapor interface, γ_{sl} = surface energy of the solid-liquid interface, and γ_{sv} = surface energy of the solid-vapor interface.

Girifalco and Good [46] assumed the interaction between a solid and a liquid could be quantified by an interaction parameter (ϕ) times the geometric mean of the surface tension of the solid and the liquid resulting in Equation (3).

$$\gamma_{sl} = \gamma_{sv} + \gamma_{lv} - 2\phi(\gamma_{sv}\gamma_{lv})^{1/2} \tag{3}$$

Fowkes [47] later postulated that the interaction energy due to wetting of solids by liquids with dispersion force interactions only, could be described by a geometric mean equation as shown in Eq. (4):

$$\gamma_{sl} = \gamma_{sv} + \gamma_{lv} + 2(\gamma_{sv}{}^d\gamma_{lv}{}^d)^{1/2} \tag{4}$$

where $\gamma_{sv}{}^d$ is the dispersion surface energy of the solid and $\gamma_{lv}{}^d$ is the dispersion surface energy of the liquid.

The interaction between solid and liquid due to polar groups has been considered by Fowkes [48] to be defined by acid-base interactions. In this model, Fowkes assumes that the interaction between two materials can be described by a component due to dispersion interactions in the form of a geometric mean relationship plus a component due to acid-base interaction. The acid-base interaction indicates the ability of a polar group on one surface to donate or accept electrons from polar groups on the other surface. The work of adhesion is then described by Eq. (5).

$$W_a{}^T = W_a{}^d + W_a{}^{ab} \tag{5}$$

where $W_a{}^d$ is the work due to dispersion forces, and $W_a{}^{ab}$ is the work due to acid-base interactions.

Kaelble et al. [49,50] have developed a technique to determine the polar and dispersion components of the surface energy of carbon fibers and other solid surfaces. In this technique, the contact angle of the fibers in several liquids of varying polar and dispersion components is measured. The work of adhesion W_a is assumed to be equal to the sum of the geometric mean of the polar components of the surface energies plus a geometric mean for dispersion surface energy components of the liquid and solid surface energies as shown in Eq. (6).

$$W_a = 2(\gamma_{sv}{}^d \gamma_{lv}{}^d)^{1/2} + 2(\gamma_{sv}{}^p \gamma_{lv}{}^p)^{1/2} = \gamma_{lv}[1 + \cos\theta] \tag{6}$$

where $\gamma_{sv}{}^p$ is the polar component of the solid surface energy and $\gamma_{lv}{}^p$ is the polar component of the liquid surface energy.

The polar dispersion components of the surface energy were calculated by dividing both sides of Eq. (6) by the square root of $2\gamma_{lv}{}^d$ as shown in Eq. (7).

$$\frac{\gamma_{lv}[1 + \cos\theta]}{2(\gamma_{lv}{}^d)^{1/2}} = (\gamma_{sv}{}^d)^{1/2} + \left(\frac{\gamma_{lv}{}^p}{\gamma_{lv}{}^d}\right)^{1/2} * (\gamma_{sv}{}^p)^{1/2} \tag{7}$$

The contact angle is measured in a series of liquids with varying polar and dispersion surface energy components. The components of the fiber surface energy can be determined by plotting the left-hand side of Eq. (7) as a function of $(\gamma_{lv}{}^p/\gamma_{lv}{}^d)^{1/2}$ of the liquid. The slope of this plot will be equal to $(\gamma_{sv}{}^d)^{1/2}$ of the solid. The intercept will be equal to $(\gamma_{sv}{}^d)^{1/2}$ of the solid.

If the surface energy is estimated for a polymer and for a solid using Kaelble's method, the work of adhesion between polymer and solid can be calculated using Eq. (6). However, it should again be noted that Fowkes [48] has argued that the geometric mean relationship to describe the polar group interaction between two materials may better be described by acid-base interactions.

3.3.2. *Contact Angle Measurement on Carbon Fiber Composites.* Contact angles of liquids on solids may be measured directly by use of a comparater microscope fitted with a goniometer scale [51]. Typically, contact angles are measured to ±3°, however, Neumann and co-workers [52] have described elegant experiments to increase the precision of contact angle measurements to ±0.3°. Contact angles of water measured against a carbon fiber/polyimide matrix composite before and after short time exposures to an oxygen plasma are shown in Figure 10 [53]. The contact angle is shown to decrease dramatically with only short exposures to an oxygen plasma. This enhanced wettability has been shown [53] to be due to a decrease in fluoropolymer surface contamination and an increase in surface oxygen functionality both brought about by plasma exposure.

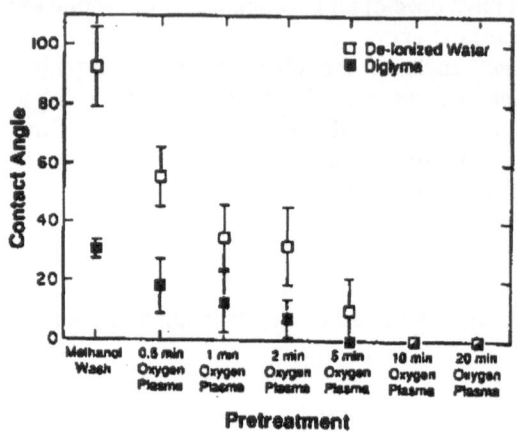

Figure 10. Effect of oxygen plasma pretreatment on contact angle [53].

3.3.3. *Contact Angle Measurement on Fibers.*

Since carbon fibers are so small, it is very difficult to measure the contact angle of a drop on a fiber. Nevertheless, several techniques have been developed to measure the contact angle of a drop on a small fiber under a microscope [54]. A simpler method for contact angle determination is to measure the wetting force of the fiber when it comes into contact with a liquid [49,50,55,56]. If the surface energy of the liquid is known, the contact angle of the liquid on the fiber can be calculated by:

$$F = \pi d \gamma_{lv} \cos\theta$$

(8)

where F is the wetting force and d is the fiber diameter.

Equation (8) assumes that the cross section of the fiber is circular. If the cross section is not circular, the πd term is replaced by the actual circumference of the fiber. The circumference may be determined by measuring the wetting force of the fiber in a liquid that completely wets the fiber [$\cos(\theta) = 0$] as described by Herb et al. [57].

3.3.4. *Inverse Gas Chromatography for Measurement of Solid-Vapor Interaction.*

The idea of putting carbon fibers in a gas chromatography column and passing probe molecules through the column to measure the fiber-liquid interaction was first used by Brooks and Scola [58]. Initial investigations using this technique, known as inverse gas chromatography (IGC), were inconclusive [59]. However, Schultz et al. [60] used IGC to show that the surface of carbon fibers obtained from Hercules, Inc., was acidic.

IGC measures the retention times of the probe molecules in the column. Molecules with a high adsorption enthalpy will take longer to pass through the column than molecules with a low adsorption enthalpy. If probe molecules of varying acid-base character are used, the acid-base properties of the fiber can be determined. This technique has also been used to determine the surface area of carbon fibers by using nonpolar probe molecules [61].

3.4.5. Calorimetric Measurement of Solid-Liquid Interaction. The thermodynamic interaction between a liquid and a solid can be measured using calorimetry. In this technique, a solid and a liquid are brought into contact in a cell with a sensitive heat detector. As the liquid wets the solid, heat is generated, which is detected by the sensitive device. This technique has been used by Rand and Robinson [62] to measure the heats of wetting of carbon fibers in acidic and basic liquids. It was found that basic probes gave a much higher heat of wetting, indicating an acidic fiber surface. Since the surface areas of carbon fibers are low, the amount of heat generated is low, and precise measurement is difficult.

4. Detection of Functional Groups on Carbon Fiber Surfaces

Some of the functional groups expected on carbon surfaces are carboxylic acids, phenols, quinones, lactones, ethers, peroxides, and esters (Fig. 11). Functional groups on carbon surfaces have been detected by such methods as polarography [63], titration [63-68], X-ray photoelectron spectroscopy [34], radioisotope labeling [34], and infrared spectroscopy [63]. Some of these techniques are discussed below.

4.1. TITRIMETRIC METHODS FOR CARBON FIBER SURFACE FUNCTIONALITY DETERMINATION

Functional groups on carbon surfaces have been identified by reacting the material with reagents that will react with specific functional groups on the carbon surface. The amount of reagent reacted is determined by titration. Several reviews of these techniques for analysis of carbon surfaces are available [63-68]. For these methods to be sensitive enough to detect the functional groups, either large surface areas or large quantities of material must be available.

4.2. X-RAY PHOTOELECTRON SPECTROSCOPY OF CARBON FIBER SURFACES

X-Ray photoelectron spectroscopy, commonly referred to as XPS or ESCA (electron spectroscopy for chemical analysis), uses the photoelectric effect to analyze the chemistry of solid surfaces [69]. In this technique (Fig. 12), a solid surface is exposed to nearly monochromatic X-rays. This exposure causes electrons to be ejected from the solid surface. These ejected electrons will have a spectrum of kinetic energies as they come from within the structure of the solid. At certain energies, the number of ejected electrons will increase. This peak in intensity is caused by ejection of electrons from atoms in the surface of the solid.

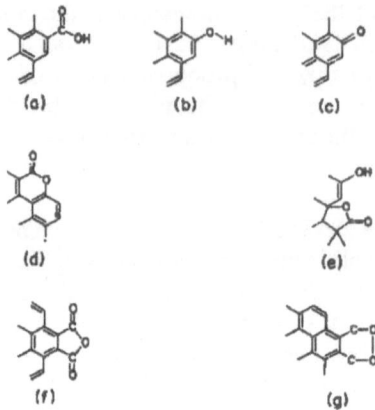

Figure 11. Predicted oxygen functional groups on carbon surfaces: (a) carboxyl, (b) phenolic hydroxyl, (c) quinone, (d) lactone, (e) fluoresceinlike lactone, (f) carboxylic acid anhydride, and (g) cyclic peroxide. (Reprinted with permission from Ref. 63. Copyright 1976 Marcel Dekker.)

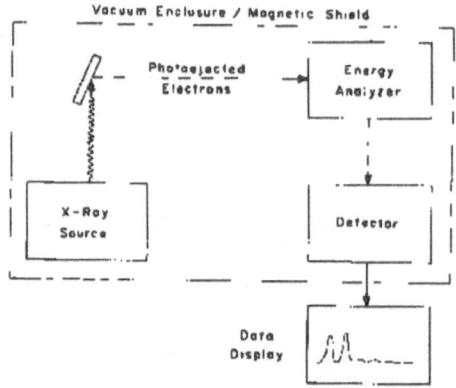

Figure 12. Schematic diagram of X-ray photoelectron spectroscopy experiment. (Reprinted with permission from Ref. 69. Copyright 1975 Elsevier.)

The energy of the electrons at the peak is indicative of the element present. Since the X-rays are nearly monochromatic, the kinetic energy distribution of these electrons will be narrow. The kinetic energy KE will be equal to the photon energy of the X-rays hv minus the binding energy of the electron in the atomic structure of the element BE minus a work function Ω:

$$KE = hv - BE - \Omega \qquad (9)$$

The intensity of the photoelectron signal for element i is proportional to the number of atoms on the solid surface N_i, the cross section of the atom to X-rays μ, the mean free path λ of the electrons in the solid (which is typically 5 nm), the X-ray energy flux F, and the geometric arrangement of the spectrometer (70). The relationship for the photoelectron peak intensity I_i for element i is:

$$I_i = \int_0^\infty FkN_i(x)\mu\exp-[\frac{d}{\lambda}]dx \qquad (10)$$

where k is a constant specific to spectrometer, x is the perpendicular distance from the surface into the sample, and d is the distance the electrons travel through the solid before exiting.

This relationship allows determination of the relative percentage of a given element on a surface from the relative peak intensities. Values for the atomic cross sections have been calculated by Scofield [71]. Empirical equations for calculating the electron mean free path have been developed by Cadman et al. [72]. Wagner [73] has determined sensitivity factors for each element to relate peak intensities to atomic concentration.

4.2.1. *Peak Shape Analysis*. If all the elements present on a solid surface were in the same bonding environment, we would expect the kinetic energy of the photoelectrons emitted from the solid surface to have a very narrow distribution. However, the actual width of the photoelectron peak is influenced by interaction of the ejected electrons with the sample as well as by the spectrometer itself.

The elements in most materials are not in just one binding state. The binding energy of electrons in the atom is influenced by the valence state of the atom. Thus the functionality of a solid surface often can be determined by observing shifts in the XPS peaks.

XPS photopeaks are typically fitted with Gaussian-shaped peaks. The peaks are assigned a width typical of the spectrometer being used and of the element being studies. The peaks are shifted in binding energy to represent the chemical environment of the element. The intensity of the fitted peak is proportional to the amount of that functional group present.

For a particular element, the number of peaks used to fit the photopeak is equal to the number of different functionalities expected for that element. Sometimes, the expected functionalities will have similar or overlapping binding energies. In this case, either several overlapping peaks will be used to fit the overall peak, or the width of the fitted peak can be increased.

Much work has been done to determine the binding energy shift caused by specific binding states [74]. The shifts in binding energy expected for carbon, oxygen, and nitrogen in organic materials are summarized below.

The photopeak from the carbon 1s atomic orbital has been widely analyzed. The peak due to C-C bonding occurs at a binding energy of 285.0 eV and is often used for instrument calibration. Many standard materials have been examined and the corresponding shift due to functional groups determined. Clark and coworkers have done numerous XPS analyses of polymer surfaces and have reported the carbon 1s binding energy shifts of many functional groups. Clark's results have been reviewed briefly by Briggs and Seah [75]. A basic trend [75] is that the R-O-C type of bond will shift the carbon 1s photopeak about + 1.5 eV, R-C=O bonds will cause about a + 3 eV binding energy shift, and $R—C\underset{\diagdown O—}{\overset{\diagup O}{}}$ bonds will cause about a + 4.5 eV binding energy shift.

Proctor and Sherwood [76] have studied the carbon 1s spectrum of carbon fiber and graphitic surfaces. They have shown that in addition to peak shifts due to functional groups, there is a peak at about + 6 eV from the main C-C peak due to interaction of ejected core electrons with plasmons from the conduction electrons in graphite. They also pointed out that certain functional groups on graphitic-type surfaces may occur at different binding energies due to aromaticity of the carbon structure to which the functional group is attached.

The binding energy shift for oxygen bound to carbon is less well defined than the shift of carbon. Most of the oxygen peaks fall in a narrow (2 eV) range centered around 533 eV. Oxygen doubly bound to carbon tends to have a lower binding energy than oxygen singly bound to carbon.

Most nitrogen associated with carbon also falls in a narrow region between 399 and 401 eV. Oxidized nitrogen shifts (6-8 eV) to higher binding energy. Clark and Harrison [75] have shown for polymers that a nitrogen binding energy of about 400 eV is due to amine groups, whereas a binding energy of about 401.5 eV is due to nitrogen bound to oxygen and/or nitrogen bound to carbon containing carbonyl groups.

4.2.2. *Derivatization*. Everhart and Reilley [77] have developed a systematic approach to the identification of functional groups on oxidized polymer surfaces. Here polymer surfaces were exposed to a series of reagents that reacted only with specific functional groups. The reagents chosen also contained an element that could easily be detected with XPS. Most of the reagents used by Everhart and Reilley were fluorine containing compounds.

This technique does have several drawbacks involving, for example, the specificity of the derivatizing reagent, the determination of the extent of the reaction, the determination of the number of functional groups that have reacted, and the stability of the reagent to X-rays. The question of whether reactions that occur in solution can be extended to a two-dimensional surface remains unanswered.

4.3. SECONDARY-ION MASS SPECTROMETRY

In secondary ion mass spectrometry (SIMS), a solid surface is bombarded with ions, causing ejection of ion fragments from the solid surface. The mass-to-charge ratios of the ejected ion fragments are analyzed in a mass spectrometer, permitting the investigator to infer the molecular and atomic structure of the solid. Trace elements also can be detected.

With high primary ion currents, material is removed from the solid surface rapidly and a depth profile is obtained. At low ion currents, the top few atomic layers of the surface are removed. Detailed information about the molecular structure of the solid surface is thus obtained.

This technique has been applied to the analysis of polymer surfaces by Briggs [78,79] and Brown and Vickerman [80]. Spectra obtained at low current observation led to detailed fingerprint spectra of polymer surfaces. Specific fragments could be assigned to either aliphatic or aromatic compounds.

One problem with analyzing polymer surfaces with SIMS is static charging caused by the primary ion current. In Briggs's work [78,79], the ion current was neutralized with an electron gun. Recent developments in the analysis of polymer surfaces by mass spectrometry have included a gun that will bombard the solid surface with neutral atom bombardment, referred to as fast atom bombardment mass spectrometry (FABMS), greatly reduces the static charging problem encountered in the normal SIMS experiment.

5. Adhesion Principles

Adhesion commonly refers to the potential for stress transfer across an interface between two materials [81]. In a fiber-reinforced composite, adhesion will result in stress transfer between fiber and matrix. The matrix thus acts to transfer stress between adjacent fibers. The adhesion between fiber and matrix will affect shear stress transfer in a composite. In addition, stress will be transferred from the ends of broken fibers to adjacent fibers through the interface and the matrix.

5.1. THEORIES OF ADHESION

To form an adhesive bond between two materials, it is necessary that they come into close molecular contact. One of the materials must be capable of flowing, wetting the other material, and solidifying [81-83].

The mechanical interlocking theory assumes that adhesion is due to irregularities on the surface into which the liquid material can penetrate. Upon solidification, the solid adhesive material is held in place by the geometry on the adsorbed layer. Mechanical interlocking is thus enhanced by increasing the surface roughness or porosity of the solid material.

The electronic attraction theory assumes adhesion to be caused by the electronic attraction between the atoms in the two materials being bonded. These forces of attraction will result from interaction of specific functional groups on the two surfaces as well as from nonlocalized electronic interactions due to the molecular structure of the materials being bonded.

For either of these mechanisms to be valid, the polymer used as the adhesive must form close contact with the solid. Huntsberger [82,83] has shown that the adhesive bond strength of polymethyl methacrylate to aluminum adherends depends on the temperature at which the bond was formed. This result was thought to be caused by inadequate molecular contact between adhesive and adherend at lower temperatures.

5.2. FORCES OF ATTRACTION ACROSS AN INTERFACE

The basic electronic forces that hold "homogeneous" materials together [84-85] include ionic bonding, dipolar interactions, covalent bonding, dispersion forces, metallic bonding, and hydrogen bonding. Ionic bonding results from the electrostatic attraction between oppositely charged ions. Dipolar bonding results from interaction of permanent dipoles within the material. Covalent bonding results from the formation of chemical bonds within the material. Dispersion force bonding results from attraction between local electron density fluctuations in the material caused by electron mobility. Metallic bonding results from attraction of metal ions to a sea of electrons. Hydrogen bonding is similar to ionic bonding and results from sharing of an adjacent hydrogen atom by two other atoms.

When two dissimilar materials are brought into contact, as is the case in an adhesive bond, the resulting electronic attraction can be caused by any combination of the interactions listed above. The attractive forces across the interface have been classified into two broad categories: dispersion and polar. These forces have been discussed by Atkins [86] and Wake [87]. The polar component results from electric dipoles associated with specific atom pairs or functional groups on the materials surface. The dispersion component results from loosely bound electrons such as those in the conduction band of metals or simply from electrons in the atoms or molecules in the material.

If the possibility of interdiffusion between the two materials does not exist, the interaction can be simplified. The attraction across an interface can thus result from dispersion interaction or dipole-dipole interactions. In addition, dipolar groups in one material can induce dipoles and thus create a dipole-induced dipole attractive force.

To predict and understand adhesion between two materials, it is necessary to understand the chemical and physical structure of the materials being bonded. A brief summary of adhesion tests for carbon fibers in polymeric matrices has been given [88].

6. Adhesion Between Carbon Fibers and Polymer Matrices

Carbon fiber surfaces are treated chemically to enhance bonding of the fiber to the resin in a composite. In surface treating, functional groups are created on the fiber surface. Many studies have been conducted to observe the effect of functional groups on carbon fiber-epoxy matrix adhesion [89-94].

Donnet and coworkers [89-91] have studied adhesion between epoxy resins and carbon acid. The number of acid groups on the fiber surface was used to neutralize strong acidic groups such as carboxyl and phenol. Sodium ethoxide was used to neutralize weaker acidic groups such as hydroxyl and carbonyl. A direct correlation was found between the number of carboxylic acid groups and the interlaminar shear strength of the composite (Figure 13).

Fitzer et al. [92] have studied the surface treatment of carbon fibers by boiling in nitric acid. By chemically blocking specific functional groups, they were able to determine which groups were most responsible for adhesion. Blocking of strong and weak acidic oxides resulted in a significant reduction of composite shear strength, leading to the conclusion that adhesion is caused by chemical bonding of the epoxy to acidic groups on the fiber surface.

In addition to functional groups being created by surface treatment, it is possible that the surface treatment will affect the molecular and morphological structure of the carbon fiber surface. Pores may be created which can enhance mechanical bonding. This has been best demonstrated by Drzal and coworkers [93,94], whose results are shown in Figure 14, where the interfacial shear strength of a carbon fiber-epoxy matrix bond is plotted against the surface oxygen content of the carbon fiber. The oxygen content is assumed to be indicative of the amount of functional groups on the surface.

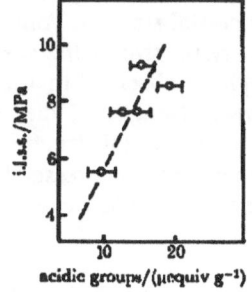

Figure 13. Effect of surface acidic groups on the interlaminar shear strength (i.l.s.s.) of carbon fiber composites. (Reprinted with permission from Ref. 89. Copyright 1980 Royal Society).

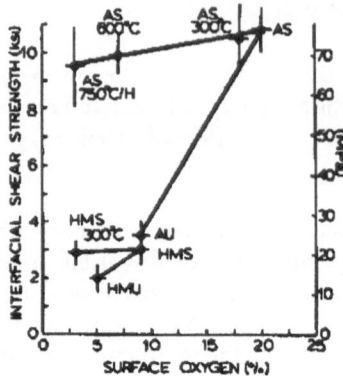

Figure 14. Effect of surface oxygen content on interfacial shear strength measured on single fiber: AU and HMU indicate no commercial surface treatment; AS and HMS indicate commercial surface treatment. Temperatures indicate heat treatment for removal of oxygen. (Reprinted with permission from Ref. 94. Copyright 1982 Gordon and Breach Science Publications.)

The AU fiber had no surface treatment and thus a low oxygen content and a low interfacial shear strength. The AS fiber had been surface treated commercially and thus had a high oxygen content and a high interfacial shear strength. Upon removal of the functional groups from the AS fiber, the oxygen content is greatly reduced, indicating loss of functional groups. However, the interfacial shear strength remains high. This indicates that adhesion promotion is not due solely to the addition of functional groups on the fiber surface, but also to the crystallinity, molecular structure, and/or possible of the fiber surface.

The crystallinity of carbon fiber surfaces has been studied with Raman spectroscopy [95-97]. The Raman spectrum of carbon fibers gives two peaks. One peak due to the graphitic nature of carbon fibers occurs at 1575 cm^{-1}. The other peak at 1355 cm^{-1} has an intensity inversely proportional to the graphite crystal size. By comparing the ratio of the 1355 cm^{-1} peak to the 1575 cm^{-1} peak, the crystal size at the surface of carbon fibers can be estimated. Tuinstra and Koenig [96] have observed that as the 1355 cm^{-1} peak decreases in intensity, the interlaminar shear strength of the composite also decreases. This indicates that bonding is affected by the crystal structure of the carbon fiber surface. Further evidence supporting this idea has been expressed by Brelant [98], who showed an inverse correlation between the thermal conductivity of the composite and its interlaminar shear strength.

To understand how the surface treatments enhance adhesion, it is necessary to understand how specific surface properties of carbon fibers. Considering the turbostratic carbon structure shown above (Fig. 4b), there are three possible locations for oxidation. Functional groups can be created at the crystal edges, between layers, or at the basal planes.

The surface treatments used by commercial producers of carbon fibers are proprietary. Indeed, much of the work on characterization of carbon fiber surfaces has been performed on fibers with proprietary surface treatments [93,94,99-103]. Although these studies show that oxygen and nitrogen functionalities are being added to the fiber surface by surface treatment, they do not give much insight into the reactions occurring during surface treatment. Nor do these studies begin to set a standard for surface treatment.

The most extensive study of carbon fiber surface treatments has been done by Sherwood and coworkers [104-108]. Some of their conclusions are outlined here. The functional groups created by anodization in sulfuric acid and in ammonium bicarbonate were dependent on the anodization potential [104]. Carbonyl (R-C=O) bonds were formed at low potential, while R-O-C bonds were formed at higher potentials [104].

Anodization in nitric acid yielded mostly $\diagdown$C=O functionality [106]. Anodization at high pH attacked primarily the edges of the graphitic planes, while anodization at neutral and low pH attacked between the carbon layers, creating a graphite oxide [107]. Nitrogen functionality can be created on the carbon fiber surface by anodization in ammonia-saturated ammonium bicarbonate solution [108].

7. Acknowledgements

The author wishes to acknowledge the efforts of Dr. T. A. DeVilbiss in compiling the thorough literature review on which this paper is based. The financial support of the NASA-Langley Research Center of work on which this review is based is appreciated.

8. References

1. M. S. Reisch, *Chem. Eng. News*, Feb. 2, 1987, p. 9.

2. J. Delmonte, *Technology of Carbon and Graphite Fiber Composites*, Van Nostrand, New York, 1981.

3. J. Diamont and R. J. Moulton, *SAMPE Q., 16(1)*, 13 (1984).

4. G. D. M. DiSalvo and S. M. Lee, *SAMPE Q, 14(2)*, 14 (1983).

5. J. Brandt and J. Warnecke, *Proceedings of the 7th Int. Conference of SAMPE, European chapter*, in K. Brunsch, H. D. Golden and D. M. Herkert, Eds., *High Tech-The Way into the Nineties*, Elsevier, Amsterdam, 1986, pp. 251-260.

6. A. F. Yee and R. A. Pearson, *J. Mater. Sci, 21*, 2462 (1986).

7. A. F. Yee and R. A. Pearson, *J. Mater. Sci, 21*, 2475 (1986).

8. D. J. Willats, *SAMPE J., 20(5)*, 6 (1984).

9. J. L. Kardos, *J. Adhes., 5*, 119 (1973).

10. T. W. Johnson and C. L. Ryan, *Proceedings of the 31st SAMPE Int. Symposium*, 1986, pp. 1537-1548.

11. D. L. Hunston, *Compos. Technol. Rev., 6*, 176 (Winter 1984).

12. J. H. Williams and P. N. Kousiounelos, *Fibre Sci. Technol., 11*, 83 (1978).

13. R. V. Subramanian, V. Sundram and A. K. Patel, *Proceedings of the 33rd Annu. Tech. Conf. SPI Reinf. Plast./Compos. Inst.*, 1978, 20F, 1.

14. R. V. Subramanian, J. J. Jakubowski, and F. D. Williams, *J. Adhes., 9*, 185 (1978).

15. L. Ying, *SAMPE Q, 14(3)*, 26 (1983).

16. S. Lehmann, C. Megerdigian, and R. Papalia, *SAMPE Q., 16(3)*, 7 (1985).

17. R. Robinson, S. Lehmann, G. Askew, D. Wilford, C. Megerdigian, and R. Papalia, *Proceedings of the 7th Int. Conference of SAMPE European chapter*, in K. Brunsch, H. D. Golden and C. M. Herkert, Eds., *High Tech-The Way into the Nineties*, Elsevier, Amsterdam, 1986, pp. 299-310.

18. J. B. Donnet and R. C. Bansal, *Carbon Fibers*, Dekker, New York, 1984, Ch. 1.

19. R. J. Diefendorf, W. C. Stevens, and S. H. Chen, *Fiber Producer, 7(6)*, 16 (1979).

20. R. J. Diefendorf and E. Tokarsky, *Polym. Eng. Sci., 15*, 150 (1975).

21. W. Watt, in W. Watt and B. V. Perov, Eds., *Strong Fibers*, Elsevier, Amsterdam, 1985.

22. P. J. Goodhew, A. J. Clarke, and J. E. Bailey, *Mater. Sci. Eng., 17*, 3, (1975).

23. M. M. Coleman and G. T. Sivy, *Carbon, 19*, 123 (1981).

24. M. M. Coleman and R. J. Petcavich, *J. Polym. Sci., Polym. Phys. Ed., 16*, 821 (1978).

25. R. Bacon, in L. F. Vosteen, N. J. Johnston, and L. A. Teichman, Eds., *Tough Composite Materials*, NASA CP-2334, 1983, p. 245.

26. J. C. Bokros, in P. L. Walker, Ed., *Chemistry and Physics of Carbon*, Vol. 5, Dekker, New York, 1969, p. 1.

27. R. Perret and W. Ruland, *J. Appl. Crystallogr., 3,* 525 (1970).

28. M. Guigon, A. Oberlin, and G. Desarmot, *Fibre Sci. Technol., 20,* 55 (1984).

29. L. T. Drzal, *Carbon, 15,* 129 (1977).

30. S. C. Bennet, "Strength Structure Relationships in Carbon Fibers," Ph.D. thesis, University of Leeds, 1976.

31. R. J. Diefendorf, in L. F. Vosteen, N. J. Johnston, and L. A. Teichman, Eds., *Tough Composite Materials,* NASA CP-2334, 1983, p. 209.

32. J. T. Paul, British Patent 1,433,712 (1976).

33. K. Saito, H. Ogawa, and T. Shigei, U.S. Patent 4,401,533 (1983).

34. G. Gynn, R. N. King, S. F. Chappell, and M. L. Deviney, "Improved Graphite Fiber Adhesion," AFWAL-TR-81-4096, 1981.

35. D. J. Pinchin and R. T. Woodhams, *J. Mater. Sci., 9,* 300 (1974).

36. D. M. Brewis, J. Comyn, J. R. Fowler, D. Briggs, and V. A. Gibson, *Fibre Sci. Technol., 12,* 41 (1979).

37. J. C. Goan, L. A. Joo, and G. E. Sharpe, *Proceedings of the 27th Annu. Tech. Conf. SPI Reinf. Plast./Compos. Inst. 21E,* 1 (1972).

38. P. Ehrburger and J. B. Donnet, in W. Watt and B. V. Perov, Eds., *Strong Fibers,* Elsevier, Amsterdam, 1985, p. 577.

39. L. T. Drzal, M. J. Rich, M. F. Koenig, and P. F. Lloyd, *J. Adhes., 16,* 133 (1983).

40. O. P. Bahl, R. B. Mathur, and T. L. Dhami, *Polym. Eng. Sci., 24,* 455 (1984).

41. E. Fitzer and R. Weiss, *Proceedings of the 16th Biennial Conference on Carbon,* July 1983, p. 473.

42. S. Brunauer, *The Adsorption of Gases and Vapors,* Princeton Univ. Press, Princeton, NJ, 1945.

43. S. J. Gregg and K. S. W. Sing, *Adsorption, Surface Area and Porosity,* 2nd edition, Academic Press, London, 1982.

44. B. O. Bateup, *Int. J. Adhes. Adhes., 1(5),* 233 (1981).

45. H. F. Webster, M. S. thesis, Virginia Polytechnic Institute and State University, Blacksburg, VA, 1985.

46. L. A. Girifalco and R. J. Good, *J. Phys. Chem*, *61*, 904 (1957).

47. F. M. Fowkes, *Ind. Eng. Chem.*, *56*, 40 (1964).

48. F. M. Fowkes in K. L. Mittal, Ed., *Physicochemical Aspects of Polymer Surfaces*, Vol. 2, Plenum Press, New York, 1983, p. 583.

49. D. H. Kaelble, P. J. Dynes, and E. H. Cirlin, *J. Adhes.*, *6*, 23 (1974).

50. D. H. Kaelble, P. J. Dynes and L. Maus, *J. Adhes.*, *6*, 239 (1974).

51. A. W. Adamson, *Physical Chemistry of Surfaces*, 5th edition, Wiley Interscience, New York, 1990, p. 389.

52. J. K. Spelt, Y. Rotenberg, D. R. Absolom and A. W. Newmann, *Colloids and Surfaces*, **24**, 127 (1987).

53. D. J. D. Moyer and J. P. Wightman, *Surf. Interf. Analysis*, **17**, 457 (1984).

54. W. C. Jones and M. C. Porter, *J. Colloid Interface Sci.*, *24*, 1 (1967).

55. G. E. Hammer and L. T. Drzal, *Appl. Surf. Sci.*, *4*, 340 (1980).

56. J. Schultz, C. Cazeneuve, M. E. R. Shanahan, and J. B. Donnet, *J. Adhes.*, *12*, 221 (1981).

57. C. A. Herb, J. L. Buckner and J. R. Overton, *J. Colloid Interface Sci.*, *94*, 14 (1983).

58. C. S. Brooks and D. A. Scola, *J. Colloid Interface Sci.*, *32*, 561 (1970).

59. J. V. Larsen, T. C. Smith, and P. W. Erickson, "Carbon Fiber Surface Treatments," NOLTR 71-165, 1971.

60. J. Schultz, L. Lavielle, and H. Simon, *Proceedings of the Intl. Symposium on Science and New Applications of Carbon Fibers*, 1984, p. 125.

61. A. S. Gozdz and H. D. Weigmann, *J. Appl. Polym. Sci.*, *29*, 3965 (1984).

62. B. Rand and R. Robinson, *Carbon*, *15*, 311 (1977).

63. J. P. Randin, in A. J. Bard, Ed., *Encyclopedia of Electrochemistry of the Elements*, Dekker, New York, 1976, p. 1.

64. J. B. Donnet, *Carbon*, *6*, 161 (1968).

65. H. P. Boehm, *Angew. Chem.*, *5*, 533 (1966).

66. D. Rivin, *Rubber Chem. Technol.*, *44*, 307 (1971).

67. D. Rivin, *Rubber Chem. Technol.*, *36*, 729 (1963).

68. D. W. McKee and V. J. Mimeault, in P. L. Walker and P. A. Thrower, Eds., *Chemistry and Physics of Carbon*, Vol. 8, Dekker, New York, 1973, p. 151.

69. W. M. Riggs and M. J. Parker, in S. P. Wolsky and A. W. Czanderna, Eds., *Methods of Surface Analysis*, Elsevier, Amsterdam, 1975, p. 103.

70. C. S. Fadley, *Prog. Solid State Chem.*, *11*, 265 (1976).

71. J. H. Scofield, *J. Electron Spectrosc. Relat. Phenom.*, *8*, 129 (1976).

72. P. Cadman, G. Gossedge, and J. D. Scott, *J. Electron Spectrosc. Relat. Phenom.*, *13*, 1 (1978).

73. C. D. Wagner, *J. Electron Spectrosc. Relat. Phenom.*, *32*, 99 (1983).

74. D. Briggs and M. P. Seah, *Practical Surface Analysis*, Wiley, New York, 1983, p. 363.

75. D. T. Clark and A. Harrison, *J. Polym. Sci., Polym. Sci. Ed.*, *19*, 1945 (1981).

76. A. Proctor and P. M. A. Sherwood, *J. Electron Spectrosc. Relat. Phenom.*, *27*, 39 (1982).

77. D. S. Everhart and C. N. Reilley, *Anal. Chem.*, *53*, 665 (1981).

78. D. Briggs, *Surf. Interface Anal.*, *4*, 151 (1982).

79. D. Briggs, *Surf. Interface Anal.*, *9*, 391 (1986).

80. A. Brown and J. C. Vickerman, *Surf. Interface Anal.*, *8*, 75 (1986).

81. L. H. Sharpe, in Notes from Adhesion Society Short Course, Savannah, Georgia, February 1985.

82. J. R. Huntsberger, in R. L. Patrick, Ed., *Treatise on Adhesion and Adhesives*, Vol. 1, Dekker, New York, 1967, p. 119.

83. J. R. Huntsberger, *J. Paint Technol.*, *30*, 199 (1967).

84. C. Kittel, *Introduction to Solid State Physics*, 5th ed., Wiley, New York, 1976, p. 71.

85. R. J. Good, in R. L. Patrick, Ed., *Treatise on Adhesion and Adhesives*, Vol. 1, Dekker, New York, 1967, p. 71.

86. P. W. Atkins, *Physical Chemistry*, 3rd ed., Freeman, New York, 1986, p. 584.

87. W. C. Wake, *Adhesion and the Formulation of Adhesives*, 2nd ed., Applied Sciences Publishers, Barking, England, 1982, p. 8.

88. J. P. Wightman, T. A. DeVilbiss and J. G. Dillard in S. M. Lee, Ed., *Intl. Encyclopedia of Composites*, Vol. 1, VCH Publishers, New York, 1990, p. 236.

89. P. Ehrburger and J. B. Donnet, *Phil. Trans. R. Soc. London, A295*, 495 (1980).

90. P. Ehrburger, J. J. Herque, and J. P. Donnet, *Proceedings of the Fifth London International Carbon Graphite Conference*, 1978, p. 398.

91. J. Lahaye, J. J. Herque, and J. B. Donnet, *Proceedings of the Fourth London International Carbon Graphite Conference*, 1976, p. 201.

92. E. Fitzer, K.-H. Geigl, and W. Huttner, *Carbon, 18*, 389 (1980).

93. L. T. Drzal, *SAMPE J., 19(5)*, 7 (1983).

94. L. T. Drzal, M. J. Rich, and P. F. Lloyd, *J. Adhes., 16*, 1 (1982).

95. F. Tuinstra and J. L. Koenig, *J. Chem. Phys., 53*, 1126 (1970).

96. F. Tuinstra and J. L. Koenig, *J. Compos. Mater., 4*, 492 (1970).

97. A. Ishitani, H. Ishida, G. Katagiri, and S. Tomita, in H. Ishida and J. L. Koenig, Eds., *Composite Interfaces*, North Holland, New York, 1986, p. 195.

98. S. Brelant, *SAMPE J., 21(5)*, 6 (1985).

99. F. Hopfgarten, *Fibre Sci. Technol, 11*, 67 (1978).

100. F. Hopfgarten, *Fibre Sci. Technol., 12*, 283 (1979).

101. A. Ishitani, *Carbon, 19*, 269 (1981).

102. T. Takahagi and A. Ishitani, *Carbon, 22*, 43 (1984).

103. K. Waltersson, *Fibre Sci. Technol., 17*, 289 (1982).

104. A. Proctor and P. M. A. Sherwood, *Carbon, 21*, 53 (1983).

105. A. Proctor and P. M. A. Sherwood, *Surf. Interface Anal., 4*, 212 (1982).

106. A. Proctor and P. M. A. Sherwood, *J. Chem. Soc., Faraday Trans. 1*, 80, 2099 (1984).

107. C. Kozlowski and P. M. A. Sherwood, *J. Chem. Soc., Faraday Trans. 1*, *8*, 2745 (1984).

108. C. Kozlowski and P. M. A. Sherwood, *Carbon*, *24*, 357 (1986).

"In Situ" composites formed with Liquid Crystalline Polymers
and
Thermoplastic Matrices

E. Amendola

CNR-ITMC c/o University of Naples, Department of Materials and Production Engineering,
Piazzale Tecchio, 80125 Napoli, Italy

L. Nicolais, and C. Carfagna

University of Naples, Department of Materials and Production Engineering,
Piazzale Tecchio, 80125 Napoli, Italy

INTRODUCTION

The modification of polymeric materials performances are commonly achieved in the industrial practice by the addition of fillers as reinforcing agents. The effect of reinforcing fibers and fillers on the mechanical properties of polymers are, in fact, well documented. (1,2)

Structural fiber composite materials with very high specific properties have been successfully utilized in many engineering application. Due to the higher strenght of the fibers compared to that of ordinary polymeric materials, they normally impart strength to whatever matrix they are in. When the fibers are all aligned in one direction, maximum strength is achieved in the composite material along the direction of the fiber length. Moreover, although these fillers are usually used to increase the stiffness of the composite, an improvement of the dimensional stability is also achieved. The extent of reinforcement depends on factors such as the shape and the amount of the reinforcing phase, interfacial filler-matrix adhesion, filler surface treatments.

However, the addition of rigid fillers to a polymeric matrix, which is a benefit in terms of properties and dimensional stability, has an adverse effect on the processability. For reasons similar to those accounting to the increase in the elastic modulus, the presence of solid fibers results in a dramatic increase in the melt viscosity.

The interest in thermotropic liquid-crystalline polymers has grown in recent years due to the their inherently high stiffness and strength, high use temperatures, excellent chemical resistance, low melt viscosity and low coefficient of thermal expansion.

G. Akovali (ed.), The Interfacial Interactions in Polymeric Composites, 387–407.
© 1993 Kluwer Academic Publishers.

Nevertheless, despite extensive research and development, LCPs have relatively few established markets. The materials tend to be expensive, primarily because of monomer high cost. The LCP have found their greatest success in the production of intricate mouldings, where the ability to fill the mould and the low thermal shrinkage are the major specifications.

Blending of LCP with thermoplastic offers the possibility of exploiting many of their desirable characteristics. The in situ formation of the LCP reinforcing agents in the shape of fibers or needles is the main reason for referring to this class of materials as "in situ composites". The first approach to materials involved the crystallization of low molecular weight compounds, molten into the polymeric matrix during processing, and crystallize as the temperature decreases. The mechanical performances of this composites did not excede the performances of the unfilled palstics.

In fact, the weak Wan der Walls forces acting in the low molecular weight crystals was not effective in inducing any reinforcing effect (3-5).

A further approach to the use of organic crystals as reinforcing agents in composites was the "in situ polymerization". In this case a monomeric filler added to the matrix, is crystallized during processing and then polymerized via solid state reaction. The final reinforcing agents is a polymer, and as a consequence a great improvement in the composite performance is expected. Nevertheless, great deal of difficulties was encountered during the solid state reaction, due to its low yield (6,7).

From a survey of the recent development in the field of self reinforcing polymers it appears clearly that the unique properties of the LCP make them primary candidates as reinforcing agent, added to engineering thermoplastic in the form of second phase. In the design of the blend several parameters have to be considered, but first let consider in close details the major features of liquid-crystalline compounds.

LIQUID-CRYSTALS: BASIC DEFINITIONS AND CHARACTERISTICS.

Although the following remarks do apply to ordinary, low molecular weight liquid-crystalline compounds, the definitions are as well suitable for the description of molecular arrangement of polymers.

The term "Liquid-Crystals" represents a number of different state of matter in which the degree of molecular order lie intermediate in between the almost perfect long range positional and orientational order characteristic of solid crystals and the statistical long range disorder found in ordinary isotropic amorphous liquids.

The earliest recognition of "Liquid-Crystals" is usually attributed to Reinitzer (8), who in 1888 noted the colour phenomena which arise when melts of cholesteryl acetate or benzoate are cooled. He observed, using a carefully purified sample of cholesteryl benzoate that this "melted at 145.5°C to give a cloudy fluid which, on further heating, suddenly clarified at 178.5°C".

Altough the term "Liquid-Crystals" remains in general use, it was vehemently opposed in the past on the basis that "Liquid-Crystals" were neither true liquids nor true crystals but indicated a new state of matter intermediate between the solid and the liquid. With the aim of an unambiguous adjective to describe this new class of materials the term mesomorphic was proposed.

The basic classification of "Liquid-Crystals" is divided into two major groups: the thermotropic and lyotropic. In the first case the structure is modified from the crystalline state to the amorphous liquid, passing trough the "Liquid-Crystals" state as the temperature is increased; in the second case the order related to the crystalline texture is decreased trough the non ordered amorphous structure as a solvent is added to the crystalline materials. In both cases the adjective mesomorphic refers to the occurence of the new state of matter in the range of stability interposed, for a particular compound, between that of the crystalline type, which prevails at low temperature or in the pure state, and that of the amorphous type which, in turn, is found at higher temperatures or in diluite solutions.

The thermotropic liquid-crystals are conventionally classified into nematics, cholesterics or smetics.

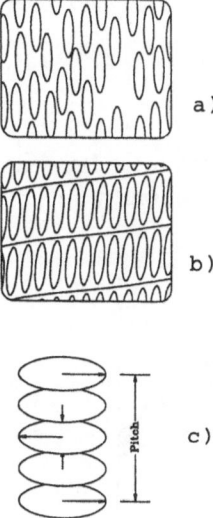

a)

b)

c)

Fig. 1   Arrangement of rod-like molecules in the liquid-crystalline state:
a) nematic phase
b) smectic phase
c) cholesteric phase.

Nematic phases can be formed by compounds that are optically inactive or by racemic mixtures, and consists of molecules that are more or less elongated in shape. The distribution of the molecule centres of gravity is without long range order, as in ordinary amorphous phases, but the long axis of the molecules are nearly parallel to each other. Domain is defined the portion of the space where the molecules lie with their long axes parallel to one preferred direction, called director. The occurence of uniform orientation and parallelism between the molecules is limited in the domain region, and there is no long range correlations between the directors of the adiacent domains. In that respect the overall orientation of the molecules in the nematic phase is determided by statistical distribution.

Cholesteric liquid-crystals are formed by optically active substances, or optically active mixtures. They are miscibile and isomorphous with nematic liquid-crystals, but while in nematics the equilibrium structure corresponds to uniform parallel orientation, in cholesterics it corresponds to the planar uniformily twisted structure. The distance for a 360°C rotation is usually referred as the pitch (P). It is generally assumed that smectic phases are to be organized in lamellar structures which are developed in addition to a long range order similar to that in nematics. Lamellar structure means that the centres of gravity of the elongated molecules lie in parallel planes. Several types of smectic phases are characterized. In the more structured textures, besides the parallel orientation of the long molecural axes the molecules lie in the smectic planes determining well defined textures. The parallel orientation of the molecules is far to be as idealized as described in fig. 1. Thermal fluctuation are very strong and large deviations from the perfect parallel orientation do exhist.

In the assumption of cylindrical molecules the average orientation can be expressed in terms of the Legendre polinomials. The second rank order parameter then results as

$$<P2> = 1/2 \ (<3\cos^2\beta> -1).$$

Empirically introduced by Zwetkoff (9-11) and named S (or Hermann orientation function) it is a measure of the fluctuations of the degree of order.

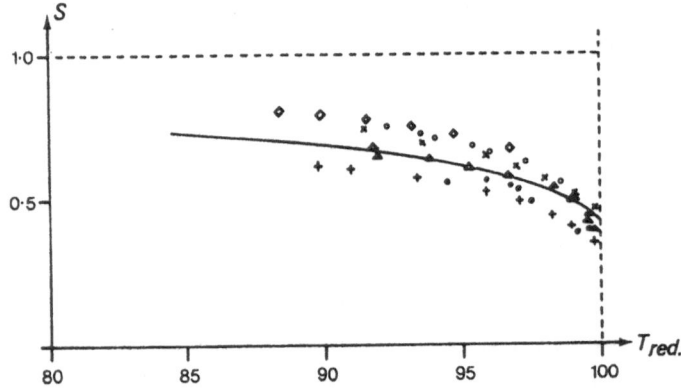

Fig. 2 Effect of the reduced temperature on the degree of order (S) for nematic liquids;

(△)	2,4-nonadienoic acid
(●) and +	4,4'-dimethoxyazoxybenzene
(○) and x	4,4'-diethoxyazoxybenzene
(◊)	4-n-hexyloxybenzoic acid.

(reprinted with permission from ref. 12)

For ideal nematic phases all molecules are perfectly parallel, and S equals to one. In amorphous isotropic liquids the average orientation of the molecules is described by the "magic angle" (54.7°) and S equals zero, while for molecules perpendicular to the reference direction S results in minus one half.

For compounds in the nematic state, the order parameter is strongly affected by the temperature. The experimental behavior is related with a theoretical prediction derived in using a simple inner field theory (12). A reduced temperature scale is used in fig.2 defined by $T_{red.} = 100T/T_c$, where T_c is the nematic to isotropic transition temperature.

The plot of S with temperature shows a first order discontinuity, indicating the occurence of a truly thermodynamic transition as the liquid-crystalline order is destroyed passing from the mesomorphous to isotropic phase. The temperature transition is often called clearing point, or isotropization, and is the "second melting" observed by Reinitzer during the pioneering study on the colestherol derivates. The temperature range between melting and clearing point is of crucial importance in describing the properties of a mesomorphous compound.

The mesomorphic behaviour is exhibited when the liquid-crystalline ordering is thermodynamically stabilized, either for structural configurations of the molecular backbone and/or for dipole interactions, in the temperature range between the melting and clearing point. In that respect the stable mesophase is termed enantiotropic. On the contrary, if the crystalline phase is stable over a temperature range overlapping the potential liquid-crystalline structure, the mesophase is named omeotropic. Only if the organization of the crystals are hindered or delayed on the basis of kynetic effects the molecules are lead to orient in the meta-stable liquid-crystalline phase. For omeotropic LCP compounds the mesogenic character is evidenced only during the cooling cycle.

The concepts introduced for low molecular weight compounds are still adequate in describing the liquid-crystalline phases exhibited by high polymers. The orientation of molecular segment, either in the molecular backbone or as pendant side groups, can give rise to the observed anysotropy. The present review deals with a well defined class of liquid-crystalline polymers, the main chain thermotropic nematic polymers (LCP). The rigid or semi-rigid molecular backbone polymers potentially exhibit interesting mechanical properties, coupled with low thermal expansion coefficients, good chemical resistance and low viscosity in the molten state. The alteration of engineering thermoplastic polymers through a physical blending with the LCP looks a challenging opportunity to take an advantage of the unique properties of this developing class of materials, overaging in the mean time the extremely anysotropic properties of the LCP itself. The strict requirement of processability of both constituent in the same temperature range determines the processing window for the blend. If the in situ formation of reinforcing phase is the desired goal, the mesogenic polymers ought to be processed in the temperature range of stability of the mesophase. It is obvious that the matrix itself is in the molten state, but it is not equally obvious the right choice for the precise processing temperature. In fact the increase in the temperature will reduce the viscosity of both component (if phase transition are not involved, i.e. the mesogenic polymer will be confined in the nematic phase). Nevertheless, different temperature dependence of the viscosity of the two components determines the viscosity ratio to be a function of the temperature. It is the viscosity ratio between the two components of the blend that play an important role in determining the final morphology for the resulting product. We will describe in more details the effect of the processing temperature on the rheology of the blend in following sections .

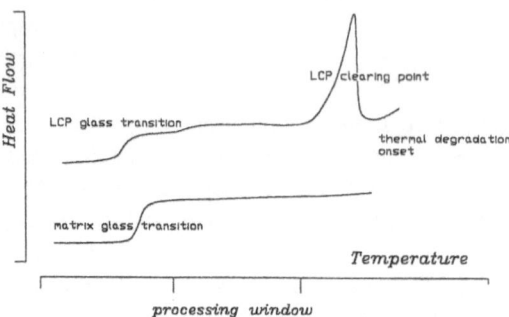

Fig. 3   Processing window for LCP/thermoplastic blend. The transition temperatures are characterized with the aim of Differential Scanning Calorimetry (DSC).

LIQUID-CRYSTALLINE POLYMERS

Since 1975 when the first thermotropic liquid-crystalline polymer was synthesized by Roviello and Sirigu (13), many mesomorphic polymers were prepared by both academic and industrial reserchers.

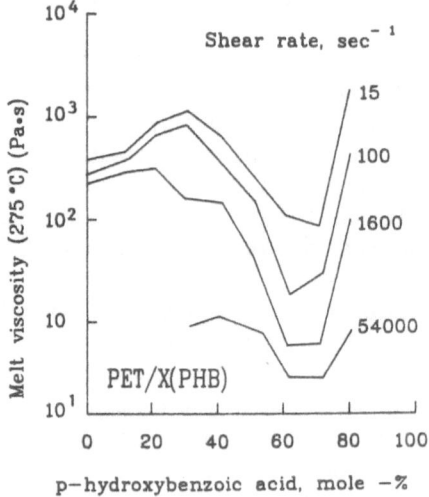

Fig. 4   Melt viscosity of poly (ethylene therephtalate) modified with p-hydroxybenzoic acid.
(reprinted with permission from "International Encyclopedia of Composites", VCH, New York (1990))

In the mean time great deal of attention was devoted to the first commercially available product, the co-polyesther produced by Tennesse Eastman Company and Synthesyzed by Jackson and coworkers (14,15). It was produced through chemical modification of polyethylene therephtalate (PET). To make stiffer the flexible molecular chains of PET, p-hydroxy benzoic acid (PHB) was used as co-monomer. The desired mesogenic behavior was found at approximatively 60 mol % of PHB co-monomer. An X-ray analysis performed on the turbid melt of the copolymer (reported as PET/PHB60) revealed that the mesophase was nematic. One of the most intriguing phenomen that is related with the appearence of nematic phases is the sudden reduction in the viscosity of the melt. The behaviour is illustred in fig. 4 for the PET/PHB copolyesther as a function of the shear rate. Wissburn has accounted for this effect by assuming that the basic morphological units in the liquid-crystalline melt are rigid rod units, which must move cohoperatively. (16).

When the macromolecules in the nematic phase are subjected to shear stresses, the rigid rod assembly line up more or less, thereby reducing the resistance to flow. In the amorphous phase physical entanglements occurring between the random coil molecules induce higher shear stresses, as compared with the nematic phase. Nevertheless, to account for the occurrence of higher shear viscosity at low shear rates rather that at high shear rates, considerable free volume must characterize the nematic phase.

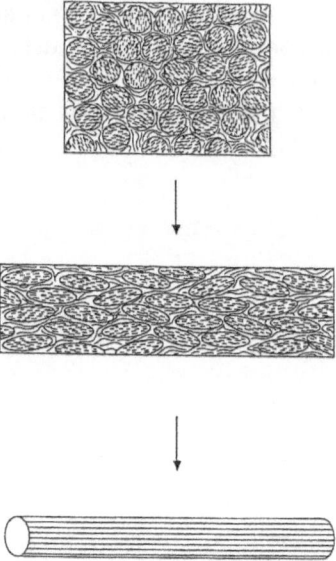

Fig. 5   Orientation of Thermotropic Material:
         model of LCP as it is beeing hot drawn.
         (reprinted with permission from ref. 26)

During the processing the nematic polymers produce a high degree of orientation in the melt state, and are able to maintain the orientation in the melt state. When the nematic liquid-crystalline compound is extruded it will undergo both shear and elongational flow. The orientation and texture upon exiting the extruder will be complex because of the variation of flow field within the extruder and at the die exit. The orientation of the liquid-crystalline extrudate will vary from highly oriented at the skin to less oriented at the core.(17,18).

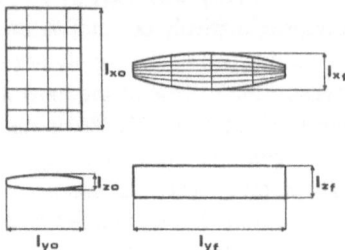

Fig. 6   Recoil of oriented films.
(reprinted with permission from *"Handbook of Polymer Science and Technology"*, Marcel Dekker, New York (1989))

When the melt is hot drawn in an extensional flow field, the nematic domains are oriented in the direction of extension and appear to be into an oriented fibrous form with a high length-to-diameter ratio. Once the fiber is hot spun and cooled down to low temperature a considerable amount of energy is frozen in the fiber as a result of the processing condition. If the temperature is increased above the glass transition of the LCP, some degree of spring back deformation results. The total deformation is generally described as formed by instantaneous elastic, viscoelastic , and viscous (plastic) components (19,21). As compared with drawn amorphous polymers the degree of spring back reversion ratio is considerably lower.

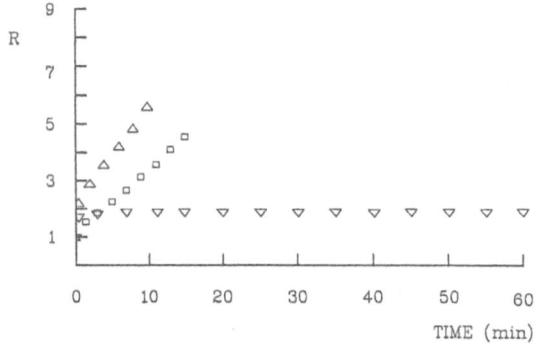

Fig. 7   R (L_0/L_{00}) vs. time for LCP (▼), PC (□) and PC/LCP10 (▲) fibers. Recoil temperature = 180°C.
(reprinted with permission from Makromol. Chem., Macromol. Symp. **23**, 253 (1989))

For what concerns the mechanical performances of liquid-crystalline polymers, many papers were published on the characterization of the LCP processed under different conditions. The major feature in determining the final properties of the extrudate or of the drawn fibers appears to be the conformation in the alligned state. It appears clearly that well oriented polymers exhibit higher values for the elastic modulus, as compared with amorphous or ordinarily semi-crystalline compounds (25). With the purpose of relating the mechanical properties of nematic oriented polymers with their molecular arrangement, a mathematical model was proposed (26). During the processing operations the molecules in the nematic domains undergo shear and elongational flow. Unlikely ordinary polymers, elevated values of molecular orientation, measured by means of the order parameter S, are achieved even for low values of drawn ratio during the spinning procedure of the fibers.

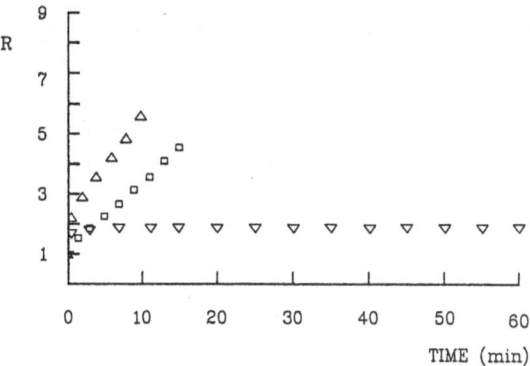

Fig. 8   R (L_0/L_{00}) vs. time for LCP (▽), PC (□) and PC/LCP10 (△) fibers. Recoil
temperature = 220°C.
(reprinted with permission from Makromol. Chem., Macromol. Symp. **23**,
253 (1989))

In spite of the promising performances of the liquid-crystalline polymers, their use is limited to few practical application. This is due to the high pronounced anisotropy of orientation and properties of LCP processed in the pure state. The apparent contraddiction can be avercome by blending the mesomorphous polymer with ordinary engineering thermoplastics.

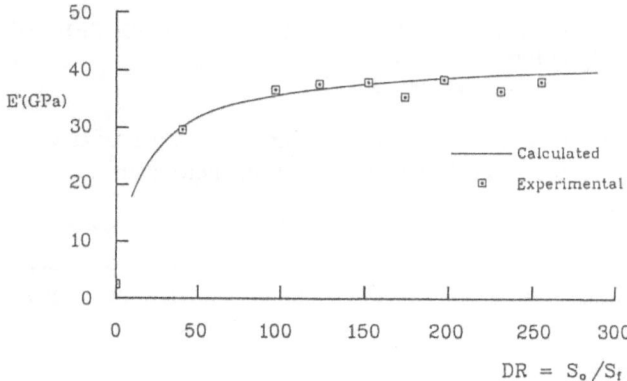

Fig. 9   Effect of draw ratio on the longitudinal elastic modulus of PET/PHB60
fibers: (□) experimantal values, ----- theoretical.
(reprinted with permission from ref. 26)

Optimizing the final performance of plastic objects via physical blending of different compounds is a quite successfull application. Nevertheless, many parameters have to be accounted for in the formulation and the processing of the resulting material.

Rheology of Blends of Thermotropic Polymers and Engineering Thermoplastic.

As previously discussed, fibers that form in situ should reduce the viscosity of the composite during processing conditions, if compared to short fiber reinforced thermoplastic. Moreover, it will be shown that the viscosity of the LCP and thermoplastic polymer will be resonably lower than the viscosity of the enginnering thermoplastic by itself, accordingly with the viscosity of deformable particles in viscous fluid.

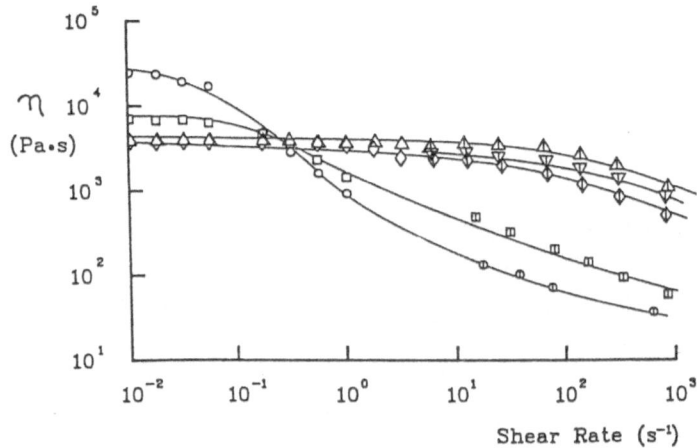

Fig. 10 Viscosity vs. shear rate at 240°C for PC (Δ), PC/LCP5 (▽), PC/LCP10 (◊), PC/LCP50 (□), LCP (○).
(reprinted with permission from ref. 35)

The reduction in the viscosity will be obtained without even plasticizing the matrix, i.e. without depleting the mechanical performances of the pure matrix.

If the desired goal is the final production of a double phase mixture, in order to reinforce the matrix with the LCP fibers, the choise of the components of the polymeric blend is limited to partially incompatible compounds. Nevertheless, some extend of chemically adhesion is necessary to promote chemucal binding of the matrix to the reinforcing agents.

If we assume that in the molten state the blend is a suspension, constituded by the suspendig medium, the matrix, and the inclusion (the LCP phase), the theories of rheology of sunspension can be applied.

Taylor (27) widely studied the effect of different parameters on the elongation and the burst of droplets in different suspensions. He extended the Einstein theory of the viscosity of rigid particles embedded in liquid matrix to the case of two immiscible liquids. Taylor suggested that, when the maximum value of pressure differences across the interface between a suspending medium and the dispersion, tending to disrupt the drops, exceeds the force due to surface tension, which holds it together, the drop will burst.

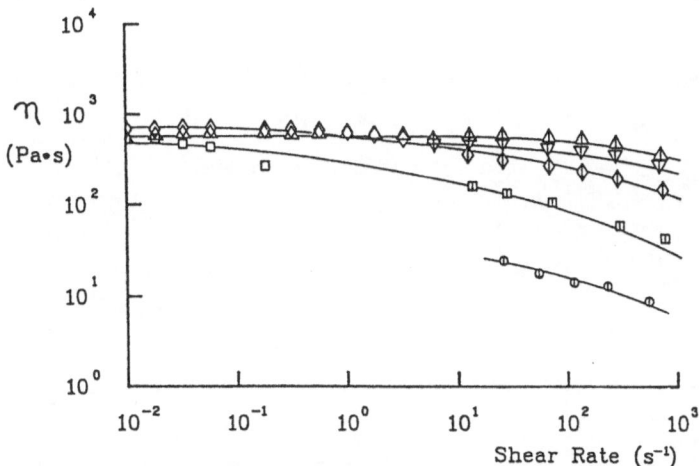

Fig. 11   Viscosity vs. shear rate at 260°C for PC (Δ), PC/LCP5 (▽),
PC/LCP10 (◊), PC/LCP50 (□), LCP (○).
(reprinted with permission from ref. 35)

Flow curves illustrating this point are shown in figg.10, 11 for the blends of polycarbonate (PC) and a liquid-crystalline copolyesther, poly(ethylene therephthalate-co-p-oxybenzoate) (PET/PHB60), at 240 and 260°C.

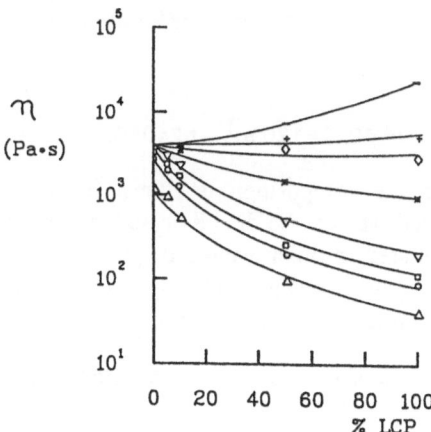

Fig. 12   Viscosity vs. LCP percentage at 240°C, at different shear rates ($\dot{\gamma}$, s^{-1}):
(-) $1 \cdot 10^{-1}$, (+) $1.7 \cdot 10^{-1}$, (◊) 0.31, (✱) 1, (▽) 12, (□) 57, (o) 130,
(Δ) 700.
(reprinted with permission from ref. 35)

The polycarbonate exhibits a typical Newtonian flow behaviour at the low shear rates investigated, while it shows the tendency to continuosly shear thin, if subjected to high shear rates. The same flow behaviour is exhibited by the blends with the lower content of the liquid-crystalline filler. At higher content of liquid-crystalline filler, the rheology of the blend is strongly affected by the presence of the second phase, whose viscosity continuously shear thins over the shear rates investigated.

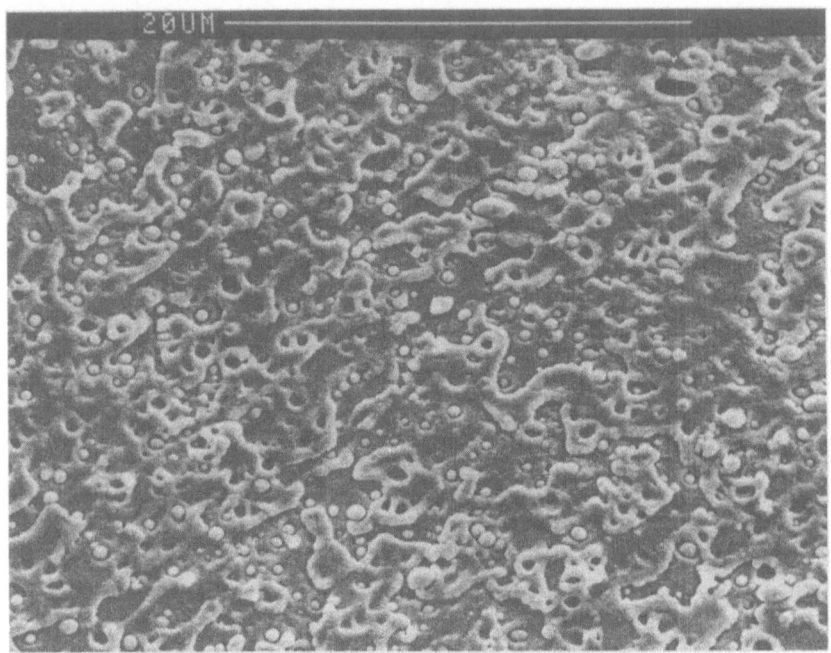

Fig. 13 SEM of PC/LCP10 extruded with L/D = 40, $\dot{\gamma}$ = 120 s^{-1}, at 240°C.

The dependence of the viscosity on the composition is shown in fig.12 for T=240°C, where the viscosity is plotted versus the liquid-crystalline percentage. Practical processing condition are simulated at high values of the shear rate. A significant drop of about 30% in the viscosity is achieved at only 5% content of the LCP filler, and it becomes 50% at 10% content of the LCP iclusion.

The morphological features of the biphasic system can be investigated by means of Electron Scanning Microscopy (SEM).

Electron micrographs of the blend with 10% content of LCP, extruded with a capillary of L/D=40 and a shear rate of 120 s^{-1} at 240 and 260°C are respectively shown in fig.13, 14.

In both cases the liquid-crystalline component forms a discrete phase of about microns in diameter, whose shape and elongation ratio is not significantly influenced either by the L/D ratio of the capillary or by the shear rate in the capillary.

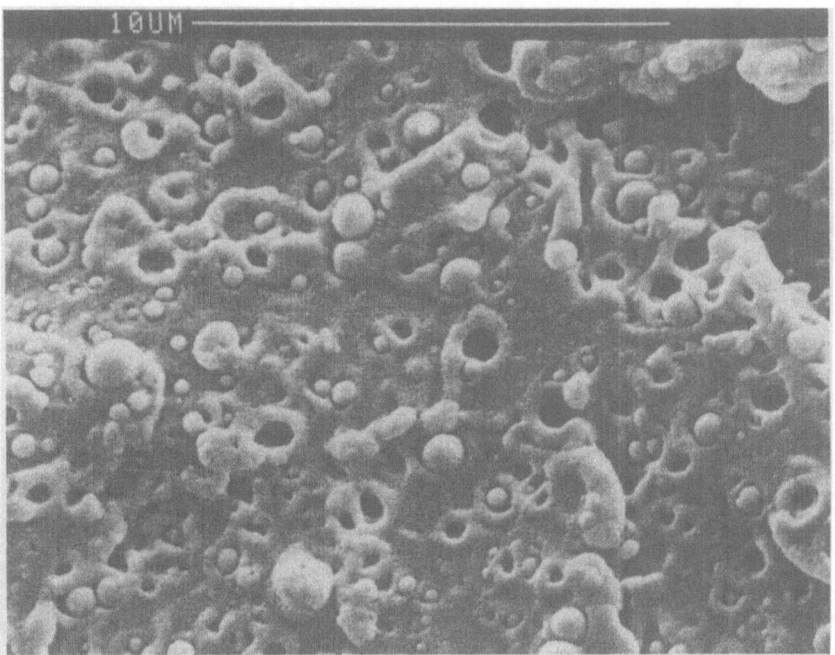

Fig. 14   SEM of PC/LCP10 extruded with L/D = 40, $\dot{\gamma}$ = 120 s^{-1}, at 260°C.

On the other hand, if the LCP content increases significantly, the second phase is no longer homogeneously distributed through the matrix. Regions with sferoidal LCP inclusion alternate with the occurence of fibrous morphology for the LCP phase.

To understand the morphology described, it must be considered that the entrance flow in a die produces an elongational flow that deforms dispersed particles into fibrils (28). The extent of particle elongation depends, among other variables, on the interfacial tension and the initial size of the particle. In the low filler content blends, small droplets of LCP were initially present which could undergo only a slight deformation. Moreover it must be considered that the morphology evidenced by means of SEM is relative to material at the exit of the die, where part of the deformation might have been recovered as a result of the elasticity of the spinning fluid. The spheroidal shape observed for inclusions in the low LCP content blend is a consequence of these combined effects.

Nevertheless the deformability of the nematic inclusion in the blend subjected to shear stresses could account for the dramatic dercreas in the viscosity for the mixture.

In the 50% blend the LCP phase could aggregate in large initial domains, that highly deform, determining the fibrillar structure evidenced in fig.15.

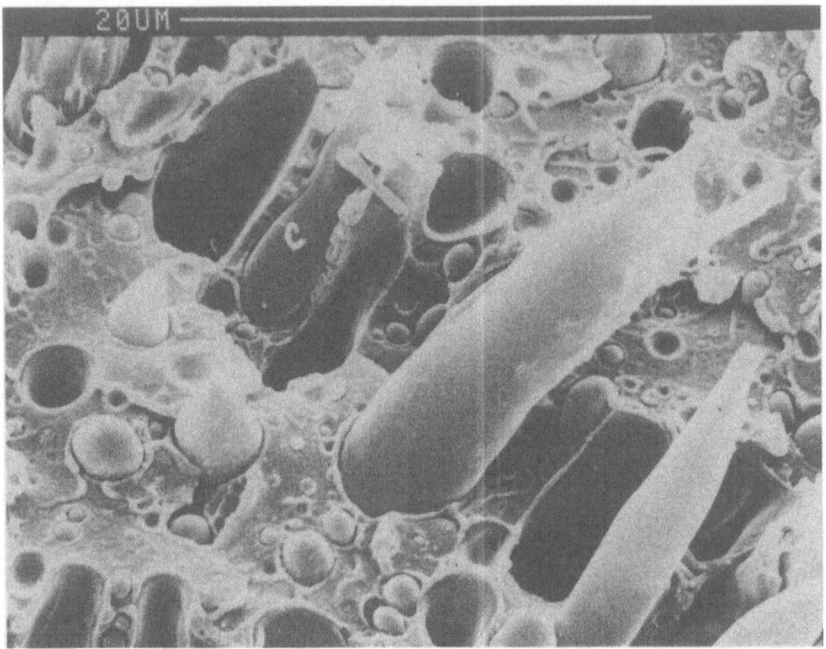

Fig. 15   SEM of PC/LCP50 extruded with L/D = 120, $\dot{\gamma}$ = 120 s^{-1}, at 260°C.

The viscosity reduction in flexible chain polymers by addition of small amounts of liquid-crystalline polymers can, therefore, be reflected in the processing parameters.

In many cases the observed reduction in the viscosity of the blend is more dramatic than the evaluated value, indicating therefore the high effectiveness of thermotropic polymers used as processing aid in the manufacture of composites.

The effect of the added liquid-crystalline polymer on the processing parameters of the polyblends is shown in tab. for some engineering thermoplastics (29).

Mechanical Properties.

The mechanical properties of the in situ composites are further considered. The tensile modulus of the blend based on an amorphous polyamide (PA) and the aromatic co-polyesther HBA/HNA by Celanese are reported (see fig.16, 17)   (30). Both the elastic modulus and the ultimate tensile strength increase rapidly with increasing LCP content.

The increase in the elastic modulus of with draw ratio for PC matrix reinforced with 10%w.w. of PET/PHB60 is shown in the following pictures for the temperatures

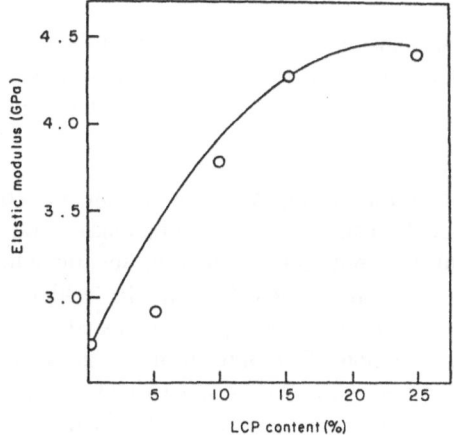

Fig. 16   Composition dependence of the Elastic Modulus of injection moulded LCP/PA blends.
(reprinted with permission from ref. 30)

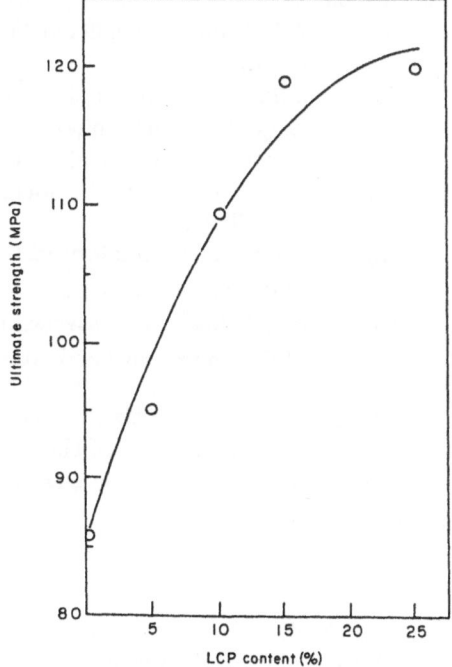

Fig. 17   Composition dependence of the Maximun Tensile Strength of injection moulded LCP/PA blends.
(reprinted with permission from ref. 30)

of 260 and 220°C respectively. At the lowest temperature an increase of about 50 % in the elastic modulus of the blend is observed, while at higher temperature the improvement of the performancese is much more reduced. In fact at the lowest temperature the viscosity ratio of the component in the blend is suitable for succesfull load transfer from the matrix through the inclusion. As a result the LCP droplets elongate into fibrills, acting as effective reinforcing agents.

The morphological analysis can account for the differences in mechanical behavior between the fibers spun at 260 and at 220°C. In fact, SEM of transverse sections of the samples processed at the higher temperature reveal droplet-like LCP regions in the PC matrix.

Therefore, since the LCP inclusion do not elongate during the spinning conditions, the modulus of the thermotropic particles can be saaumed to be close to the modulus of the undrawn LCP, which in turn is not too different to the modulus of the polycarbonate. As a consequence, no reinforcing effect can be achieved for the amorphous matrix from the addition of the LCP phase. If the temperature determines the right flow conditions during the spinning of the molten blend, the LCp elongate into fibrils, acting as reinforcing agents in the matrix of polycarbonate. The modulus of the needle-like inclusion is increased dramatically, reaching the plateau value of 37 GPa, as compared with the value of modulus of about 3 GPa for the undrawn LCP.

It is well known that the elastic modulus of short fiber reinforced composite can be evaluated by means of the Halpin-Tsai equation:

$$\frac{(E_J)_F}{E_{A_J}} = \frac{1 + ABX_N}{1 - BX_N}$$

$$B = \frac{(E_N/E_A)_J - 1}{(E_N/E_A)_J + A}.$$

A = 2*(l/d) for the longitudinal direction $(J=L)$

A = 2 for the transversal direction $(J=T)$

l/d = length-to-diameter ratio of the oriented liquid crystalline domains

Xn = volume fraction of the oriented liquid crystalline domains

E_{NJ} = longitudinal and transversal moduli of the nematic domains

E_{AJ} = longitudinal and transversal moduli of the amorphous matrix domains

If we assume the values of 3 and 37 GPa respectively for the elastic modulus of PC and LCP in the blend spun at 220°C, with an average aspect ratio of about 10 for the fibers included in the matrix, the averall value of 4.3 result for the elastic modulus, which in turn is very close to the experimantal observation.

Dimensional Stability.

Processing of thermoplastic materials often leads to parts that are partially oriented. In these operations the materials undergo large multiaxial deformations, which remain as frozen in stresses once the material is cooled under its glass transition. (31,32).

If the temperature is further increased a poor dimensional stability results, as a consequence of the relaxation of the entropic driving force back to equilibrium conditions.

Particulate fillers are often used to improve the dimensional stability of polymeric manufacts. Several papers have been published on the effect of the draw ratio and glass filler contents on the recoil kinetics of unfilled and composites specimens (32,33). The presence of the filler, however, do not modify the recoil kinetics of composite samples, but significantly reduces the total extent of the lenght reversion ratio. This can be attributed to the orientation of the fibers in the direction of drawing, that act as a constrain for the matrix surrounding the fibers; the fiber-filled specimens consequequently undergo partial sterss relaxation and partial creep.

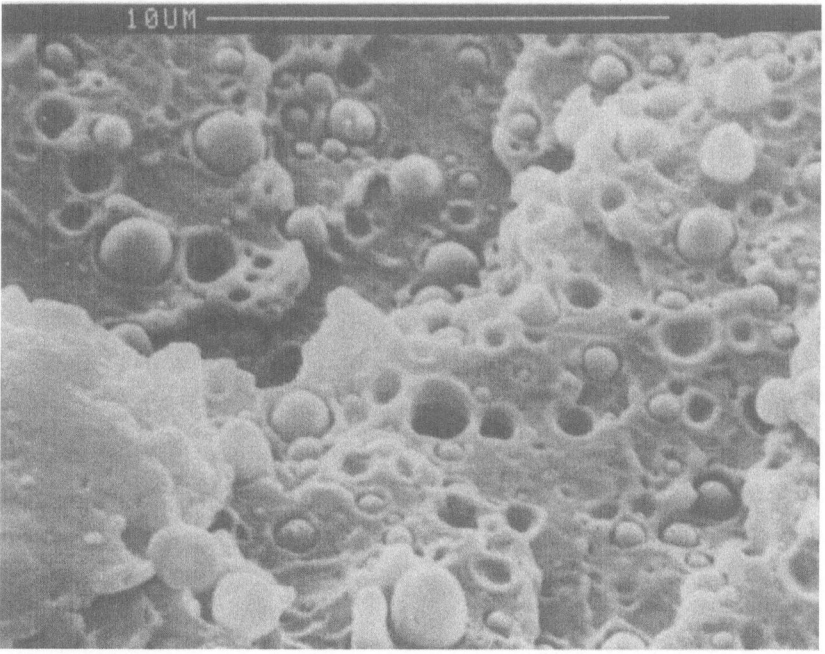

Fig. 18   SEM of PC/LCP10 fibers drawn at 260°C with $S_o/S_f = 200$.

However, many difficulties are encountered during the procesing of the fiber-reinforced composites.

The feature of ultra-high modulus and elevated dimensional stability of nematogenic LCP seems to be suitable to determine great dimensional stability in the composite.

Ther semirigid macromolecules in the nematic state can be easily oriented in the shear field during the processing operations. This leads to the formation of higly oriented fibrils that can enhance the dimensional stability and the mechanical properties of the specimens.

The data regarding polystyrene reinforced with liquid-crystalline polymers clearly show the effect of the filler on the process of recoil.

In the range of temperature where the mesophasic polymer is in the solid state, the blends behaves as a solid fiber reinforced composite. In fact, during the extrusion and subsequent drawing, the LCP inclusion are molten and easily drawn in the directio of the flow (34).

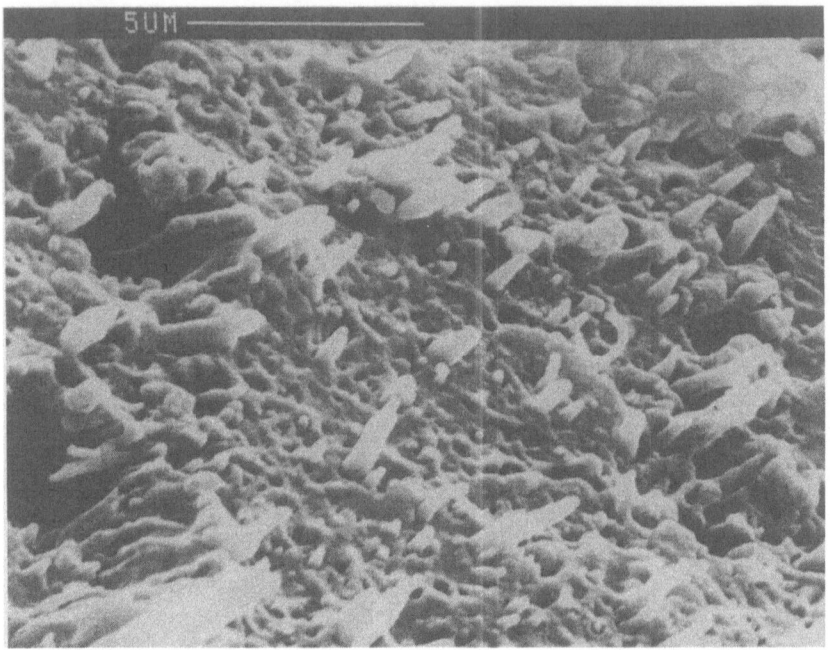

Fig. 19   SEM of PC/LCP10 fibers drawn at 220°C with $S_0/S_f = 200$.

In the uniaxial deformation field the LCP droplets are oriented in fibrils, and upon cooling of the blend, the reinforcing face take place.

While the presence of the LCP inclusion strongly reduces the degree of spring back shrinkage, the recoil kinetics has not been altered of the liquid-crystalline filler.

The reinforcing effect of the liquid-crystalline phase is evidenced up to the temperature of the melting or softening of the fbrillar inclusions. In fact, once the melting temperature has been attained, the liquid-crystalline polymer cannot act anymore as a mechanical constrain for the surrounding matrix. As a consequence the equilibrium length reversion ratio of the blend is close to the values obtained for unfilled matrix.

In the case of polycarbonate the addition of PET/PHB60 reduces, at high temperature, the dimensional stability of the composite blend. In fact the temperature of the shrinkage experiments is higher than the glass transition temperature of the LCP (35). In this case the nematic inclusions cannot act as solid fibers, but on the contrary act as internal lubricant particles, due to the low viscosity of the nematic phase, allowing then the complete relaxation of the amorphous matrix.

CONCLUSIONS

Many advantages can be derived by the use of liquid-crystalline compound as reinforcing filler in producing blends with engineering thermoplastics.

Several apects of the blend behaviour has to be considered in order to fully appreciate the effect of liquid crystalline polymers added to thermoplastic matrices.

In first instance the viscosity of the molten phase is dramatically reduced by the addition of small amount of nematic polymers. The deformation of nematic domains, acting as lubricating particles, is claimed for dramatic decrease in the viscosity of the molten phase. In fact, the presence of the LCP domains as a second phase in the thermoplastic matrix do not plasticize the host matrix, and hence the mechanical properties of the hosting polymer are not reduced, but eventually increased by the addition of the second phase. Besides, if the amount of the LCP is significant and the processing conditions of the blend are effective in inducing high mechanical properties to the LCP fiber, several reinforcing effects can take place, resembling the performances of solid fibers reinforced composites.

Therefore the high cost of the liquid-cristalline polymers can be balanced by the several advantages that can be achieved with the use of nematic polymers as second phase.

REFERENCES

1   L.E. Nielsen, *"Mechanical properties of Polymers and Composites"*, vol. 2, Marcel Dekker, New York (1974)

2   J.A. Manson and L.H. Sperling, *"Polymer Blends and Composites"*, Plenum Pres, New York (1974)

3   J.R. Joseph, J.L. Kardos, and L.E. Nielsen, J. Appl. Polym. Sci., **12**, 1151 (1986)

4   J.L. Kardos, W.L. McDonnell, and J. Raisoni, J. Macromol. Sci. Phys., **86 (2)**, 397 (1972)

5   A. Siegmann, M. Narkis, M. Putermann, and A.T. Di Benedetto, Polymer, **20**, 89 (1979).

6   G. Kiss, A.J. Kovacs, and J.C. Wittmann, J. Appl. Polym. Sci., **26**, 2665 (1981)

7   I. Voigt-Martin, Makromol. Chem., **175**, 2669 (1974)

8   Reinitzer, F. Mh. Chem. **9,**421(1888)

9   G.R. Luckhurst and G.W. Gray eds. *"The Molecular Physics of Liquid-Crystals"*, Academic (1979)

10   P.G. De Gennes, *"The Physics of Liquid-Crystals"*, Oxford U.P. (1974)

11   A. De Vries, J. Chem. Phys., **56**, 4489 (1972)

12   Saupe, A. Angew. Chem. Int. Edn. **7,** 97 (1968)

13   A. Roviello, and A. Sirigu, J.Polm.Sci., Polym. Lett. Ed., **13,**455 (1975)

14   W.J. Jackson, Jr., and H.K. Kuhfuss, J. Polym. Sci. A, **14**, 2043 (1976)

15   F. N. Cogswell, in L.C. Chapoy, Ed., *"Recent Advances in Liquid-Crystalline Polymers"*, Elsevier, London (1985), pp. 165-175

16   K.F. Wissburn, Br. Polym. J., **December**, 163 (1980)

17   Y. Ide, and T.S. Chung, J. Macromol. Sci., Phys., **B-23(4-6)**, 497 (1984-85)

18   D.G. Baird , and G.L. Wilkes, Polym. Eng. Sci., **23**, 633 (1983)

19   A. Peterlin, ed., *"Plastic Deformation of Polymers"*, Marcell Dekker, New York (1971)

20   R.J. Samuels, *"Structured Polymer Properties"*, Wiley-Interscience, New York (1974)

21   A.Ram, Z. Tadmor, and M. Schwartz, Int. J. Polym. Mat., **6**, 57 (1977)

22   L.C. Sayer, M. Jaffe, J. Mat. Sci., **21**, 1897 (1986)

23   C. Carfagna, E. Amendola, M.R. Nobile, and L. Nicolais, J. Mat. Sci. Lett., **7**, 563 (1988).

24   H. Muramatsu, W.R. Krigbaum, J. Polym. Sci.,Polym. Phys., **25**, 2303 (1987)

25   L.E. Nielsen, *"Mechanical Properties of Polymers and of Composites"*, Vol.2, Marcel Dekker Inc., New York (1974)

26   A.T. DiBenedetto, L. Nicolais, E. Amendola, C. Carfagna, andM.R. Nobile, Polym. Eng. Sci., **29**, 153 (1989)

27   G.I. Taylor, Proc. Roy. Soc., **A146**, 501 (1934)

28   D.R. Paul and N.S. Newmann, *"Polymer Blends"*,.Academic Press, New York (1978)

29   G. Kiss, Polym. Eng. Sci., **27**, 410 (1987)

30   A. Siegmann, A. Dagan, and S. Kenig, Polymer, **26**, 1325 (1985)

31   K.M. Kulkarni, Polym. Eng. Sci., **19**, 474 (1979)

32   A. Apicella, L. Nicodemo, L. Nicolais, Rheol. Acta, **19**, 291 (1980)

33   L. Nicodemo, L. Nicolais, and A. Apicella, J. Appl. Polym. Sci., **26**, 129 (1981)

34   L. Nicolais, L. Nicodemo, P. Masi, and A.T. DiBenedetto, Polym. Eng. Sci., **19**, 1046 (1984)

35   M.R. Nobile, E. Amendola, L. Nicolais, D. Acierno, and C. Carfagna, Polym. Eng. Sci., **29**, 244 (1989)

SOME

SHORT COMMUNICATIONS

OF

PARTICIPANTS

OPEN QUESTIONS ON EFFECTS OF FIBRE–MATRIX INTERACTIONS ON COMPOUND AND COMPOSITE PROPERTIES

Boudewijn J.R. Scholtens
Research Fellow Interfacial Phenomena
DSM Research, P.O. Box 18, 6160 MD Geleen, The Netherlands

Compounds are short ($\sim$ 0.3 mm) or long (> 0.6 mm) fibre reinforced thermoplastic or thermoset polymeric materials, which are processed automatically (injection or compression moulding), have good (mechanical) properties (for automotive, electric and electronic applications) and are relatively cheap. Composites contain continuous fibres (rovings, fabrics or mats), usually combined with thermosets, have excellent mechanical (structural) properties, but are very expensive because lack of an industrial process (mainly used in aerospace and aircraft industry).
It is a part of DSM's present business strategy to strengthen its position as a polymer materials supplier, in particular for the automotive, transport, electric and electronic industries. In these applications glass fibres are the preferred reinforcing material as a result of its favourable price/performance characteristics.

Glass fibres are always sized with a protective coating or sizing. Such sizing is necessary to make the fibres' production and processing possible [1]. This sizing is applied to the fibres as an aqueous emulsion, which forms a film during drying. It consists of a film former and coupling agent (silane) as the main ingredients (after drying about 80 and 15 mass %, respectively) and of an emulsifier (of the film former), a lubricant and an antistatic agent, which are processing aids [1]. Glass fibres normally contain about 1 mass % of such a protective sizing, which determines their processing, strand integrity, wetting and dispersion of the fibres by/in the matrix, fibre strength, fibre–matrix adhesive strength and moisture resistance to a high degree. As a consequence, the type of sizing greatly determines the mechanical properties of the compound/composite. Some typical results for injection moulded thermoplastic PP, reinforced with 30 mass % glass fibres (average length $\sim$ 0.3 mm), having two different sizings and with additionally 1 mass % of maleic anhydride modified PP in the matrix are presented below.

mechanical property (23°C)	PP	PP + 30 m % glass fibres (0.3 mm)		
		typical size	good size	good size + 1 m % coupling agent
tensile modulus (GPa)	1.5	2.3/4.5	2.5/5.3	2.6/5.1
tensile strength (MPa)	23	34	65	92
flexural strength (MPa)	49	43/58	51/87	74/123
elongation at break (%)	480	7	4	7
Izod (kJ/m²)	2	4/5	4/5	7/7
falling dart energy (J)	–	3	3	5
HDT at 250 MPa (°C)	80	138	157	158

perpendicular/parallel to injection moulding direction

G. Akovali (ed.), The Interfacial Interactions in Polymeric Composites, 411–413.
© 1993 Kluwer Academic Publishers.

The following techniques are useful for analysis and charaterization: for the sizing on the fibres: SEM, IR and GPC of extracts, IGC and DMA (T_g); for the fibre-matrix bond: micro bond, micro failure with SEM [2], confocal laser scanning microscopy [3]; DMA, flexural strength, ILSS and impact of the uni-directional composites (fixed fibre length, orientation and volume fraction distributions). After this detailed analysis, the fibres are tested in their application, which is an injection moulded compound for thermoplasts, where the microstructure (fibre length, fibre orientation and fibre-fibre distance distributions) as well as the fibre- matrix adhesion determine the mechanical properties.

Fibre-matrix adhesive strength is unimportant for the modulus of compounds, which is determined by the volume fraction of fibres, the moduli of fibres and matrix, the fibre aspect ratio and the fibre orientation. The modulus is a small-strain property and the matrix shrinkage is usually sufficiently high to ensure the modulus increase desired without any true adhesion [4].

Interfaces play, however, a dominant part in the strength of compounds (besides the parameters mentioned above). The adhesive strength is expressed by the critical aspect ratio, $(l/d)_c = \sigma_f/(2\tau)$, where σ_f is the fibre strength (at length l) and τ the interfacial (or matrix) shear strength. High compound strength values can be obtained with large $(l/d)/(l/d)_c$ values. This is achieved either by small $(l/d)_c$ values, so by high fibre-matrix adhesion, or by large (l/d) values, so by very careful processing with minimum fibre breakage. However, interfacial defects such as voids, flaws, fibre (bundle) ends, poor fibre wetting and dispersion, residual shrinkage stresses and nonuniformities in τ and σ_f seriously limit the ultimate compound strength in a way not yet amenable to theoretical treatment.

The theory of fracture toughness of compounds and composites is still insufficiently developed, probably due to the very complex micromechanical processes, which are governed by complicated combinations of the microstructural and interfacial properties. As a general rule, the fracture toughness is enlarged by increasing the active volume of damage and by addressing high-energy dissipating mechanisms [5,6]. The following recipes of controlled interfaces are claimed to give higher fracture toughness: well-bonded, thin elastomeric sizings [7], weak fibre-matrix adhesion [6], optimum utilisation of fibre pull-out [5], hybrid composites [6], intermittent bonding concept [6,8] and long fibres [9]. At present, most recipes indicate that maximum toughness and maximum strength are mutually exclusive, but much basic research on well characterized systems (microstructure and fiber-matrix adhesion) is still to be done.

The three most important aspects to be characterized in fibre reinforced compounds are: fibre aspect ratio, l/d, fibre orientation and fibre-matrix adhesion, $(l/d)_c$. An experimental difficulty in studying the processing-structure-properties relationships in these materials is the strongly coupled interrelations of interfaces and processing on the microstructure.

In our opinion, some important issues of future research on fibre reinforced compounds are included in the following concluding questions:

1. How can the magnitude and concentration of defects, like incomplete wetting or fibre bundles, be taken into account in a strength theory?

2. What combination(s) of microstructure and fibre-matrix adhesion can provide high strength and high toughness?

3. Is there a relationship between the fibre-thermoplast adhesion in the melt and l/d after processing?

4. Which processing route(s) enable(s) industrial production of long or continuous glass fibre compounds/composites?

REFERENCES

1. Loewenstein, K.L. (1983), "The Manufacturing Technology of Continuous Glass Fibres", Glass Science and Technology 6, Elsevier, Amsterdam.
2. Sato, N., Kurauchi, T., Sato, S. and Kamigaito, O. (1991), J. Mater. Sci., 26, 3891.
3. Thomason, J.L. and Knoester, A. (1990), J. Mater. Sci. Lett., 9, 258.
4. Kardos, J.L. (1985), in "Molecular Characterization of Composite Interfaces", H. Ishida and G. Kumar, eds., Plenum Press, New York, p.1.
5. Lauke, B., Schultrich, B., and Pompe, W. (1990), Polym. Plast. Technol. Eng., 29, 607.
6. Kim, J.K. and Mai, Y.M. (1991), Comp. Sci. Technol., 41, 333.
7. McGarry, F.J. (1989), in: "Rubber-Toughened Plastics", Advances in Chemistry Series 222, C.K. Riew, ed., American Chem. Soc., Washington, DC, ch. 7, p. 173.
8. Atkins, A.G. (1975), J. Mater. Sci., 10, 819.
9. Ward, S. and Bailey, R. (1991), in "International Encyclopedia of Composites", Vol. 4, S.M. Lee, ed., VCH, New York, p. 433.

Interface Stabilization in Polymer Blends by Means of Block and Graft Copolymers

M. Fischer

Deutsches Kunststoff-Institut (DKI)

Schloßgartenstraße 6, 6100 Darmstadt, FRG

Among polymeric materials polymer blends are becoming an increasing importance as they permit tailoring of new properties easily. Unfortunately, many blends are heterogeneous and exhibit poor mechanical properties because of high interfacial tensions and a lack of phase adhesion. Such instable blends can be improved by the addition of copolymers that stabilize the phase borders kinetically and thermodynamically. Stabilizing copolymers can have different architectures. In block copolymers chemically different blocks are linked together linearly, whereas in graft copolymers, one type of block is linked laterally to a chemically different main chain.

This paper deals with the conditions of an effective stabilization. The miscibility behaviour of blends $A/\alpha\beta$ and $A/B/\alpha\beta$ where **A and B are homopolymers and $\alpha\beta$ is a symmetric two-block copolymer of the same monomer units as A respectively B or a graft copolymer** was analyzed theoretically by model calculations, and experimentally by electron microscopy and X-ray scattering. Due to the uniformity of the two-block copolymers, (P(S-b-MMA)) blends with a high degree of structural control are obtained. The morphology of the blends of one homopolymer and one block copolymer depends on the chain length ratio λ between the homopolymers and the block copolymer.

$$\lambda = \frac{V_{A,B}}{V_{\alpha\beta}} \qquad (1)$$

$V_{A,B}$: chain volumina of the homopolymers

$V_{\alpha\beta}$: chain volumina of the block copolymers

415

G. Akovali (ed.), The Interfacial Interactions in Polymeric Composites, 415–416.
© 1993 Kluwer Academic Publishers.

For λ « 1 microphase seperation (micelles or lamallaes depending on the composition of the blend) is observed whereas values of λ » 1 lead to macrophase separation. In the range of λ ≃ 0.50 a new structure was found , which is a micelle aggregation. The compatibilizing effect of the two-block copolymers αβ in blends with two immiscible homopolymers A,B also strongly depends on the chain length ratio λ. It was found that a large amount of the block copolymer does not go into the interface and forms micelles in the pure homopolymer phases.

It could be shown that , in spite of their nonuniformity , radically synthesized graft copolymers based on a polybutadiene backbone are also able to build up well defined and particularly highly ordered blend structures. The structure formation follows the same rules as described in the case of blends with the two-block copolymers. Besides that, graft copolymers offer the possibility to modify the chemistry of the side chains very easily by radical copolymerisation.This was illustrated by the copolymerisation of styrene and cyclohexylemethacrylate in the graft chains. A great variability of phase morphologies in blends with po lystyrene (λ » 1) is observed which is controlled by the chemical composition of the grafts. This seems to make graft copolymers proper for the use as effective and easily available compatibilizers.

INTERFACIAL CHEMICAL INTERACTIONS IN CONDENSATION POLYMERS AND THEIR BLENDS

S. Fakirov, M. Evstatiev
Sofia University,
Laboratory ob Structure and
Properties of Polymers,
1126 Sofia, Bulgaria

ABSTRACT. Chemical interactions (additional condensation and transreactions) on the interfaces of condensation polymers (nylons and polyesters) are discussed. In addition to the recently discovered phenomenon of chemical healing, a new type of composites from polymer blends - the microfibrillar reinforced ones are introduced. It is demonstrated that interfacial chemical interactions result in the formation of a new interphase representing a copolymer of the blended components and playing the role of selfcompatibilizer.

1. INTRODUCTION

Chemical reactions (additional condensation and transreactions) between linear polycondensates are well known to occur in their melts [1] as well as in the bulk solid state [2]. One can therefore expect that they will take place on the contact surface, too. If so, two pieces of bulk polycondensates held in good contact at a temperature below the melting point will be welded, i.e. the phase boundary between them will disappear as a result of solid state chemical reactions. This phenomenon was called chemical healing [3] in contrast to the well studied phenomenon of physical healing where bonding is due to mutual seld-diffusion of macromolecules across the contact interface. Depending on the chemical composition of the healing partners, one differentiates between homochemical healing [3,4] and heterochemical healing [5]. Basically, the performance of chemical reactions at elevated temperatures does not exclude the occurrence of diffusion. By elimination or strong restriction of the chain mobility using chemical cross-linking in the amorphous areas, it was concluded that the bonding effect observed can be considered as a result mainly of chemical reactions [6]. Healing experiments with partially cross-linked polyamides [6] led further to the conclusion that parallel to the healing effect a new type of diffusion - chemically released diffusion takes place. The mass transfer in this case is connected with chemical reactions, i.e. one deals with chemotransfer [2,5].
 The described phenomena are of particular importance for incompatible blends of condensation polymers as far as interfaces are available. This is demonstrated on the recently discovered [7] microfibrillar reinforced composites.

2. EXPERIMENTAL

Microfibrillar reinforced composites can be obtained from incompatible polymer blends by extrusion, drawing (with good orientation, e.g. zone drawing) and annealing above the melting temperature of the low melting component(s) and below that of the high melting one at constant strain. Drawing and eventual annealing below T_m of all components results in the improvement of orientation and perfection of the structure. Upon heating at higher temperatures, isotropization of the molten component(s) takes place together with preservation of the orientation and morphological characteristics of the high melting component, i.e. microfibrillar reinforced composites are obtained. In the case of poly(ethylene terephthalate) (PET) and Nylon 6 (PA 6) blend, this temperature is 240°C.

G. Akovali (ed.), The Interfacial Interactions in Polymeric Composites, 417–420.
© 1993 *Kluwer Academic Publishers*.

3. RESULTS AND DISCUSSION

The microfibrillar reinforced composites (MFC) represent a new class of composites and can be situated between the two extreme cases (in respect of the dimensions of the reinforcing elements) of molecular composites (blends with liquid crystalline polymers) and fibre reinforced composites. The mechanical properties (Young modulus and tensile strength) of MFC are higher by 30-50% than the weight average values of the components [7-9] and comparable to those of glass fibre reinforced composites having the same matrix [9].

In addition to isotropization, during short (several hours) thermal treatment chemical interactions (additional condensation and exchange reactions) take place at the interfaces, resulting in the formation of copolycondensates, i.e. an interphase appears. This interphase plays the role of compatibilizer. i.e. one deals with a <u>self-compatibilization effect</u> as far as one does not introduce in the blend extra synthesized copolymer of the blend's components according to the common approach [10].

At prolonged annealing (10-20 hours), the interphase grows involving the entire amount of the isotropic (molten) component(s) and the amorphous portion of the fibrillized one in block copolymers. This process leads to the complete transformation of the homopolymeric matrix into a copolymeric one:

$$(A)_n + (B)_n \longrightarrow \quad \cdots\cdot(A)_n\text{-}(B)_n\cdots\cdot$$

Since $(A)_n$ and $(B)_n$ are crystallizable, the block copolymers crystallize, too.

With the progress of the chemical interactions the block copolymers randomize and convert themselves into statistic ones:

$$\cdots\cdot AABBBBBBAAAAAAB\cdots\cdot \longrightarrow \cdots\cdot ABABBABAA\cdots\cdot$$

An inportant result of the randomization is the lost of crystallization ability of the matrix, as demonstrated by DSC, WAXS, SAXS and solubility measurements performed on binary and ternary blends of PET, PA 6 and poly-butylene terephthalate (PBT) 7-9 .

Evidences of the described physical processes occuring during the preparation of MFC as well as some of the steps of the interfacial chemical interactions can be seen in the WAXS patterns displayed in Fig. 1

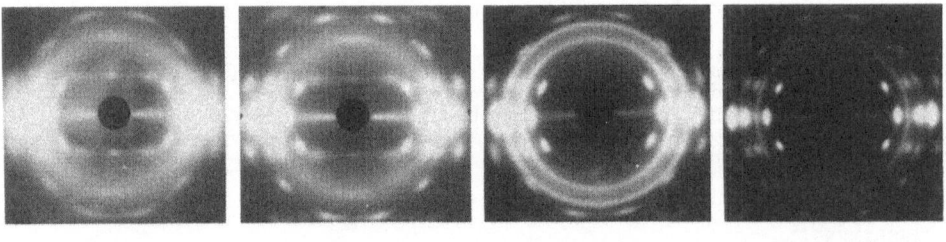

| (a) | (b) | (c) | (d) |
| Zone drawn | $T_a=220°C, t_a=5$ h | $T_a=240°C, t_a=5$ h | $T_a=240°C, t_a=25$ h |

Fig. 1   WAXS patterns of a PET/PA 6 blend (1:1 by wt.) after being subjected to different treatments.

Zone drawing leads to the orientation of both components (Fig. 1a), subsequent annealing below T_m of the components improves both the orientation and crystal perfection (Fig. 1b). Annealing for a shorter time above T_m

of PA 6 and below that of PET results in the isotropization of PA 6 (Fig. 1c) while during prolonged annealing (25 hours) at the same T_a randomization of the block copolymers takes place and PA 6 disappears as a crystalline phase (Fig. 1d). Fig. 1c demonstrates in the best way the existence of the new type of composites - the microfibrillar reinforced ones, and Fig. 1d proves the drastic changes occurring in the blend of condensation polymers as a result of interfacial chemical interactions - the interphase representing initially a thin layer grows up to the complete conversion of the starting homo-PA 6 matrix into an intermediate PA 6-PET block copolymer and finally into a statistical, non crystallizable copolymer.

Evidences of the occurrence of chemical interactions can be found also in the DSC traces of the PET/PA 6 blend. Fig. 2 shows curves of the second melting of blends crystallized from the melt of isotropic polymers and then subjected to zone drawing and annealing at different T_a and t_a.

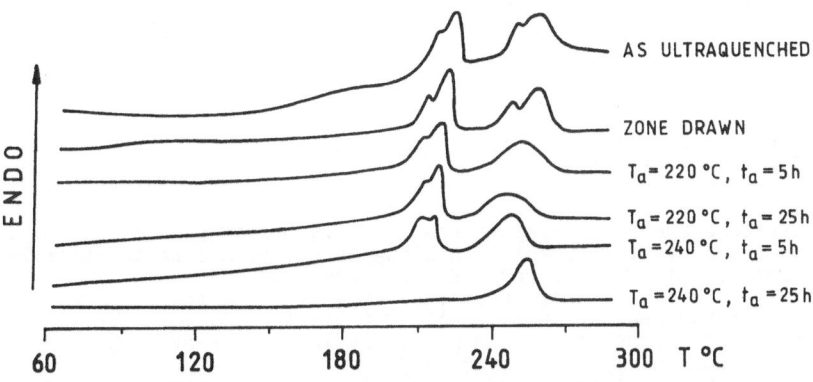

Fig. 2   DSC curves of PET/PA 6 blend (1:1 by wt.) taken in second heating mode. The sample pretreatment is indicated to the curves.

These experimental results indicate the loss of the ability of the polyamide component to crystallize in the blend. This loss is in turn an indication of chemical interaction and formation of a copolymer involving the entire initial amount of PA 6. Such behaviour has already been observed; Golovoy et al. report the nonappearance of the crystalline phase in PET upon cooling after a 30 min retention at 300°C of molten PET/polyacrylate and PET/polycarbonate blends [11]. It should be noted, however, that in the above case the chemical interactions take place in the melt, at a temperature 40°C above the melting point of the crystallizable polymer (the second component being noncrystallizable). In the case of the PET/PA 6 blend only the polyamide component is in the molten state at 240°C, while PET retains its microfibrillar structure (Fig. 1). For this reason, the copolymer formed under these conditions is likely built up of short PET blocks (originating solely from the amorphous fraction of this polymer) and of relatively long PA 6 blocks (since the entire PA 6, being in the molten state, participates in the chemical reactions). The presence of longer PA 6 blocks leads to the formation upon cooling of small and imperfect PA 6 regions of some ordering that follow to some extent the orientation of the PET component 8 . The preservation of the ability of PET to crystallize in the last sample shown in Fig. 2 is related to the retention of its fibrillar structure, i.e. to the fact that the chain segments building up the crystallites do not participate in chemical reactions upon annealing at 240°C and thus entire macromolecules or long PET segments remain free of PA 6.

It is important to note here that the observed compatibilizing effect resulting from the formation of copolymer layers at the interface between the two components of the blend can be effective only at the initial stages of chemical interaction. With a larger degree of exchange (via higher annealing temperature and/or annealing duration), one of the components (PA 6 in the present case) can be completely converted into a copolymer. This trend would result in the total change of the chemical composition of the microfibrillar reinforced composite, the initial matrix of homo-PA 6 being now replaced by a new one, a copolymer of PA 6 and PET. Such a change in the chemical composition of the matrix can influence the entire behaviour of the microfibrillar reinforced composite.

4. CONCLUSION

On the basis of the above considerations, one can conclude that by following the approach described a new type of composites - microfibrillar reinforced ones can be obtained from all thermoplastic polymers. Using blends of condensation polymers, a strong self-compatibilizing effect can be achieved due to interfacial chemical interactions. Opportunities are also available for continuous controllable changes of the chemical composition and crystallization ability of the matrix. The formation of an interphase drastically improves the mechanical integrity of drawn blends.

5. REFERENCES

[1] Flory, J. P. (1953) Principles of Polymer Chemistry, Cornell University Press, Ithaca.
[2] Fakirov, S. (1990) 'Solid state reactions in linear polycondensates' in J. M. Schultz and S. Fakirov (eds.), Solid State Behavior of Linear Polyesters and Polyamides, Prentice Hall, Englewood Cliffs, New Jersey.
[3] Fakirov, S. (1984) 'Chemical healing in poly(ethylene terephthalate)', J. Polym. Sci., Polym. Phys. Ed. 22, 2095-2104.
[4] Fakirov, S. (1984) 'Effect of the temperature on the chemical healing of poly(ethylene terephthalate)', Makromol. Chem. 185, 1607-1611.
[5] Fakirov, S. (1985) 'Heterochemical healing in linear polycondensates', Polymer Communications 26, 137-139.
[6] Fakirov, S. and Avramova, N. (1987) 'Chemical healing in partially cross-linked polyamides', J. Polym. Sci., Polym. Phys. Ed. 25, 1331-39.
[7] Evstatiev, M. and Fakirov, S. (1992) 'Microfibrillar reinforcement of polymer blends', Polymer 33, 877-880.
[8] Fakirov, S., Evstatiev, M. and Schultz, J. M. (1992) 'Microfibrillar reinforced composite from drawn PET/PA-6 blend', Polymer, in press.
[9] Evstatiev, M., Petrovich, S. and Fakirov, S. (1992) 'Microfibrillar reinforced composites from binary and ternary blends of polyesters and Nylon 6', Macromolecules, submitted.
[10] Fischer, M. (1992) 'Interface stabilization in polymer blends by means of block and graft copolymers' NATO-A.S.I. on 'The Interfacial Interactions in Polymeric Composites', 15-26 June, Antalya/Kemer, Türkiye.
[11] Golovoy, A., Cheung, M.-F., Carduner, K. R. and Rokosz, M. J. (1989) 'The influence of aging on the effectiveness of an organophosphite in suppressing transesterification in polymer blends', Polymer Bulletin 21, 327-334.

INTERFACIAL POLARIZATION AND ITS DIAGNOSTIC SIGNIFICANCE IN POLYMERIC COMPOSITES

GYÖRGY BANHEGYI
Furukawa Electric Institute of Technology,
H-1158, Cservenka Miklós út 86,
Budapest,
Hungary

ABSTRACT, The phenomenon of interfacial polarization appearing in composite materials due to the hindered movement of mobile charge carriers is described. Models of composite dielectrics, including bilayer dielectric, statistical mixtures and matrix-filler type composites are discussed together with theoretical methods used to describe their effective dielectric parameters. Applications of dielectric spectroscopy to polymeric composites are demonstrated using mainly data obtained in the author's laboratory. Miscibility and phase sepration in polymeric blends can be studied well, as demonstrated for a polyimide/polysulfone blend exhibiting apparent (non-equilibrium) miscibility. Polymer-filler interactions can be monitored in filled/reinforced composites by comparing the glass tansition temperatures of the matrix polymer in neat and in reinforced state. Origin of low frequency polarization processes can also be identified by this comparison. Example is shown for wollastonite and quartz filled epoxies. The effect of adsorbed water, surface treatment on AC dielectric loss and on thermally stimulated currents is shown for $CaCO_3$ filled polyolefinic composites. Finally the effect of insulating surface layers in polycrystalline ceramic materials is discussed.

Introduction

As the main purpose of this conference is to understand phenomena occuring at the interfaces in polymeric composites and because dielectric methods of investigation are relatively less utilized by composite scientists (perhaps with the exception of cure montioring in thermoset composites) it seems worthwhile to introduce briefly a dielectric relaxation mechanism specific to composite materials, *interfacial polarization*, and to discuss its applicability to solve practical problems.

G. Akovali (ed.), The Interfacial Interactions in Polymeric Composites, 421–430.
© 1993 *Kluwer Academic Publishers.*

1. Interfacial Polarization

Interfacial or Maxwell-Wagner polarization is a special mechanism of dielectric polarization caused by charge build-up at the interfaces of different phases, characterized by different permittivities and conductivities. The simplest model is the bilayer dielectric [1,2], (see Fig. 1.) where this mechanism can be described by a simple Debye response (exponential current decay). The effective dielectric parameters (unrelaxed and relaxed permittivities, relaxation time and static conductivity) of the bilayer dielectric are functions of the dielectric parameters and of the relative amount of the constituent phases:

$$E^* = E_\infty + \frac{E_s - E_\infty}{1 + i\omega\tau} - \frac{i\Sigma}{\epsilon_0\omega} \tag{1}$$

$$E_\infty = \frac{\epsilon_1\epsilon_2}{\epsilon_1 v_2 + \epsilon_2 v_1} \tag{2}$$

$$E_s = \frac{\epsilon_1 v_1 \sigma_2{}^2 + \epsilon_2 v_2 \sigma_1{}^2}{[\sigma_1 v_2 + \sigma_2 v_1]^2} \tag{3}$$

$$\Sigma = \frac{\sigma_1\sigma_2}{\sigma_1 v_2 + \sigma_2 v_1} \tag{4}$$

$$\tau = \frac{\epsilon_0(\epsilon_1 v_2 + \epsilon_2 v_1)}{\sigma_1 v_2 + \sigma_2 v_1} \tag{5}$$

where E and ϵ_i denote the permittivities, Σ and σ_i the conductivities of the composite and of phase i respectively, ϵ_0 is the permittivity of free space, τ is the relaxation time of the interfacial relaxation process, v_i-s are the volume fractions ($v_i = d_i/d$, where d is the total thickness). As shown in [1,2], if the components exhibit own dielectric relaxation processes, they appear together with the interfacial process, but with different relaxation strength and at a slightly shifted frequency.

In the more general case, which is more complicated than the bilayer dielectric, the components are modelled by ellipsoidal elements, and the shape factors ascribed to the phases are also to be taken into account. One can use one- and two-phase elemntary units in the calculations [3].

Composites can be divided into two subgroups: statistical mixtures and matrix-inclusion type composites. The effective dielectric function of the first subgroup can be calculated by equations, which are symmetrical with respect to phase indices. Statistical mixtures exhibit the so called percolation phenomenon which is extremely important in conductor-insulator composites. Percolation threshold is a critical concentration limit, above which the conducting filler renders the previously insulating composite conducting. Some basic methods used for the calculation of effective dielectric parameters (effective medium theroy, mean field method, integral methods etc.) are discussed in [3,4], the resulting equations are compared with published data on various model composites. The Debye approximation valid for the bilayer dielectric remains applicable for matrix-inclusion type composites in the low concentration limit (of course, with different functional form). In the high concentration limit the response function can be no more approximated by an exponential, but the qualitative behavior remains the same. Large interfacial polarization effects can be expected if a conducting inclusion is embedded in an insulating medium. The relaxation strength of the interfacial polarization process increases if the inclusion is elongated along the field direction.

2. Applications

First application is studying <u>miscibility and phase separation</u> in polymeric blends. A series of measurements is presented for polyimide-polyethersulfone blends [5]. The solvent cast films exhibit a composition dependent T_g (Fig. 2.), which is usually an indicator of miscibility. If, however, the films are heated above this temperature, the system immediately phase separates, i.e. the miscibility was only an apparent one, phase separation was prevented by high viscosty during the casting process. If the lower T_g, higher conductivity, component (PES) forms discrete minority phase on phase separation, a well defined interfacial polarization process appears, which can be modeled using the dielectric parameters of the neat components only if it is assumed, that the phase separation is nontotal: a few % of PES remains distributed in PI.

The second, obvious application is <u>studying polymer-filler interactions</u> in filled and reinforced composites. Some data are presented for wollastonite and quartz filled Bisphenol-A based and cycloaliphatic epoxies [6]. These data show, that T_g shifts observed by different relaxation methods (dielectric spectroscopy, DSC, thermomechanical measurements) are not necessarily the same (Table 1.), they depend on the effective frequency, changes in activation energy have also to be taken into account. Correlations between T_g shift and polymer adsorption can be understood using Lipatov's theory [7]. Positive T_g shift usually indicates strong adhesion, while negative T_g shift can be explained by the fact that the adsorbed polymer layer forms a looser structure than that of the bulk material. If both the neat resins and their composites are studied dielectrically, the origin of the low-frequency, high temperature dispersion process (whether it is a matrix/electrode or a matrixfiller interfacial polarization) can be identified. If the neat resin does not show this dispersion process, and the composites does, it is most probably a matrix/filler interfacial process (Fig.3.).

The effect of interphases, adsorbed layers, surface conductivity can be easily simulated theoretically using two-phase elementary units (Bilboul model for confocal ellipsoids [8]). In the case of highly conductive, or high permittivty interphases their presence can influence the dielectric properties of the composites, in spite of their low con-centraion. Some data are presented for PE/CaCO$_3$ composites, which have been conditioned in atmospheres of different humdity [9]. The data can be superimposed into a master curve by shifting along the log(frequency) scale (Fig.4.). In the case of titanate treated CaCO$_3$ fillers dielectric losses increase systematically with filler content and relatve humidity, and decrease with increasing frequency. The observed dielectric behavior can be well approximated by linear Cole-Cole plot, instead of a semicircular one characteristic of Debye processes. If a special composite, containing 48 wt% CaCO$_3$ filler surface treated by polymerization catalysts, is studied, a qualitatively similar dielectric response is observed, but the loss is one order of magnitude higher. In the case of these catalyst treated fillers the dielectric response is determined by the fine details of the catalyst treatment. One may suppose, that this treatment influences the electrical behavior of the adsorbed water layer. In well defined PE/glass bead composites [10] the adsorbed water layer results in normal interfacial loss peaks, which can be analyzed in terms of theoretical models.

The effect of surface treatment on dielectric behavior can be well demonstrated in PP/CaCO$_3$ composites as well [11]. The dielectric loss of three composites containing 30 wt% CaCO$_3$ filler having different average particle diameters (i.e. different specific surfaces) were compared with that of a composite containing the same amount of stearate treated CaCO$_3$ filler (Fig.5.). On heating all composites exhibited a loss maximum, which increased as filler particle diameter de-creased, which disappeared on cooling. The dielectric loss was significantly reduced on cooling in all cases, but remained an order of magnitude higher in stearate treated samples. This behavior can be well understood taking into account the desorption of adsorbed water, and the non-volatile nature of stearic acid.

Surface conditions influence partial discharge behavior as well. Thermally stimulated polarization (TSP) and depolar-ization (TSD) properties of PP/CaCO$_3$/nonionic surfactant systems [12] have been studied. TSP curve of a simple PP/CaCO$_3$ composite, containing no surfactant, stored under ambient contions, shows a maximum above room temperature, analogous to the AC loss properties of the same composite. If, however, the same composite is dried thorughly, the TSP curve cannot be registrated because of frequently appearing 100 pA current pulses (the usual noise level of the amplifier with non-polarized sample was 0.01 pA). In spite of the relatively low field strength applied (1 kV/mm) those current pulses prove the presence of partial discharges. If the sample is stored again under ambient conditions, the original dielectric properties are regained, i.e. the adsorbed water layer "quenches" these partial discharges. If noninic surfactants are also used, the dried sample also shows these current spikes around room temparture, which, however, disappear at higher temperatures. One can speculate, that at this temperature the conductivity of the surfactant layer is high enogh to prevent partial discharges.

Finally examples are mentioned from the field of <u>polycrystalline ceramic composites</u> used for capacitor production [13]. Barrier layer capacitors made of $BaTiO_3$ or $SrTiO_3$ (the latter is non-ferroelectric) utilize the interfacial relaxation process to produce high dielectric constants. The ceramic materials are first fired in air, then, by heating it in a reducing forming gas, the whole body is converted into semiconducting state and finally, by a slight oxidation, an insulating surface layer is produced. This structure results in fairly high effective dielectric constants at all frequencies below the GHz range.

426

3. Conclusion

Heterogenous dielectric systems exhibit some special properties, among which most notable is interfacial polarization caused by hindered charge movement at the interfaces. This effect makes dielectric spectroscopy (AC and DC methods), in combination with other methods of structure determination, a usefol tool in studying the following phenomena in composite materials:
- miscibility/phase separation
- interactions between polymeric/non-polymeric components
- monitoring of cure/post cure
- monitoring of water adsorption
- non-destructive testing (partial discharges)
- selection of coupling agents for insulating composites
- morphological studies in phase separated systems
- connectivty/processing in conducting composites
- detection of barrier layers etc.
I hope that the examples presented prove the usefulness of dielectric methods in this field and encourage other scientists and technologists to use them.

4. References:

1.) G. Bánhegyi: Colloid and Polymer Science, **262**, 956-966, (1984)
2.) G. Bánhegyi: Colloid and Polymer Science, **262**, 967-977, (1984)
3.) G. Bánhegyi: Colloid and Polymer Science, **264**, 1030-IID, (1986)
4.) G. Bánhegyi: Colloid and Polymer Science, **266**, 11-28, (1988)
5.) G. Bánhegyi, L. Wu, F.E. Karasz, W.J. MacKnight: to be published
6.) G. Bánhegyi, P. Szaplonczay, G. Frojimovics, F.E. Karasz: Polymer Composites, **11**, 133-143, (1990)
7.) Yu. S. Lipatov: Physical Chemistry of Filled Polymers í(Russian), Khimiya, Moscow, 1977
8.) R.R. Bilboul: J.Phys.D., **2**, 921, (1969)
9.) G. Bánhegyi, P. Hedvig, F.E. Karasz: Colloid and Polymer Science, **266**, 701-715, (1988)
10.) P.A.M. Steetman, F.H.J. Maurer: Collid and Polymer Science, **268**, 315, (1990)
11.) G. Bánhegyi, F.E. Karasz, Z. Petrovic: Polymer Engineering and Science, **30**, 374-383, (1990)
12.) G. Bánhegyi, G. Marosi, G. Bertalan, F.E. Karasz: Colloid and Polymer Science, **270**, 113, (1992)
13.) R.C. Buchanan (ed.): Ceramic Materials for Electronics, M. Dekker, New York, 1986

Table 1. Glass transition temperatures measured by DSC, thermomechanical and AC dielectric methods on filled and non-filled epoxy resins (after Ref.[6]).

Sample	T_g ($^{\circ}$C) measured by		
	DSC	Thermomechanical	AC dielectric (1 kHz)
Cycloaliphatic epoxy			
- non-filled	116	122	147
- filled with 60% wollastonite	105	124	135
- filled with 60% quartz	112	118	134
Bisphenol-A base epoxy			
- non-filled	96	98	116
- filled with 60% wollastonite	95	114	120

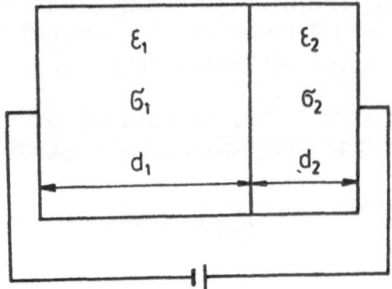

Fig. 1. Bilayer dielectric, the simplest model of composite materials exhibiting interfacial polarization.

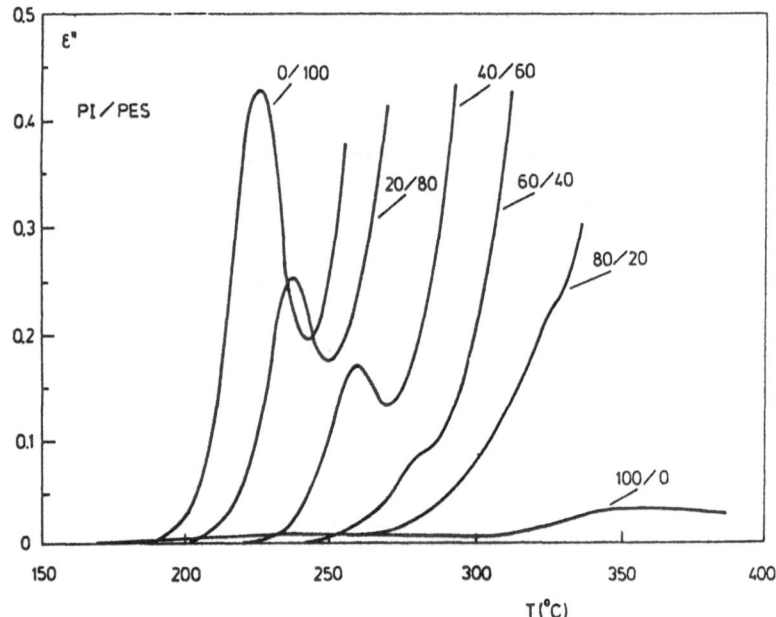

Fig. 2. 100 kHz dielectric loss spectra of polyimide – poly(ether-sulfone) blends, together with the spectra of the pure constituents, in the first heating run. Samples were solvent cast films.

429

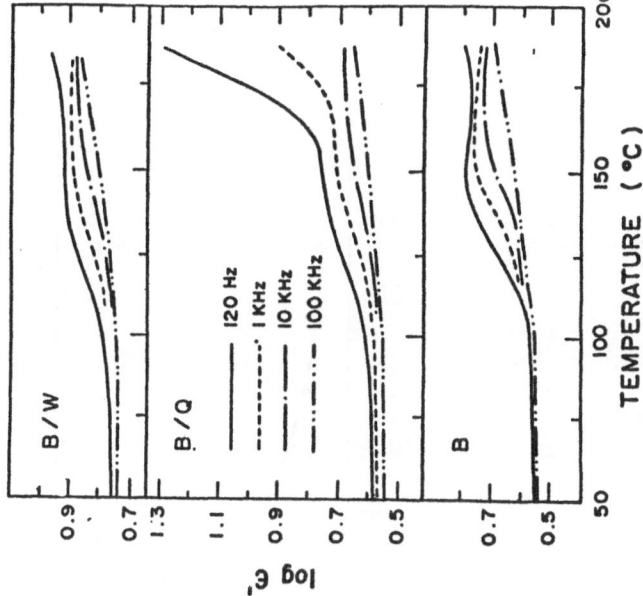

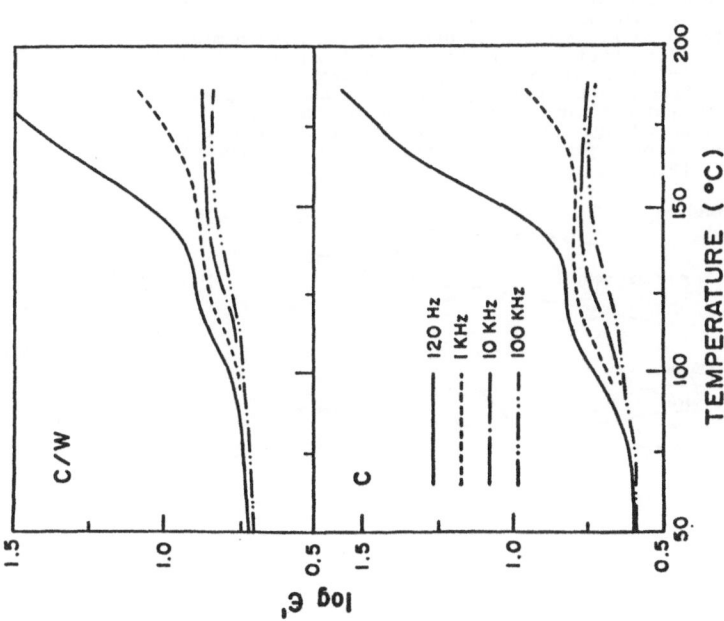

Fig. 3. Dielectric permittivities of a non-filled Bisphenol-A
based epoxy resin (C) and its composite filled with 60 wt%
wollastonite (C/W, left) and of a non-filled cycloaliphatic
epoxy resin (B) and its composites filled with 60 wt%
wollastonite (B/W) and quartz (B/Q, right). The high-tempera-
ture, low frequency polarization process is most probably due
to the matrix/electrode interface in the case of resin C,
while it is related to the matrix/filler interface in the case
of resin B. (After Ref. [6]).

430

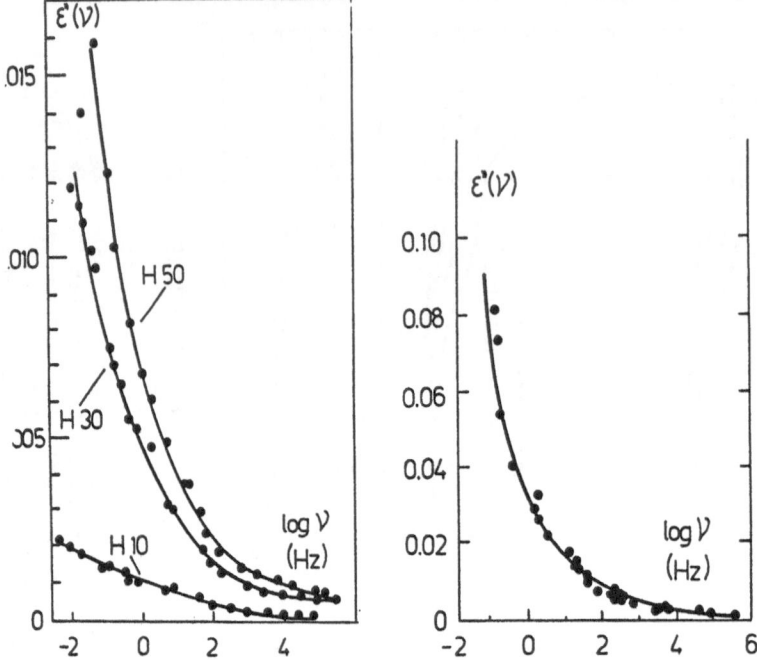

Fig. 4. Dielectric loss master curves obtained by horizontal shifting from data measured on $CaCO_3$ filled polyethylene samples stored in atmospheres of different relative humidities. Left: Data obtained on mechanical mixtures containing 10, 30 and 50 wt% titanate treated (hydrophobized) Ca-carbonate fillers. Right: Data obtained on a special composite prepared by polymerizing ethylene onto the surface of a polymerization catalyst treated Ca-carbonate filler (48 wt%). Note the order of magnitude difference between the samples. (After Ref. [9]).

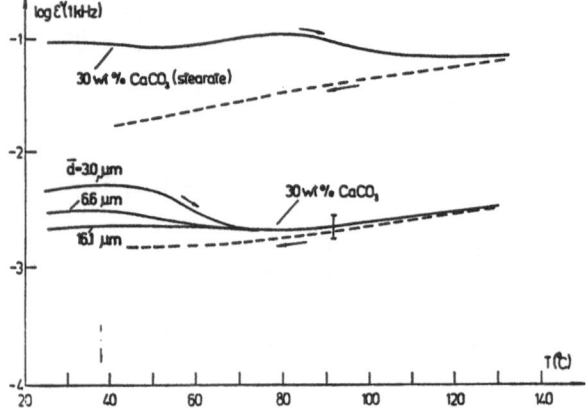

Fig. 5. 1 kHz dielectric losses measured on heating (continuous line) and on cooling (dashed line) on $CaCO_3$ filled polypropylene samples conating 30 wt% filler. Lower curves: non-treated filler with different average particle diameter values (higher specific area - higher loss peak attributed to water desoprtion). Upper curves: stearate treated filler (higher residual loss plus water desorption). (After Ref. [11]).

ON THE PHYSICAL NATURE OF INTERFACIAL LAYER IN POLYMER COATINGS

M.R.KISELEV and V.M.STARSEV

Institute of Physical Chemistry, Russian Academy of Sciences, Moscow

Polymer coatings are widely utilized for protection and decoration of various substrates. They may be considered as simplified model of composite materials. Interest in composite sistems stimulates investigation of properties of interfacial layers (IL), it's thickness (r) and volume fraction (V_f). r of IL may be calculated for dispersive polymer composite (DPC) and for fibrous PC, /1/.

Unfortunately, the physical nature of the IL, i.e. its origin, is not discussed in papers and they do not provide an adequate approach to the calculation of thickness and volume fraction in polymeric coatings. Meanwhile the influence of surface energy of substrate on morphology and thermal behavior of polymeric coatings was revealed /2,3/.

In present paper polymeric coating is considered as a three—phase system, i.e. as a PC model, where the substrate is modelling as a filler with an infinitely large radious. For this model the following equality will be true:

$$V_s + V_m + V_i = 1 \qquad (1)$$

where volume fractions of substrate, IL and matrix are shown as V_s , V_i , V_m ; respectively. They may be expressed in terms of thethickness (h) of these phases:

$$V_s = \frac{h_s}{h_s + h_c} ; \quad V_i = \frac{h_i}{h_s + h_c} ; \quad V_m = \frac{h_c - h_i}{h_s + h_c} = \frac{h_m}{h_s + h_c} \qquad (2)$$

where $h_c = h_m + h_c$ (3) is the thickness of the coating.

The coating on the substrate ($h_s + h_c$) will have thermal capacity ($C_p^\cdot$) different from C_p^m due to the presence of the IL. Based on the proportionality between jumps C_p and the volumes of the phases one can write:

$$\frac{V_m}{V_c} = \frac{C_p^m}{C_p^c} \qquad (4), \text{ but } V_m = V_c - V_i \qquad (5)$$

G. Akovali (ed.), The Interfacial Interactions in Polymeric Composites, 431–432.

and we obtain
$$V_1 = (\frac{C_p^c - C_p^m}{C_p^c}) V_c \qquad (6).$$

If by analogy of /1/ the expression in brackets is designated by J then; $V_1 = J.V_c$ (7) and $h_1 = J.h_c$ (8)

Suggested formulas (7) and (8) were verified on polyimide coatings of different thicknesses from 3 to 48 micrometers. Calculation should be done, using C_p^m of polymer matrix or free film.

Free films were produced on mercury, dried at room temperature, then removed from the mercury and in a free state was thermally cycled.

The large thickness and relaxation character of the IL in samples used favoured the "mechanical" origin of the IL during the formation of coatings. A "mechanical" IL is not important, but a sone of a polymer, excited by a substrate, i.e. it is a result of the resistance of the solid phase, whose volume is constant, to the shrinkage stress in a polymer component. On the other hand a "mechanical" IL is a sone of the localization of inner stretching stresses at the interface.

References

1. Yu.S.Lipatov. Physical Chemistry of Polymers. Moscow. Chemistry. 1977. p.89.

2. M.R.Kiselev, P.I.Zubov. Electronmicroscopic Study of the Structure of Polymeric Coatings. The VII Internat. Congress on Electron Microscopy. 1970 Grenoble, France.

3. Ch.A.Kirakov, M.R.Kiselev, P.I.Zubov. Peculiarity of thermal extension of curing epoxy resin. Doclady AN USSR. 1980. v.251. N.5. p.1160.

THE EFFECT OF CORONA MODIFICATION ON THE COMPOSITE INTERFACES

S.SAPIEHA

Pulp and Paper Research Institute of Canada and
Chemical Engineering Department, Ecole Polytechnique
P.O.Box 6079, Station "A", Montreal, Que., H3C 3A7 Canada

ABSTRACT

Corona treatment represents one of the most versatile methods for surface modification of natural and synthetic polymeric materials and it is widely used in industry for the improvement of adhesion and printability (1,2). Treatment of cellulose fibers produces several physical and chemical effects. Among other things, such a treatment improves remarkably the interfiber bonding in paper (3-4), which may be related to the increased surface acidity and basicity (5), and a formation of low-molecular-weight degradation products during the treatment (4). It has been shown for composite materials based on cellulose fibers and thermoplastics that corona treatment of composite components results in remarkable increase in its strength and, at the same time, reduces the melt viscosity thus lowering the energy expenses during processing (6).

The interactions at fiber/thermoplastic interfaces depend on the surface properties of the composite components. Therefore, to produce reliable composite materials one needs efficient, rapid and reproducible method for the evaluation of the treatment level of cellulose surfaces. In this study it has been shown that electrical conductance of the distilled water suspensions of treated cellulose fibers is a reliable measure of the surface treatment and it relates directly to the physical properties of composites.

Corona treatment of cellulose fibers was performed in a treatment cell presented schematically in Figure 1. Metal electrodes are separated by an air-gap and 1.5 mm thick quartz plate. Corona system operates at 3.5 kHz and the treatment current can be varied within a wide range. The experimental variables include: fiber concentration, corona treatment current, treatment time, volume of treated fibers, and controlled ratio of the fresh-air supply. Several types of cellulose fibers were investigated, high-molecular-weight cellulose (above 99.5% of α-cellulose), cotton, and bleached softwood kraft pulp (Q90). Fibers were conditioned at 50% RH before the treatment. Electrical conductance measurements were performed using the Omega Engineering Inc. conductance meter model CDH-70. Typical fiber concentration in distilled water was 0.5 wt%.

The addition of treated fibers to the distilled water produces almost instantaneous response in electrical conductance. A linear relationship exists between electrical conductance of suspension and a concentration of all types of fibers investigated. The level of conductance is a strong function of treatment and post-treatment storage conditions. For any given current and treatment time the conductance level depends on

433

G. Akovali (ed.), The Interfacial Interactions in Polymeric Composites, 433–434.
© 1993 *Kluwer Academic Publishers.*

434

the amount of oxygen available for oxidative reactions; it was found to be highest for a continuous air-flow configuration. Electrical conductance of suspensions corresponds to the mechanical properties of composites made of these fibers and polypropylene as a matrix material (Figure 2). These results suggest that electrical conductance is a useful method for determination of the treatment level of fibers, permitting thus the control of the mechanical properties of resulting composites.

The low-molecular-weight, water soluble degradation products of cellulose are responsible for the increased conductance; their chemistry seems to be similar to that observed during the ozone bleaching of wood fibers. They include water soluble acids and hydroxy acids such as acetic, formic, glyceric, adipic, tartaric, succinic, glutaric, lactic, gluconic, and α-hydroxybutyric. These products can be removed completely; after washing and immersing fibers again in distilled water, the conductance is equal to that of untreated cellulose fibers suspensions.

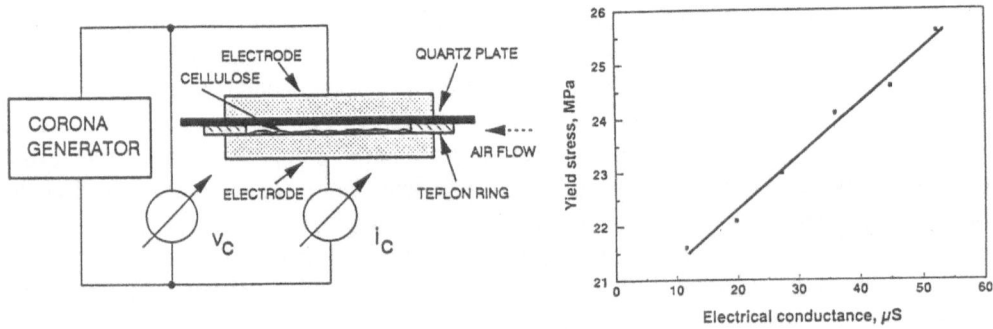

Figure 1. Schematic diagram of the corona treatment cell.

Figure 2. Yield stress of TC/PP composite as a function of electrical conductance of the corona-treated fibers. Fiber content: 30%. Corona current: 25mA. After (7).

REFERENCES

1. S.Wu, Polymer Interface and Adhesion, Marcel Dekker, Inc. New York (1982)
2. D.A.I. Goring., Pulp Paper Mag. Can. 68(8), T372 (1967).
3. H. Nishimura, T. Nakano, T.Uehara and S. Yano, Tappi J., 73(10), 275 (1990).
4. T.Uehara and I.Sakata, J.Appl.Polym.Sci., 41, 1695 (1990)
5. S.Dong and S.Sapieha, 13th Canadian Seminar on Surfaces, Peterborough, May 28-29, 1991
6. S.Dong, S.Sapieha and H.P.Schreiber, Polym.Eng.Sci., in press.
7. N.M.Belgacem, P.Bataille and S.Sapieha, to be published

PERCULIARITY OF FILM-FORMING AND HYDROLYTIC DECAY OF TIN-CONTAINING POLYMER COATINGS

Zakir M.O.RZAYEV, Rafik R.ABDULLAYEV and Viktor A.ZUBOV

Azerbaijan Academy of Sciences, Department of Chemical Sciences, Institute of Polymer Materials, 37000I BAKU, Azerbaijan

The investigation results of Sn-containing carboxylate polymer coatings influence on mechanism of their hydrolytic splitting showed that microstructure of organotin macromolecules defines the geometrical accessability of hydrolyzing particles to carboxylate groups, i.e. changes its destruction rate. It was shown that the decay rate of polymers with isotactic location of hydrolyzing functional groups in the chains is sigificantly higher than the decay rate of polymers with syndiotactic location of these groups.

Hydrolysis of organotin polymer coatings with isolated (syndiotactical structure) organotin groups is a result of nucleophile attack of carboxyle group of electrophile carbon atom or tin (which electrophile capacity is specified by a vacant d-orbital) by water or hydroxide-ion. Further charges distribution or nucleophile substitution causes separation of side groups from polymer chain. For polymers with isotactic structure hydrolysis is being complicated by nearby organotin groups which effect each other greatly.

The orgraph of decay of bioresistant organotin polymer coatings of the methacrylate and maleate types in the sea water is described. The diffusion of water into macromolecular coils to the hydrolytically unstable bonds of biocides with a polymer matrix is the limiting stage of the decay process. The calculated value of the coefficient of diffusion of water is of the same order with that obtained from experimental data on the flow of organotin biocides for a coating in the stationary regime.

Distribution of internal stresses in organotin coatings is defined by formation of two phase boundaries, i.e. gas-liquid (polymer, as highviscous liquid), liquid-solid body (Fig.I). In this model the internal stresses are presented as balls with dimensions proportional to stress value and arrows show direction of forces. The coating damage starts from the surface and boundary layers.

In Tin-containing coatings there are devoloping only internal compression stresses, i.e. all values are positive. Data given in Fig.2 show that the internal stresses in boundary layer dominates the stresses in surface and middle layers. The curves peculiarity is the starting point of time of internal stresses calculation which is in compliance with the time of boundary layer forming at high-energy surfaces.

G. Akovali (ed.), The Interfacial Interactions in Polymeric Composites, 435–436.

436

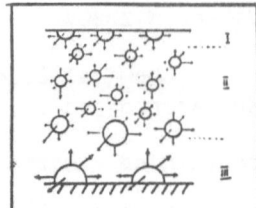

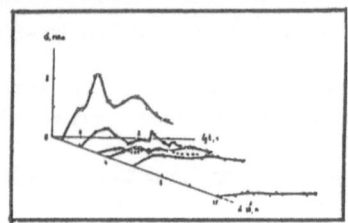

Fig.I. Distribution of the internal stresses in polymer coating.Models I,II and III are surface, middle and boundary layers, respectively.

Fig.II. Relaxation spectra of the internal stresses in coatings formed of tin-containing polymers (in Xylol, $2I \pm I^{o}$)

Absorption from concentrated solutions is identicial to absorption as macromolecules in the form of statistical balls or ball aggregates. Convergence of parallelepipedes (Fig.3) in this case will be accompanied by their deformation and due to cooperative effect in the system a definite ball rigidness for steel plate bending is observed.

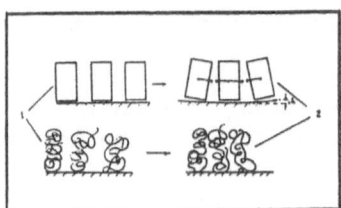

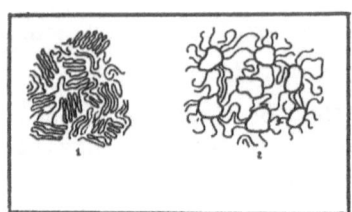

Fig.3. Model and scheme show formation on the internal compression stresses in boundary layer of organotin polymer coatings: I-the obsorption balls, 2-final result of balls interaction (ball agglomeration).

Fig.4. Scheme of macromolecule surface packing:
I- fringemiccelar one
2- associative one

Upon reconstruction the adsorption layer structure is formed as boundary layer structure (Fig.4). The last one made of organoting polymer concentrated solutions is well described by associative model which specifies interaction of polymer balls by separate chain sections being outside of ball. Oranotin polymer coatings form various structures independent of molecular formation from associates to crystalls. And if films structural defects can exert a significant effect on reproduction of hydrohytic decay interactions proceed in interphase stage.

DIFFUSION OF METAL IONS IN CARBOXYLIC POLYMER SORBENTS OF DIFFERENT MORPHOLOGIC STRUCTURE

A.A.EFENDIEV
Azerbaijan Academy of Sciences
Institute of Polymer Materials
373204 Sumgait
Azerbaijan Republic

The sorption mechanism and diagnostic equations for uptake and binding of metal ions by ion exchange sorbents, particularly by those with complex-forming functional groups was studied by numerous authors and contradictory opinions were expressed. In papers published the equations based on quasi-homogeneous model of the sorbent were used in majority of cases.

In this report we aim to show that any understanding of the mechanism of ion uptake where complexes are formed must seek to connect the kinetic data with structure and morphology of the polymer sorbent.

Zerolit 236 carboxylic ion exchanger and copper ions as sorbing metal were chosen for investigation. It is well known that carboxylic ion exchangers easily form complexes with copper ions. Nonionized H-form and ionized Na-form of the sorbent were investigated. Stereoscanning electron microscope "Cambridge Stereoscan S4" with a fixture for X-ray microanalysis was used to investigate the morphologic structure of the sorbents and character of ion diffusion into the beads of different forms of sorbent. Both forms of sorbent beads were dried at room temperature followed by bringing to constant weight in dessicator over moisture-absorbing agent, at critical point and by freeze drying with centrifugation.

It was found from the observation made with stereoscanning electron microscope that sorbent in nonionized H-form is seen to have a coiled granular structure and to be made up of globules. The external appearance of

437

the globules differs a little depending on the method of
drying but in all cases the diameter of the globules is
within a range of 0.1-0.2 micron.

In case of ionized Na-form of the sorbent globular
structure disappears and developed fibrillar structure
may be observed.

In order to investigate the character of the diffusion of ions we took specimens of the H- and Na-form of
the sorbent that had sorbed different amount of copper
ions. It was found from the observation made with X-ray
microanalysis that copper ions initially take up accessible sites in the surface layers of the bead. Sorption
zone forms a clear boundary which moves from the periphery to the centre of the bead as the sorption degree
increases. Completely different distribution pattern is
found for the copper ions when they are sorbed by H-form
of the sorbent. The copper ions may be seen to be almost
uniformly distributed throughout the entire volume of
the bead even at a very low degree of sorption, pene-
trating from the periphery to the centre through interglobular space. The density of the copper ions increases
uniformly throughout the entire volume of the bead as
the degree of sorption increases. This proves that diffusion occurs much more rapidly in the interglobular
space than within the globules.

Thus, a comparison of the stereoscanning electron
microscope results with data of X-ray microanalysis indicates that the different types of morphologic structure determine the different character of ions diffusion into the sorbent beads.

So, it appears that sorption of copper ions by Na-
form of the sorbent may be analysed in terms of quasi-
homogeneous model of the sorbent because diffusion coefficients in the regions between the fibrillas and
within them are comparable whereas the H-form has heterogeneous structure with two diffusion coefficients.

ADSORPTION OF METHYLENE BLUE ON PVC-DOP-NATURAL ZEOLITE COMPOSITES

Devrim Balköse; Semra Ülkü
Ege University
Department of Chemical Engineering
Bornova-İzmir-TÜRKİYE

ABSTRACT. Rates of methylene blue adsorption from 0.02 gdm^{-3} aquous solution on composites prepared from 100 phr PVC, 60 phr DOP, 0-40 phr natural zeolite (clinoptillolite) were studied. The adsorption of methylene blue was negligible for composites having no zeolite additive, whereas extensive adsorption was observed for composites having 5-40 phr zeolite.

1. INTRODUCTION

Microporous materials from PVC are used as battery seperators, filters, supports for enzymatic catalyts, breathable liners and paper substitutes. While the mechanical properties of PVC composites were studied extensively little work exists on adsorptive properties(2,3). Thus adsorption of methylene blue from aquous solutions was chosen to be studied as the representative of adsorptive properties in this work.

2. MATERIALS AND METHODS

PVC(Petvinyl E 36/71), dioctylphtalate (DOP)(Sankim),natural zeolite (clinoptillolite, from Bigadiç-Türkiye), were used in this work. 60 phr DOP and 100 phr PVC were mixed in Rowenta dough kneader. Clinoptillolite was added in the range of 5 to 40 phr to the mixture. Plastic plates of 2.5 mm thickness were obtained by casting at 160°C for 15 minutes. 3.8 cm$_3$ diameter pellets obtained from plates were contacted with 0.02 gdm^{-3} methylene blue solution in 1/100 solid liquid ratio. Solution concentrations were determined by visible spectroscopy using Jasco 7800 spectrophotometer measuring absorbance at 678 nm. Equilibrium uptakes were determined after 5 weeks contact time.

3. RESULTS AND DISCUSSION

While the control samples having no zeolites did not adsorb methylene blue, composites having zeolites adsorbed it significantly as shown in Table 1.

Table 1. Equilibrium uptake of methylene blue by composites having different zeolite concentration

phr zeolite	0	5	10	20	40
equilibrium uptake	0	1.0	0.4	1.0	1.0

Although the composites are heterogenous in microscopic scale they can be assumed to be homogeneous in macroscopic scale(4). The

439

G. Akovali (ed.), The Interfacial Interactions in Polymeric Composites, 439–440.

adsorption of methylene blue is a diffusion controlled process as indicated by linear dependence of fractional uptake on square root of time in Figure 1. The diffusion coefficient (D) was calculated from experimental uptake data using Equation 1

$$M_t/M = 4/1 \ (Dt/ \quad)^{1/2} \qquad (1)$$

where M_t amount adsorbed at time t; M , amount adsorbed at equilibrium; 1, half of the layer thickness. It is found that it was $3.7 \times 10^{-9} \ m^2 s^{-1}$ and it was nearly independent of zeolite concentrations.

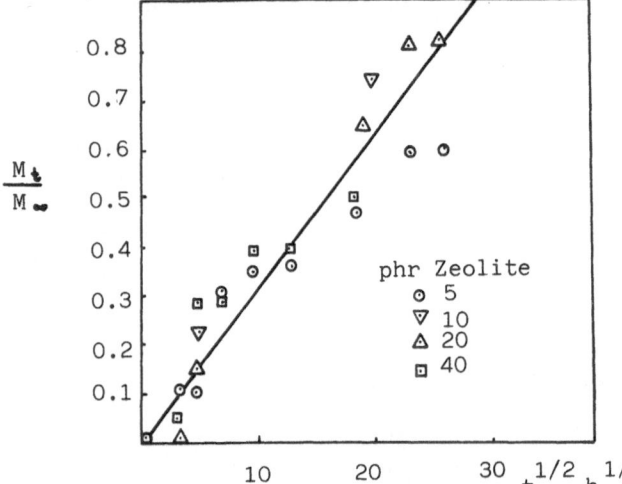

Figure 1.Fractional uptake of methylene blue versus square root of time

The degree of fusion of the plastisol and the extent of wetting of the zeolite particles by the plastic matrix affects the rate of diffusion and extent of adsorption of methylene blue whic are preferantially adsorbed on zeolites. Experimental work indicated that addition of zeolites to PVC plastisols created an interface accessible to aquous solutions.

4. REFERENCES

1. Wagner,M.P.(1978)'Natural and synthetic silicas in plastics' in R.Deanin(ed), Additives for Plastics,Academic Press,London pp9–28
2. Ajji A.,Schreiber HP (1987),'Rates of property change in plasticized filled PVC compounds', J.App.Polym.Sci.,33, 2493–2501
3. Kokta B.V.,Maldas D.,Deneault T., Béland P.(1990), 'Composites of Poly(vinyl chloride) and wood Fibers', Polymer Composites, 11, 84–89
4. Park G.S.(1986) 'Transport principles–solution,diffusion and permeation in polymer membranes' in P.M.Bungay (ed), Synthetic Membranes: Science,Engineering and Application, D.Reidel Pub.Com.,London,pp 57–107

Optically Transparent Glass Fiber Reinforced Poly(methylmethacrylate) Laminates

R.K. Six, J.O. Stoffer, and D.E. Day
Dept. of Chemsitry
142 Schrenk Hall
University of Missouri at Rolla
Rolla, MO 65401

Abstract:

Transparent composites have been made at the University of Missouri-Rolla, USA. The refractive index of optical glass fibers was adjusted to match the refractive index of poly(methyl methacrylate) for the middle of the visible spectrum at 25 °C. A 1.5 mm thick composite made in this manner containing 8% glass fiber has 80% transmission and forms clear images of distant objects. Dissimilar thermooptic coefficients (dn/dT) between the glass and PMMA cause a decrease in transmission as a function of temperature change. However there is good transmission of light over a broad temperature range (approx. ±25 °C from the temperature at which the refractive index was matched between the glass and the PMMA). These composites are made by two processes: a polymerized in place method, and by hot pressing glass fiber/PMMA prepregs to form laminates. The hot pressing process allows fiber to be placed on the surfaces of a cast sheet of PMMA core. Since bending stresses occur at or near the surfaces this results in increased strength with less fiber and therefore a more highly transparent composite.

G. Akovali (ed.), The Interfacial Interactions in Polymeric Composites, 441.
© 1993 *Kluwer Academic Publishers.*

The effect of recycling on the properties of thermoplastics composites

C.A.Bernardo, A.M.Cunha and M.J.Oliveira

Department of Polymer Engineering, University of Minho

4719 Braga Codex - PORTUGAL

The **recycling of plastics** waste has received an increased interest in the last few years, due to an increasing perception of the environmental impact of plastics, namely as packaging materials. In particular, the **primary recycling** of plastics waste (which is done directly in the industry to produce parts whose properties are similar to those of the original products), has been the subject of various recent publications [1-4], some of which devoted to the deduction of algorithms to predict the properties (and the economics) of mixtures of virgin and recycled polymers[1,3]. The interest of these studies, specially with **engineering plastics**, is obvious, as in this case it is critical to compatibilize the maximum incorporation of recycled material with adequate values of the specified property. Moreover, this type of research may help to elucidate the nature of the degradation mechanism.

Figure 1 represents, in a schematic way, a continuous operation of plastics processing, such as injection moulding, incorporating a recycling/granulation step.

In the figure, R, V, F and O represent, respectively, the stream of recycled, virgin, feed and output material, and P, P_o, P_n and P_n the values of the corresponding properties (the subscript n refers to the nth cycle). In the derivation of the algorithms we assumed that the properties of the mixtures of virgin and recycled polymers obey one of two general laws. These are:

$$P_n = kP_o + rP_r \qquad \text{linear law (additive mixture)}$$

443

G. Akovali (ed.), The Interfacial Interactions in Polymeric Composites, 443–448.

444

and

$$\ln P_n = k \ln P_o + r \ln P_r \qquad\qquad \text{logarithmic law}$$

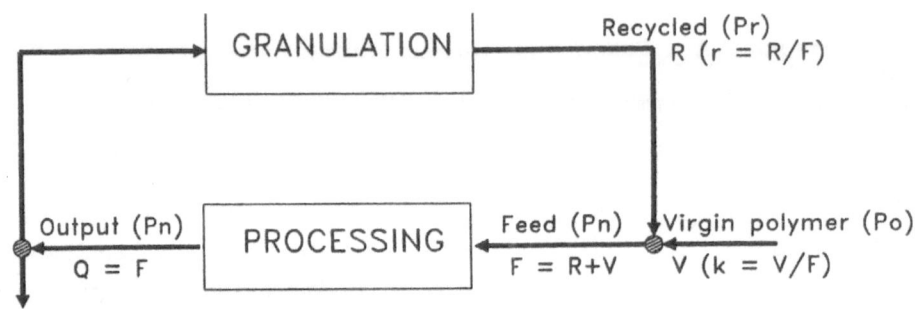

Fig.1 - Schematic representation of a processing/recycling operation

The algorithms also imply that the degradation of the properties of the material in each processing cycle (considered to be the set of all the steps between two consecutive feeds), may be described by a general equation, usually **different** for **each** material and **each** property, that is integrated in the expression that gives the properties of the mixture. Various degradation equations have been described in the literature [1,3]. In this work, we used only 3 of these equations, which, in our experience, are applicable to a large range of materials and properties. The corresponding algorithms are:

a) Degradation equation in the form of an exponential decay $(P_i = P_o e^{-bi})$ and additive mixture:

$$\frac{P_n}{P_o} = k + k\,\frac{(1-k)\,e^{-b} - (1-k)^n\,e^{-bn}}{1-(1-k)\,e^{-b}} + (1-k)^n\,e^{-bn} \qquad\qquad [\text{model 1}]$$

b) Degradation equation in the form of an exponential decay to an assimptotic value, P_a $(P_i = a_o\,e^{-bi} + P_a)$ and additive mixture:

$$\frac{P_n}{P_o} = k + \frac{a_o}{P_o}\,k\,\frac{(1-k)\,e^{-b} - (1-k)^n\,e^{-bn}}{1-(1-k)\,e^{-b}} + \frac{a_o}{P_o}(1-k)^n\,e^{-bn} + \frac{P_a}{P_o}\,(1-k) \qquad [\text{model 2}]$$

c) Logarithmic law of mixtures and loss of property **in each processing step** obeying a power law ($P_i=cP_o{}^d$):

$$P_n = C_*{}^{(1-(rd)^n)/(1-rd)} \cdot P_o \qquad \text{[model 3]}$$

with,

$$C_* = c\, P_o{}^{d-1}$$

In the derivation of the above algorithms, r (and k = 1-r) is supposed to be the same in all cycles, and P_i represents the value of the property in the output of the ith processing. The property loss in the granulation process is not normally considered but, when significant, can be easily included by incorporating its effect in the degradation equation.

In the case of thermoplastics composites reinforced with glass or carbon fibres, besides the equations that represent the degration of the mechanical properties (or the changes in the molecular weight of the polymer in the matrix), the decrease in the length of the fibres must also be known. This can be done by burning the polymer after each processing/granulation step, and observing, by optical microscopy, the resulting ashes.

In the present work we report the study of the effect of recycling on the properties of two engineering plastics, a glass fibre reinforced polypropylene, produced by ICI (Propathene HW60 GR20; 20% W/W) and a glass fibre reinforced polycarbonate, produced by GEP (Lexan 500R;10%W/W). Some of the results obtained with the polypropylene are presented in Table 1 and Figures 2 and 3.

Table 1 – Variation of the second moment of the distribution of fibre lengths (L_w) with the number of cycles (dimensions in μm)

n	0	1	2	3	4	5	6	7	8	9
L_w(Polypropylene)	710	610	540	480	380	330	260	280	250	259
L_w(Polycarbonate)	270	260	167	149	133	108	140	99	–	–

446

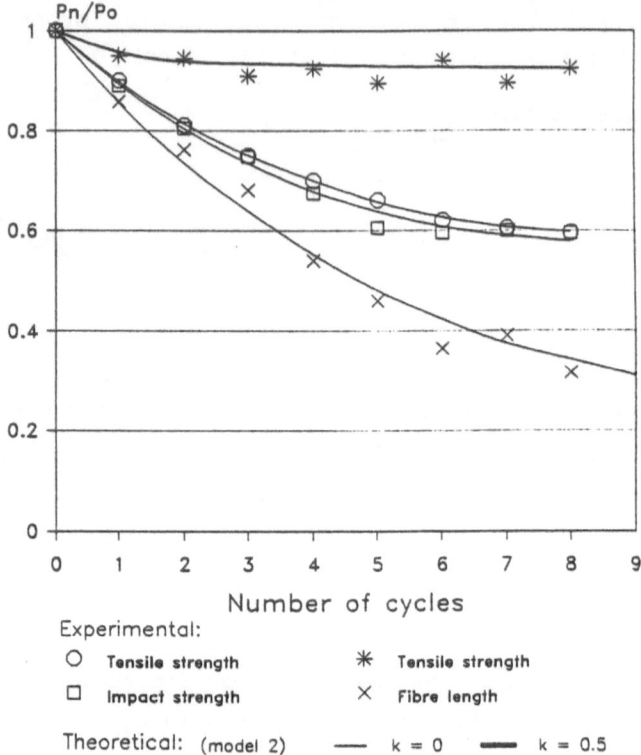

Fig.2 - Variation of the mechanical properties and fibre lengt
        PP/20% GFR with the number of processing/granulation cy

In Figure 2 each property - tensile strength, impact strengt}
fibre length - is normalised with the value of the one correspo
to the virgin polymer. The lines correspond to model 2, which
the best fit to the experimental values. Although the length o:
fibres (determined as the second moment of the distributic
lengths, measured in the optical micrographs) decreased to 3(
that of the virgin polymer after 9 cycles, it remained always h
than 250 μm. It can be observed that, even when no virgin mat.
was added between cycles (k=0), the values of the mechal
properties of the composite stabilize around 58% of those of
original material. When the fraction of virgin polymer in the
increases to 50% (k=0.5) the tensile strength after 3 cycles i
of that of the original material and remains constant from the
SEM observations of fracture surfaces show that most of the f:
are broken close to the surface, without debonding from the ma

In this case, the effect of the fibres in the perfomance of the composite is additive, as predicted in various theoretical formulations [5], and confirmed by the present results.

It can also be observed that model 2 decribes adequately the variation of both the mechanical properties and the length of the fibres with the number of cycles, and can thus be used to predict the properties of the mixtures.

The loss of impact strength of the 10%(W/W) glass fibre reinforced polycarbonate with the number of processing/granulation cycles is presented in Figure 3, which also shows the variation of the length of the fibres with recycling. It can be observed that, unlike the previous case, the value of the property decreases pratically to zero after only 4 cycles. Although the length of the fibres stabilizes after 3 cycles, its value diminishes from ca 270 μm (n=0) to 130 μm (n=4). This value is apparently lower than that of L_{cr} (critical fibre length), below which, when the composite fails, the fibre's debond from the matrix without breakage. In fact, values of L_{cr} between 125 and 200 μm have been reported in the literature [6,7]. Hence, in this case, the fibres do not contribute towards the overall mechanical properties of the composite, but act instead as stress concentrators, leading to catastrophic failure. It can also be observed that one of the algorithms (model 3) describes quite well both the loss of impact strength and the reduction of fibre length.

In this work we presented a methodology that can be used to predict the properties of thermoplastics composites made with mixtures of recycled and virgin polymers. The algorithms developped perform equally well when recycling leads to a catastrophic failure of the properties or when these properties stabilize after a number of cycles.

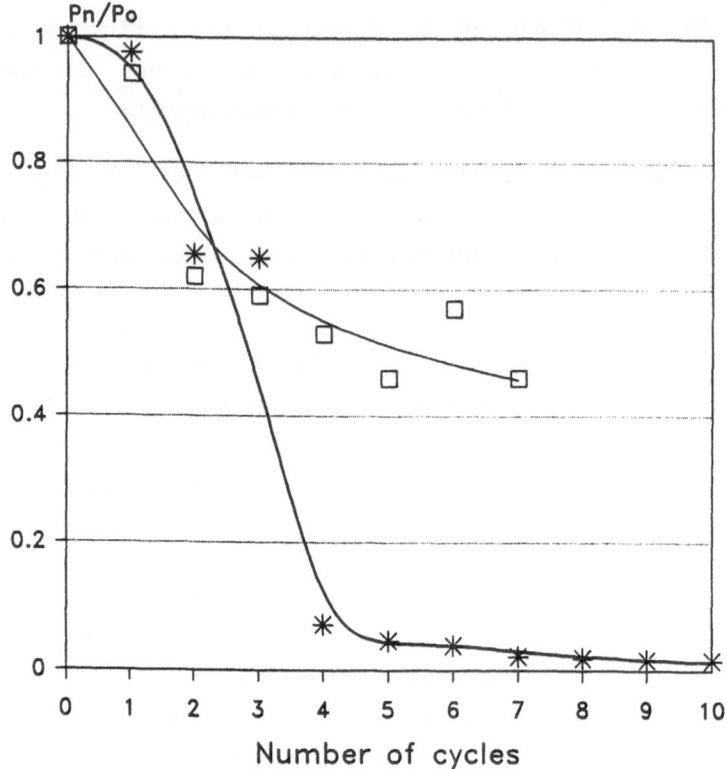

Fig.3 - Variation of the mechanical properties and fibre length of
PC/10% GFR with the number of processing/granulation cycles

References

[1] Throne, J.L., Advances in Polymer Technology **7**,347 (1987)

[2] Wandhal,W., Proc. Conf. *Plastics Recycling 88*, Copenhagen, comm.
24.1, SPE- Scandinavian Section (1988)

[3] Bernardo, C.A., Tecnometal **70**, 13 (1990)

[4] La Mantia,F.P., Macplas, 53, May (1990)

[5] Yam,K.L., Gogoi,B.K., Lai,C.C. and Selke,S.L., Polym.Eng.Sci.
30, 693 (1990)

[6] Filbert,W.C., SPE Techical Papers **14**,3 94(1968).

[7] Yang,H.W., Farris,R.and Chien, J.C., J.Appl.Polym.Sci.**23**, 11,
(1979)

CARBON FIBERS FROM METHANE

M. TERESA SOUSA, J.L. FIGUEIREDO
Faculdade de Engenharia (LCM/DEQ)
4099 Porto codex, Portugal

Carbon fibers were produced from methane by a CVD process in two stages:
1- Growth of carbon filaments by hydrocarbon decomposition catalysed by metal particles;
2- Thickening of these filaments by pyrolytic carbon deposition.

The mechanism proposed for the initial stage of filament growth is the following [1]:
a) Adsorption of the carbonaceous reactant at the metal surface, followed by decomposition reactions leading to chemisorbed carbon species; b) Carbon dissolution in, and diffusion through, the metal particles to active growth areas (such as grain boundaries or metal-support interfaces) where carbon precipitates out. As a result, metal particles are detached from the surface and transported on top of the growing filaments; c) Alternatively, the carbon species may react on the surface of the metal to originate a film of "encapsulating" carbon; consequently, the catalyst deactivates and filament growth ceases.

The filaments grow with a diameter which is close to that of the catalyst particles at their tips. In the second stage, carried out at higher temperature, the filaments stop growing as a result of catalyst deactivation, but they thicken by carbon deposition due to the pyrolysis of the hydrocarbon, becoming fibers.

The experiments were conducted in a mixture of 30% methane in hydrogen. Iron supported on grafoil (® Union Carbide) was used as catalyst, submitted to an appropriate temperature programme [2]. Fibers were produced with diameters in the range 5-15 μm and lengths up to 5 cm. Characterization of these fibers is in progress, using the available techniques for determination of physical and chemical properties. The most important parameters are total surface area, active surface area (ASA), porous texture, surface energy and surface functional groups.

Carbon fibers are used as reinforcement material in composites, the mechanical properties of which are dependent upon the matrix/fiber adhesion. Surface modification such as dry or wet oxidation of the carbon fibers may be used to improve those characteristics [3].

Oxidation in boiling HNO_3 and in air is being studied and the results have been evaluated in terms of their influence on structure and surface oxygen concentration. Wet oxidation (HNO_3) does not affect the fibers surface, but an significant increase in the surface oxygen concentration (~ 7 %) is detected as shown by the XPS spectrum presented in Figure 1. Dry oxidation experiments were carried out in air at 600 °C to different extents of burn-off. Even at low burn-off level the surface morphology is strongly affected showing pits (Figure 2).

The results of these experiments as well as those of anodic oxidation will be analysed in terms of total surface area, ASA and strength of the carbon fibers.

G. Akovali (ed.), The Interfacial Interactions in Polymeric Composites, 449–450.
© 1993 Kluwer Academic Publishers.

450

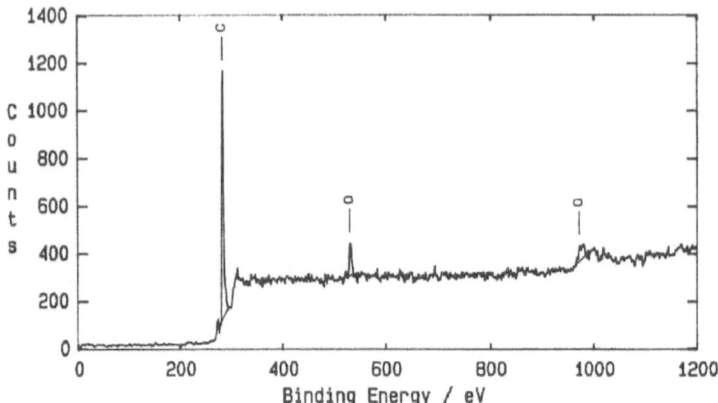

Figure 1 : XPS spectrum of carbon fibers oxidised in boiling HNO₃ during 12 hrs

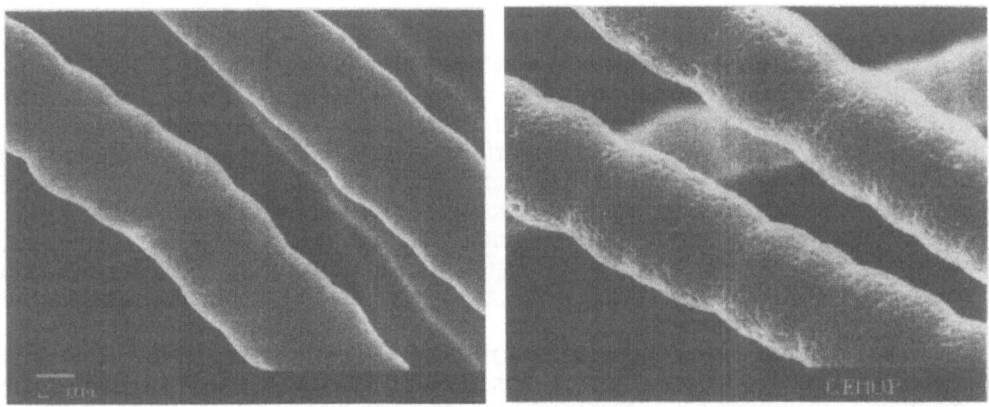

Figure 2 : Carbon fibers : left original, right after oxidation in air at 600 °C, B.O. = 4 %

REFERENCES

1. Figueiredo, J. L. and Bernardo, C. A. (1990) "Filamentous carbon formation on metal and alloys " in J. L. Figueiredo, C. A. Bernardo, R. T. K. Baker and K. J. Hüttinger (eds), Carbon Fibers Filaments and Composites, Kluwer Academic Publishers, Dordrecht, pp. 441-457.
2. Benissad, F., Gadelle, P. , Coulon, M. and Bonnetain, L. (1988) " Formation de fibres de carbone a partir du methane: I-Croissance catalytique et epaississement pyrolytique ", Carbon 26, 61-69.
3. Fitzer, E. and Weiss, R. (1987) "Effect of surface treatment and sizing of C-fibers on the mechanical properties of CFR thermosetting and thermoplastic polymers ", Carbon 25, 455-467.

INDEX

A

B

C

D

Debonding; 142
Diffusion Theory; 8

E

Electrostatic Theory; 7
Electron Microscopy
    Scanning; 127
    Scanning Transmission; 128
    Auger; 131
Electron Spin Resonance (ESR); 297
ESCA; 164

F

Failure; 15, 139
Fracture; 61
Fragmentation; 89
Friction; 72
Filament Winding; 324
Fluoronation; 12

G

Gutman Probe; 30
Grazing Incidence Diffraction (GID); 297

H

Healing; 61
Hooking Theory; 7

I

Interface; 1, 42, 61, 82, 107, 160, 171, 415, 417, 431